ENGINEERING MATHEMATICS–I

ENGINEERING MATHEMATICS–I

SUBJECT REVIEWERS

I would like to thank all those reviewers who took out time to review the manuscript and gave important suggestions. Their names are given below.

Dr. Abdul Majeed

Dr. Loka Pavani	Osmania University College for Women, Hyderabad
Dr. Gitti Narsimlu	Chaitanya Bharathi Institute of Tech., Hyderabad
C. Naga Anuradha	Vasavi College of Engineering, Hyderabad
Dr. K. Ramesh Babu	MVSR Engineering College, Hyderabad
Dr. G. Krishna Kumari	Vidya Jyothi Institute of Tech, Hyderabad
Saroj M Revankar	Stanley Engg College for Womens, Hyderabad
D. Swamy	Methodist College of Engg and Technology, Hyderabad
Naveen Voruganti	Methodist College of Engg and Technology, Hyderabad
M.A. Rawoof Sayeed	Muffakham Jah College of Engg and Tech, Hyderabad
Syed Nisar Ahmed	Deccan College of Engg and Tech, Hyderabad
Dr. T. Nagaiah	Kakatiya University, Warangal

ENGINEERING MATHEMATICS–I

For

B.E., 1st Year/1st semester

Strictly according to the latest revised syllabi 2016

FOR OSMANIA UNIVERSITY, HYDERABAD (TELANGANA)

By

Dr. Abdul Majeed
Head Dept. of Mathematics
Muffakham Jah College of
Engineering and Technology,
Banjara Hills, Hyderabad
Telangana

N.P. Bali
Former Principal
S.B. College, Gurugram
Haryana

UNIVERSITY SCIENCE PRESS
(An Imprint of Laxmi Publications Pvt. Ltd.)
An ISO 9001:2008 Company

BENGALURU • CHENNAI • COCHIN • GUWAHATI • HYDERABAD
JALANDHAR • KOLKATA • LUCKNOW • MUMBAI • RANCHI • NEW DELHI
BOSTON (USA) • ACCRA (GHANA) • NAIROBI (KENYA)

ENGINEERING MATHEMATICS-I

Printed and bound in India
Typeset at Goswami Associates, Delhi
First Edition : 2015; Second Edition : 2017
ISBN 978-93-5138-207-2

Published in India by

UNIVERSITY SCIENCE PRESS
(An Imprint of Laxmi Publications Pvt. Ltd.)
An ISO 9001:2008 Company
113, GOLDEN HOUSE, DARYAGANJ,
NEW DELHI - 110002, INDIA
Telephone : 91-11-4353 2500, 4353 2501
Fax : 91-11-2325 2572, 4353 2528
www.laxmipublications.com info@laxmipublications.com

Branches

✆	Bengaluru	080-26 75 69 30	
✆	Chennai	044-24 34 47 26,	24 35 95 07
✆	Cochin	0484-237 70 04,	405 13 03
✆	Guwahati	0361-254 36 69,	251 38 81
✆	Hyderabad	040-27 55 53 83,	27 55 53 93
✆	Jalandhar	0181-222 12 72	
✆	Kolkata	033-22 27 43 84	
✆	Lucknow	0522-220 99 16	
✆	Mumbai	022-24 91 54 15,	24 92 78 69
✆	Ranchi	0651-220 44 64	

C—
Printed at: Repro Knowledgecast Limited, Thane

CONTENTS

ACKNOWLEDGEMENTS

We are indebted to many of our former teachers and senior professors who have directly or indirectly helped us in preparing this book.

We have great pleasure to express our sincere thanks and deep sense of gratitude to **Prof. T. Srinivas, Prof. Shabbir Ahmed** and **Prof. M.H. Muddebihal** for their valuable suggestions and encouragement.

Our special and cordial thanks go to the highly esteemed Chairman, Hon'ble Secretary, Governing council members and Board of Governors of Muffakham Jah College of Engineering and Technology, Hyderabad for their cooperation and motivation in preparing this book. We are very much grateful to **Dr. Basheer Ahmed**, Advisor-cum-Director, MJCET for his kind encouragement.

We also thank the Principal, Dean (Academics), Dean (Administration), Heads of the department and colleagues of MJCET for their great help and effort during the periods of preparing the manuscript and publishing the book.

Lastly, we offer our love and gratitude to our beloved parents, family members for their continued encouragement and inspiration without which this book could not have taken the shape.

—Authors

PREFACE TO THE SECOND EDITION

There is no dearth of books on Engineering Mathematics–I. The students find it difficult to solve most of the problems in the exercises in the absence of an adequate number of solved examples. This inspired us to write the book Engineering Mathematics–I. The contents of the book are based on revised syllabus prescribed by Osmania University. An outstanding and distinguishing feature of the book is the large number of workedout examples followed by well graded problems. Numerous examples and problems have been given from previous question papers of Osmania University.

The selection, arrangement, and presentation of the text have been made with the greatest care, based on past and present teaching experience. The presentation is detailed, to help readers avoid frequent references to other books. We have endeavoured to present the fundamental concepts in a comprehensive and lucid manner.

Intimations of errors and suggestions for improvement will be highly appreciated and gratefully acknowledged.

—Authors

PREFACE TO THE FIRST EDITION

There is no dearth of books on Engineering Mathematics. The students find difficulty to solve most of the problems in the exercises in the absence of an adequate number of solved examples. This inspired us to write this book Engineering Mathematics-I . The contents of the book is based on revised syllabus prescribed by Osmania University. An outstanding and distinguishing feature of the book is the large number of typical solved examples followed by well graded problems. Numerous of examples and problems have been given from previous question papers of Osmania University.

The selection, arrangement, and presentation of the text has been made with the greatest care, based on past and present teaching experience. The presentation is detailed, to avoid readers by frequent references to details in other books. The examples are simple.

We have endeavoured to present the fundamental concepts in a comprehensive and lucid manner.

Intimations of errors and suggestions for improvement will be highly appreciated and gratefully acknowledged.

—Authors

Syllabus

Engineering Mathematics–I

(Common to All Branches)

UNIT-I: Linear Algebra

Introduction to matrices, Elementary row and column operations, Rank of a Matrix, Echelon form, System of linear equations, Eigenvalues, Eigen vectors, Cayley-Hamilton theorem, Diagonalization, Quadratic forms, Signature and Index.

UNIT-II: Infinite Series

Sequences, Infinite series, Convergence and Divergence, P-Series test, Geometric series test, Comparison tests, D'Alembert's Ratio test, Raabe's test, Cauchy's nth root test, Alternating series, Leibnitz's test, Absolute convergence, Conditional convergence.

UNIT-III: Differential Calculus

Rolle's theorem, Lagrange's and Cauchy's mean value theorems, Taylor's series, Curvature, Radius of curvature, Envelopes, Evolutes and Involutes, Asymptotes of a curve, Curve sketching (Cartesian).

UNIT-IV: Functions of Several Variables

Limits and Continuity of Functions of two variables, Partial derivatives, Total differentials and derivatives, Derivatives of composite and implicit functions, Higher order partial derivatives, Taylor's theorem for functions of two variables, Maxima and minima of functions of two variables, Jacobian, Change of variables.

UNIT-V: Vector Calculus

Scalar and vector fields, Vector differentiation, Gradient of a scalar field, Directional derivative, Divergence and Curl of a vector field, Line, Surface and Volume integrals, Green's theorem in a plane, Gauss's divergence theorem, Stoke's theorem and their applications.

UNIT 1
Linear Algebra

1.1. INTRODUCTION

Linear algebra comprises of the theory and applications of linear system of equations, linear transformations, eigen values and eigen vector problems. Determinants were first introduced for solving linear system of equations and have important engineering applications in systems of differential equations, electrical networks, eigen value problems and many more. Many complicated expressions occurring in electrical and mechanical systems can be simplified by expressing them in the form of determinants.

Matrices originated as mere stores of information but, at present, have found very wide application. They play a very vital role not only in mathematics but also in communication theory, network analysis, theory of structures, quantum mechanics, biology, sociology, economics, psychology, statistics etc.

Determinants and matrices play the key role in providing suitable criteria for testing the consistency of a system of linear equations.

In this unit, we first deal with determinants and then matrices.

DETERMINANTS

1.2. DEFINITIONS

Let us eliminate x and y from the equations

$$a_1x + b_1y = 0 \quad \text{and} \quad a_2x + b_2y = 0$$

The eliminant is $\dfrac{a_1}{b_1} = \dfrac{a_2}{b_2}$ $\qquad \left(\text{each} = -\dfrac{y}{x}\right)$

or $\qquad a_1b_2 - a_2b_1 = 0$

which is conveniently written in the compact form $\begin{vmatrix} a_1 & b_1 \\ a_2 & b_2 \end{vmatrix} = 0$

In other words, we have $\begin{vmatrix} a_1 & b_1 \\ a_2 & b_2 \end{vmatrix} = a_1b_2 - a_2b_1$

The expression $\begin{vmatrix} a_1 & b_1 \\ a_2 & b_2 \end{vmatrix}$ is called a **determinant of the second order.** The numbers a_1, b_1, a_2, b_2 are called the **elements** of the determinant. $a_1b_2 - a_2b_1$ is called the **expansion** or the **value** of the determinant.

We have

$$\begin{vmatrix} a_1 & b_1 & c_1 \\ a_2 & b_2 & c_2 \\ a_3 & b_3 & c_3 \end{vmatrix} = a_1(b_2c_3 - b_3c_2) + b_1(c_2a_3 - c_3a_2) + c_1(a_2b_3 - a_3b_2) \quad \text{...(1)}$$

The expression $\begin{vmatrix} a_1 & b_1 & c_1 \\ a_2 & b_2 & c_2 \\ a_3 & b_3 & c_3 \end{vmatrix}$ is called a **determinant of the third order** and the RHS of (1) is called the **expansion** or the **value** of the determinant.

A determinant of the nth order is denoted by

$$\Delta = \begin{vmatrix} a_{11} & a_{12} & a_{13} \text{......} a_{1n} \\ a_{21} & a_{22} & a_{23} \text{......} a_{2n} \\ \text{..............................} \\ \text{..............................} \\ a_{n1} & a_{n2} & a_{n3} \text{......} a_{nn} \end{vmatrix}$$

The elements $a_{11}, a_{22}, a_{33}, ..., a_{nn}$ (*i.e.*, the elements a_{ij} with $i = j$) are called the **diagonal elements.** The diagonal through the left hand top corner along which the diagonal elements lie is called the **leading or principal diagonal.** Also $a_{11}, a_{22}, a_{33}, ..., a_{nn}$ is called the **leading term.**

1.3. MINORS AND CO-FACTORS

The **minor** of an element in a determinant is the determinant obtained by **omitting** the row and the column in which that element lies.

For example, consider the determinant, $\Delta = \begin{vmatrix} a_1 & b_1 & c_1 \\ a_2 & b_2 & c_2 \\ a_3 & b_3 & c_3 \end{vmatrix}$

To find the minor of c_2 which occurs in the second row and third column, we **omit** the second row and the third column.

Thus, minor of $c_2 = \begin{vmatrix} a_1 & b_1 \\ a_3 & b_3 \end{vmatrix}$

Similarly, minor of $a_3 = \begin{vmatrix} b_1 & c_1 \\ b_2 & c_2 \end{vmatrix}$

The **co-factor** of an element in a determinant is its minor with proper sign and is usually denoted by the corresponding capital letter. The co-factor of the element which lies in the ith row and jth column is $(-1)^{i+j}$ times the minor of the element.

Thus, the co-factor of $c_2 = (-1)^{2+3} \times$ minor of c_2 *i.e.*, $C_2 = -\begin{vmatrix} a_1 & b_1 \\ a_3 & b_3 \end{vmatrix}$

Similarly, A_3 = co-factor of $a_3 = (-1)^{3+1} \times$ minor of $a_3 = \begin{vmatrix} b_1 & c_1 \\ b_2 & c_2 \end{vmatrix}$.

1.4. EXPANSION OF A DETERMINANT

(*a*) For a determinant of second order, we have seen that $\begin{vmatrix} a_1 & b_1 \\ a_2 & b_2 \end{vmatrix} = a_1b_2 - a_2b_1$

Sign of a product remains unchanged with downward arrow while it changes with upward arrow.

Thus, $\begin{vmatrix} -2 & 5 \\ -3 & 4 \end{vmatrix} = (-2)(4) - (-3)(5) = 7$

(*b*) For a determinant of third order, we have seen that

$$\begin{vmatrix} a_1 & b_1 & c_1 \\ a_2 & b_2 & c_2 \\ a_3 & b_3 & c_3 \end{vmatrix} = a_1(b_2c_3 - b_3c_2) + b_1(c_2a_3 - c_3a_2) + c_1(a_2b_3 - a_3b_2)$$

$$= a_1\begin{vmatrix} b_2 & c_2 \\ b_3 & c_3 \end{vmatrix} - b_1\begin{vmatrix} a_2 & c_2 \\ a_3 & c_3 \end{vmatrix} + c_1\begin{vmatrix} a_2 & b_2 \\ a_3 & b_3 \end{vmatrix} = a_1A_1 + b_1B_1 + c_1C_1$$

= the sum of the products of elements in the first row with their respective co-factors.

In fact, there is nothing particular about the first row. We can expand a determinant in terms of any row or column as follows :

Multiply the elements of any row (or column) with their respective co-factors and add up.

Thus, $\Delta = \begin{vmatrix} a_1 & b_1 & c_1 \\ a_2 & b_2 & c_2 \\ a_3 & b_3 & c_3 \end{vmatrix}$

$= a_1A_1 + b_1B_1 + c_1C_1$ (expanding by first row)

$= a_2A_2 + b_2B_2 + c_2C_2$ (expanding by second row)

$= a_3A_3 + b_3B_3 + c_3C_3$ (expanding by third row)

$= a_1A_1 + a_2A_2 + a_3A_3$ (expanding by first column)

$= b_1B_1 + b_2B_2 + b_3B_3$ (expanding by second column)

$= c_1C_1 + c_2C_2 + c_3C_3$ (expanding by third column)

These expansions are called **Laplace's expansions** and the technique can be adopted for determinants of any order.

However, the sum of the products of the elements of any row (or column) by the co-factors of any other row (or column) is zero.

Thus, $a_1A_3 + b_1B_3 + c_1C_3 = a_1\begin{vmatrix} b_1 & c_1 \\ b_2 & c_2 \end{vmatrix} - b_1\begin{vmatrix} a_1 & c_1 \\ a_2 & c_2 \end{vmatrix} + c_1\begin{vmatrix} a_1 & b_1 \\ a_2 & b_2 \end{vmatrix}$

$$= a_1(b_1c_2 - b_2c_1) - b_1(a_1c_2 - a_2c_1) + c_1(a_1b_2 - a_2b_1)$$

$$= a_1b_1c_2 - a_1b_2c_1 - a_1b_1c_2 + a_2b_1c_1 + a_1b_2c_1 - a_2b_1c_1 = 0$$

In general, $a_iA_j + b_iB_j + c_iC_j = \begin{cases} \Delta & \text{if } i = j \\ 0 & \text{if } i \neq j \end{cases}$

Example. *Find the value of the determinant* $\Delta = \begin{vmatrix} 2 & -4 & 3 \\ 3 & 1 & 2 \\ 7 & 6 & 1 \end{vmatrix}$ *by expanding (i) from second row (ii) from third column.*

Sol. (*i*) Expanding from second row,

$$\Delta = 3\left(-\begin{vmatrix} -4 & 3 \\ 6 & 1 \end{vmatrix}\right) + 1\begin{vmatrix} 2 & 3 \\ 7 & 1 \end{vmatrix} + 2\left(-\begin{vmatrix} 2 & -4 \\ 7 & 6 \end{vmatrix}\right)$$

$$= -3(-4 - 18) + (2 - 21) - 2(12 + 28) = 66 - 19 - 80 = -33$$

(*ii*) Expanding from third column,

$$\Delta = 3\begin{vmatrix} 3 & 1 \\ 7 & 6 \end{vmatrix} + 2\left(-\begin{vmatrix} 2 & -4 \\ 7 & 6 \end{vmatrix}\right) + 1\begin{vmatrix} 2 & -4 \\ 3 & 1 \end{vmatrix}$$

$$= 3(18 - 7) - 2\,(12 + 28) + (2 + 12) = 33 - 80 + 14 = -33.$$

1.5. PROPERTIES OF DETERMINANTS

The following properties of determinants simplify their evaluation. We shall verify these properties for determinants of third order only, however, they hold good for determinants of any order.

Property 1. *The value of a determinant remains unaltered when rows are changed into columns and columns into rows.*

Property 2. *If two parallel lines (rows or columns) of a determinant are interchanged, the determinant retains its numerical value but changes in sign.*

Property 3. *If two parallel lines (rows or columns) of a determinant are identical, the determinant vanishes.*

Property 4. *If all the elements of a line (row or column) of a determinant are multiplied by a common factor, the whole determinant gets multiplied by that factor.*

Property 5. *If each element of a line of a determinant is the sum of n terms, the determinant can be expressed as the sum of n determinants of the same order.*

Property 6. *If each element of a line (row or column) of a determinant be added equimultiples of the corresponding elements of one or more parallel lines, the determinant remains unaltered.*

Note 1. Property 6 is used most frequently to simplify and evaluate determinants.

Note 2. If all the elements in a line of a determinant are zero, the value of the determinant is zero.

Note 3. Always expand by that line which has maximum number of zeros.

Note 4. If all the elements on one side of the leading diagonal are zero, then the determinant is called a *triangular determinant* and its value is equal to the leading term, *i.e.*, the product of elements in the leading diagonal.

For example,

$$\begin{vmatrix} a_1 & 0 & 0 & 0 \\ b_1 & b_2 & 0 & 0 \\ c_1 & c_2 & c_3 & 0 \\ d_1 & d_2 & d_3 & d_4 \end{vmatrix} = a_1\begin{vmatrix} b_2 & 0 & 0 \\ c_2 & c_3 & 0 \\ d_2 & d_3 & d_4 \end{vmatrix} \quad \text{(Expanding along } R_1\text{)}$$

$$= a_1 b_2 \begin{vmatrix} c_3 & 0 \\ d_3 & d_4 \end{vmatrix} \quad \text{(Expanding along } R_1\text{)}$$

$$= a_1 b_2 c_3 d_4 = \text{the leading term.}$$

Note 5. While operating, several parallel lines of a determinant may be altered simultaneously but one line must be left unaltered at every stage.

Property 7. (Factor Theorem). *If the elements of a determinant Δ are rational integral functions of x and two parallel lines become identical when x = a, then (x – a) is a factor of Δ.*

Example. *Evaluate* $\begin{vmatrix} 1^2 & 2^2 & 3^2 & 4^2 \\ 2^2 & 3^2 & 4^2 & 5^2 \\ 3^2 & 4^2 & 5^2 & 6^2 \\ 4^2 & 5^2 & 6^2 & 7^2 \end{vmatrix}$.

Sol. Let

$$\Delta = \begin{vmatrix} 1 & 4 & 9 & 16 \\ 4 & 9 & 16 & 25 \\ 9 & 16 & 25 & 36 \\ 16 & 25 & 36 & 49 \end{vmatrix} \quad \text{Operating } C_4 - C_3,\ C_3 - C_2,\ C_2 - C_1$$

$$= \begin{vmatrix} 1 & 3 & 5 & 7 \\ 4 & 5 & 7 & 9 \\ 9 & 7 & 9 & 11 \\ 16 & 9 & 11 & 13 \end{vmatrix} \quad \text{Operating } C_4 - C_3,\ C_3 - C_2$$

$$= \begin{vmatrix} 1 & 3 & 2 & 2 \\ 4 & 5 & 2 & 2 \\ 9 & 7 & 2 & 2 \\ 16 & 9 & 2 & 2 \end{vmatrix}$$

= 0 since C_3 and C_4 are identical.

MATRICES

1.6. DEFINITIONS

A set of *mn* numbers (real or complex) arranged in a rectangular array having *m* rows (horizontal lines) and *n* columns (vertical lines), the numbers being enclosed by brackets [ ] or ( ), is called an *m* × *n* **matrix** (read as "*m* by *n* matrix").

An *m* × *n* matrix is also called a matrix of order *m* × *n*. Each of the *mn* numbers is called an **element** of the matrix.

For example, $\begin{bmatrix} 2 & -1 & 5 \\ 3 & 0 & 4 \end{bmatrix}$ is a 2 × 3 matrix or matrix of order 2 × 3. It has two rows and three columns. The numbers 2, – 1, 5, 3, 0, 4 are its elements.

An *m* × *n* matrix is usually written as $\begin{bmatrix} a_{11} & a_{12} & a_{13} & \dots & a_{1n} \\ a_{21} & a_{22} & a_{23} & \dots & a_{2n} \\ a_{31} & a_{32} & a_{33} & \dots & a_{3n} \\ \dots & \dots & \dots & \dots & \dots \\ a_{m1} & a_{m2} & a_{m3} & \dots & a_{mn} \end{bmatrix}$

Here each element has two suffixes. The first suffix indicates the row and the second suffix indicates the column in which the element lies. Thus, a_{23} is the element lying in the second row and third column, a_{ij} is the element lying in the ith row and jth column.

For brevity, a matrix is usually denoted by a single capital letter A or B or C etc.

Thus, an $m \times n$ matrix A may be written as

$$A = [a_{ij}]_{m \times n} \quad \text{or} \quad A = [a_{ij}], \quad \text{where } i = 1, 2, 3, \ldots\ldots, m \,;\, j = 1, 2, 3, \ldots\ldots, n.$$

Real Matrix. A matrix is said to be real if all its elements are real numbers.

e.g., $\begin{bmatrix} \sqrt{5} & -3 & 1 \\ 0 & -\sqrt{2} & 7 \end{bmatrix}$ is a real matrix.

Square Matrix. A matrix in which the number of rows is equal to the number of columns is called a **square matrix,** otherwise, it is said to be a **rectangular matrix.**

Thus, a matrix $A = [a_{ij}]_{m \times n}$ is a square matrix if $m = n$ and a rectangular matrix if $m \neq n$.

A square matrix having n rows and n columns is called "a square matrix of order n" or "an n-rowed square matrix",

e.g., $\begin{bmatrix} a_{11} & a_{12} & a_{13} \\ a_{21} & a_{22} & a_{23} \\ a_{31} & a_{32} & a_{33} \end{bmatrix}$ is a 3-rowed square matrix.

The elements a_{11}, a_{22}, a_{33} of a square matrix are called its **diagonal elements** and the diagonal along which these elements lie is called the **principal diagonal.**

In a square matrix $A = [a_{ij}]$,

(*i*) for elements along the principal diagonal, $i = j$

(*ii*) for elements above the principal diagonal, $i < j$

(*iii*) for elements below the principal diagonal, $i > j$

(*iv*) for non-diagonal elements, $i \neq j$

The sum of the diagonal elements of a square matrix is called its **trace.** Thus, trace of the n rowed square matrix $A = [a_{ij}]$ is, $a_{11} + a_{22} + a_{33} + \ldots\ldots + a_{nn} = \sum_{i=1}^{n} a_{ij}$.

Row Matrix. A matrix having only one row and any number of columns *i.e.*, a matrix of order $1 \times n$ is called a **row matrix.** *e.g.*, $[2 \quad 5 \quad -3 \quad 0]$ is a row matrix.

Column Matrix. A matrix having only one column and any number of rows *i.e.*, a matrix of order $m \times 1$ is called a **column matrix.** *e.g.*, $\begin{bmatrix} \sqrt{2} \\ 0 \\ -1 \end{bmatrix}$ is a column matrix.

Null Matrix. A matrix in which each element is zero is called a **null matrix** or a **zero matrix.** A null matrix of order $m \times n$ is denoted by $O_{m \times n}$.

$$\begin{bmatrix} 0 & 0 \\ 0 & 0 \\ 0 & 0 \end{bmatrix} = O_{3 \times 2}, \quad \begin{bmatrix} 0 & 0 & 0 & 0 \\ 0 & 0 & 0 & 0 \end{bmatrix} = O_{2 \times 4}$$

Sub-matrix. A matrix obtained from a given matrix A by deleting some of its rows or column or both is called a **sub-matrix of A.**

Thus, $B = \begin{bmatrix} 3 & 0 \\ 1 & 4 \end{bmatrix}$ is a sub-matrix of $A = \begin{bmatrix} 0 & -1 & 2 & 5 \\ 3 & 5 & 0 & 7 \\ 1 & 6 & 4 & -2 \end{bmatrix}$

obtained by deleting the first row, second and fourth columns of A.

Diagonal Matrix. A square matrix in which all non-diagonal elements are zero is called a **diagonal matrix.**

Thus, $A = [a_{ij}]_{n \times n}$ is a diagonal matrix if $a_{ij} = 0$ for $i \neq j$.

For example, $A = \begin{bmatrix} 2 & 0 & 0 \\ 0 & -1 & 0 \\ 0 & 0 & 0 \end{bmatrix}$ is a diagonal matrix.

An n-rowed diagonal matrix is briefly written as diag. $[d_1, d_2, \ldots\ldots, d_n]$, where $d_1, d_2, \ldots\ldots, d_n$ are the diagonal elements. Thus, the above diagonal matrix A can be written as diag. [2, – 1, 0].

Scalar Matrix. A diagonal matrix in which all the diagonal elements are equal to a scalar, say k, is called a **scalar matrix.**

Thus a scalar matrix is a square matrix in which all non-diagonal elements are zero and all diagonal elements are equal to a scalar, say k.

i.e., $A = [a_{ij}]_{n \times n}$ is a scalar matrix if $a_{ij} = \begin{cases} 0 & \text{when } i \neq j \\ k & \text{when } i = j \end{cases}$

For example, $\begin{bmatrix} 2 & 0 & 0 \\ 0 & 2 & 0 \\ 0 & 0 & 2 \end{bmatrix} \begin{bmatrix} -5 & 0 & 0 & 0 \\ 0 & -5 & 0 & 0 \\ 0 & 0 & -5 & 0 \\ 0 & 0 & 0 & -5 \end{bmatrix}$ are scalar matrices.

Unit Matrix or Identity Matrix. A scalar matrix in which each diagonal element is unity (*i.e.*, 1) is called a **unit matrix or identity matrix.**

Thus, a unit matrix is a square matrix in which all non-diagonal elements are zero and all diagonal elements are equal to 1.

i.e., $A = [a_{ij}]_{n \times n}$ is a unit matrix if $a_{ij} = \begin{cases} 0 & \text{when } i \neq j \\ 1 & \text{when } i = j \end{cases}$

A unit matrix of order n is denoted by I_n. If the order is evident, it may be simply denoted by I.

Thus, $\begin{bmatrix} 1 & 0 \\ 0 & 1 \end{bmatrix} = I_2, \begin{bmatrix} 1 & 0 & 0 \\ 0 & 1 & 0 \\ 0 & 0 & 1 \end{bmatrix} = I_3$

Upper Triangular Matrix. A square matrix in which all the elements *below* the principal diagonal are zero is called an **upper triangular matrix.**

Thus, $A = [a_{ij}]_{n \times n}$ is an upper triangular matrix if $a_{ij} = 0$ for $i > j$.

For example, $\begin{bmatrix} 2 & 3 & 4 \\ 0 & -1 & 5 \\ 0 & 0 & 3 \end{bmatrix}$ is an upper triangular matrix.

Lower Triangular Matrix. A square matrix in which all the elements *above* the principal diagonal are zero is called a **lower triangular matrix.**

Thus, $A = [a_{ij}]_{n \times n}$ is a lower triangular matrix if $a_{ij} = 0$ for $i < j$.

For example, $\begin{bmatrix} -1 & 0 & 0 \\ 5 & 6 & 0 \\ 3 & 2 & 0 \end{bmatrix}$ is a lower triangular matrix.

Triangular Matrix. A square matrix in which all the elements either below or above the principal diagonal are zero is called a triangular matrix. Thus, a triangular matrix is either upper triangular or lower triangular.

Equal Matrices. Two matrices A and B are said to be equal (written as A = B) if and only if they have the same order and their corresponding elements are equal.

Thus, if $A = [a_{ij}]_{m \times n}$ and $B = [b_{ij}]_{p \times q}$, then A = B if and only if

(*i*) $m = p$ and $n = q$ (*ii*) $a_{ij} = b_{ij}$ for all i and j.

1.7. MATRIX MULTIPLICATION

Two matrices A and B are said to be conformable for the product AB **(in this very order of A and B)** if the number of columns in A (called the pre-factor) is equal to the number of rows in B (called the post-factor).

Thus, if the orders of A and B are $m \times n$ and $p \times q$ respectively, then

(*i*) AB is defined if number of columns in A = number of rows in B, *i.e.*, if $n = p$.

(*ii*) BA is defined if number of columns in B = number of rows in A, *i.e.*, if $q = m$.

Let $A = [a_{ij}]_{m \times n}$ and $B = [b_{ij}]_{n \times p}$ be two matrices conformable for the product AB, then AB is defined as the matrix $C = [c_{ij}]_{m \times p}$, where $c_{ij} = a_{i1} b_{1j} + a_{i2}b_{2j} + \ldots\ldots + a_{in}b_{nj} = \sum_{k=1}^{n} a_{ik}b_{kj}$

i.e., (*i, j*)th element of AB = sum of the products of the elements of *i*th row of A with the corresponding elements of *j*th column of B.

The rule for multiplication of two conformable matrices is called **row-by-column method.**

Consider $A = \begin{bmatrix} a_{11} & a_{12} & a_{13} \\ a_{21} & a_{22} & a_{23} \\ a_{31} & a_{32} & a_{33} \end{bmatrix}$ and $B = \begin{bmatrix} b_{11} & b_{12} \\ b_{21} & b_{22} \\ b_{31} & b_{32} \end{bmatrix}$

Orders of A and B are 3 × 3 and 3 × 2 respectively. AB is defined and is of order 3 × 2.

$$AB = C = \begin{bmatrix} c_{11} & c_{12} \\ c_{21} & c_{22} \\ c_{31} & c_{32} \end{bmatrix}$$

where c_{11} = sum of products of elements of 1st row of A and 1st column of B

$$= [a_{11}\ a_{12}\ a_{13}] \begin{bmatrix} b_{11} \\ b_{21} \\ b_{31} \end{bmatrix} = a_{11}b_{11} + a_{12}b_{21} + a_{13}b_{31}$$

c_{12} = sum of products of elements of 1st row of A and 2nd column of B

$$= [a_{11}\ a_{12}\ a_{13}] \begin{bmatrix} b_{12} \\ b_{22} \\ b_{32} \end{bmatrix} = a_{11}b_{12} + a_{12}b_{22} + a_{13}b_{32}$$

c_{21} = sum of products of elements of 2nd row of A and 1st column of B

$$= [a_{21}\ a_{22}\ a_{23}] \begin{bmatrix} b_{11} \\ b_{21} \\ b_{31} \end{bmatrix} = a_{21}b_{11} + a_{22}b_{21} + a_{23}b_{31}$$

c_{22} = sum of products of elements of 2nd row of A and 2nd column of B

$$= [a_{21}\ a_{22}\ a_{23}] \begin{bmatrix} b_{12} \\ b_{22} \\ b_{32} \end{bmatrix} = a_{21}b_{12} + a_{22}b_{22} + a_{23}b_{32}$$

c_{31} = sum of products of elements of 3rd row of A and 1st column of B

$$= [a_{31}\ a_{32}\ a_{33}] \begin{bmatrix} b_{11} \\ b_{21} \\ b_{31} \end{bmatrix} = a_{31}b_{11} + a_{32}b_{21} + a_{33}b_{31}$$

c_{32} = sum of products of elements of 3rd row of A and 2nd column of B

$$= [a_{31}\ a_{32}\ a_{33}] \begin{bmatrix} b_{12} \\ b_{22} \\ b_{32} \end{bmatrix} = a_{31}b_{12} + a_{32}b_{22} + a_{33}b_{32}$$

Thus,
$$AB = \begin{bmatrix} a_{11}b_{11} + a_{12}b_{21} + a_{13}b_{31} & a_{11}b_{12} + a_{12}b_{22} + a_{13}b_{32} \\ a_{21}b_{11} + a_{22}b_{21} + a_{23}b_{31} & a_{21}b_{12} + a_{22}b_{22} + a_{23}b_{32} \\ a_{31}b_{11} + a_{32}b_{21} + a_{33}b_{31} & a_{31}b_{12} + a_{32}b_{22} + a_{33}b_{32} \end{bmatrix}$$

Note. Another useful notation to remember matrix multiplication.

$$A = \begin{bmatrix} a_{11} & a_{12} & a_{13} \\ a_{21} & a_{22} & a_{23} \\ a_{31} & a_{32} & a_{33} \end{bmatrix} = \begin{bmatrix} R_1 \\ R_2 \\ R_3 \end{bmatrix},$$

where R_1 = 1st Row, R_2 = 2nd Row, R_3 = 3rd Row

$$B = \begin{bmatrix} b_{11} & b_{12} \\ b_{21} & b_{22} \\ b_{31} & b_{32} \end{bmatrix} = [C_1\ C_2],$$

where C_1 = 1st column, C_2 = 2nd column

$$AB = \begin{bmatrix} R_1 \\ R_2 \\ R_3 \end{bmatrix} [C_1 \ C_2] = \begin{bmatrix} R_1C_1 & R_1C_2 \\ R_2C_1 & R_2C_2 \\ R_3C_1 & R_3C_2 \end{bmatrix}.$$

Example. *If* $A = \begin{bmatrix} 1 & -2 & 3 \\ 2 & 3 & -1 \\ -3 & 1 & 2 \end{bmatrix}$ *and* $B = \begin{bmatrix} 1 & 0 & 2 \\ 0 & 1 & 2 \\ 1 & 2 & 0 \end{bmatrix}$, *form the products AB and BA and show that AB ≠ BA.*

Sol. $AB = \begin{bmatrix} 1 & -2 & 3 \\ 2 & 3 & -1 \\ -3 & 1 & 2 \end{bmatrix} \begin{bmatrix} 1 & 0 & 2 \\ 0 & 1 & 2 \\ 1 & 2 & 0 \end{bmatrix}$

$$= \begin{bmatrix} 1(1) - 2(0) + 3(1) & 1(0) - 2(1) + 3(2) & 1(2) - 2(2) + 3(0) \\ 2(1) + 3(0) - 1(1) & 2(0) + 3(1) - 1(2) & 2(2) + 3(2) - 1(0) \\ -3(1) + 1(0) + 2(1) & -3(0) + 1(1) + 2(2) & -3(2) + 1(2) + 2(0) \end{bmatrix} = \begin{bmatrix} 4 & 4 & -2 \\ 1 & 1 & 10 \\ -1 & 5 & -4 \end{bmatrix}$$

$$BA = \begin{bmatrix} 1 & 0 & 2 \\ 0 & 1 & 2 \\ 1 & 2 & 0 \end{bmatrix} \begin{bmatrix} 1 & -2 & 3 \\ 2 & 3 & -1 \\ -3 & 1 & 2 \end{bmatrix}$$

$$= \begin{bmatrix} 1(1) + 0(2) + 2(-3) & 1(-2) + 0(3) + 2(1) & 1(3) + 0(-1) + 2(2) \\ 0(1) + 1(2) + 2(-3) & 0(-2) + 1(3) + 2(1) & 0(3) + 1(-1) + 2(2) \\ 1(1) + 2(2) + 0(-3) & 1(-2) + 2(3) + 0(1) & 1(3) + 2(-1) + 0(2) \end{bmatrix} = \begin{bmatrix} -5 & 0 & 7 \\ -4 & 5 & 3 \\ 5 & 4 & 1 \end{bmatrix}$$

Orders of AB and BA are the same but their corresponding elements are not equal.

Hence, $AB \neq BA$.

1.8. PROPERTIES OF MATRIX MULTIPLICATION

(*i*) Matrix multiplication is not commutative in general *i.e.*, $AB \neq BA$.

(*ii*) Matrix multiplication is associative *i.e.*, $(AB)C = A(BC)$.

(*iii*) Matrix multiplication is distributive with respect to matrix addition. *i.e.*,

$$A(B + C) = AB + AC.$$

(*iv*) If A and I are square matrices of the same order, then $AI = IA = A$.

(*v*) If A is a square matrix of order n, then $A \times A = A^2$, $A \times A \times A = A^3$, $A \times A \times A$ ..... m times $= A^m$ are all square matrices of order n.

Also, we define $A^0 = I$

(*vi*) For any positive integer n, $I^n = I$.

Note. If A is a square matrix such that $A^2 = I$, then A is called an **involutory matrix.**

1.9. TRANSPOSE OF A MATRIX

Given a matrix A, then the matrix obtained from A by changing its rows into columns and columns into rows is called the **transpose** of A and is denoted by A′ or A^T.

For example, if $A = \begin{bmatrix} 1 & 0 & 2 & 5 \\ 2 & -1 & 3 & 7 \end{bmatrix}$, then $A' = \begin{bmatrix} 1 & 2 \\ 0 & -1 \\ 2 & 3 \\ 5 & 7 \end{bmatrix}$

Clearly, (*i*) if the order of A is $m \times n$ then order of A′ is $n \times m$.

(*ii*) (*i, j*)th element of A′ = (*j, i*)th element of A.

In symbols, if $A = [a_{ij}]_{m \times n}$, then $A' = [b_{ij}]_{n \times m}$, where $b_{ij} = a_{ji}$.

1.10. PROPERTIES OF TRANSPOSE OF A MATRIX

If A′ and B′ denote the transposes of A and B respectively, then

(*a*) **(A′)′ = A** *i.e.*, **the transpose of a matrix is the matrix itself.**

(*b*) **(A + B)′ = A′ + B′** *i.e.*, **the transpose of the sum of two matrices is equal to the sum of their transposes.**

(*c*) **(AB)′ = B′A′** *i.e.*, **the transpose of the product of two matrices is equal to the product of their transposes taken in the reverse order.**

1.11. SYMMETRIC MATRIX

A square matrix $A = [a_{ij}]$ is said to be symmetric if **A′ = A** *i.e.*, if the transpose of the matrix is equal to the matrix itself.

Thus, for a symmetric matrix $A = [a_{ij}]$, $a_{ij} = a_{ji}$

i.e., (*i, j*)th element of A = (*j, i*)th element of A

For example, $\begin{bmatrix} -1 & 2 & 3 \\ 2 & 5 & 6 \\ 3 & 6 & 4 \end{bmatrix}$, $\begin{bmatrix} a & h & g \\ h & b & f \\ g & f & c \end{bmatrix}$ are symmetric matrices.

1.12. SKEW-SYMMETRIC MATRIX (OR ANTI-SYMMETRIC MATRIX)

A square matrix $A = [a_{ij}]$ is said to be skew-symmetric if **A′ = – A** *i.e.*, if the transpose of the matrix is equal to the negative of the matrix.

Thus, for a skew-symmetric matrix $A = [a_{ij}]$, $a_{ij} = -a_{ji}$

i.e., (*i, j*)th element of A = negative of (*j, i*)th element of A

Putting $j = i$, $a_{ii} = -a_{ii} \Rightarrow 2a_{ii} = 0$ or $a_{ii} = 0$ for all i.

Thus, all diagonal elements of a skew-symmetric matrix are zero.

For example, $\begin{bmatrix} 0 & 2 & -3 \\ -2 & 0 & 1 \\ 3 & -1 & 0 \end{bmatrix}$, $\begin{bmatrix} 0 & h & -g \\ -h & 0 & f \\ g & -f & 0 \end{bmatrix}$ are skew-symmetric matrices.

1.13. ORTHOGONAL MATRIX

A square matrix A is called an orthogonal matrix if $AA' = I = A'A$ or $A^{-1} = A'$.

1.14. ADJOINT OF A SQUARE MATRIX

The adjoint of a square matrix is the transpose of the matrix obtained by replacing each element of A by its co-factor in A.

Adjoint of A is briefly written as adj A.

Thus, if $A = \begin{bmatrix} a_1 & b_1 & c_1 \\ a_2 & b_2 & c_2 \\ a_3 & b_3 & c_3 \end{bmatrix}$ then adj A = transpose of $\begin{bmatrix} A_1 & B_1 & C_1 \\ A_2 & B_2 & C_2 \\ A_3 & B_3 & C_3 \end{bmatrix} = \begin{bmatrix} A_1 & A_2 & A_3 \\ B_1 & B_2 & B_3 \\ C_1 & C_2 & C_3 \end{bmatrix}$

where the capital letters denote the co-factors of corresponding small letters in

$$A = \begin{bmatrix} a_1 & b_1 & c_1 \\ a_2 & b_2 & c_2 \\ a_3 & b_3 & c_3 \end{bmatrix}.$$

1.15. SINGULAR AND NON-SINGULAR MATRICES

A square matrix A is said to be singular if $|A| = 0$ and non-singular if $|A| \neq 0$.

1.16. INVERSE (OR RECIPROCAL) OF A SQUARE MATRIX

Let A be an n-rowed square matrix. If there exists an n-rowed square matrix B such that

$$AB = BA = I$$

then the matrix **A is said to be invertible** and B is called the **inverse** (or **reciprocal**) of A.

Note 1. Only square matrices can be invertible.

Note 2. From the definition, it is clear that if B is the inverse of A, then A is the inverse of B.

Note 3. Inverse of A is denoted by A^{-1}, thus $B = A^{-1}$ and $AA^{-1} = A^{-1}A = I$.

Note 4. From the above theorem, we conclude that A has inverse if and only if $|A| \neq 0$ and then $A^{-1} = \dfrac{\text{adj } A}{|A|}$.

Example. *If* $A = \begin{bmatrix} 3 & -3 & 4 \\ 2 & -3 & 4 \\ 0 & -1 & 1 \end{bmatrix}$, *find* A^{-1}.

Sol.

$$A = \begin{bmatrix} 3 & -3 & 4 \\ 2 & -3 & 4 \\ 0 & -1 & 1 \end{bmatrix} = \begin{bmatrix} a_1 & b_1 & c_1 \\ a_2 & b_2 & c_2 \\ a_3 & b_3 & c_3 \end{bmatrix}$$

$$|A| = \begin{vmatrix} 3 & -3 & 4 \\ 2 & -3 & 4 \\ 0 & -1 & 1 \end{vmatrix} \quad \text{operating } C_2 + C_3$$

$$= \begin{vmatrix} 3 & 1 & 4 \\ 2 & 1 & 4 \\ 0 & 0 & 1 \end{vmatrix} \quad \text{expanding by third row}$$

$$= 1\begin{vmatrix} 3 & 1 \\ 2 & 1 \end{vmatrix} = 3 - 2 = 1$$

$$|A| = 1 \neq 0 \quad \therefore \quad A^{-1} \text{ exists.}$$

$$A_1 = \begin{vmatrix} -3 & 4 \\ -1 & 1 \end{vmatrix} = 1, \quad A_2 = -\begin{vmatrix} -3 & 4 \\ -1 & 1 \end{vmatrix} = -1, \quad A_3 = \begin{vmatrix} -3 & 4 \\ -3 & 4 \end{vmatrix} = 0$$

$$B_1 = -\begin{vmatrix} 2 & 4 \\ 0 & 1 \end{vmatrix} = -2, \quad B_2 = \begin{vmatrix} 3 & 4 \\ 0 & 1 \end{vmatrix} = 3, \quad B_3 = -\begin{vmatrix} 3 & 4 \\ 2 & 4 \end{vmatrix} = -4$$

$$C_1 = \begin{vmatrix} 2 & -3 \\ 0 & -1 \end{vmatrix} = -2, \quad C_2 = -\begin{vmatrix} 3 & -3 \\ 0 & -1 \end{vmatrix} = 3, \quad C_3 = \begin{vmatrix} 3 & -3 \\ 2 & -3 \end{vmatrix} = -3$$

$$\therefore \quad \text{adj A} = \text{transpose of} \begin{bmatrix} A_1 & B_1 & C_1 \\ A_2 & B_2 & C_2 \\ A_3 & B_3 & C_3 \end{bmatrix} = \begin{bmatrix} A_1 & A_2 & A_3 \\ B_1 & B_2 & B_3 \\ C_1 & C_2 & C_3 \end{bmatrix} = \begin{bmatrix} 1 & -1 & 0 \\ -2 & 3 & -4 \\ -2 & 3 & -3 \end{bmatrix}$$

$$A^{-1} = \frac{1}{|A|} \text{adj A} = \begin{bmatrix} 1 & -1 & 0 \\ -2 & 3 & -4 \\ -2 & 3 & -3 \end{bmatrix} \qquad (\because \ |A| = 1)$$

Note 1. If A is invertible, then so is A^{-1} and $(A^{-1})^{-1} = A$.

Note 2. If A and B be two non-singular square matrices of the same order, then

$$(AB)^{-1} = B^{-1} A^{-1}.$$

Note 3. If A is a non-singular square matrix, then so is A′ and $(A')^{-1} = (A^{-1})'$.

Note 4. If A and B are two non-singular square matrices of the same order, then

$$\text{adj (AB)} = (\text{adj B}) (\text{adj A}).$$

1.17. ELEMENTARY TRANSFORMATIONS (OR OPERATIONS)

Any one of the following operations on a matrix is called an elementary transformation (or E-operation).

(*i*) **Interchange of two rows or two columns.**

The interchange of *i*th and *j*th rows is denoted by R_{ij}.

The interchange of *i*th and *j*th columns is denoted by C_{ij}.

(*ii*) **Multiplication of (each element of) a row or column by a non-zero number k.**

The multiplication of *i*th row by *k* is denoted by kR_i.

The multiplication of *i*th column by *k* is denoted by kC_i.

(*iii*) **Addition of k times the elements of a row (or column) to the corresponding elements of another row (or column), k ≠ 0.**

The addition of *k* times the *j*th row to the *i*th row is denoted by $R_i + kR_j$.

The addition of *k* times the *j*th column to the *i*th column is denoted by $C_i + kC_j$.

If a matrix B is obtained from a matrix A by one or more E-operations, then B is said to be **equivalent** to A.

Two equivalent matrices A and B are written as A ~ B.

1.18. ELEMENTARY MATRICES

The matrix obtained from a unit matrix I by subjecting it to one of the E-operations is called an elementary matrix.

For example, Let $I = \begin{bmatrix} 1 & 0 & 0 \\ 0 & 1 & 0 \\ 0 & 0 & 1 \end{bmatrix}$

(*i*) Operating R_{23} or C_{23} on I, we get the same elementary matrix $\begin{bmatrix} 1 & 0 & 0 \\ 0 & 0 & 1 \\ 0 & 1 & 0 \end{bmatrix}$.

It is denoted by E_{23}. Thus, the E-matrix obtained by either of the operations R_{ij} or C_{ij} on I is denoted by E_{ij}.

(*ii*) Operating $5R_2$ or $5C_2$ on I, we get the same elementary matrix $\begin{bmatrix} 1 & 0 & 0 \\ 0 & 5 & 0 \\ 0 & 0 & 1 \end{bmatrix}$.

It is denoted by $5E_2$. Thus, the E-matrix obtained by either of the operations kR_i or kC_i is denoted by kE_i.

(*iii*) Operating $R_2 + 4R_3$ on I, we get the elementary matrix $\begin{bmatrix} 1 & 0 & 0 \\ 0 & 1 & 4 \\ 0 & 0 & 1 \end{bmatrix}$.

It is denoted by $E_{23}(4)$. Thus, the E-matrix obtained by the operation $R_i + kR_j$ is denoted by $E_{ij}(k)$.

(*iv*) Operating $C_2 + 4C_3$ on I, we get the elementary matrix $\begin{bmatrix} 1 & 0 & 0 \\ 0 & 1 & 0 \\ 0 & 4 & 1 \end{bmatrix}$, which is the transpose of $\begin{bmatrix} 1 & 0 & 0 \\ 0 & 1 & 4 \\ 0 & 0 & 1 \end{bmatrix} = E_{23}(4)$ and is therefore, denoted by $E'_{23}(4)$. Thus, the E-matrix obtained by the operation $C_i + kC_j$ is denoted by $E_{ij}'(k)$.

1.19. RANK OF A MATRIX

(OU-Nov. 2006)

Let A be any $m \times n$ matrix. It has square sub-matrices of different orders. The determinants of these square sub-matrices are called minors of A. If **all** minors of order $(r + 1)$ are zero but there **is at least one** non-zero minor of order r, then r is called the rank of A. Symbolically, rank of A = r is written as $\rho(A) = r$.

From the definition of the rank of a matrix A, it follows that :

(*i*) If A is a null matrix, then $\rho(A) = 0$.

[$\because$ every minor of A has zero value.]

(*ii*) If A is not a null matrix, then $\rho(A) \geq 1$.

(*iii*) If A is a non-singular $n \times n$ matrix, then $\rho(A) = n$

[$\because$ $|A| \neq 0$ is the largest minor of A.]

If I_n is the $n \times n$ unit matrix, then $|I_n| = 1 \neq 0 \Rightarrow \rho(I_n) = n$.

(*iv*) If A is an $m \times n$ matrix, then $\rho(A) \leq$ minimum of m and n.

(*v*) If all minors of order r are equal to zero, then $\rho(A) < r$.

To determine the rank of a matrix A, we adopt the following different methods :

METHOD 1: Start with the highest order minor (or minors) of A. Let their order be r. **If any one of them** is non-zero, then $\rho(A) = r$.

If **all of them** are zero, start with minors of next lower order $(r - 1)$ and so on till you get a non-zero minor. The order of that minor is the rank of A.

This method **usually** involves a lot of computational work since we have to evaluate several determinants.

1.20. METHOD 2: ECHELON FORM OR TRIANGULAR FORM

By using row and column operations, any non-zero matrix A can be reduced to Echelon form in which all the elements below the principal diagonal elements are zeros.

Note 1. $\rho(A)$ = Rank of A = Number of non-zero rows in Echelon form of A.

Note 2. Non-zero row means a row which contains at least one non-zero element.

Example 1. *Find the rank of the matrix* $\begin{bmatrix} 1 & 2 & 3 & 2 \\ 2 & 3 & 5 & 1 \\ 1 & 3 & 4 & 5 \end{bmatrix}$ *by reducing into Echelon form.*

Sol. Let
$$A = \begin{bmatrix} 1 & 2 & 3 & 2 \\ 2 & 3 & 5 & 1 \\ 1 & 3 & 4 & 5 \end{bmatrix}$$

Applying $R_2 \to R_2 - 2R_1$, $R_3 \to R_3 - R_1$
$$\sim \begin{bmatrix} 1 & 2 & 3 & 2 \\ 0 & -1 & -1 & -3 \\ 0 & 1 & 1 & 3 \end{bmatrix}$$

Applying $R_3 \to R_3 + R_2$
$$\sim \begin{bmatrix} 1 & 2 & 3 & 2 \\ 0 & -1 & -1 & -3 \\ 0 & 0 & 0 & 0 \end{bmatrix}$$

This is an Echelon form.

$\therefore$ Rank = Number of non-zero rows = 2.

Example 2. *Use elementary transformations to reduce the following matrix A to triangular form and hence find the rank of A.*
$$A = \begin{bmatrix} -1 & 2 & 3 & -2 \\ 2 & -5 & 1 & 2 \\ 3 & -8 & 5 & 2 \\ 5 & -12 & -1 & 6 \end{bmatrix}.$$

Sol. Given
$$A = \begin{bmatrix} -1 & 2 & 3 & -2 \\ \mathbf{2} & -5 & 1 & 2 \\ \mathbf{3} & -8 & 5 & 2 \\ \mathbf{5} & -12 & -1 & 6 \end{bmatrix}$$

Applying $R_2 \to R_2 + 2R_1$, $R_3 \to R_3 + 3R_1$, $R_4 \to R_4 + 5R_1$
$$\sim \begin{bmatrix} -1 & 2 & 3 & -2 \\ 0 & -1 & 7 & -2 \\ 0 & -\mathbf{2} & 14 & -4 \\ 0 & -\mathbf{2} & 14 & -4 \end{bmatrix}$$

Applying $R_3 \to R_3 - 2R_2$, $R_4 \to R_4 - 2R_2$

$$\sim \begin{bmatrix} -1 & 2 & 3 & -2 \\ 0 & -1 & 7 & -2 \\ 0 & 0 & 0 & 0 \\ 0 & 0 & 0 & 0 \end{bmatrix}$$

is a triangular form.

$\therefore$ Rank = 2.

Example 3. *Find the rank of the matrix*

$$A = \begin{bmatrix} 2 & 3 & 1 & 0 & 4 \\ 3 & 1 & 2 & -1 & 1 \\ 4 & -1 & 3 & -2 & -2 \\ 5 & 4 & 3 & -1 & 5 \end{bmatrix}.$$ (OU-Dec. 2013)

Sol. Given
$$A = \begin{bmatrix} 2 & 3 & 1 & 0 & 4 \\ \mathbf{3} & 1 & 2 & -1 & 1 \\ \mathbf{4} & -1 & 3 & -2 & -2 \\ \mathbf{5} & 4 & 3 & -1 & 5 \end{bmatrix}$$

Applying $R_2 \to 2R_2 - 3R_1$, $R_3 \to R_3 - 2R_1$, $R_4 \to 2R_4 - 5R_1$

$$\sim \begin{bmatrix} 2 & 3 & 1 & 0 & 4 \\ 0 & -7 & 1 & -2 & -10 \\ 0 & -\mathbf{7} & 1 & -2 & -10 \\ 0 & -\mathbf{7} & 1 & -2 & -10 \end{bmatrix}$$

Now, applying $R_3 \to R_3 - R_2$, $R_4 \to R_4 - R_2$

$$\sim \begin{bmatrix} 2 & 3 & 1 & 0 & 4 \\ 0 & -7 & 1 & -2 & -10 \\ 0 & 0 & 0 & 0 & 0 \\ 0 & 0 & 0 & 0 & 0 \end{bmatrix}$$

which is an Echelon form.

Hence, Rank = 2.

Example 4. *Determine rank of the matrix* $\begin{bmatrix} 5 & 6 & 7 & 8 \\ 6 & 7 & 8 & 9 \\ 11 & 12 & 13 & 14 \\ 16 & 17 & 18 & 19 \end{bmatrix}$. (OU-Apr. 2007)

Sol. Let
$$A = \begin{bmatrix} 5 & 6 & 7 & 8 \\ \mathbf{6} & 7 & 8 & 9 \\ \mathbf{11} & 12 & 13 & 14 \\ \mathbf{16} & 17 & 18 & 19 \end{bmatrix}$$

Applying $R_2 \to 5R_2 - 6R_1$, $R_3 \to 5R_3 - 11R_1$, $R_4 \to 5R_4 - 16R_1$

$$\sim \begin{bmatrix} 5 & 6 & 7 & 8 \\ 0 & -1 & -2 & -3 \\ 0 & -\mathbf{6} & -12 & -18 \\ 0 & -\mathbf{11} & -22 & -33 \end{bmatrix}$$

Applying $R_3 \to R_3 - 6R_2$, $R_4 \to R_4 - 11R_2$

$$\sim \begin{bmatrix} 5 & 6 & 7 & 8 \\ 0 & -1 & -2 & -3 \\ 0 & 0 & 0 & 0 \\ 0 & 0 & 0 & 0 \end{bmatrix}$$

This is an Echelon form.

$\therefore$ Rank = 2.

Example 5. *Find the rank of the matrix*

$$A = \begin{bmatrix} 2 & 1 & 5 & -1 \\ -1 & 2 & 5 & 3 \\ 3 & 2 & 9 & -1 \end{bmatrix}.$$ (OU-Dec. 2008)

Sol. Given
$$A = \begin{bmatrix} 2 & 1 & 5 & -1 \\ -1 & 2 & 5 & 3 \\ 3 & 2 & 9 & -1 \end{bmatrix}$$

Applying $R_1 \leftrightarrow R_2$

$$\sim \begin{bmatrix} -1 & 2 & 5 & 3 \\ \mathbf{2} & 1 & 5 & -1 \\ \mathbf{3} & 2 & 9 & -1 \end{bmatrix}$$

Applying $R_2 \to R_2 + 2R_1$, $R_3 \to R_3 + 3R_1$

$$\sim \begin{bmatrix} -1 & 2 & 5 & 3 \\ 0 & 5 & 15 & 5 \\ 0 & 8 & 24 & 8 \end{bmatrix}$$

Applying $R_2 \to \dfrac{R_2}{5}$, $R_3 \to \dfrac{R_3}{8}$

$$\sim \begin{bmatrix} -1 & 2 & 5 & 3 \\ 0 & 1 & 3 & 1 \\ 0 & \mathbf{1} & 3 & 1 \end{bmatrix}$$

Applying $R_3 \to R_3 - R_2$

$$\sim \begin{bmatrix} -1 & 2 & 5 & 3 \\ 0 & 1 & 3 & 1 \\ 0 & 0 & 0 & 0 \end{bmatrix}$$

This is an Echelon form.

$\therefore$ Rank = 2.

Example 6. *Find the rank of the matrix*

$$A = \begin{bmatrix} 1 & 3 & 4 & 3 \\ 3 & 9 & 12 & 3 \\ 5 & 15 & 20 & 9 \end{bmatrix}.$$ (OU-June 2011)

Sol.
$$A = \begin{bmatrix} 1 & 3 & 4 & 3 \\ \mathbf{3} & 9 & 12 & 3 \\ \mathbf{5} & 15 & 20 & 9 \end{bmatrix}$$

Applying $R_2 \to R_2 - 3R_1$, $R_3 \to R_3 - 5R_1$

$$\sim \begin{bmatrix} 1 & 3 & 4 & 3 \\ 0 & 0 & 0 & -6 \\ 0 & 0 & 0 & -\mathbf{6} \end{bmatrix}$$

Applying $R_3 \to R_3 - R_2$

$$\sim \begin{bmatrix} 1 & 3 & 4 & 3 \\ 0 & 0 & 0 & -6 \\ 0 & 0 & 0 & 0 \end{bmatrix}$$

This is an Echelon form.

$\therefore$ Rank = 2.

Example 7. *Determine the rank of the matrix* $\begin{bmatrix} 1 & 3 & 4 & 3 \\ 3 & 9 & 12 & 3 \\ 1 & 3 & 4 & 1 \end{bmatrix}$. (OU-Dec. 2007)

Sol. Let
$$A = \begin{bmatrix} 1 & 3 & 4 & 3 \\ \mathbf{3} & 9 & 12 & 3 \\ \mathbf{1} & 3 & 4 & 1 \end{bmatrix}$$

Applying $R_2 \to R_2 - 3R_1$, $R_3 \to R_3 - R_1$

$$\sim \begin{bmatrix} 1 & 3 & 4 & 3 \\ 0 & 0 & 0 & -6 \\ 0 & 0 & 0 & -\mathbf{2} \end{bmatrix}$$

Applying $R_3 \to 3R_3 - R_2$

$$\sim \begin{bmatrix} 1 & 3 & 4 & 3 \\ 0 & 0 & 0 & -6 \\ 0 & 0 & 0 & 0 \end{bmatrix}$$

This is an Echelon form.

$\therefore$ Rank = 2.

Example 8. *Find the rank of the matrix* $A = \begin{bmatrix} 1 & -2 & 1 & 2 \\ 2 & -2 & 0 & 6 \\ 4 & 2 & 0 & 2 \\ 1 & -1 & 0 & 3 \end{bmatrix}$. (OU-Nov. 2006)

Sol. Applying $R_2 \to R_2 - 2R_1$, $R_3 \to R_3 - 4R_1$, $R_4 \to R_4 - R_1$

$$\sim \begin{bmatrix} 1 & -2 & 1 & 2 \\ 0 & 2 & -2 & 2 \\ 0 & \mathbf{10} & -4 & -6 \\ 0 & \mathbf{1} & -1 & 1 \end{bmatrix}$$

Applying $R_3 \to R_3 - 5R_2$, $R_4 \to 2R_4 - R_2$

$$\sim \begin{bmatrix} 1 & -2 & 1 & 2 \\ 0 & 2 & -2 & 2 \\ 0 & 0 & 6 & -16 \\ 0 & 0 & 0 & 0 \end{bmatrix}$$

This is an Echelon form.

$\therefore$ Rank = 3.

1.21. METHOD 3: NORMAL FORM OR CANONICAL FORM (This Topic is for Student Reference Only)

By using row and column operations, any non-zero matrix A can be reduced to one of the following four forms, called the normal form of A

(*i*) I_r (*ii*) $[I_r \;\; 0]$

(*iii*) $\begin{bmatrix} I_r \\ 0 \end{bmatrix}$ (*iv*) $\begin{bmatrix} I_r & 0 \\ 0 & 0 \end{bmatrix}$

then rank of A = r(A) = *r*.

Note. $\rho(A)$ = The number of non-zero rows in normal form.

Example 1. *Find the rank of the matrix* $\begin{bmatrix} 2 & 3 & -1 & -1 \\ 1 & -1 & -2 & -4 \\ 3 & 1 & 3 & -2 \\ 6 & 3 & 0 & -7 \end{bmatrix}$.

Sol. Let $$A = \begin{bmatrix} 2 & 3 & -1 & -1 \\ 1 & -1 & -2 & -4 \\ 3 & 1 & 3 & -2 \\ 6 & 3 & 0 & -7 \end{bmatrix}$$

$\therefore$ $\rho(A) \le 4$

Applying $R_1 \leftrightarrow R_2$

$$\sim \begin{bmatrix} 1 & -1 & -2 & -4 \\ \mathbf{2} & 3 & -1 & -1 \\ \mathbf{3} & 1 & 3 & -2 \\ \mathbf{6} & 3 & 0 & -7 \end{bmatrix}$$

Applying $R_2 \to R_2 - 2R_1$, $R_3 \to R_3 - 3R_1$, $R_4 \to R_4 - 6R_1$

$$\sim \begin{bmatrix} 1 & \mathbf{-1} & \mathbf{-2} & \mathbf{-4} \\ 0 & 5 & 3 & 7 \\ 0 & 4 & 9 & 10 \\ 0 & 9 & 12 & 17 \end{bmatrix}$$

Applying $C_2 \to C_2 + C_1$, $C_3 \to C_3 + 2C_1$, $C_4 \to C_4 + 4C_1$

$$\sim \begin{bmatrix} 1 & 0 & 0 & 0 \\ 0 & 5 & 3 & 7 \\ 0 & \mathbf{4} & 9 & 10 \\ 0 & \mathbf{9} & 12 & 17 \end{bmatrix}$$

Applying $R_3 \to 5R_3 - 4R_2$, $R_4 \to 5R_4 - 9R_2$

$$\sim \begin{bmatrix} 1 & 0 & 0 & 0 \\ 0 & 5 & 3 & 7 \\ 0 & 0 & 33 & 22 \\ 0 & 0 & 33 & 22 \end{bmatrix}$$

Applying $R_3 \to \dfrac{R_3}{11}$, $R_4 \to \dfrac{R_4}{11}$

$$\sim \begin{bmatrix} 1 & 0 & 0 & 0 \\ 0 & 5 & 3 & 7 \\ 0 & 0 & 3 & 2 \\ 0 & 0 & 3 & 2 \end{bmatrix}$$

Applying $C_2 \to \dfrac{C_2}{5}$, $C_3 \to \dfrac{C_3}{3}$

$$\sim \begin{bmatrix} 1 & 0 & 0 & 0 \\ 0 & 1 & \mathbf{1} & \mathbf{7} \\ 0 & 0 & 1 & 2 \\ 0 & 0 & 1 & 2 \end{bmatrix}$$

Applying $C_3 \to C_3 - C_2$, $C_4 \to C_4 - 7C_2$

$$\sim \begin{bmatrix} 1 & 0 & 0 & 0 \\ 0 & 1 & 0 & 0 \\ 0 & 0 & 1 & 2 \\ 0 & 0 & \mathbf{1} & 2 \end{bmatrix}$$

Applying $R_4 \to R_4 - R_3$

$$\sim \begin{bmatrix} 1 & 0 & 0 & 0 \\ 0 & 1 & 0 & 0 \\ 0 & 0 & 1 & \mathbf{2} \\ 0 & 0 & 0 & 0 \end{bmatrix}$$

Applying $C_4 \to C_4 - 2C_3$

$$\sim \begin{bmatrix} 1 & 0 & 0 & 0 \\ 0 & 1 & 0 & 0 \\ 0 & 0 & 1 & 0 \\ 0 & 0 & 0 & 0 \end{bmatrix} \sim \begin{bmatrix} I_3 & 0 \\ 0 & 0 \end{bmatrix}$$

This is the normal form.

$\therefore$ Rank of A = $\rho(A) = 3$.

Example 2. *Reduce the matrix* $\begin{bmatrix} 2 & 1 & -6 & -3 \\ 3 & -3 & 2 & 1 \\ 1 & 1 & 2 & 1 \end{bmatrix}$ *to normal form and find its rank.*

(OU-May 2012)

Sol. Let $A = \begin{bmatrix} 2 & 1 & -6 & -3 \\ 3 & -3 & 2 & 1 \\ 1 & 1 & 2 & 1 \end{bmatrix}$

Applying $R_1 \leftrightarrow R_3$

$$\sim \begin{bmatrix} 1 & 1 & 2 & 1 \\ \mathbf{3} & -3 & 2 & 1 \\ \mathbf{2} & 1 & -6 & -3 \end{bmatrix}$$

Applying $R_2 \to R_2 - 3R_1$, $R_3 \to R_3 - 2R_1$

$$\sim \begin{bmatrix} 1 & \mathbf{1} & \mathbf{2} & \mathbf{1} \\ 0 & -6 & -4 & -2 \\ 0 & -1 & -10 & -5 \end{bmatrix}$$

Applying $C_2 \to C_2 - C_1$, $C_3 \to C_3 - 2C_1$, $C_4 \to C_4 - C_1$

$$\sim \begin{bmatrix} 1 & 0 & 0 & 0 \\ 0 & -6 & -4 & -2 \\ 0 & -1 & -10 & -5 \end{bmatrix}$$

Applying $R_2 \to \dfrac{R_2}{-2}$, $R_3 \to (-1)R_3$

$$\sim \begin{bmatrix} 1 & 0 & 0 & 0 \\ 0 & 3 & 2 & 1 \\ 0 & \mathbf{1} & 10 & 5 \end{bmatrix}$$

Applying $R_3 \to 3R_3 - R_2$

$$\sim \begin{bmatrix} 1 & 0 & 0 & 0 \\ 0 & 3 & \mathbf{2} & \mathbf{1} \\ 0 & 0 & 28 & 14 \end{bmatrix}$$

Applying $C_3 \to 3C_3 - 2C_2$, $C_4 \to 3C_4 - C_2$

$$\sim \begin{bmatrix} 1 & 0 & 0 & 0 \\ 0 & 3 & 0 & 0 \\ 0 & 0 & 84 & 42 \end{bmatrix}$$

Applying $C_2 \to \dfrac{C_2}{3}$, $C_3 \to \dfrac{C_3}{84}$, $C_4 \to \dfrac{C_4}{42}$

$$\sim \begin{bmatrix} 1 & 0 & 0 & 0 \\ 0 & 1 & 0 & 0 \\ 0 & 0 & 1 & \mathbf{1} \end{bmatrix}$$

Applying $C_4 \to C_4 - C_3$

$$\sim \begin{bmatrix} 1 & 0 & 0 & 0 \\ 0 & 1 & 0 & 0 \\ 0 & 0 & 1 & 0 \end{bmatrix} \sim [I_3 \quad 0]$$

This is a normal form.

$\therefore \qquad \rho(A) = 3.$

Example 3. *Reduce the matrix to normal form and find rank* $\begin{bmatrix} 0 & 1 & 2 & -2 \\ 4 & 0 & 2 & 6 \\ 2 & 1 & 3 & 1 \end{bmatrix}$.

Sol. Given
$$A = \begin{bmatrix} 0 & 1 & 2 & -2 \\ 4 & 0 & 2 & 6 \\ 2 & 1 & 3 & 1 \end{bmatrix}$$

Applying $C_1 \leftrightarrow C_2$, we get

$$\sim \begin{bmatrix} 1 & 0 & 2 & -2 \\ 0 & 4 & 2 & 6 \\ \mathbf{1} & 2 & 3 & 1 \end{bmatrix}$$

Applying $R_3 \to R_3 - R_1$, we get

$$\sim \begin{bmatrix} 1 & 0 & \mathbf{2} & \mathbf{-2} \\ 0 & 4 & 2 & 6 \\ 0 & 2 & 1 & 3 \end{bmatrix}$$

Applying $C_3 \to C_3 - 2C_1$, $C_4 \to C_4 + 2C_1$, we get

$$\sim \begin{bmatrix} 1 & 0 & 0 & 0 \\ 0 & 4 & 2 & 6 \\ 0 & 2 & 1 & 3 \end{bmatrix}$$

Applying $R_3 \to 2R_3 - R_2$, we get

$$\sim \begin{bmatrix} 1 & 0 & 0 & 0 \\ 0 & 4 & 2 & 6 \\ 0 & 0 & 0 & 0 \end{bmatrix}$$

Applying $C_2 \to \dfrac{C_2}{4}$, $C_3 \to \dfrac{C_3}{2}$, $C_4 \to \dfrac{C_4}{6}$

$$\sim \begin{bmatrix} 1 & 0 & 0 & 0 \\ 0 & 1 & 1 & 1 \\ 0 & 0 & 0 & 0 \end{bmatrix}$$

Applying $C_3 \to C_3 - C_2$, $C_4 \to C_4 - C_2$

$$\sim \begin{bmatrix} 1 & 0 & 0 & 0 \\ 0 & 1 & 0 & 0 \\ 0 & 0 & 0 & 0 \end{bmatrix} \sim \begin{bmatrix} I_2 & 0 \\ 0 & 0 \end{bmatrix}$$

This is the normal form

Hence the rank of matrix A is 2.

Example 4. *Find the rank of the matrix* $A = \begin{bmatrix} 2 & -2 & 0 & 6 \\ 4 & 2 & 0 & 2 \\ 1 & -1 & 0 & 3 \\ 1 & -2 & 1 & 2 \end{bmatrix}$ *by reducing it to canonical form.*

Sol. Given

$$A = \begin{bmatrix} 2 & -2 & 0 & 6 \\ 4 & 2 & 0 & 2 \\ 1 & -1 & 0 & 3 \\ 1 & -2 & 1 & 2 \end{bmatrix}$$

Applying $R_1 \leftrightarrow R_3$

$$\sim \begin{bmatrix} 1 & -1 & 0 & 3 \\ 4 & 2 & 0 & 2 \\ 2 & -2 & 0 & -6 \\ 1 & -2 & 1 & 2 \end{bmatrix}$$

Applying $R_2 \to R_2 - 4R_1$, $R_3 \to R_3 - 2R_1$, $R_4 \to R_4 - R_1$

$$\sim \begin{bmatrix} 1 & -1 & 0 & 3 \\ 0 & 6 & 0 & -10 \\ 0 & 0 & 0 & 0 \\ 0 & -1 & 1 & -1 \end{bmatrix}$$

Applying $C_2 \to C_2 + C_1$, $C_4 \to C_4 - 3C_1$

$$\sim \begin{bmatrix} 1 & 0 & 0 & 0 \\ 0 & 6 & 0 & -10 \\ 0 & 0 & 0 & 0 \\ 0 & -1 & 1 & -1 \end{bmatrix}$$

Applying $R_3 \to 6R_4 + R_2$

$$\sim \begin{bmatrix} 1 & 0 & 0 & 0 \\ 0 & 6 & 0 & -10 \\ 0 & 0 & 0 & 0 \\ 0 & 0 & 6 & -16 \end{bmatrix}$$

Applying $C_2 \to \dfrac{C_2}{6}$, $C_3 \to \dfrac{C_3}{6}$, $C_4 \to \dfrac{C_4}{-2}$

$$\sim \begin{bmatrix} 1 & 0 & 0 & 0 \\ 0 & 1 & 0 & 5 \\ 0 & 0 & 0 & 0 \\ 0 & 0 & 1 & 8 \end{bmatrix}$$

Applying $C_4 \to C_4 - 5C_2$

$$\sim \begin{bmatrix} 1 & 0 & 0 & 0 \\ 0 & 1 & 0 & 0 \\ 0 & 0 & 0 & 0 \\ 0 & 0 & 1 & 8 \end{bmatrix}$$

Applying $R_3 \leftrightarrow R_4$

$$\sim \begin{bmatrix} 1 & 0 & 0 & 0 \\ 0 & 1 & 0 & 0 \\ 0 & 0 & 1 & 8 \\ 0 & 0 & 0 & 0 \end{bmatrix}$$

Applying $C_4 \to C_4 - 8C_3$

$$\sim \begin{bmatrix} 1 & 0 & 0 & 0 \\ 0 & 1 & 0 & 0 \\ 0 & 0 & 1 & 0 \\ 0 & 0 & 0 & 0 \end{bmatrix} \sim \begin{bmatrix} I_3 & 0 \\ 0 & 0 \end{bmatrix}$$

This is a canonical form.

$\therefore$ Rank of A is 3.

Example 5. *Find the rank of the matrix*

(*i*) $\begin{bmatrix} 1 & 2 & 3 & 4 \\ -2 & 0 & 5 & 7 \end{bmatrix}$ (*ii*) $\begin{bmatrix} 1 & 2 & 3 \\ 1 & 4 & 2 \\ 2 & 6 & 5 \end{bmatrix}$

(*iii*) $\begin{bmatrix} 2 & 3 & 4 & -1 \\ 5 & 2 & 0 & -1 \\ -4 & 5 & 12 & -1 \end{bmatrix}$ (*iv*) $\begin{bmatrix} 1 & 2 & 3 & 0 \\ 2 & 4 & 3 & 2 \\ 3 & 2 & 1 & 3 \\ 6 & 8 & 7 & 5 \end{bmatrix}$.

Sol. (*i*) Here $A = \begin{bmatrix} 1 & 2 & 3 & 4 \\ -2 & 0 & 5 & 7 \end{bmatrix}$ is a 2×4 matrix.

$\therefore$ $\rho(A) \le 2$, the smaller of 2 and 4.

The second order minor $\begin{vmatrix} 1 & 2 \\ -2 & 0 \end{vmatrix} = 4 \neq 0$ $\therefore$ $\rho(A) = 2$.

(*ii*) Here $A = \begin{bmatrix} 1 & 2 & 3 \\ 1 & 4 & 2 \\ 2 & 6 & 5 \end{bmatrix}$ is a 3×3 matrix. $\therefore$ $\rho(A) \le 3$

Operating $R_2 - R_1$, $R_3 - 2R_1$

$$A \sim \begin{bmatrix} 1 & 2 & 3 \\ 0 & 2 & -1 \\ 0 & 2 & -1 \end{bmatrix} \text{ Operating } R_3 - R_2$$

$$\sim \begin{bmatrix} 1 & 2 & 3 \\ 0 & 2 & -1 \\ 0 & 0 & 0 \end{bmatrix}$$

The only third order minor is zero, but the second order minor

$$\begin{vmatrix} 1 & 2 \\ 0 & 2 \end{vmatrix} = 2 \neq 0 \quad \therefore \quad \rho(A) = 2.$$

(*iii*) Here $A = \begin{bmatrix} 2 & 3 & 4 & -1 \\ 5 & 2 & 0 & -1 \\ -4 & 5 & 12 & -1 \end{bmatrix}$ is a 3×4 matrix.

$\therefore$ $\rho(A) \le 3$

Operating C_{14}

$$A \sim \begin{bmatrix} -1 & 3 & 4 & 2 \\ -1 & 2 & 0 & 5 \\ -1 & 5 & 12 & -4 \end{bmatrix} \text{ Operating } R_2 - R_1, R_3 - R_1$$

$$\sim \begin{bmatrix} -1 & 3 & 4 & 2 \\ 0 & -1 & -4 & 3 \\ 0 & 2 & 8 & -6 \end{bmatrix} \text{ Operating } R_1 + 3R_2,\ R_3 + 2R_2$$

$$\sim \begin{bmatrix} -1 & 0 & -8 & 11 \\ 0 & -1 & -4 & 3 \\ 0 & 0 & 0 & 0 \end{bmatrix}$$

All the third order minors are zero but the second order minor

$$= \begin{vmatrix} -1 & 0 \\ 0 & -1 \end{vmatrix} = 1 \neq 0 \quad \therefore \quad \rho(A) = 2.$$

(*iv*) Here $\quad A = \begin{bmatrix} 1 & 2 & 3 & 0 \\ 2 & 4 & 3 & 2 \\ 3 & 2 & 1 & 3 \\ 6 & 8 & 7 & 5 \end{bmatrix}$ is a 4 × 4 matrix.

$\therefore \quad \rho(A) \leq 4$

Operating $R_4 - (R_3 + R_2 + R_1)$

$$A \sim \begin{bmatrix} 1 & 2 & 3 & 0 \\ 2 & 4 & 3 & 2 \\ 3 & 2 & 1 & 3 \\ 0 & 0 & 0 & 0 \end{bmatrix} \quad \text{Operating } R_2 - 2R_1,\ R_3 - 3R_1$$

$$\sim \begin{bmatrix} 1 & 2 & 3 & 0 \\ 0 & 0 & -3 & 2 \\ 0 & -4 & -8 & 3 \\ 0 & 0 & 0 & 0 \end{bmatrix} \quad \text{Operating } R_{23}$$

$$\sim \begin{bmatrix} 1 & 2 & 3 & 0 \\ 0 & -4 & -8 & 3 \\ 0 & 0 & -3 & 2 \\ 0 & 0 & 0 & 0 \end{bmatrix}$$

The only fourth order minor is zero. Since the third order minor

$$\begin{vmatrix} 1 & 2 & 3 \\ 0 & -4 & -8 \\ 0 & 0 & -3 \end{vmatrix} = (1)(-4)(-3) = 12 \neq 0 \qquad \therefore \quad \rho(A) = 3$$

Note. If $\rho(A) = n$ then determinant of A of order higher than n should be vanishes.

Example 6. *If the rank of the matrix* $A = \begin{bmatrix} 1 & 2 \\ a & 3 \end{bmatrix}$ *is one, then find 'a'.*

(OU-June 2013)

Sol. Since $\rho(A) = 1$

$\Rightarrow \quad |A|_{2\times 2} = 0$

$\Rightarrow \quad \begin{vmatrix} 1 & 2 \\ a & 3 \end{vmatrix} = 0 \Rightarrow 3 - 2a = 0$

$\Rightarrow \quad a = \dfrac{3}{2}.$

Example 7. *Find all values of* μ *for which the rank of the matrix* $A = \begin{bmatrix} \mu & -1 & 0 & 0 \\ 0 & \mu & -1 & 0 \\ 0 & 0 & \mu & -1 \\ -6 & 11 & -6 & 1 \end{bmatrix}$ *is 3.*

(OU-May 2008)

Sol. Since $\rho(A) = 3 \Rightarrow |A|_{4\times4} = 0$

$$\Rightarrow \begin{vmatrix} \mu & -1 & 0 & 0 \\ 0 & \mu & -1 & 0 \\ 0 & 0 & \mu & -1 \\ -6 & 11 & -6 & 1 \end{vmatrix} = 0$$

$$\Rightarrow \mu\begin{vmatrix} \mu & -1 & 0 \\ 0 & \mu & -1 \\ 11 & -6 & 1 \end{vmatrix} + 1\begin{vmatrix} \mu & -1 & 0 \\ 0 & \mu & -1 \\ -6 & -6 & 1 \end{vmatrix} = 0$$

$$\Rightarrow \mu[\mu(\mu - 6) + 1(0 + 11)] + 1[0 + 1(0 - 6) + 0] = 0$$

$$\Rightarrow \mu^2(\mu - 6) + 11\mu - 6 = 0$$

$$\Rightarrow \mu^3 - 6\mu^2 + 11\mu - 6 = 0$$

$$\Rightarrow (\mu - 1)(\mu - 2)(\mu - 3) = 0$$

$$\Rightarrow \mu = 1, 2, 3.$$

Example 8. *For which value of 'b' the rank of the matrix* $A = \begin{bmatrix} 1 & 5 & 4 \\ 0 & 3 & 2 \\ b & 13 & 10 \end{bmatrix}$ *is 2.*

Sol. Since $\rho(A) = 2 \Rightarrow |A|_{3\times3} = 0$

$$\Rightarrow \begin{vmatrix} 1 & 5 & 4 \\ 0 & 3 & 2 \\ b & 13 & 10 \end{vmatrix} = 0$$

$$\Rightarrow 1(30 - 26) - 5(0 - 2b) + 4(0 - 3b) = 0$$

$$\Rightarrow 4 + 10b - 12b = 0$$

$$\Rightarrow 4 = 2b$$

$$\Rightarrow b = 2.$$

EXERCISE 1

1. Find the rank of the matrices by reducing into normal form.

(*i*) $\begin{bmatrix} 2 & -1 & 0 & 5 \\ 0 & 3 & 1 & 4 \end{bmatrix}$ (*ii*) $\begin{bmatrix} 1 & 2 & 3 \\ 2 & 4 & 7 \\ 3 & 6 & 10 \end{bmatrix}$ (*iii*) $\begin{bmatrix} 1 & 3 & 4 & 5 \\ 1 & 2 & 6 & 7 \\ 1 & 5 & 0 & 10 \end{bmatrix}$.

2. Determine the rank of the matrix $A = \begin{bmatrix} 6 & 1 & 3 & 8 \\ 4 & 2 & 6 & -1 \\ 10 & 3 & 9 & 7 \\ 16 & 4 & 12 & 15 \end{bmatrix}$ by reducing it to the normal form.

3. Find the rank of the matrices by reducing into Echelon form.

(*i*) $\begin{bmatrix} 3 & -1 & 2 \\ -6 & 2 & 4 \\ -3 & 1 & 2 \end{bmatrix}$ (*ii*) $\begin{bmatrix} 1 & 2 & 3 & 2 \\ 2 & 3 & 5 & 1 \\ 1 & 3 & 4 & 5 \end{bmatrix}$

(*iii*) $\begin{bmatrix} 1 & -1 & 2 & -3 \\ 4 & 1 & 0 & 2 \\ 0 & 3 & 1 & 4 \\ 0 & 1 & 0 & 2 \end{bmatrix}$ (*iv*) $\begin{bmatrix} 0 & 1 & -3 & -1 \\ 1 & 0 & 1 & 1 \\ 3 & 1 & 0 & 2 \\ 1 & 1 & -2 & 0 \end{bmatrix}$.

Answers

1. (*i*) 2 (*ii*) 2 (*iii*) 3 **2.** 2

3. (*i*) 2 (*ii*) 2 (*iii*) 4 (*iv*) 2

1.22. SOLUTION OF A SYSTEM OF LINEAR EQUATIONS

Consider the system of equations $\left.\begin{aligned} a_1x + b_1y + c_1z = d_1 \\ a_2x + b_2y + c_2z = d_2 \\ a_3x + b_3y + c_3z = d_3 \end{aligned}\right\}$ (3 equations in 3 unknowns)

In matrix notation, these equations can be written as

$$\begin{bmatrix} a_1x + b_1y + c_1z \\ a_2x + b_2y + c_2z \\ a_3x + b_3y + c_3z \end{bmatrix} = \begin{bmatrix} d_1 \\ d_2 \\ d_3 \end{bmatrix}$$

or
$$\begin{bmatrix} a_1 & b_1 & c_1 \\ a_2 & b_2 & c_2 \\ a_3 & b_3 & c_3 \end{bmatrix}\begin{bmatrix} x \\ y \\ z \end{bmatrix} = \begin{bmatrix} d_1 \\ d_2 \\ d_3 \end{bmatrix}$$

or
$$AX = B$$

where $A = \begin{bmatrix} a_1 & b_1 & c_1 \\ a_2 & b_2 & c_2 \\ a_3 & b_3 & c_3 \end{bmatrix}$ is called the co-efficient matrix, $X = \begin{bmatrix} x \\ y \\ z \end{bmatrix}$ is the column matrix of unknowns, $B = \begin{bmatrix} d_1 \\ d_2 \\ d_3 \end{bmatrix}$ is the column matrix of constants.

If $d_1 = d_2 = d_3 = 0$, then B = O and the matrix equation AX = B reduces to AX = O.

Such a system of equations is called a system of **homogeneous linear equations.**

If atleast one of d_1, d_2, d_3 is non-zero, then B ≠ O.

Such a system of equations is called a system of **non-homogeneous linear equations.**

Solving the matrix equation AX = B means finding X, *i.e.*, finding a column matrix $\begin{bmatrix} \alpha \\ \beta \\ \gamma \end{bmatrix}$ such that

$$X = \begin{bmatrix} x \\ y \\ z \end{bmatrix} = \begin{bmatrix} \alpha \\ \beta \\ \gamma \end{bmatrix}. \text{ Then } x = \alpha, y = \beta, z = \gamma.$$

The matrix equation AX = B need not always have a solution. It may have no solution or a unique solution or an infinite number of solutions.

A system of equations having **no solution** is called an **inconsistent** system of equations.

A system of equations having **one or more solution** is called **a consistent** system of equations.

For a system of non-homogeneous linear equations AX = B.

(*i*) if $\rho[A \vdots B] \neq \rho(A)$, the system is inconsistent.

(*ii*) if $\rho[A \vdots B] = \rho(A)$ = number of unknowns, the system has a unique solution.

(*iii*) $\rho[A \vdots B] = \rho(A)$ < number of unknowns, the system has an infinite number of solutions.

The matrix $[A \vdots B]$ in which the elements of A and B are written side by side is called the **augmented matrix.**

For a system of homogeneous linear equations AX = O

(*i*) X = O is always a solution. This solution in which each unknown has the value zero is called the **Null Solution** or the **Trivial Solution.** Thus a homogeneous system is always consistent.

A system of homogeneous linear equations has either the trivial solution or an infinite number of solutions.

(*ii*) if ρ (A) = number of unknowns, the system has only the trivial solution.

(*iii*) if ρ (A) < number of unknowns, the system has an infinite number of non-trivial solutions.

ILLUSTRATIVE EXAMPLES

Example 1. *Solve the equations :*

$$x_1 + 3x_2 + 2x_3 = 0, \quad 2x_1 - x_2 + 3x_3 = 0, \quad 3x_1 - 5x_2 + 4x_3 = 0, x_1 + 17x_2 + 4x_3 = 0.$$

Sol. In matrix notation, the given system of equations can be written as

$$AX = O$$

where $A = \begin{bmatrix} 1 & 3 & 2 \\ 2 & -1 & 3 \\ 3 & -5 & 4 \\ 1 & 17 & 4 \end{bmatrix}, X = \begin{bmatrix} x_1 \\ x_2 \\ x_3 \end{bmatrix}$

Operating $R_2 - 2R_1, R_3 - 3R_1, R_4 - R_1$

$$A \sim \begin{bmatrix} 1 & 3 & 2 \\ 0 & -7 & -1 \\ 0 & -14 & -2 \\ 0 & 14 & 2 \end{bmatrix} \quad \text{Operating } R_3 - 2R_2, R_4 + 2R_2$$

$$\sim \begin{bmatrix} 1 & 3 & 2 \\ 0 & -7 & -1 \\ 0 & 0 & 0 \\ 0 & 0 & 0 \end{bmatrix} \quad \text{Operating } R_1 + 2R_2$$

$$\sim \begin{bmatrix} 1 & -11 & 0 \\ 0 & -7 & -1 \\ 0 & 0 & 0 \\ 0 & 0 & 0 \end{bmatrix}$$

$\rho(A) = 2 <$ number of unknowns

$\Rightarrow$ The system has an infinite number of non-trivial solutions given by

$$x_1 - 11x_2 = 0, \quad -7x_2 - x_3 = 0$$

i.e., $x_1 = 11k$, $x_2 = k$, $x_3 = -7k$, where k is any number.

Different values of k give different solutions.

Example 2. *Solve:* $x + y + z = 3$, $x + 2y + 3z = 4$, $x + 4y + 9z = 6$.

Sol. Augmented matrix $[A \vdots B] = \begin{bmatrix} 1 & 1 & 1 & \vdots & 3 \\ 1 & 2 & 3 & \vdots & 4 \\ 1 & 4 & 9 & \vdots & 6 \end{bmatrix}$

Operating $R_2 - R_1$, $R_3 - R_1$

$$\sim \begin{bmatrix} 1 & 1 & 1 & \vdots & 3 \\ 0 & 1 & 2 & \vdots & 1 \\ 0 & 3 & 8 & \vdots & 3 \end{bmatrix} \text{ Operating } R_3 - 3R_2$$

$$\sim \begin{bmatrix} 1 & 1 & 1 & \vdots & 3 \\ 0 & 1 & 2 & \vdots & 1 \\ 0 & 0 & 2 & \vdots & 0 \end{bmatrix} \text{ Operating } (\tfrac{1}{2})R_3$$

$$\sim \begin{bmatrix} 1 & 1 & 1 & \vdots & 3 \\ 0 & 1 & 2 & \vdots & 1 \\ 0 & 0 & 1 & \vdots & 0 \end{bmatrix} \text{ Operating } R_1 - R_3,\ R_2 - 2R_3$$

$$\sim \begin{bmatrix} 1 & 1 & 0 & \vdots & 3 \\ 0 & 1 & 0 & \vdots & 1 \\ 0 & 0 & 1 & \vdots & 0 \end{bmatrix} \text{ Operating } R_1 - R_2$$

$$\sim \begin{bmatrix} 1 & 0 & 0 & \vdots & 2 \\ 0 & 1 & 0 & \vdots & 1 \\ 0 & 0 & 1 & \vdots & 0 \end{bmatrix}$$

$\therefore$ $\rho(A) = \rho(A \vdots B) = 3 =$ number of unknowns

$\Rightarrow$ The given system of equations is consistent and solution is

$$x = 2, y = 1, z = 0.$$

Example 3. *Solve the system of equations* :

$$2x_1 + x_2 + 2x_3 + x_4 = 6, \qquad 6x_1 - 6x_2 + 6x_3 + 12x_4 = 36$$

$$4x_1 + 3x_2 + 3x_3 - 3x_4 = -1, \qquad 2x_1 + 2x_2 - x_3 + x_4 = 10.$$

Sol. In matrix notation, the given system of equations can be written as AX = B

where $A = \begin{bmatrix} 2 & 1 & 2 & 1 \\ 6 & -6 & 6 & 12 \\ 4 & 3 & 3 & -3 \\ 2 & 2 & -1 & 1 \end{bmatrix}, X = \begin{bmatrix} x_1 \\ x_2 \\ x_3 \\ x_4 \end{bmatrix}, B = \begin{bmatrix} 6 \\ 36 \\ -1 \\ 10 \end{bmatrix}$

Augmented matrix [A : B]

$$= \left[\begin{array}{cccc:c} 2 & 1 & 2 & 1 & 6 \\ 6 & -6 & 6 & 12 & 36 \\ 4 & 3 & 3 & -3 & -1 \\ 2 & 2 & -1 & 1 & 10 \end{array}\right] \quad \text{Operating } R_2 - 3R_1,\ R_3 - 2R_1,\ R_4 - R_1$$

$$\sim \left[\begin{array}{cccc:c} 2 & 1 & 2 & 1 & 6 \\ 0 & -9 & 0 & 9 & 18 \\ 0 & 1 & -1 & -5 & -13 \\ 0 & 1 & -3 & 0 & 4 \end{array}\right] \quad \text{Operating } -\frac{1}{9} R_2$$

$$\sim \left[\begin{array}{cccc:c} 2 & 1 & 2 & 1 & 6 \\ 0 & 1 & 0 & -1 & -2 \\ 0 & 1 & -1 & -5 & -13 \\ 0 & 1 & -3 & 0 & 4 \end{array}\right] \quad \text{Operating } R_1 - R_2,\ R_3 - R_2,\ R_4 - R_2$$

$$\sim \left[\begin{array}{cccc:c} 2 & 0 & 2 & 2 & 8 \\ 0 & 1 & 0 & -1 & -2 \\ 0 & 0 & -1 & -4 & -11 \\ 0 & 0 & -3 & 1 & 6 \end{array}\right] \quad \text{Operating } R_4 - 3R_3,\ \tfrac{1}{2}R_1$$

$$\sim \left[\begin{array}{cccc:c} 1 & 0 & 1 & 1 & 4 \\ 0 & 1 & 0 & -1 & -2 \\ 0 & 0 & -1 & -4 & -11 \\ 0 & 0 & 0 & 13 & 39 \end{array}\right] \quad \text{Operating } R_1 + R_3,\ \frac{1}{13} R_4$$

$$\sim \left[\begin{array}{cccc:c} 1 & 0 & 0 & -3 & -7 \\ 0 & 1 & 0 & -1 & -2 \\ 0 & 0 & -1 & -4 & -11 \\ 0 & 0 & 0 & 1 & 3 \end{array}\right] \quad \text{Operating } R_1 + 3R_4,\ R_2 + R_4,\ R_3 + 4R_4$$

$$\sim \left[\begin{array}{cccc:c} 1 & 0 & 0 & 0 & 2 \\ 0 & 1 & 0 & 0 & 1 \\ 0 & 0 & -1 & 0 & 1 \\ 0 & 0 & 0 & 1 & 3 \end{array}\right] \quad \text{Operating } (-1)R_3$$

$$\sim \left[\begin{array}{cccc:c} 1 & 0 & 0 & 0 & 2 \\ 0 & 1 & 0 & 0 & 1 \\ 0 & 0 & 1 & 0 & -1 \\ 0 & 0 & 0 & 1 & 3 \end{array}\right]$$

$\therefore$ $\rho(A) = \rho(A : B) = 4$ = number of unknowns

$\Rightarrow$ The given system of equations is consistent and solution is

$x_1 = 2,\ x_2 = 1,\ x_3 = -1,\ x_4 = 3.$

Example 4. *Using matrix method, show that the equations :*

$3x + 3y + 2z = 1, \quad x + 2y = 4, \quad 10y + 3z = -2, \quad 2x - 3y - z = 5$

are consistent and hence obtain the solutions for x, y and z.

Sol. In matrix notation, the given system of equations can be written as

$$AX = B$$

where $A = \begin{bmatrix} 3 & 3 & 2 \\ 1 & 2 & 0 \\ 0 & 10 & 3 \\ 2 & -3 & -1 \end{bmatrix}, X = \begin{bmatrix} x \\ y \\ z \end{bmatrix}, B = \begin{bmatrix} 1 \\ 4 \\ -2 \\ 5 \end{bmatrix}$

Augmented matrix $[A \vdots B]$

$$= \begin{bmatrix} 3 & 3 & 2 & \vdots & 1 \\ 1 & 2 & 0 & \vdots & 4 \\ 0 & 10 & 3 & \vdots & -2 \\ 2 & -3 & -1 & \vdots & 5 \end{bmatrix} \quad \text{Operating } R_{12}$$

$$\sim \begin{bmatrix} 1 & 2 & 0 & \vdots & 4 \\ 3 & 3 & 2 & \vdots & 1 \\ 0 & 10 & 3 & \vdots & -2 \\ 2 & -3 & -1 & \vdots & 5 \end{bmatrix} \quad \text{Operating } R_2 - 3R_1,\ R_4 - 2R_1$$

$$\sim \begin{bmatrix} 1 & 2 & 0 & \vdots & 4 \\ 0 & -3 & 2 & \vdots & -11 \\ 0 & 10 & 3 & \vdots & -2 \\ 0 & -7 & -1 & \vdots & -3 \end{bmatrix} \quad \text{Operating } R_3 + 3R_2,\ R_4 - 2R_2$$

$$\sim \begin{bmatrix} 1 & 2 & 0 & \vdots & 4 \\ 0 & -3 & 2 & \vdots & -11 \\ 0 & 1 & 9 & \vdots & -35 \\ 0 & -1 & -5 & \vdots & 19 \end{bmatrix} \quad \text{Operating } R_1 - 2R_3,\ R_2 + 3R_3,\ R_4 + R_3$$

$$\sim \begin{bmatrix} 1 & 0 & -18 & \vdots & 74 \\ 0 & 0 & 29 & \vdots & -116 \\ 0 & 1 & 9 & \vdots & -35 \\ 0 & 0 & 4 & \vdots & -16 \end{bmatrix} \quad \text{Operating } R_{23},\ \tfrac{1}{4}R_4$$

$$\sim \begin{bmatrix} 1 & 0 & -18 & \vdots & 74 \\ 0 & 1 & 9 & \vdots & -35 \\ 0 & 0 & 29 & \vdots & -116 \\ 0 & 0 & 1 & \vdots & -4 \end{bmatrix} \quad \text{Operating } R_1 + 18R_4,\ R_2 - 9R_4,\ R_3 - 29R_4$$

$$\sim \begin{bmatrix} 1 & 0 & 0 & \vdots & 2 \\ 0 & 1 & 0 & \vdots & 1 \\ 0 & 0 & 0 & \vdots & 0 \\ 0 & 0 & 1 & \vdots & -4 \end{bmatrix} \quad \text{Operating } R_{34}$$

$$\sim \begin{bmatrix} 1 & 0 & 0 & \vdots & 2 \\ 0 & 1 & 0 & \vdots & 1 \\ 0 & 0 & 1 & \vdots & -4 \\ 0 & 0 & 0 & \vdots & 0 \end{bmatrix}$$

$\rho(A) = \rho(A \vdots B) = 3$ = number of unknowns

$\Rightarrow$ The given system of equations is consistent and the solution is $x = 2, y = 1, z = -4$.

Example 5. *Test for consistency and solve*

$$5x + 3y + 7z = 4$$
$$3x + 26y + 2z = 9$$
$$7x + 2y + 10z = 5.$$

(OU-Dec. 2008)

Sol. Write the given equations in the form AX = B, where

$$A = \begin{bmatrix} 5 & 3 & 7 \\ 3 & 26 & 2 \\ 7 & 2 & 10 \end{bmatrix}, X = \begin{bmatrix} x \\ y \\ z \end{bmatrix}, B = \begin{bmatrix} 4 \\ 9 \\ 5 \end{bmatrix}$$

Consider $$[A : B] = \begin{bmatrix} 5 & 3 & 7 & 4 \\ \mathbf{3} & 26 & 2 & 9 \\ \mathbf{7} & 2 & 10 & 5 \end{bmatrix}$$

Applying $R_2 \to 5R_2 - 3R_1$, $R_3 \to 5R_3 - 7R_1$, we get

$$\sim \begin{bmatrix} 5 & 3 & 7 & 4 \\ 0 & 121 & -11 & 33 \\ 0 & -11 & 1 & -3 \end{bmatrix}$$

Applying $R_2 \to \frac{R_2}{11}$, $R_3 \to (-1)R_3$, we get

$$\sim \begin{bmatrix} 5 & 3 & 7 & 4 \\ 0 & 11 & -1 & 3 \\ 0 & \mathbf{11} & -1 & 3 \end{bmatrix}$$

Applying $R_3 \to R_3 - R_2$, we get

$$\sim \begin{bmatrix} 5 & 3 & 7 & 4 \\ 0 & 11 & -1 & 3 \\ 0 & 0 & 0 & 0 \end{bmatrix} \qquad ...(1)$$

Here, $\rho(A) = 2$, $\rho[A : B] = 2$, number of unknowns = 3

Since $\rho(A) = \rho(A : B) <$ number of unknowns.

$\therefore$ The given system of equations have an infinite number of solutions.

From (1), we get $5x + 3y + 7z = 4$ and $11y - z = 3$.

Let $z = k$ then $$y = \frac{3+k}{11}$$

and $$5x = 4 - 3y - 7z = 4 - 3\left(\frac{3+k}{11}\right) - 7k = \frac{44 - 9 - 3k - 77k}{11}$$

$$x = \frac{35 - 80k}{11 \times 5} = \frac{7 - 16k}{11}$$

$\therefore$ $$x = \frac{7-16k}{11}, y = \frac{3+k}{11}, z = k \text{ is a required solution.}$$

Example 6. *Show that the system of equations*

$$\begin{bmatrix} 4 & 9 & 3 \\ 2 & 3 & 1 \\ 2 & 6 & 2 \end{bmatrix} \begin{bmatrix} x \\ y \\ z \end{bmatrix} = \begin{bmatrix} 6 \\ 2 \\ 7 \end{bmatrix}$$ *is inconsistent.* (OU-May 2008)

Sol. Here $A = \begin{bmatrix} 4 & 9 & 3 \\ 2 & 3 & 1 \\ 2 & 6 & 2 \end{bmatrix}, \quad B = \begin{bmatrix} 6 \\ 2 \\ 7 \end{bmatrix}$

Consider $[A : B] = \begin{bmatrix} 4 & 9 & 3 & 6 \\ 2 & 3 & 1 & 2 \\ 2 & 6 & 2 & 7 \end{bmatrix}$

Applying $R_2 \to 2R_2 - R_1$, $R_3 \to 2R_3 - R_1$

$$\sim \begin{bmatrix} 4 & 9 & 3 & 6 \\ 0 & -3 & -1 & -2 \\ 0 & 3 & 1 & 8 \end{bmatrix}$$

Applying $R_3 \to R_3 + R_2$

$$\sim \begin{bmatrix} 4 & 9 & 3 & 6 \\ 0 & -3 & -1 & -2 \\ 0 & 0 & 0 & 0 \end{bmatrix}$$

is an Echelon form

Here $\rho(A) = 2, \rho(A : B) = 3$

Since $\rho(A) \neq \rho(A : B)$, the given system of equations is inconsistent and there is no solution.

Example 7. *For what values of* λ *and* μ *do the system of equations :*

$$x + y + z = 6, \quad x + 2y + 3z = 10, \quad x + 2y + \lambda z = \mu$$

have (i) no solution (ii) unique solution (iii) more than one solutions ?

(OU-Dec. 2013, June 2013, Apr. 2007)

Sol. In matrix notation, the given system of equations can be written as

$$AX = B$$

where $A = \begin{bmatrix} 1 & 1 & 1 \\ 1 & 2 & 3 \\ 1 & 2 & \lambda \end{bmatrix}, X = \begin{bmatrix} x \\ y \\ z \end{bmatrix}, B = \begin{bmatrix} 6 \\ 10 \\ \mu \end{bmatrix}$

Augmented matrix $[A \vdots B]$

$$= \begin{bmatrix} 1 & 1 & 1 & \vdots & 6 \\ 1 & 2 & 3 & \vdots & 10 \\ 1 & 2 & \lambda & \vdots & \mu \end{bmatrix} \qquad \text{Operating } R_2 - R_1, R_3 - R_1$$

$$\sim \begin{bmatrix} 1 & 1 & 1 & \vdots & 6 \\ 0 & 1 & 2 & \vdots & 4 \\ 0 & 1 & \lambda - 1 & \vdots & \mu - 6 \end{bmatrix} \qquad \text{Operating } R_1 - R_2, R_3 - R_2$$

$$\sim \begin{bmatrix} 1 & 0 & -1 & \vdots & 2 \\ 0 & 1 & 2 & \vdots & 4 \\ 0 & 0 & \lambda - 3 & \vdots & \mu - 10 \end{bmatrix}$$

Case I: If $\lambda = 3, \mu \neq 10$

$$\rho(A) = 2, \rho(A \vdots B) = 3$$

$\because \qquad \rho(A) \neq \rho(A \vdots B)$

$\therefore$ The system has no solution.

Case II: If $\lambda \neq 3$, μ may have any value

$$\rho(A) = \rho(A \vdots B) = 3 = \text{number of unknowns}$$

$\therefore$ The system has unique solution.

Case III: If $\lambda = 3$, $\mu = 10$

$$\rho(A) = \rho(A \vdots B) = 2 < \text{number of unknowns}$$

$\therefore$ The system has an infinite number of solutions.

Example 8. *For what values of k, the equations $x + y + z = 1$, $2x + y + 4z = k$, $4x + y + 10z = k^2$ have a solution and solve them completely in each case?* (OU-Apr. 2007, May 2009)

Sol. The matrix form of given system of equations is AX = B

$$\Rightarrow \qquad \begin{bmatrix} 1 & 1 & 1 \\ 2 & 1 & 4 \\ 4 & 1 & 10 \end{bmatrix} \begin{bmatrix} x \\ y \\ z \end{bmatrix} = \begin{bmatrix} 1 \\ k \\ k^2 \end{bmatrix}$$

$$[A : B] = \begin{bmatrix} 1 & 1 & 1 & 1 \\ 2 & 1 & 4 & k \\ 4 & 1 & 10 & k^2 \end{bmatrix}$$

Convert into Echelon Form:

Applying $R_2 \to R_2 - 2R_1$, $R_3 \to R_3 - 4R_1$, we get

$$\sim \begin{bmatrix} 1 & 1 & 1 & 1 \\ 0 & -1 & 2 & k-2 \\ 0 & -3 & 6 & k^2-4 \end{bmatrix}$$

Applying $R_3 \to R_3 - 3R_2$, we get

$$\sim \begin{bmatrix} 1 & 1 & 1 & 1 \\ 0 & -1 & 2 & k-2 \\ 0 & 0 & 0 & k^2-3k+2 \end{bmatrix}$$

Since the given system of equations have a solution, $\rho(A) = \rho(A : B)$.

To satisfy this condition, we should consider

$$k^2 - 3k + 2 = 0$$

$$\Rightarrow \qquad (k-1)(k-2) = 0$$

$$\Rightarrow \qquad k = 1, 2$$

Case I: If $k = 1$ then we have

$$A = \begin{bmatrix} 1 & 1 & 1 \\ 0 & -1 & 2 \\ 0 & 0 & 0 \end{bmatrix}$$

$$\Rightarrow \qquad \rho(A) = 2$$

$$[A : B] = \begin{bmatrix} 1 & 1 & 1 & 1 \\ 0 & -1 & 2 & -1 \\ 0 & 0 & 0 & 0 \end{bmatrix} \Rightarrow \rho(A : B) = 2$$

$\therefore$ System of equations is consistent. From [A : B] we can write linear equations as

$$x + y + z = 1 \text{ and } -y + 2z = -1$$

Let $z = k_1$ then $y = 2k + 1$ and

$$x = 1 - y - z = 1 - (2k_1 + 1) - k_1 = -3k_1$$

$$\therefore \quad \begin{bmatrix} x \\ y \\ z \end{bmatrix} = \begin{bmatrix} -3k_1 \\ 2k_1 + 1 \\ k_1 \end{bmatrix}$$

where k_1 is a parameter.

In this case the system of equations have an infinite number of solutions.

Case II: If $k = 2$ then $[A : B] = \begin{bmatrix} 1 & 1 & 1 & 0 \\ 0 & -1 & 2 & 0 \\ 0 & 0 & 0 & 0 \end{bmatrix}$.

Here, $x + y + z = 1, -y + 2z = 0$

Take $z = k_2$ then $\quad y = 2k_2$

and $\quad x = 1 - y - z = 1 - 2k_2 - k_2 = 1 - 3k_2$

$$\therefore \quad \begin{bmatrix} x \\ y \\ z \end{bmatrix} = \begin{bmatrix} 1 - 3k_2 \\ 2k_2 \\ k_2 \end{bmatrix}$$

where k_2 is a parameter.

In this case also the given system of equations have an infinite number of solutions.

Example 9. *Discuss for all values of k, the system of equations* $x + y + 4z = 6$, $x + 2y - 2z = 6$, $kx + y + z = 6$ *with regard to consistency.*

Sol. Take $\begin{bmatrix} 1 & 1 & 4 \\ 1 & 2 & -2 \\ k & 1 & 1 \end{bmatrix} \begin{bmatrix} x \\ y \\ z \end{bmatrix} = \begin{bmatrix} 6 \\ 6 \\ 6 \end{bmatrix}$

Applying $R_2 \to R_2 - R_1$, $R_3 \to R_3 - kR_1$, we get

$$\begin{bmatrix} 1 & 1 & 4 \\ 0 & 1 & -6 \\ 0 & 1-k & 1-4k \end{bmatrix} \begin{bmatrix} x \\ y \\ z \end{bmatrix} = \begin{bmatrix} 6 \\ 0 \\ 6 - 6k \end{bmatrix}$$

For the above system to have a unique solution we must have

rank of A = 3 = number of unknowns

$$\Rightarrow \quad \text{det. } A \neq 0 \Rightarrow (1 - 4k) + (6 - 6k) \neq 0$$

$$\Rightarrow \quad k \neq \frac{7}{10}$$

$\therefore$ If $k \neq \frac{7}{10}$ then the given system of equations is consistent and the solution is unique.

If $k = \frac{7}{10}$ then we get $[AB] \sim \begin{bmatrix} 1 & 1 & 4 & 6 \\ 0 & 1 & -6 & 0 \\ 0 & \frac{3}{10} & -\frac{18}{10} & \frac{18}{10} \end{bmatrix}$.

Applying $R_3 \to R_3 - \frac{3}{10} R_2$, we get

$$[A : B] \sim \begin{bmatrix} 1 & 1 & 4 & 6 \\ 0 & 1 & -6 & 0 \\ 0 & 0 & 0 & \frac{18}{10} \end{bmatrix}$$

$\Rightarrow$ $\rho(A) = 2, \rho(A : B) = 3$

Since $\rho(A) \neq \rho(A : B)$, the given system of equations is inconsistent and there is no solution.

Example 10. *Find the values of a and b for which the equations $x + y + z = 3$, $x + 2y + 2z = 6$, $x + ay + 3z = b$ have (i) no solution (ii) a unique solution (iii) infinite number of solutions.*

Sol. Given system of equations can be expressed in the matrix form AX = B *i.e.,*

$$\begin{bmatrix} 1 & 1 & 1 \\ 1 & 2 & 2 \\ 1 & a & 3 \end{bmatrix} \begin{bmatrix} x \\ y \\ z \end{bmatrix} = \begin{bmatrix} 3 \\ 6 \\ b \end{bmatrix}$$

Consider an augmented matrix

$$[A : B] = \begin{bmatrix} 1 & 1 & 1 & 3 \\ 1 & 2 & 2 & 6 \\ 1 & a & 3 & b \end{bmatrix}$$

Applying $R_2 \to R_2 - R_1$, $R_3 \to R_3 - R_1$

$$\sim \begin{bmatrix} 1 & 1 & 1 & 3 \\ 0 & 1 & 1 & 3 \\ 0 & a-1 & 2 & b-3 \end{bmatrix}$$

Applying $R_3 \to R_3 - 2R_2$

$$\sim \begin{bmatrix} 1 & 1 & 1 & 3 \\ 0 & 1 & 1 & 3 \\ 0 & a-3 & 0 & b-9 \end{bmatrix}$$

Case I: If $a \neq 3$, then $\rho(A) = 3$, $\rho(A : B) = 3$, $n = 3$.

Hence, $\rho(A) = \rho(A : B) = n$

$\therefore$ The given system of equations is consistent and there is a unique solution.

Case II: If $a = 3$, $b = 9$ then $\rho(A) = 2$, $\rho(A : B) = 2$, $n = 3$.
Hence, $\rho(A) = \rho(A : B) < n$

$\therefore$ The given system of equations is consistent and there are infinite number of solutions.

Case III: If $a = 3$, $b \neq 9$ then $\rho(A) = 2$, $\rho[A : B] = 3$.
Since $\rho(A) \neq \rho(A : B)$, the system is inconsistent and there is no solution.

EXERCISE 2

1. Write the following equations in matrix form AX = B and solve:

   (*i*) $2x - 2y + z = 1$, $x + 2y + 2z = 2$, $2x + y - 2z = 7$

   (*ii*) $2x_1 - x_2 + x_3 = 4$, $x_1 + x_2 + x_3 = 1$, $x_1 - 3x_2 - 2x_3 = 2$.

2. Using the loop current method on a circuit, the following equations were obtained :
$$7i_1 - 4i_2 = 12, \; -4i_1 + 12i_2 - 6i_3 = 0, \; -6i_2 + 14i_3 = 0.$$
Solve for i_1, i_2 and i_3.

3. Solve the following systems of equations:

   (*i*) $x + y + z = 8$, $x - y + 2z = 6$, $3x + 5y - 7z = 14$

   (*ii*) $x + y + z = 6$, $x - y + 2z = 5$, $3x + y + z = 8$

   (*iii*) $x + 2y + 3z = 1$, $2x + 3y + 2z = 2$, $3x + 3y + 4z = 1$.

4. Show that the equations $x + 2y - z = 3$, $3x - y + 2z = 1$, $2x - 2y + 3z = 2$, $x - y + z = -1$ are consistent and solve them.

5. Test for consistency the equations $2x - 3y + 7z = 5$, $3x + y - 3z = 13$, $2x + 19y - 47z = 32$.

6. Solve the equations $x + 3y + 2z = 0$, $2x - y + 3z = 0$, $3x - 5y + 4z = 0$, $x + 17y + 4z = 0$.

7. For what values of a and b do the equations $x + 2y + 3z = 6$, $x + 3y + 5z = 9$, $2x + 5y + az = b$ have (*i*) no solution, (*ii*) a unique solution, and (*iii*) more than one solution ?

8. Find the value of k so that the equations $x + y + 3z = 0$, $4x + 3y + kz = 0$, $2x + y + 2z = 0$ have a non-trivial solution.

9. Show that the equations $3x + 4y + 5z = a$, $4x + 5y + 6z = b$, $5x + 6y + 7z = c$ do not have a solution unless $a + c = 2b$.

10. Investigate the values of λ and μ so that the equations $2x + 3y + 5z = 9$, $7x + 3y - 2z = 8$, $2x + 3y + \lambda z = \mu$ have (*i*) no solution, (*ii*) a unique solution, and (*iii*) an infinite number of solutions.

11. Determine the values of λ for which the following set of equations may possess non-trivial solution. $3x_1 + x_2 - \lambda x_3 = 0$, $4x_1 - 2x_2 - 3x_3 = 0$, $2\lambda x_1 + 4x_2 + \lambda x_3 = 0$. For each permissible value of λ, determine the general solution.

12. Investigate for consistency of the following equations and if possible find the solutions :
$$4x - 2y + 6z = 8, \quad x + y - 3z = -1, \quad 15x - 3y + 9z = 21.$$

13. For what values of k the equations $x + y + z = 1$, $2x + y + 4z = k$, $4x + y + 10z = k^2$ have a solution and solve them completely in each case.

14. Show that if $\lambda \neq -5$, the system of equations $3x - y + 4z = 3$, $x + 2y - 3z = -2$, $6x + 5y + \lambda z = -3$ have a unique solution. If $\lambda = -5$, show that the equations are consistent. Determine the solutions in each case.

15. Find the values of a and b for which the equations $x + ay + z = 3$, $x + 2y + 2z = b$, $x + 5y + 3z = 9$ are consistent. When will these equations have a unique solution ?

Answers

1. (i) $x = 2, y = 1, z = -1$ (ii) $x_1 = 1, x_2 = -1, x_3 = 1$ 2. $i_1 = \frac{396}{175}, i_2 = \frac{24}{25}, i_3 = \frac{72}{175}$
3. (i) $x = 5, y = \frac{5}{3}, z = \frac{4}{3}$ (ii) $x = 1, y = 2, z = 3$ (iii) $x = -\frac{3}{7}, y = \frac{8}{7}, z = -\frac{2}{7}$
4. $x = -1, y = 4, z = 4$ 5. Inconsistent
6. $x = 11k, y = k, z = -7k$, where k is arbitrary
7. (i) $a = 8, b \neq 15$ (ii) $a \neq 8$, b may have any value (iii) $a = 8, b = 15$
8. $k = 8$ 10. (i) $\lambda = 5, \mu \neq 9$ (ii) $\lambda \neq 5$, μ arbitrary (iii) $\lambda = 5, \mu = 9$
11. $\lambda = 1, -9$. For $\lambda = 1$, solution is $x = k, y = -k, z = 2k$. For $\lambda = -9$, solution is $x = 3k, y = 9k, z = -2k$.
12. Consistent ; $x = 1, y = 3k - 2, z = k$ where k is arbitrary.
13. $k = 1, x = -3\lambda, y = 1 + 2\lambda, z = \lambda$; $k = 2, x = 1 - 3\lambda, y = 2\lambda, z = \lambda$, where λ is arbitrary.
14. $\lambda \neq -5, x = \frac{4}{7}, y = -\frac{9}{7}, z = 0$; $\lambda = -5, x = \frac{4-5k}{7}, y = \frac{13k-9}{7}, z = k$, where k is arbitrary.
15. If $a = -1, b = 6$, equations will be consistent and have infinitely many solutions
If $a = -1, b \neq 6$, equations will be inconsistent.
If $a \neq -1$, b is arbitrary, equations will be consistent and have a unique solution.

1.23. CHARACTERISTIC EQUATION

If A is a square matrix of order n, we can form the matrix $A - \lambda I$, where λ is a scalar and I is the unit matrix of order n. The determinant of this matrix equated to zero, *i.e.*,

$$|A - \lambda I| = \begin{vmatrix} a_{11}-\lambda & a_{12} & \cdots & a_{1n} \\ a_{21} & a_{22}-\lambda & \cdots & a_{2n} \\ \cdots & \cdots & \cdots & \cdots \\ a_{n1} & a_{n2} & \cdots & a_{nn}-\lambda \end{vmatrix} = 0$$

is called the *characteristic equation* of A.

1.24. EIGEN VALUES OR CHARACTERISTIC ROOTS OR LATENT ROOTS

(OU-June 2013)

On expanding the determinant, the characteristic equation can be written as a polynomial equation of degree n in λ of the form $(-1)^n \lambda^n + k_1 \lambda^{n-1} + k_2 \lambda^{n-2} + \ldots\ldots + k_n = 0$.

The roots of this equation are called the *characteristic roots* or *latent roots* or *eigen values* of A.

Note. The sum of the eigen-values of a matrix A is equal to trace of A.
[The trace of a square matrix is the sum of its diagonal elements.]

1.25. EIGEN VECTORS OR CHARACTERISTIC VECTORS OR LATENT VECTORS

(OU-June 2013)

Consider the linear transformation $Y = AX$ ...(1)

which transforms the column vector X into the column vector Y. In practice, we are often required to find those vectors X which transform into scalar multiples of themselves.

Let X be such a vector which transforms into λX(λ being a non-zero scalar) by the transformation (1).

Then $$Y = \lambda X \quad ...(2)$$

From (1) and (2), $$AX = \lambda X \Rightarrow AX - \lambda IX = O \Rightarrow (A - \lambda I)X = O \quad ...(3)$$

This matrix equation gives n homogeneous linear equations

$$\left.\begin{array}{llll}(a_{11} - \lambda)x_1 & + a_{12}\, x_2 & \ldots\ldots + a_{1n}\, x_n & = 0 \\ a_{21}\, x_1 & + (a_{22} - \lambda)x_2 & \ldots\ldots + a_{2n}\, x_n & = 0 \\ \ldots\ldots\ldots & \ldots\ldots\ldots & \ldots\ldots\ldots & \\ a_{n1}\, x_1 & + a_{n2}\, x_2 & \ldots\ldots + (a_{nn} - \lambda)x_n & = 0\end{array}\right] \quad ...(4)$$

These equations will have a non-trivial solution only if the co-efficient matrix $A - \lambda I$ is singular

i.e., if $$| A - \lambda I | = 0 \quad ...(5)$$

This is the characteristic equation of the matrix A and has n roots which are the eigen-values of A. Corresponding to each root of (5), the homogeneous system (3) has a non-zero solution

$$X = \begin{bmatrix} x_1 \\ x_2 \\ \vdots \\ x_n \end{bmatrix},$$ which is called an *eigen vector* or *latent vector*.

Note. If X is a solution of (3), then so is kX, where k is an arbitrary constant. Thus, the eigen vector corresponding to an eigen-value is not unique.

ILLUSTRATIVE EXAMPLES

Example 1. *Find the eigen values and eigen vectors of the matrix* $A = \begin{bmatrix} 1 & -2 \\ -5 & 4 \end{bmatrix}$.

Sol. The characteristic equation of the given matrix is

$$| A - \lambda I | = 0 \quad \text{or} \quad \begin{vmatrix} 1-\lambda & -2 \\ -5 & 4-\lambda \end{vmatrix} = 0$$

or $$(1 - \lambda)(4 - \lambda) - 10 = 0 \quad \text{or} \quad \lambda^2 - 5\lambda - 6 = 0$$

or $$(\lambda - 6)(\lambda + 1) = 0 \quad \therefore \quad \lambda = 6, -1.$$

Thus, the eigen values of A are 6, – 1.

Corresponding to $\lambda = 6$, the eigen vectors are given by $(A - 6I)X = O$

or $$\begin{bmatrix} 1-6 & -2 \\ -5 & 4-6 \end{bmatrix}\begin{bmatrix} x_1 \\ x_2 \end{bmatrix} = O \quad \text{or} \quad \begin{bmatrix} -5 & -2 \\ -5 & -2 \end{bmatrix}\begin{bmatrix} x_1 \\ x_2 \end{bmatrix} = O$$

We get only one independent equation $-5x_1 - 2x_2 = 0$

$\therefore \quad \dfrac{x_1}{2} = \dfrac{x_2}{-5}$ giving the eigen vector $k_1 \begin{bmatrix} 2 \\ -5 \end{bmatrix}$.

Corresponding to $\lambda = -1$, the eigen vectors are given by $\begin{bmatrix} 2 & -2 \\ -5 & 5 \end{bmatrix}\begin{bmatrix} x_1 \\ x_2 \end{bmatrix} = O$

We get only one independent equation $2x_1 - 2x_2 = 0$.

$\therefore \quad x_1 = x_2$ giving the eigen vector $k_2\begin{bmatrix}1\\1\end{bmatrix}$.

Example 2. *Find the eigen values and eigen vectors of the matrix* $A = \begin{bmatrix}-2 & 2 & -3\\ 2 & 1 & -6\\ -1 & -2 & 0\end{bmatrix}$.

Sol. The characteristic equation of the given matrix is $|A - \lambda I| = 0$

or
$$\begin{vmatrix}-2-\lambda & 2 & -3\\ 2 & 1-\lambda & -6\\ -1 & -2 & -\lambda\end{vmatrix} = 0$$

or
$$(-2-\lambda)[-\lambda(1-\lambda) - 12] - 2[-2\lambda - 6] - 3[-4 + 1(1-\lambda)] = 0$$

or
$$\lambda^3 + \lambda^2 - 21\lambda - 45 = 0$$

By trial, $\lambda = -3$ satisfies it.

$\therefore \quad (\lambda + 3)(\lambda^2 - 2\lambda - 15) = 0 \Rightarrow (\lambda + 3)(\lambda + 3)(\lambda - 5) = 0 \Rightarrow \lambda = -3, -3, 5$

Thus, the eigen values of A are $-3, -3, 5$.

Corresponding to $\lambda = -3$, the eigen vectors are given by

$$(A + 3I)X = O \quad \text{or} \quad \begin{bmatrix}1 & 2 & -3\\ 2 & 4 & -6\\ -1 & -2 & 3\end{bmatrix}\begin{bmatrix}x_1\\x_2\\x_3\end{bmatrix} = O$$

We get only one independent equation $x_1 + 2x_2 - 3x_3 = 0$

Choosing $x_2 = 0$, we have $x_1 - 3x_3 = 0$

$\therefore \quad \dfrac{x_1}{3} = \dfrac{x_2}{0} = \dfrac{x_3}{1}$ giving the eigen vector $k_1\begin{bmatrix}3\\0\\1\end{bmatrix}$

Choosing $x_3 = 0$, we have $x_1 + 2x_2 = 0$

$\therefore \quad \dfrac{x_1}{2} = \dfrac{x_2}{-1} = \dfrac{x_3}{0}$ giving the eigen vector $k_2\begin{bmatrix}2\\-1\\0\end{bmatrix}$

Any other eigen vector corresponding to $\lambda = -3$ will be a linear combination of these two.

Corresponding to $\lambda = 5$, the eigen vectors are given by $\begin{bmatrix}-7 & 2 & -3\\ 2 & -4 & -6\\ -1 & -2 & -5\end{bmatrix}\begin{bmatrix}x_1\\x_2\\x_3\end{bmatrix} = O$

$$\Rightarrow \quad \begin{aligned}-7x_1 + 2x_2 - 3x_3 &= 0\\ 2x_1 - 4x_2 - 6x_3 &= 0\\ -x_1 - 2x_2 - 5x_3 &= 0\end{aligned}$$

Since only two of them are independent, we can omit one of them.

From first two equations, we have $\dfrac{x_1}{-12-12} = \dfrac{x_2}{-6-42} = \dfrac{x_3}{28-4}$

or
$$\frac{x_1}{1} = \frac{x_2}{2} = \frac{x_3}{-1} \text{ giving the eigen vector } k_3\begin{bmatrix}1\\2\\-1\end{bmatrix}.$$

Example 3. *Find the eigen values and eigen vectors of the matrix* $\begin{bmatrix} 5 & 4 \\ 1 & 2 \end{bmatrix}$.

(OU-May 2009)

Sol. Let $$A = \begin{bmatrix} 5 & 4 \\ 1 & 2 \end{bmatrix}$$

Its characteristic equation is $|A - \lambda I| = 0$

$$\Rightarrow \quad \begin{vmatrix} 5-\lambda & 4 \\ 1 & 2-\lambda \end{vmatrix} = 0$$

$$\Rightarrow \quad (5 - \lambda)(2 - \lambda) - 4 = 0$$

$$\Rightarrow \quad 10 - 5\lambda - 2\lambda + \lambda^2 - 4 = 0$$

$$\Rightarrow \quad \lambda^2 - 7\lambda + 6 = 0$$

$$\Rightarrow \quad (\lambda - 1)(\lambda - 6) = 0$$

$$\Rightarrow \quad \lambda = 1, 6 \text{ are eigen values.}$$

Now consider $$[A - \lambda I]X = 0$$

$$\Rightarrow \quad \begin{bmatrix} 5-\lambda & 4 \\ 1 & 2-\lambda \end{bmatrix}\begin{bmatrix} x \\ y \end{bmatrix} = \begin{bmatrix} 0 \\ 0 \end{bmatrix} \quad \text{...(1)}$$

Case I: Put $\lambda = 1$ in (1), we get

$$\begin{bmatrix} 4 & 4 \\ 1 & 1 \end{bmatrix}\begin{bmatrix} x \\ y \end{bmatrix} = \begin{bmatrix} 0 \\ 0 \end{bmatrix} \text{ Applying } R_2 \to 4R_2 - R_1$$

$$\begin{bmatrix} 4 & 4 \\ 0 & 0 \end{bmatrix}\begin{bmatrix} x \\ y \end{bmatrix} = \begin{bmatrix} 0 \\ 0 \end{bmatrix}, \text{ which is an Echelon form.}$$

$$\Rightarrow \quad 4x + 4y = 0 \quad \Rightarrow \quad x = -y$$

$$\Rightarrow \quad \frac{x}{1} = \frac{y}{-1} = k_1$$

Hence $x = k_1, y = -k_1$

$$\therefore \quad X_1 = \begin{bmatrix} k_1 \\ -k_1 \end{bmatrix} = k_1\begin{bmatrix} 1 \\ -1 \end{bmatrix}, \text{ where } k_1 \neq 0 \text{ is a scalar.}$$

Hence $\begin{bmatrix} 1 \\ -1 \end{bmatrix}$ is an eigen vector of A corresponding to $\lambda = 1$.

Case II: Put $\lambda = 6$ in (1)

$$\begin{bmatrix} -1 & 4 \\ 1 & -4 \end{bmatrix}\begin{bmatrix} x \\ y \end{bmatrix} = \begin{bmatrix} 0 \\ 0 \end{bmatrix}$$

Applying $R_2 \to R_2 + R_1$

$$\begin{bmatrix} -1 & 4 \\ 0 & 0 \end{bmatrix}\begin{bmatrix} x \\ y \end{bmatrix} = \begin{bmatrix} 0 \\ 0 \end{bmatrix}$$

This is an Echelon form.

$$\Rightarrow \quad -x + 4y = 0$$

$$\Rightarrow \qquad x = 4y$$

$$\Rightarrow \qquad \frac{x}{4} = \frac{y}{1} = k_2$$

$$\Rightarrow \qquad x = 4k_2, y = k_2$$

$$\therefore \qquad X_2 = \begin{bmatrix} 4k_2 \\ k_2 \end{bmatrix} = k_2\begin{bmatrix} 4 \\ 1 \end{bmatrix}, \text{ where } k_2 \neq 0 \text{ is a scalar.}$$

Hence $\begin{bmatrix} 4 \\ 1 \end{bmatrix}$ is an eigen vector corresponding to $\lambda = 6$.

Example 4. *Find the sum of the eigen values of the matrix* $A = \begin{bmatrix} 2 & 5 & 3 & 1 \\ 1 & 6 & 3 & 2 \\ 3 & 4 & 1 & 2 \\ 2 & 0 & 0 & 2 \end{bmatrix}$.

(OU-May 2012)

Sol. We have

Sum of the eigen values of the matrix A = Trace of A

= Sum of its diagonal elements

= 2 + 6 + 1 + 2 = 11.

Example 5. *Find the sum of the eigen values of the matrix* $A = \begin{bmatrix} 1 & 2 & -1 \\ 0 & 2 & 2 \\ 0 & 0 & -2 \end{bmatrix}$.

Sol. We know that

Sum of the eigen values of A = Trace of A

= Sum of its diagonal elements

= 1 + 2 – 2 = 1.

Note. Product of the eigen values of A = determinant of A.

Example 6. *Find the sum and product of the eigen values of the matrix* $A = \begin{bmatrix} 1 & 1 & 1 \\ 1 & 1 & 1 \\ 1 & 1 & 1 \end{bmatrix}$.

Sol. Since sum of the eigen values of A = TrA = 1 + 1 + 1 = 3

and product of eigen values of A $= |A| = \begin{vmatrix} 1 & 1 & 1 \\ 1 & 1 & 1 \\ 1 & 1 & 1 \end{vmatrix} = 0.$

Example 7. *Find the product of the eigen values of the matrix* $A = \begin{bmatrix} 3 & -1 & 1 \\ -1 & 5 & -1 \\ 1 & -1 & 3 \end{bmatrix}$.

Sol. Since product of eigen values of A = | A |

$$= \begin{vmatrix} 3 & -1 & 1 \\ -1 & 5 & -1 \\ 1 & -1 & 3 \end{vmatrix}$$

$$= 3(15 - 1) + 1(-3 + 1) + 1(1 - 5) = 42 - 2 - 4 = 36.$$

Example 8. *Find the eigen values and eigen vectors of the matrix* $A = \begin{bmatrix} 6 & -2 & 2 \\ -2 & 3 & -1 \\ 2 & -1 & 3 \end{bmatrix}$.

(OU-Apr. 2007)

Sol. Characteristic equation is $| A - \lambda I | = 0$

$$\Rightarrow \begin{vmatrix} 6-\lambda & -2 & 2 \\ -2 & 3-\lambda & -1 \\ 2 & -1 & 3-\lambda \end{vmatrix} = 0$$

$$\Rightarrow (6 - \lambda)[(3 - \lambda)^2 - 1] + 2[-2(3 - \lambda) + 2] + 2[2 - 2(3 - \lambda)] = 0$$

$$\Rightarrow (6 - \lambda)[\lambda^2 - 6\lambda + 8] + 2[2\lambda - 4] + 2[2\lambda - 4] = 0$$

$$\Rightarrow 6\lambda^2 - 36\lambda + 48 - \lambda^3 + 6\lambda^2 - 8\lambda + 4\lambda - 8 + 4\lambda - 8 = 0$$

$$\Rightarrow -\lambda^3 + 12\lambda^2 - 36\lambda + 32 = 0$$

$$\Rightarrow \lambda^3 - 12\lambda^2 + 36\lambda - 32 = 0$$

By synthetic division

$\lambda = 2$	1	– 12	36	– 32
	0	2	– 20	32
	1	– 10	16	0

$$\lambda^2 - 10\lambda + 16 = 0$$

$$\Rightarrow (\lambda - 2)(\lambda - 8) = 0$$

$\therefore$ $\lambda = 2, 2, 8$ are eigen values of A.

Consider $[A - \lambda I]X = 0$

$$\Rightarrow \begin{bmatrix} 6-\lambda & -2 & 2 \\ -2 & 3-\lambda & -1 \\ 2 & -1 & 3-\lambda \end{bmatrix} \begin{bmatrix} x \\ y \\ z \end{bmatrix} = \begin{bmatrix} 0 \\ 0 \\ 0 \end{bmatrix} \quad \text{...(1)}$$

Case I: Put $\lambda = 2$ in (1)

$$\begin{bmatrix} 4 & -2 & 2 \\ -2 & 1 & -1 \\ 2 & -1 & 1 \end{bmatrix} \begin{bmatrix} x \\ y \\ z \end{bmatrix} = \begin{bmatrix} 0 \\ 0 \\ 0 \end{bmatrix}$$

Applying $R_2 \to 2R_2 + R_1$, $R_3 \to 2R_3 - R_1$

$$\begin{bmatrix} 4 & -2 & 2 \\ 0 & 0 & 0 \\ 0 & 0 & 0 \end{bmatrix} \begin{bmatrix} x \\ y \\ z \end{bmatrix} = \begin{bmatrix} 0 \\ 0 \\ 0 \end{bmatrix}$$

which is an Echelon form.

$$\Rightarrow \qquad 4x - 2y + 2z = 0$$

Let $y = k_1$, $z = k_2$. Then $\qquad x = \dfrac{k_1}{2} - \dfrac{k_2}{2}$

$$\therefore \qquad X = \begin{bmatrix} x \\ y \\ z \end{bmatrix} = \begin{bmatrix} \dfrac{k_1}{2} - \dfrac{k_2}{2} \\ k_1 \\ k_2 \end{bmatrix} = k_1 \begin{bmatrix} 1/2 \\ 1 \\ 0 \end{bmatrix} + k_2 \begin{bmatrix} -1/2 \\ 0 \\ 1 \end{bmatrix}$$

is an eigen vector of A where k_1, k_2 are arbitrary constants.

Case II: Put $\lambda = 8$ in (1), we get

$$\begin{bmatrix} -2 & -2 & 2 \\ -2 & -5 & -1 \\ 2 & -1 & -5 \end{bmatrix} \begin{bmatrix} x \\ y \\ z \end{bmatrix} = \begin{bmatrix} 0 \\ 0 \\ 0 \end{bmatrix}$$

Applying $R_2 \to R_2 - R_1$, $R_3 \to R_3 + R_1$

$$\Rightarrow \qquad \begin{bmatrix} -2 & -2 & 2 \\ 0 & -3 & -3 \\ 0 & -3 & -3 \end{bmatrix} \begin{bmatrix} x \\ y \\ z \end{bmatrix} = \begin{bmatrix} 0 \\ 0 \\ 0 \end{bmatrix}$$

Applying $R_3 \to R_3 - R_2$

$$\Rightarrow \qquad \begin{bmatrix} -2 & -2 & 2 \\ 0 & -3 & -3 \\ 0 & 0 & 0 \end{bmatrix} \begin{bmatrix} x \\ y \\ z \end{bmatrix} = \begin{bmatrix} 0 \\ 0 \\ 0 \end{bmatrix}$$

$$\Rightarrow \qquad -2x - 2y + 2z = 0$$

$$\Rightarrow \qquad x + y - z = 0$$

and $\qquad -3y - 3z = 0$

$$\Rightarrow \qquad y + z = 0$$

Let $z = k_3$, then $y = -k_3$ and $x = 2k_3$

$$\therefore \qquad X = \begin{bmatrix} 2k_3 \\ -k_3 \\ k_3 \end{bmatrix} = k_3 \begin{bmatrix} 2 \\ -1 \\ 1 \end{bmatrix} \text{ is an eigen vector.}$$

Corresponding to $\lambda = 8$ where k_3 is any non-zero arbitrary constant.

Note. Eigen values of a triangular matrix A = diagonal elements of A.

Example 9. *The characteristic roots of* $A = \begin{bmatrix} 1 & 2 & 3 \\ 0 & -4 & 2 \\ 0 & 0 & 6 \end{bmatrix}$ *are ...* (OU-Nov. 2006)

Sol. 1, – 4, 6 since the characteristic roots of a triangular matrix are just the diagonal elements of that matrix.

Example 10. *Find the characteristic roots and characteristic vectors of the matrix*

$$A = \begin{bmatrix} 8 & -6 & 2 \\ -6 & 7 & -4 \\ 2 & -4 & 3 \end{bmatrix}. \qquad \text{(OU-May 2009)}$$

Sol. Characteristic equation is $|A - \lambda I| = 0$

$$\Rightarrow \quad \begin{vmatrix} 8-\lambda & -6 & 2 \\ -6 & 7-\lambda & -4 \\ 2 & -4 & 3-\lambda \end{vmatrix} = 0$$

$$\Rightarrow \quad (8-\lambda)[(7-\lambda)(3-\lambda) - 16] + 6[-6(3-\lambda) + 8] + 2[24 - 2(7-\lambda)] = 0$$

$$\Rightarrow \quad \lambda^3 - 18\lambda^2 + 45\lambda = 0$$

$$\Rightarrow \quad \lambda(\lambda^2 - 18\lambda + 45) = 0$$

$$\Rightarrow \quad \lambda(\lambda - 3)(\lambda - 15) = 0$$

$\Rightarrow$ $\lambda = 0, 3, 15$ are eigen values of A

Consider $[A - \lambda I]X = 0$

$$\Rightarrow \quad \begin{bmatrix} 8-\lambda & -6 & 2 \\ -6 & 7-\lambda & -4 \\ 2 & -4 & 3-\lambda \end{bmatrix} \begin{bmatrix} x \\ y \\ z \end{bmatrix} = \begin{bmatrix} 0 \\ 0 \\ 0 \end{bmatrix} \qquad ...(1)$$

Case I: Put $\lambda = 0$ in (1)

$$\begin{bmatrix} 8 & -6 & 2 \\ -6 & 7 & -4 \\ 2 & -4 & 3 \end{bmatrix} \begin{bmatrix} x \\ y \\ z \end{bmatrix} = \begin{bmatrix} 0 \\ 0 \\ 0 \end{bmatrix}$$

Applying $R_1 \to \frac{R_1}{2}$

$$\begin{bmatrix} 4 & -3 & 1 \\ -6 & 7 & -4 \\ 2 & -4 & 3 \end{bmatrix} \begin{bmatrix} x \\ y \\ z \end{bmatrix} = \begin{bmatrix} 0 \\ 0 \\ 0 \end{bmatrix}$$

Applying $R_2 \to 2R_2 + 3R_1$, $R_3 \to 2R_3 - R_1$

$$\Rightarrow \quad \begin{bmatrix} 4 & -3 & 1 \\ 0 & 5 & -5 \\ 0 & -5 & 5 \end{bmatrix} \begin{bmatrix} x \\ y \\ z \end{bmatrix} = \begin{bmatrix} 0 \\ 0 \\ 0 \end{bmatrix}$$

Applying $R_3 \to R_3 + R_2$

$$\Rightarrow \quad \begin{bmatrix} 4 & -3 & 1 \\ 0 & 5 & -5 \\ 0 & 0 & 0 \end{bmatrix} \begin{bmatrix} x \\ y \\ z \end{bmatrix} = \begin{bmatrix} 0 \\ 0 \\ 0 \end{bmatrix}$$

This is an Echelon form.

We have $4x - 3y + z = 0$ and

$5y - 5z = 0 \Rightarrow y = z$

Let $z = k_1$, then $y = k_1$ and $x = \frac{k_1}{2}$

$$X = k_1 \begin{bmatrix} 1/2 \\ 1 \\ 1 \end{bmatrix}$$

$\therefore \quad \begin{bmatrix} 1/2 \\ 1 \\ 1 \end{bmatrix}$ is an eigen vector corresponding to $\lambda = 0$.

Similarly, eigen vectors corresponding to $\lambda = 3$ and 15 are $\begin{bmatrix} -2 \\ -1 \\ 2 \end{bmatrix}$ and $\begin{bmatrix} 2 \\ -2 \\ 1 \end{bmatrix}$ respectively.

Example 11. *Find the sum of the eigen values of the matrix* $\begin{bmatrix} 8 & 9 & 1 \\ 4 & 5 & 6 \\ 1 & 4 & 3 \end{bmatrix}$.

(OU-Jan. 2011)

Sol. We have

Sum of eigen values of A = Tr A = 8 + 5 + 3 = 16

Example 12. *Show that if $\lambda_1, \lambda_2, \ldots\ldots, \lambda_n$ are the latent roots of the matrix A, then A^3 has the latent roots $\lambda_1^3, \lambda_2^3, \ldots\ldots, \lambda_n^3$.*

Sol. Let λ be a latent root of the matrix A. Then there exists a non-zero vector X such that

$$AX = \lambda X \qquad \ldots(1)$$

$$\Rightarrow \quad A^2(AX) = A^2(\lambda X) \quad \Rightarrow \quad A^3X = \lambda(A^2X)$$

But $\quad A^2X = A(AX) = A(\lambda X)$ | Using (1)

$$= \lambda(AX) = \lambda(\lambda X) = \lambda^2 X$$

$$\therefore \quad A^3X = \lambda(\lambda^2 X) = \lambda^3 X$$

$\Rightarrow \quad \lambda^3$ is a latent root of A^3.

$\therefore \quad$ If $\lambda_1, \lambda_2, \ldots\ldots, \lambda_n$ are the latent roots of A, then $\lambda_1^3, \lambda_2^3, \ldots\ldots, \lambda_n^3$ are the latent roots of A^3.

Example 13. *If λ be an eigen value of a non-singular matrix A, show that*

(i) λ^{-1} is an eigen value of A^{-1}.

(ii) $\dfrac{|A|}{\lambda}$ is an eigen value of adj. A.

Sol. *(i)* λ is an eigen value of A.

$\Rightarrow \quad$ There exists a non-zero matrix X such that $AX = \lambda X$

$$\Rightarrow \quad X = A^{-1}(\lambda X) \quad \Rightarrow \quad X = \lambda(A^{-1}X)$$

$$\Rightarrow \quad \frac{1}{\lambda}X = A^{-1}X \quad \Rightarrow \quad A^{-1}X = \lambda^{-1}X$$

$\Rightarrow \quad \lambda^{-1}$ is an eigen value of A^{-1}.

(ii) λ is an eigen value of A

$\Rightarrow \quad$ There exists a non-zero matrix X such that $AX = \lambda X$

$$\Rightarrow \quad (\text{adj } A)(AX) = (\text{adj } A)(\lambda X)$$

$$\Rightarrow \quad ((\text{adj } A)A)\, X = \lambda\, (\text{adj } A)\, X$$

$$\Rightarrow \qquad |A|\,IX = \lambda\,(\text{adj } A)\,X \qquad [\because \ (\text{adj } A)\,A = |A|\,I]$$

$$\Rightarrow \qquad |A|\,X = \lambda\,(\text{adj } A)\,X \qquad \Rightarrow \qquad \frac{|A|}{\lambda}\,X = (\text{adj } A)\,X$$

$$\Rightarrow \qquad (\text{adj } A)\,X = \frac{|A|}{\lambda}\,X \qquad \Rightarrow \qquad \frac{|A|}{\lambda} \text{ is an eigen value of adj A.}$$

Example 14. *Let a 3 × 3 matrix A have eigen values 1, 2, – 1. Find the trace of the matrix* $B = A - A^{-1} + A^2$. (OU-Dec. 2012)

Sol. Since eigen values of A are 1, 2, – 1,

eigen values of A^{-1} are 1, $\frac{1}{2}$, – 1 and

eigen values of A^2 are 1, 4, 1

$\therefore$ eigen values of $B = A - A^{-1} + A^2$ are

$$(1 - 1 + 1), \left(2 - \frac{1}{2} + 4\right), (-1 + 1 + 1)$$

i.e., eigen values of B are 1, $\frac{11}{2}$, 1

Trace of B = Tr B = Sum of eigen values of B

$$= 1 + \frac{11}{2} + 1 = \frac{15}{2}.$$

1.26. CAYLEY-HAMILTON THEOREM (*OU-Dec. 2008, 2007*)

Statement. *Every square matrix satisfies its own characteristic equation.*

i.e., if the characteristic equation of the nth order square matrix A is

$$|A - \lambda I| = (-1)^n \lambda^n + k_1 \lambda^{n-1} + k_2 \lambda^{n-2} + \ldots\ldots + k_n = 0 \qquad \ldots(1)$$

then

$$(-1)^n A^n + k_1 A^{n-1} + k_2 A^{n-2} + \ldots\ldots + k_n I = O \qquad \ldots(2)$$

Proof. Let $P = \text{adj } (A - \lambda I)$

Since the elements of $A - \lambda I$ are at most of first degree in λ, the elements of P = adj $(A - \lambda I)$ are polynomials in λ of degree $(n - 1)$ or less. We can, therefore, split up P into a number of matrices each containing the same power of λ and write

$$P = P_0 \lambda^{n-1} + P_1 \lambda^{n-2} + \ldots\ldots + P_{n-2} \lambda + P_{n-1} \qquad \ldots(3)$$

Also, we know that if M is a square matrix, then M (adj M) = | M | I $\left(\because \ M^{-1} = \frac{\text{adj } M}{|M|}\right)$

$$\therefore \qquad (A - \lambda I)P = |A - \lambda I|\, I$$

By (1) and (3), we have

$$(A - \lambda I)(P_0 \lambda^{n-1} + P_1 \lambda^{n-2} + \ldots\ldots + P_{n-2}\lambda + P_{n-1}) = [(-1)^n \lambda^n + k_1 \lambda^{n-1} + k_2 \lambda^{n-2} + \ldots\ldots + k_{n-1}\lambda + k_n]I$$

Equating co-efficients of like powers of λ on both sides, we have

Coefficients of λ^n are $-P_0 = (-1)^n I \ \Rightarrow \ -P_0 A^n = (-1)^n A^n$ $\qquad [\because \ IP_0 = P_0]$

λ^{n-1} coefficients $\Rightarrow \quad AP_0 - P_1 = k_1 I \quad \Rightarrow \quad A^n P_0 - A^{n-1} P_1 = k_1 A^{n-1}$

λ^{n-2} coefficients $\Rightarrow \quad AP_1 - P_2 = k_2 I \quad \Rightarrow \quad A^{n-1} P_1 - A^{n-2} P_2 = k_2 A^{n-2}$

....

λ coefficients $\Rightarrow \quad AP_{n-2} - P_{n-1} = k_{n-1} I \quad \Rightarrow \quad A^2 P_{n-2} - AP_{n-1} = k_{n-1} A$

λ° coefficients are $\Rightarrow \quad AP_{n-1} = k_n I \quad \Rightarrow \quad AP_{n-1} = k_n I$

i.e., Pre-multiplying these equations by A^n, A^{n-1}, A^{n-2}, ..., A, I respectively and adding, we get

$$O = (-1)^n A^n + k_1 A^{n-1} + k_2 A^{n-2} + \ldots + k_{n-1} A + k_n I$$

[terms on the LHS cancel in pairs]

or
$$(-1)^n A^n + k_1 A^{n-1} + k_2 A^{n-2} + \ldots + k_{n-1} A + k_n I = O \qquad \ldots(4)$$

which proves the theorem.

Note 1. Multiplying (4) by A^{-1}, we have $(-1)^n A^{n-1} + k_1 A^{n-2} + \ldots + k_{n-1} I + k_n A^{-1} = O$

$$\Rightarrow \quad A^{-1} = -\frac{1}{k_n}\left[(-1)^n A^{n-1} + k_1 A^{n-2} + \ldots + k_{n-1} I\right]$$

Thus Cayley-Hamilton theorem gives another method for computing the inverse of a matrix. Since this method expresses the inverse of a matrix of order n in terms of $(n-1)$ powers of A, it is most suitable for computing inverses of large matrices.

Note 2. If m be a positive integer such that $m > n$, then multiplying (4) by A^{m-n}, we get

$$(-1)^n A^m + k_1 A^{m-1} + k_2 A^{m-2} + \ldots + k_{n-1} A^{m-n+1} + k_n A^{m-n} = O$$

showing that any positive integral power A^m $(m > n)$ of A is linearly expressible in terms of those of lower degree.

ILLUSTRATIVE EXAMPLES

Example 1. *Verify Cayley-Hamilton theorem for the matrix* $A = \begin{bmatrix} 2 & -1 & 1 \\ -1 & 2 & -1 \\ 1 & -1 & 2 \end{bmatrix}$.

Hence compute A^{-1}.

Sol. The characteristic equation of A is

$$|A - \lambda I| = 0 \quad i.e., \quad \begin{vmatrix} 2-\lambda & -1 & 1 \\ -1 & 2-\lambda & -1 \\ 1 & -1 & 2-\lambda \end{vmatrix} = 0$$

or
$$\lambda^3 - 6\lambda^2 + 9\lambda - 4 = 0 \quad \text{(on simplification)}$$

To verify Cayley-Hamilton theorem, we have to show that

$$A^3 - 6A^2 + 9A - 4I = O \qquad \ldots(1)$$

Now,
$$A^2 = \begin{bmatrix} 2 & -1 & 1 \\ -1 & 2 & -1 \\ 1 & -1 & 2 \end{bmatrix} \times \begin{bmatrix} 2 & -1 & 1 \\ -1 & 2 & -1 \\ 1 & -1 & 2 \end{bmatrix} = \begin{bmatrix} 6 & -5 & 5 \\ -5 & 6 & -5 \\ 5 & -5 & 6 \end{bmatrix}$$

$$A^3 = A^2 \times A = \begin{bmatrix} 6 & -5 & 5 \\ -5 & 6 & -5 \\ 5 & -5 & 6 \end{bmatrix} \begin{bmatrix} 2 & -1 & 1 \\ -1 & 2 & -1 \\ 1 & -1 & 2 \end{bmatrix} = \begin{bmatrix} 22 & -21 & 21 \\ -21 & 22 & -21 \\ 21 & -21 & 22 \end{bmatrix}$$

$$\therefore \quad A^3 - 6A^2 + 9A - 4I = \begin{bmatrix} 22 & -21 & 21 \\ -21 & 22 & -21 \\ 21 & -21 & 22 \end{bmatrix} - 6\begin{bmatrix} 6 & -5 & 5 \\ -5 & 6 & -5 \\ 5 & -5 & 6 \end{bmatrix} + 9\begin{bmatrix} 2 & -1 & 1 \\ -1 & 2 & -1 \\ 1 & -1 & 2 \end{bmatrix} - 4\begin{bmatrix} 1 & 0 & 0 \\ 0 & 1 & 0 \\ 0 & 0 & 1 \end{bmatrix}$$

$$= \begin{bmatrix} 0 & 0 & 0 \\ 0 & 0 & 0 \\ 0 & 0 & 0 \end{bmatrix} = O$$

This verifies Cayley-Hamilton theorem.

Now, multiplying both sides of (1) by A^{-1}, we have

$$A^2 - 6A + 9I - 4A^{-1} = O$$

$$\Rightarrow \quad 4A^{-1} = A^2 - 6A + 9I$$

$$= \begin{bmatrix} 6 & -5 & 5 \\ -5 & 6 & -5 \\ 5 & -5 & 6 \end{bmatrix} - 6\begin{bmatrix} 2 & -1 & 1 \\ -1 & 2 & -1 \\ 1 & -1 & 2 \end{bmatrix} + 9\begin{bmatrix} 1 & 0 & 0 \\ 0 & 1 & 0 \\ 0 & 0 & 1 \end{bmatrix} = \begin{bmatrix} 3 & 1 & -1 \\ 1 & 3 & 1 \\ -1 & 1 & 3 \end{bmatrix}$$

$$\therefore \quad A^{-1} = \frac{1}{4}\begin{bmatrix} 3 & 1 & -1 \\ 1 & 3 & 1 \\ -1 & 1 & 3 \end{bmatrix}.$$

Example 2. *Find the characteristic equation of the matrix* $A = \begin{bmatrix} 2 & 1 & 1 \\ 0 & 1 & 0 \\ 1 & 1 & 2 \end{bmatrix}$ *and, hence, find the matrix represented by* $A^8 - 5A^7 + 7A^6 - 3A^5 + A^4 - 5A^3 + 8A^2 - 2A + I$.

Sol. The characteristic equation of A is

$$|A - \lambda I| = \begin{vmatrix} 2-\lambda & 1 & 1 \\ 0 & 1-\lambda & 0 \\ 1 & 1 & 2-\lambda \end{vmatrix} = 0 \quad \text{or} \quad \lambda^3 - 5\lambda^2 + 7\lambda - 3 = 0$$

By Cayley-Hamilton theorem, $A^3 - 5A^2 + 7A - 3I = O$ ...(1)

Now, $A^8 - 5A^7 + 7A^6 - 3A^5 + A^4 - 5A^3 + 8A^2 - 2A + I$

$$= A^5(A^3 - 5A^2 + 7A - 3I) + A(A^3 - 5A^2 + 7A - 3I) + (A^2 + A + I)$$

$$= A^2 + A + I \qquad | \text{ Using (1)}$$

$$= \begin{bmatrix} 2 & 1 & 1 \\ 0 & 1 & 0 \\ 1 & 1 & 2 \end{bmatrix}\begin{bmatrix} 2 & 1 & 1 \\ 0 & 1 & 0 \\ 1 & 1 & 2 \end{bmatrix} + \begin{bmatrix} 2 & 1 & 1 \\ 0 & 1 & 0 \\ 1 & 1 & 2 \end{bmatrix} + \begin{bmatrix} 1 & 0 & 0 \\ 0 & 1 & 0 \\ 0 & 0 & 1 \end{bmatrix}$$

$$= \begin{bmatrix} 5 & 4 & 4 \\ 0 & 1 & 0 \\ 4 & 4 & 5 \end{bmatrix} + \begin{bmatrix} 2 & 1 & 1 \\ 0 & 1 & 0 \\ 1 & 1 & 2 \end{bmatrix} + \begin{bmatrix} 1 & 0 & 0 \\ 0 & 1 & 0 \\ 0 & 0 & 1 \end{bmatrix} = \begin{bmatrix} 8 & 5 & 5 \\ 0 & 3 & 0 \\ 5 & 5 & 8 \end{bmatrix}.$$

Example 3. *Using Cayley-Hamilton theorem, find the inverse of* $A = \begin{bmatrix} 1 & 1 & 3 \\ 1 & 3 & -1 \\ 2 & -4 & -4 \end{bmatrix}$.

(OU-May 2012)

Sol. The characteristic equation of A is $|A - \lambda I| = 0$

$$\Rightarrow \quad \begin{vmatrix} 1-\lambda & 1 & 3 \\ 1 & 3-\lambda & -1 \\ 2 & -4 & -4-\lambda \end{vmatrix} = 0$$

$$(1-\lambda)[(3-\lambda)(-4-\lambda)-4] - 1[1(-4-\lambda)+2] + 3[-4-2(3-\lambda)] = 0$$

$$\Rightarrow \quad (1-\lambda)[\lambda^2+\lambda-16] - [-\lambda-2] + 3[2\lambda-10] = 0$$

$$\Rightarrow \quad \lambda^2+\lambda-16-\lambda^3-\lambda^2+16\lambda+\lambda+2+6\lambda-30 = 0$$

$$\Rightarrow \quad -\lambda^3+24\lambda-44 = 0$$

$$\Rightarrow \quad \lambda^3-24\lambda+44 = 0$$

By Cayley-Hamilton theorem, we get

$$A^3 - 24A + 44I = 0$$

Multiplying by A^{-1} on both sides

$$A^2 - 24I + 44A^{-1} = 0 \qquad (\because \; A \cdot A^{-1} = I)$$

$$A^{-1} = \frac{1}{44}[24I - A^2] \qquad \ldots(1)$$

Now,

$$A^2 = A \cdot A = \begin{bmatrix} 1 & 1 & 3 \\ 1 & 3 & -1 \\ 2 & -4 & -4 \end{bmatrix}\begin{bmatrix} 1 & 1 & 3 \\ 1 & 3 & -1 \\ 2 & -4 & -4 \end{bmatrix}$$

$$= \begin{bmatrix} 1+1+6 & 1+3-12 & 3-1-12 \\ 1+3-2 & 1+9+4 & 3-3+4 \\ 2-4-8 & 2-12+16 & 6+4+16 \end{bmatrix} = \begin{bmatrix} 8 & -8 & -10 \\ 2 & 14 & 4 \\ -10 & 6 & 26 \end{bmatrix}$$

From (1),

$$A^{-1} = \frac{1}{44}\left[24\begin{bmatrix} 1 & 0 & 0 \\ 0 & 1 & 0 \\ 0 & 0 & 1 \end{bmatrix} - \begin{bmatrix} 8 & -8 & -10 \\ 2 & 14 & 4 \\ -10 & 6 & 26 \end{bmatrix}\right]$$

$$= \frac{1}{44}\begin{bmatrix} 16 & 8 & 10 \\ -2 & 10 & -4 \\ 10 & -6 & -2 \end{bmatrix} = \frac{1}{22}\begin{bmatrix} 8 & 4 & 5 \\ -1 & 5 & -2 \\ 5 & -3 & -1 \end{bmatrix}.$$

Example 4. *Find the inverse of the matrix* $\begin{bmatrix} 1 & -1 & 0 \\ 0 & 1 & 1 \\ 2 & 1 & 2 \end{bmatrix}$ *by using Cayley-Hamilton theorem.*

Sol. Let $$A = \begin{bmatrix} 1 & -1 & 0 \\ 0 & 1 & 1 \\ 2 & 1 & 2 \end{bmatrix}$$

Its characteristic equation is $| A - \lambda I | = 0$

$$\Rightarrow \quad \begin{vmatrix} 1-\lambda & -1 & 0 \\ 0 & 1-\lambda & 1 \\ 2 & 1 & 2-\lambda \end{vmatrix} = 0$$

$\Rightarrow \quad \lambda^3 - 4\lambda^2 + 4\lambda + 1 = 0$ on simplification

By Cayley-Hamilton theorem, we get

$$A^3 - 4A^2 + 4A + I = 0$$

Multiply by A^{-1} on both sides

$$A^2 - 4A + 4I + A^{-1} = 0$$

$$\Rightarrow \quad A^{-1} = - A^2 + 4A - 4I$$

$$= - \begin{bmatrix} 1 & -2 & -1 \\ 2 & 2 & 3 \\ 6 & 1 & 5 \end{bmatrix} + 4 \begin{bmatrix} 1 & -1 & 0 \\ 0 & 1 & 1 \\ 2 & 1 & 2 \end{bmatrix} - 4 \begin{bmatrix} 1 & 0 & 0 \\ 0 & 1 & 0 \\ 0 & 0 & 1 \end{bmatrix}$$

$$= \begin{bmatrix} -1 & -2 & 1 \\ -2 & -2 & 1 \\ 2 & 3 & -1 \end{bmatrix}.$$

Example 5. *Verify Cayley-Hamilton theorem for the matrix* $A = \begin{bmatrix} 1 & 2 & 3 \\ 2 & 4 & 5 \\ 3 & 5 & 6 \end{bmatrix}$. *Hence find* A^{-1}.

Sol. The characteristic equation of A is $| A - \lambda I | = 0$

$$\Rightarrow \quad \begin{vmatrix} 1-\lambda & 2 & 3 \\ 2 & 4-\lambda & 5 \\ 3 & 5 & 6-\lambda \end{vmatrix} = 0$$

$\Rightarrow \quad \lambda^3 - 11\lambda^2 - 4\lambda + 1 = 0$ on simplification

By Cayley-Hamilton theorem, we have

$$A^3 - 11A^2 - 4A + I = 0 \quad \quad ...(1)$$

Multiply by A^{-1}, we get

$$A^2 - 11A - 4I + A^{-1} = 0$$

$$\Rightarrow \quad A^{-1} = - A^2 + 11A + 4I \quad \quad ...(2)$$

$$A^2 = A \cdot A = \begin{bmatrix} 1 & 2 & 3 \\ 2 & 4 & 5 \\ 3 & 5 & 6 \end{bmatrix} \cdot \begin{bmatrix} 1 & 2 & 3 \\ 2 & 4 & 5 \\ 3 & 5 & 6 \end{bmatrix} = \begin{bmatrix} 14 & 25 & 31 \\ 25 & 45 & 56 \\ 31 & 56 & 70 \end{bmatrix}$$

$$A^3 = A \cdot A^2 = \begin{bmatrix} 1 & 2 & 3 \\ 2 & 4 & 5 \\ 3 & 5 & 6 \end{bmatrix} \begin{bmatrix} 14 & 25 & 31 \\ 25 & 45 & 56 \\ 31 & 56 & 70 \end{bmatrix} = \begin{bmatrix} 157 & 283 & 353 \\ 283 & 510 & 636 \\ 353 & 636 & 793 \end{bmatrix}$$

From (1), $A^3 - 11A^2 - 4A + I$

$$= \begin{bmatrix} 157 & 283 & 353 \\ 283 & 510 & 636 \\ 353 & 636 & 793 \end{bmatrix} - 11 \begin{bmatrix} 14 & 25 & 31 \\ 25 & 45 & 56 \\ 31 & 56 & 70 \end{bmatrix} - 4 \begin{bmatrix} 1 & 2 & 3 \\ 2 & 4 & 5 \\ 3 & 5 & 6 \end{bmatrix} + \begin{bmatrix} 1 & 0 & 0 \\ 0 & 1 & 0 \\ 0 & 0 & 1 \end{bmatrix}$$

$$= \begin{bmatrix} 0 & 0 & 0 \\ 0 & 0 & 0 \\ 0 & 0 & 0 \end{bmatrix} = 0$$

$\therefore$ Cayley-Hamilton theorem is verified.

From (2), we have

$$A^{-1} = - A^2 + 11A + 4I$$

$$= - \begin{bmatrix} 14 & 25 & 31 \\ 25 & 45 & 56 \\ 31 & 56 & 70 \end{bmatrix} + 11 \begin{bmatrix} 1 & 2 & 3 \\ 2 & 4 & 5 \\ 3 & 5 & 6 \end{bmatrix} + 4 \begin{bmatrix} 1 & 0 & 0 \\ 0 & 1 & 0 \\ 0 & 0 & 1 \end{bmatrix}$$

$$= \begin{bmatrix} 1 & -3 & 2 \\ -3 & 3 & -1 \\ 2 & -1 & 0 \end{bmatrix}.$$

Example 6. *Using Cayley-Hamilton theorem, find A^8 if $A = \begin{bmatrix} 1 & 2 \\ 2 & -1 \end{bmatrix}$.*

Sol. Characteristic equation of A is $| A - \lambda I | = 0$

$$\Rightarrow \quad \begin{vmatrix} 1-\lambda & 2 \\ 2 & -1-\lambda \end{vmatrix} = 0$$

$$\Rightarrow \quad \lambda^2 - 5 = 0$$

By Cayley-Hamilton theorem, A satisfied the above equation.

Hence, we get $A^2 - 5I = 0 \Rightarrow A^2 = 5I$

$$\Rightarrow \quad A^8 = 5A^6 I = 5(A^2)^3 = 5(5I)^3 = 625I$$

$$\Rightarrow \quad A^8 = \begin{bmatrix} 625 & 0 \\ 0 & 625 \end{bmatrix}.$$

Example 7. *If $A = \begin{bmatrix} 1 & 0 & 0 \\ 1 & 0 & 1 \\ 0 & 1 & 0 \end{bmatrix}$ show that $A^n = A^{n-2} + A^2 - I$ for $n \geq 3$ using Cayley-Hamilton theorem.* (OU-June 2011)

Sol. Characteristic equation of A is

$$| A - \lambda I | = 0$$

$$\Rightarrow \quad \begin{vmatrix} 1-\lambda & 0 & 0 \\ 1 & -\lambda & 1 \\ 0 & 1 & -\lambda \end{vmatrix} = 0$$

$$\Rightarrow \quad (1 - \lambda)(\lambda^2 - 1) = 0$$

$$\Rightarrow \quad \lambda^2 - 1 - \lambda^3 + \lambda = 0$$

$$\Rightarrow \quad \lambda^3 = \lambda^2 + \lambda - 1$$

Using Cayley-Hamilton theorem, we get $A^3 = A^2 + A - I$

$$\Rightarrow \quad A^3 - A^2 = A - I$$

$$\Rightarrow \quad A^4 - A^3 = A^2 - A$$

$$\vdots \quad \vdots \quad \vdots \quad \vdots$$

Similarly, $A^n - A^{n-1} = A^{n-2} - A^{n-3}$

Adding all above equations, we get

$$A^n = A^{n-2} + A^2 - I \text{ for } n \geq 3.$$

1.27. REDUCTION OF A MATRIX TO DIAGONAL FORM

If a square matrix A of order n has n linearly independent eigen vectors, then a matrix B can be found such that $B^{-1} AB$ is a diagonal matrix.

[We prove the result for a square matrix of order 3. The proof can be easily extended to matrices of higher order.]

Let A be a square matrix of order 3. Let $\lambda_1, \lambda_2, \lambda_3$ be its eigen values and

$$X_1 = \begin{bmatrix} x_1 \\ y_1 \\ z_1 \end{bmatrix}, \quad X_2 = \begin{bmatrix} x_2 \\ y_2 \\ z_2 \end{bmatrix}, \quad X_3 = \begin{bmatrix} x_3 \\ y_3 \\ z_3 \end{bmatrix}$$ be the corresponding eigen vectors.

Since eigen vectors are non-trivial solutions of the matrix equation $AX = \lambda X$, we have

$$AX_1 = \lambda_1 X_1, \; AX_2 = \lambda_2 X_2, \; AX_3 = \lambda_3 X_3$$

Let
$$B = \begin{bmatrix} x_1 & x_2 & x_3 \\ y_1 & y_2 & y_3 \\ z_1 & z_2 & z_3 \end{bmatrix} = [X_1 X_2 X_3]$$

Consider $AB = A[X_1, X_2, X_3] = [AX_1, AX_2, AX_3] = [\lambda_1 X_1, \lambda_2 X_2, \lambda_3 X_3]$

$$= \begin{bmatrix} \lambda_1 x_1 & \lambda_2 x_2 & \lambda_3 x_3 \\ \lambda_1 y_1 & \lambda_2 y_2 & \lambda_3 y_3 \\ \lambda_1 z_1 & \lambda_2 z_2 & \lambda_3 z_3 \end{bmatrix} = \begin{bmatrix} x_1 & x_2 & x_3 \\ y_1 & y_2 & y_3 \\ z_1 & z_2 & z_3 \end{bmatrix} \begin{bmatrix} \lambda_1 & 0 & 0 \\ 0 & \lambda_2 & 0 \\ 0 & 0 & \lambda_3 \end{bmatrix}$$

$$= BD, \text{ where D is the diagonal matrix } \begin{bmatrix} \lambda_1 & 0 & 0 \\ 0 & \lambda_2 & 0 \\ 0 & 0 & \lambda_3 \end{bmatrix}$$

$$\Rightarrow \quad AB = BD \quad \Rightarrow \quad B^{-1} AB = B^{-1} BD = ID = D, \text{ which proves the theorem.}$$

Note. The matrix B which diagonalises A is called the **modal matrix** of A and is obtained by grouping the eigen vectors of A into a square matrix.

Example 1. *Reduce the matrix* $A = \begin{bmatrix} -1 & 2 & -2 \\ 1 & 2 & 1 \\ -1 & -1 & 0 \end{bmatrix}$ *to the diagonal form.*

Sol. The characteristic equation of A is

$$|A - \lambda I| = \begin{vmatrix} -1-\lambda & 2 & -2 \\ 1 & 2-\lambda & 1 \\ -1 & -1 & -\lambda \end{vmatrix} = 0 \quad \text{or} \quad \lambda^3 - \lambda^2 - 5\lambda + 5 = 0 \text{ (on simplification)}$$

Solving it, we have $\lambda = 1, \pm\sqrt{5}$.

When $\lambda = 1$, the corresponding eigen vector is given by

$$-2x_1 + 2y_1 - 2z_1 = 0$$
$$x_1 + y_1 + z_1 = 0$$
$$x_1 + y_1 + z_1 = 0$$

Solving the first two $\frac{x_1}{4} = \frac{y_1}{0} = \frac{z_1}{-4}$ giving the eigen vector $k_1\begin{bmatrix} 1 \\ 0 \\ -1 \end{bmatrix}$.

When $\lambda = \sqrt{5}$, the corresponding eigen vector is given by

$$(-1 - \sqrt{5})x_2 + 2y_2 - 2z_2 = 0$$
$$x_2 + (2 - \sqrt{5})y_2 + z_2 = 0$$
$$-x_2 - y_2 - \sqrt{5}z_2 = 0$$

Solving the last two $\frac{x_2}{6 - 2\sqrt{5}} = \frac{y_2}{-1+\sqrt{5}} = \frac{z_2}{1-\sqrt{5}}$

or $$\frac{x_2}{(\sqrt{5}-1)^2} = \frac{y_2}{\sqrt{5}-1} = \frac{z_2}{1-\sqrt{5}} \quad \text{or} \quad \frac{x_2}{\sqrt{5}-1} = \frac{y_2}{1} = \frac{z_2}{-1}$$

giving the eigen vector $k_2\begin{bmatrix} \sqrt{5}-1 \\ 1 \\ -1 \end{bmatrix}$

Similarly, the eigen vector corresponding to $\lambda = -\sqrt{5}$ is $k_3\begin{bmatrix} \sqrt{5}+1 \\ -1 \\ 1 \end{bmatrix}$.

$\therefore$ The modal matrix $$B = \begin{bmatrix} 1 & \sqrt{5}-1 & \sqrt{5}+1 \\ 0 & 1 & -1 \\ -1 & -1 & 1 \end{bmatrix}$$

and $$D = \begin{bmatrix} \lambda_1 & 0 & 0 \\ 0 & \lambda_2 & 0 \\ 0 & 0 & \lambda_3 \end{bmatrix} = \begin{bmatrix} 1 & 0 & 0 \\ 0 & \sqrt{5} & 0 \\ 0 & 0 & -\sqrt{5} \end{bmatrix}$$

is the diagonal matrix, obtained by diagonalising A.

Example 2. *Diagonalise the matrix* $A = \begin{bmatrix} 8 & -8 & -2 \\ 4 & -3 & -2 \\ 3 & -4 & 1 \end{bmatrix}$.

Sol. The characteristic equation of A is $|A - \lambda I| = 0$

$$\Rightarrow \quad \begin{vmatrix} 8-\lambda & -8 & -2 \\ 4 & -3-\lambda & -2 \\ 3 & -4 & 1-\lambda \end{vmatrix} = 0$$

$$\Rightarrow \quad \lambda^3 - 6\lambda^2 + 11\lambda - 6 = 0 \text{ on simplification}$$

By synthetic division

$\lambda = 1$	1	−6	11	−6
	0	1	−5	+6
	1	−5	6	0

$\lambda^2 - 5\lambda + 6 = 0 \Rightarrow (\lambda - 2)(\lambda - 3) = 0$

Hence, $\lambda = 1, 2, 3$ are eigen values.

Now, consider $[A - \lambda I]X = 0$

$$\Rightarrow \quad \begin{bmatrix} 8-\lambda & -8 & -2 \\ 4 & -3-\lambda & -2 \\ 3 & -4 & 1-\lambda \end{bmatrix} \begin{bmatrix} x \\ y \\ z \end{bmatrix} = \begin{bmatrix} 0 \\ 0 \\ 0 \end{bmatrix} \qquad ...(1)$$

Case I: Put $\lambda = 1$ in (1), we get

$$\begin{bmatrix} 7 & -8 & -2 \\ 4 & -4 & -2 \\ 3 & -4 & 0 \end{bmatrix} \begin{bmatrix} x \\ y \\ z \end{bmatrix} = \begin{bmatrix} 0 \\ 0 \\ 0 \end{bmatrix}$$

Applying $R_2 \to 7R_2 - 4R_1$, $R_3 \to 7R_3 - 3R_1$

$$\begin{bmatrix} 7 & -8 & -2 \\ 0 & 4 & -6 \\ 0 & -4 & 6 \end{bmatrix} \begin{bmatrix} x \\ y \\ z \end{bmatrix} = \begin{bmatrix} 0 \\ 0 \\ 0 \end{bmatrix}$$

Applying $R_3 \to R_3 + R_2$

$$\begin{bmatrix} 7 & -8 & -2 \\ 0 & 4 & -6 \\ 0 & 0 & 0 \end{bmatrix} \begin{bmatrix} x \\ y \\ z \end{bmatrix} = \begin{bmatrix} 0 \\ 0 \\ 0 \end{bmatrix}$$

$$\Rightarrow \quad 7x - 8y - 2z = 0$$

and $\quad 4xy - 6z = 0 \Rightarrow 2y - 3z = 0$

Let $z = k_1$, then $\quad y = \frac{3k_1}{2}$ and $7x = 8\left(\frac{3k_1}{2}\right) + 2k_1 = 14k_1$

$$\Rightarrow \quad x = 2k_1$$

$$\therefore \qquad X_1 = k_1 \begin{bmatrix} 2 \\ 3/2 \\ 1 \end{bmatrix} \text{ or } \begin{bmatrix} 4 \\ 3 \\ 2 \end{bmatrix} \qquad (i.e., \text{ multiply by } 2)$$

is an eigen vector. Similarly, we get

$$X_2 = \begin{bmatrix} 3 \\ 2 \\ 1 \end{bmatrix} \text{ and } X_3 = \begin{bmatrix} 2 \\ 1 \\ 1 \end{bmatrix}$$

$\therefore$ Modal matrix

$$P = [X_1 \cdot X_2 \cdot X_3] = \begin{bmatrix} 4 & 3 & 2 \\ 3 & 2 & 1 \\ 2 & 1 & 1 \end{bmatrix}$$

Therefore, we get

$$P^{-1}AP = \begin{bmatrix} 1 & 0 & 0 \\ 0 & 2 & 0 \\ 0 & 0 & 3 \end{bmatrix} = \text{diag. } (1, 2, 3) = D.$$

Example 3. *Determine the modal matrix P for* $A = \begin{bmatrix} 1 & 1 & 3 \\ 1 & 5 & 1 \\ 3 & 1 & 1 \end{bmatrix}$ *and hence diagonalize A.*

Sol. The characteristic equation of A is $| A - \lambda I | = 0$

$$\Rightarrow \qquad \begin{vmatrix} 1-\lambda & 1 & 3 \\ 1 & 5-\lambda & 1 \\ 3 & 1 & 1-\lambda \end{vmatrix} = 0$$

$$\Rightarrow \qquad \lambda = -2, 3, 6 \text{ are eigen values.}$$

The corresponding eigen vectors are

$$X_1 = \begin{bmatrix} -1 \\ 0 \\ 1 \end{bmatrix}, X_2 = \begin{bmatrix} 1 \\ -1 \\ 1 \end{bmatrix}, X_3 = \begin{bmatrix} 1 \\ 2 \\ 1 \end{bmatrix}$$

$$X_1^T \cdot X_2 = [-1 \quad 0 \quad 1] \begin{bmatrix} 1 \\ -1 \\ 1 \end{bmatrix} = [-1 - 0 + 1] = 0$$

$$X_2^T \cdot X_3 = [1 \quad -1 \quad 1] \begin{bmatrix} 1 \\ 2 \\ 1 \end{bmatrix} = [1 - 2 + 1] = 0$$

$$X_3^T \cdot X_1 = [1 \quad 2 \quad 1] \begin{bmatrix} -1 \\ 0 \\ 1 \end{bmatrix} = [-1 + 0 + 1] = 0$$

$\therefore$ X_1, X_2, X_3 are pairwise orthogonal.

Hence, $P^{-1} = P^T$.

$$P = [X_1 \quad X_2 \quad X_3] = \begin{bmatrix} -1 & 1 & 1 \\ 0 & -1 & 2 \\ 1 & 1 & 1 \end{bmatrix} \text{ is a modal matrix of A.}$$

In normalized form

$$P = [e_1 \quad e_2 \quad e_3] = \begin{bmatrix} -\frac{1}{\sqrt{2}} & \frac{1}{\sqrt{3}} & \frac{1}{\sqrt{6}} \\ 0 & -\frac{1}{\sqrt{3}} & \frac{1}{\sqrt{6}} \\ \frac{1}{\sqrt{2}} & \frac{1}{\sqrt{3}} & \frac{1}{\sqrt{6}} \end{bmatrix}$$

Hence, we get

$$P^{-1}AP = P^TAP = \begin{bmatrix} -2 & 0 & 0 \\ 0 & 3 & 0 \\ 0 & 0 & 6 \end{bmatrix} = D.$$

Example 4. *Determine the modal matrix P of* $A = \begin{bmatrix} -2 & 2 & -3 \\ 2 & 1 & -6 \\ -1 & -2 & 0 \end{bmatrix}$. *Verify that $P^{-1}AP$ is a diagonal matrix.*

Sol. The characteristic equation of A is $|A - \lambda I| = 0$

$$\Rightarrow \quad \begin{vmatrix} -2-\lambda & 2 & -3 \\ 2 & 1-\lambda & -6 \\ -1 & -2 & -\lambda \end{vmatrix} = 0$$

$$\Rightarrow \quad (-2-\lambda)[(1-\lambda)(-\lambda) - 12] - 2[-2\lambda - 6] - 3[-4 + 1(1-\lambda)] = 0$$

on simplification, we get

$$\lambda = 5, -3, -3 \text{ are eigen values.}$$

To find eigen vectors we consider $[A - \lambda I]X = 0$

$$\Rightarrow \quad \begin{bmatrix} -2-\lambda & 2 & -3 \\ 2 & 1-\lambda & -6 \\ -1 & -2 & -\lambda \end{bmatrix} \begin{bmatrix} x \\ y \\ z \end{bmatrix} = \begin{bmatrix} 0 \\ 0 \\ 0 \end{bmatrix} \qquad \ldots(1)$$

Case I: Put $\lambda = 5$ in (1), then, we get

Eigen vector $\quad X_1 = \begin{bmatrix} 1 \\ 2 \\ -1 \end{bmatrix}$

Case II: Put $\lambda = -3$ in (1), then by taking $z = 0$, we get

$$X_2 = \begin{bmatrix} 2 \\ -1 \\ 0 \end{bmatrix}$$

Case III: Put $\lambda = -3$ in (1) then by taking $y = 0$, we get

$$X_3 = \begin{bmatrix} 3 \\ 0 \\ 1 \end{bmatrix}$$

$\therefore$ X_2 and X_3 are linearly independent eigen vectors.

Here X_1, X_2, X_3 are not orthogonal but they are linearly independent.

Hence, a modal matrix $P = [X_1 \quad X_2 \quad X_3] = \begin{bmatrix} 1 & 2 & 3 \\ 2 & -1 & 0 \\ -1 & 0 & 1 \end{bmatrix}$

$$|P| = -8$$

$$P^{-1} = \frac{1}{|P|} \text{adj } P = \frac{1}{-8}\begin{bmatrix} -1 & -2 & 3 \\ -2 & 4 & 6 \\ -1 & -2 & -5 \end{bmatrix}$$

$$P^{-1}AP = \begin{bmatrix} 5 & 0 & 0 \\ 0 & -3 & 0 \\ 0 & 0 & -3 \end{bmatrix} = D$$

Therefore, $P^{-1}AP$ is a diagonal matrix.

Example 5. *Diagonalize* $A = \begin{bmatrix} 1 & 0 & -1 \\ 1 & 2 & 1 \\ 2 & 2 & 3 \end{bmatrix}$. *Hence calculate* A^4.

Sol. $|A - \lambda I| = 0$

$\Rightarrow$ $\lambda = 1, 2, 3$ are eigen values.

The corresponding eigen vectors are

$$X_1 = \begin{bmatrix} 1 \\ -1 \\ 0 \end{bmatrix}, X_2 = \begin{bmatrix} -2 \\ 1 \\ 2 \end{bmatrix}, X_3 = \begin{bmatrix} -1 \\ 1 \\ 2 \end{bmatrix}$$

Hence, $P = \begin{bmatrix} 1 & -2 & -1 \\ -1 & 1 & 1 \\ 0 & 2 & 2 \end{bmatrix}$

$$P^{-1} = \frac{1}{|P|} \text{adj } P = \frac{1}{-2}\begin{bmatrix} 0 & 2 & -1 \\ 2 & 2 & 0 \\ -2 & -2 & -1 \end{bmatrix}$$

$$P^{-1}AP = \begin{bmatrix} 1 & 0 & 0 \\ 0 & 2 & 0 \\ 0 & 0 & 3 \end{bmatrix} = D$$

Now, $$A^4 = PD^4P^{-1} = \begin{bmatrix} 1 & -2 & -1 \\ -1 & 1 & 1 \\ 0 & 2 & 2 \end{bmatrix}\begin{bmatrix} 1 & 0 & 0 \\ 0 & 16 & 0 \\ 0 & 0 & 81 \end{bmatrix}\begin{bmatrix} 0 & -1 & \frac{1}{2} \\ -1 & 1 & 0 \\ 1 & 1 & \frac{1}{2} \end{bmatrix}$$

$$= \begin{bmatrix} -49 & -50 & -40 \\ 65 & 66 & 40 \\ 130 & 130 & 81 \end{bmatrix}.$$

EXERCISE 3

1. Find the eigen values and eigen vectors of the matrix $\begin{bmatrix} 5 & 4 \\ 1 & 2 \end{bmatrix}$.
2. Find the eigen values and eigen vectors of the matrices

   (*i*) $\begin{bmatrix} 8 & -6 & 2 \\ -6 & 7 & -4 \\ 2 & -4 & 3 \end{bmatrix}$ (*ii*) $\begin{bmatrix} 6 & -2 & 2 \\ -2 & 3 & -1 \\ 2 & -1 & 3 \end{bmatrix}$

   (*iii*) $\begin{bmatrix} 2 & 0 & 1 \\ 0 & 2 & 0 \\ 1 & 0 & 2 \end{bmatrix}$ (*iv*) $\begin{bmatrix} 2 & 1 & 1 \\ 1 & 2 & 1 \\ 0 & 0 & 1 \end{bmatrix}$ (*v*) $\begin{bmatrix} 3 & 1 & 4 \\ 0 & 2 & 0 \\ 0 & 0 & 5 \end{bmatrix}$.
3. Prove that the characteristic roots of a diagonal matrix are the diagonal elements of the matrix.
4. Show that 0 is a characteristic root of a matrix if and only if the matrix is singular.
5. Show that if λ is a characteristic root of the matrix A, then $\lambda + k$ is a characteristic root of the matrix $A + kI$.
6. (*a*) If $\lambda_1, \lambda_2, \ldots\ldots., \lambda_n$ are the eigen values of a matrix A, then A^m has the eigen values $\lambda_1^m, \lambda_2^m, \ldots\ldots., \lambda_n^m$ (m being a positive integer).

   (*b*) $m = 2$
7. Find the characteristic equation of the matrix $A = \begin{bmatrix} 1 & 3 & 7 \\ 4 & 2 & 3 \\ 1 & 2 & 1 \end{bmatrix}$. Show that the equation is satisfied by A and hence obtain the inverse of the given matrix.
8. Find the characteristic equation of the matrix $A = \begin{bmatrix} 1 & 2 & 3 \\ 2 & -1 & 4 \\ 3 & 1 & -1 \end{bmatrix}$. Show that the equation is satisfied by A.
9. Using Cayley-Hamilton theorem, find the inverse of

   (*i*) $\begin{bmatrix} 2 & 3 \\ 3 & 5 \end{bmatrix}$ (*ii*) $\begin{bmatrix} 7 & -1 & 3 \\ 6 & 1 & 4 \\ 2 & 4 & 8 \end{bmatrix}$.

10. Diagonalise the following matrices and obtain the modal matrix in each case :

(i) $\begin{bmatrix} -1 & 1 & 2 \\ 0 & -2 & 1 \\ 0 & 0 & -3 \end{bmatrix}$ (ii) $\begin{bmatrix} 3 & -1 & 1 \\ -1 & 5 & -1 \\ 1 & -1 & 3 \end{bmatrix}$.

Answers

1. 1, 6 ; $k_1\begin{bmatrix} 1 \\ -1 \end{bmatrix}, k_2\begin{bmatrix} 4 \\ 1 \end{bmatrix}$

2. (i) 0, 3, 15 ; $k_1\begin{bmatrix} 1 \\ 2 \\ 2 \end{bmatrix}, k_2\begin{bmatrix} 2 \\ 1 \\ -2 \end{bmatrix}, k_3\begin{bmatrix} 2 \\ -2 \\ 1 \end{bmatrix}$ (ii) 2, 2, 8 ; $k_1\begin{bmatrix} 1 \\ 0 \\ -2 \end{bmatrix}, k_2\begin{bmatrix} 1 \\ 2 \\ 0 \end{bmatrix}, k_3\begin{bmatrix} 2 \\ -1 \\ 1 \end{bmatrix}$

(iii) 1, 2, 3 ; $k_1\begin{bmatrix} 1 \\ 0 \\ -1 \end{bmatrix}, k_2\begin{bmatrix} 0 \\ 1 \\ 0 \end{bmatrix}, k_3\begin{bmatrix} 1 \\ 0 \\ 1 \end{bmatrix}$ (iv) 1, 1, 3 ; $k_1\begin{bmatrix} 1 \\ -2 \\ 1 \end{bmatrix}, k_2\begin{bmatrix} 1 \\ 1 \\ 0 \end{bmatrix}, k_3\begin{bmatrix} 1 \\ 1 \\ 0 \end{bmatrix}$

(v) 2, 3, 5 ; $k_1\begin{bmatrix} 1 \\ -1 \\ 0 \end{bmatrix}, k_2\begin{bmatrix} 1 \\ 0 \\ 0 \end{bmatrix}, k_3\begin{bmatrix} 2 \\ 0 \\ 1 \end{bmatrix}$

7. $\lambda^3 - 4\lambda^2 - 20\lambda - 35 = 0, \frac{1}{35}\begin{bmatrix} -4 & 11 & -5 \\ -1 & -6 & 25 \\ 6 & 1 & -10 \end{bmatrix}$

8. $\lambda^3 + \lambda^2 - 18\lambda - 40 = 0$ **9.** (i) $\begin{bmatrix} 5 & -3 \\ -3 & 2 \end{bmatrix}$ (ii) $\frac{1}{50}\begin{bmatrix} -8 & 20 & -7 \\ -40 & 50 & -10 \\ 22 & -30 & 13 \end{bmatrix}$

10. (i) $\begin{bmatrix} -1 & 0 & 0 \\ 0 & -2 & 0 \\ 0 & 0 & -3 \end{bmatrix}, \begin{bmatrix} 1 & 1 & 1 \\ 0 & -1 & 2 \\ 0 & 0 & -2 \end{bmatrix}$ (ii) $\begin{bmatrix} 2 & 0 & 0 \\ 0 & 3 & 0 \\ 0 & 0 & 6 \end{bmatrix}, \begin{bmatrix} 1 & 1 & 1 \\ 0 & 1 & -2 \\ -1 & 1 & 1 \end{bmatrix}$.

1.28. QUADRATIC FORMS

Definition. A homogeneous polynomial of second degree in any number of variables is called a quadratic form. For example,

(i) $ax^2 + 2hxy + by^2$ (ii) $ax^2 + by^2 + cz^2 + 2hxy + 2gzx + 2fyz$ and

(iii) $ax^2 + by^2 + cz^2 + dw^2 + 2hxy + 2gyz + 2fzx + 2lxw + 2myw + 2nzw$

are quadratic forms in two, three and four variables.

In n-variables $x_1, x_2, \ldots\ldots, x_n$, the general quadratic form is $\sum_{j=1}^{n}\sum_{i=1}^{n} b_{ij}x_ix_j$

In the expansion, the co-efficient of $x_ix_j = (b_{ij} + b_{ji})$.

Suppose $2a_{ij} = b_{ij} + b_{ij}$, where $a_{ij} = a_{ji}$ and $a_{ii} = b_{ii}$

$$\therefore \quad \sum_{j=1}^{n}\sum_{i=1}^{n} b_{ij}x_ix_j = \sum_{j=1}^{n}\sum_{i=1}^{n} a_{ij}x_ix_j, \text{ where } a_{ij} = \frac{1}{2}(b_{ij} + b_{ji}).$$

Hence every quadratic form can be written as $\sum_{j=1}^{n} \sum_{i=1}^{n} a_{ij}x_i x_j = X'AX$, so that the matrix A is always symmetric, where $A = [a_{ij}]$ and $X' = [x_1, x_2, ..., x_n]$.

Now writing the above said examples of quadratic forms in matrix form, we get

$$(i)\ ax^2 + 2hxy + by^2 = [x\ y]\begin{bmatrix} a & h \\ h & b \end{bmatrix}\begin{bmatrix} x \\ y \end{bmatrix} \Rightarrow A = \begin{bmatrix} a & h \\ h & b \end{bmatrix}$$

$$(ii)\ ax^2 + by^2 + cz^2 + 2hxy + 2gzx + 2fyz = [x\ y\ z]\begin{bmatrix} a & h & g \\ h & b & f \\ g & f & c \end{bmatrix}\begin{bmatrix} x \\ y \\ z \end{bmatrix} \Rightarrow A = \begin{bmatrix} a & h & g \\ h & b & f \\ g & f & c \end{bmatrix}$$

and $(iii)\ ax^2 + by^2 + cz^2 + dw^2 + 2hxy + 2gyz + 2fzx + 2lxw + 2myw + 2nzw$

$$= [x\ y\ z\ w]\begin{bmatrix} a & h & f & l \\ h & b & g & m \\ f & g & c & n \\ l & m & n & d \end{bmatrix}\begin{bmatrix} x \\ y \\ z \\ w \end{bmatrix} \Rightarrow A = \begin{bmatrix} a & h & f & l \\ h & b & g & m \\ f & g & c & n \\ l & m & n & d \end{bmatrix}$$

1.29. LINEAR TRANSFORMATION OF A QUADRATIC FORM

Let X′AX be a quadratic form in n-variables and let X = PY ...(1)

where P is a non-singular matrix, be the non-singular transformation.

From (1), $X' = (PY)' = Y'P'$ and hence

$$X'AX = Y'P'APY = Y'(P'AP)Y = Y'BY \quad ...(2)$$

where B = P′AP. Therefore Y′BY is also a quadratic form in n-variables. Hence it is a linear transformation of the quadratic form X′AX under the linear transformation X = PY and B = P′AP.

Note. (*i*) Here $B' = (P'AP)' = P'AP = B$ (*ii*) $\rho(B) = \rho(A)$.

∴ A and B are congruent matrices.

1.30. CANONICAL FORM

If a real quadratic form be expressed as a sum or difference of the squares of new variables by means of any real non-singular linear transformation, then the latter quadratic expression is called a *canonical form* of the given quadratic form.

i.e., if the quadratic form $X'AX = \sum_{j=1}^{n} \sum_{i=1}^{n} a_{ij}x_i x_j$ can be reduced to the quadratic form

$Y'BY = \sum_{i=1}^{n} \lambda_i {y_i}^2$ by a non-singular linear transformation

X = PY then Y′BY is called the canonical form of the given one.

∴ If $B = P'AP = \text{diag}(\lambda_1, \lambda_2, ..., \lambda_n)$ then $X'AX = Y'BY = \sum_{j=1}^{n} \lambda_i {y_i}^2$.

Note 1. Here some of λ_i (eigen values) may be positive or negative or zero.

Note 2. A quadratic form is said to be real if the elements of the symmetric matrix are real.

Note 3. If $\rho(A) = r$, then the quadratic form X′AX will contain only r terms.

1.31. INDEX AND SIGNATURE OF THE QUADRATIC FORM

The number p of positive terms in the canonical form is called the *index* of the quadratic form or Index = The number of positive eigen values

(The number of positive terms)–(The number of negative terms) *i.e.*, $p - (r - p) = 2p - r$ is called *signature* of the quadratic form, where $\rho(A) = r$ or Signature = (The number of positive eigen values) – (The number of negative eigen values).

1.32. DEFINITE, SEMI-DEFINITE AND INDEFINITE REAL QUADRATIC FORMS

Let X′AX be a real quadratic form in n-variables $x_1, x_2, \ldots\ldots, x_n$ with rank r and index p. Then we say that the quadratic form is

(*i*) *positive definite* if $r = n, p = r$ (or) *if all eigenvalues are positive*

(*ii*) *negative definite* if $r = n, p = 0$ (or) *if all eigenvalues are negative*

(*iii*) *positive semi-definite* if $r < n, p = r$ (or) *if at least one eigenvalue is zero and other eigen values are positive*

and (*iv*) *negative semi-definite* if $r < n, p = 0$ (or) *if at least one eigenvalue is zero and other eigenvalues are negative*

If the canonical form has both positive and negative eigen values, the quadratic form is said to be *indefinite*.

Note. If X′AX is positive definite, then $|A| > 0$.

ILLUSTRATIVE EXAMPLES

Example 1. *Reduce* $3x^2 + 3z^2 + 4xy + 8xz + 8yz$ *into canonical form.*

Sol. The given quadratic form can be written as X′AX, where X′ = [x, y, z] and the symmetric matrix

$$A = \begin{bmatrix} 3 & 2 & 4 \\ 2 & 0 & 4 \\ 4 & 4 & 3 \end{bmatrix}.$$

Let us reduce A into diagonal matrix. We know that $A = I_3 A I_3$

i.e.,
$$\begin{bmatrix} 3 & 2 & 4 \\ 2 & 0 & 4 \\ 4 & 4 & 3 \end{bmatrix} = \begin{bmatrix} 1 & 0 & 0 \\ 0 & 1 & 0 \\ 0 & 0 & 1 \end{bmatrix} A \begin{bmatrix} 1 & 0 & 0 \\ 0 & 1 & 0 \\ 0 & 0 & 1 \end{bmatrix}$$

Operating $R_2(-2/3)R_1$, $R_3(-4/3)R_1$ (for A on LHS and pre-factor on RHS), we get

$$\begin{bmatrix} 3 & 2 & 4 \\ 0 & -\frac{4}{3} & \frac{4}{3} \\ 0 & \frac{4}{3} & -\frac{7}{3} \end{bmatrix} = \begin{bmatrix} 1 & 0 & 0 \\ -\frac{2}{3} & 1 & 0 \\ -\frac{4}{3} & 0 & 1 \end{bmatrix} A \begin{bmatrix} 1 & 0 & 0 \\ 0 & 1 & 0 \\ 0 & 0 & 1 \end{bmatrix}$$

Operating $C_2(-2/3)\ C_1$, $C_3(-4/3)\ C_1$ (for A on LHS and post-factor on RHS), we get

$$\begin{bmatrix} 3 & 0 & 0 \\ 0 & -\frac{4}{3} & \frac{4}{3} \\ 0 & \frac{4}{3} & -\frac{7}{3} \end{bmatrix} = \begin{bmatrix} 1 & 0 & 0 \\ -\frac{2}{3} & 1 & 0 \\ -\frac{4}{3} & 0 & 1 \end{bmatrix} A \begin{bmatrix} 1 & -\frac{2}{3} & -\frac{4}{3} \\ 0 & 1 & 0 \\ 0 & 0 & 1 \end{bmatrix}$$

Operating $R_3 + R_2$, we get

$$\begin{bmatrix} 3 & 0 & 0 \\ 0 & -\frac{4}{3} & \frac{4}{3} \\ 0 & 0 & -1 \end{bmatrix} = \begin{bmatrix} 1 & 0 & 0 \\ -\frac{2}{3} & 1 & 0 \\ -2 & 1 & 1 \end{bmatrix} A \begin{bmatrix} 1 & -\frac{2}{3} & -\frac{4}{3} \\ 0 & 1 & 0 \\ 0 & 0 & 1 \end{bmatrix}$$

Operating $C_3 + C_2$, we get

$$\begin{bmatrix} 3 & 0 & 0 \\ 0 & -\frac{4}{3} & 0 \\ 0 & 0 & -1 \end{bmatrix} = \begin{bmatrix} 1 & 0 & 0 \\ -\frac{2}{3} & 1 & 0 \\ -2 & 1 & 1 \end{bmatrix} A \begin{bmatrix} 1 & -\frac{2}{3} & -2 \\ 0 & 1 & 0 \\ 0 & 0 & 1 \end{bmatrix}$$

or $$\text{Diag}\left(3, -\frac{4}{3}, -1\right) = P'AP$$

∴ The canonical form of the given quadratic form is

$$Y'(P'AP)Y = [y_1, y_2, y_3] \begin{bmatrix} 3 & 0 & 0 \\ 0 & -\frac{4}{3} & 0 \\ 0 & 0 & -1 \end{bmatrix} \begin{bmatrix} y_1 \\ y_2 \\ y_3 \end{bmatrix} = 3{y_1}^2 - \frac{4}{3}{y_2}^2 - {y_3}^2.$$

Here $\rho(A) = 3$, index = 1, signature = 1 – 2 = – 1.

Note 1. In this problem the non-singular transformation which reduces the given quadratic form into the canonical form is X = PY *i.e.*, $\begin{bmatrix} x \\ y \\ z \end{bmatrix} = \begin{bmatrix} 1 & -\frac{2}{3} & -2 \\ 0 & 1 & 0 \\ 0 & 0 & 1 \end{bmatrix} \begin{bmatrix} y_1 \\ y_2 \\ y_3 \end{bmatrix}$ *i.e.*, $x = y_1 - \frac{2}{3} y_2 - 2y_3$, $y = y_2 + y_3$, $z = y_3$.

Note 2. The above example can also be questioned as 'Diagonalise' the quadratic form $3x^2 + 3z^2 + 4xy + 8xz + 8yz$ by linear transformation and write the linear transformation.

Or

Reduce the quadratic form $3x^2 + 3z^2 + 4xy + 8xz + 8yz$ into the sum of squares.

Example 2. *Reduce the quadratic form* $x^2 - 4y^2 + 6z^2 + 2xy - 4xz + 2w^2 - 6zw$ *into sum of squares.*

Sol. The matrix form of the given quadratic form is X′AX, where $X' = [x\ y\ z\ w]$

and
$$A = \begin{bmatrix} 1 & 1 & -2 & 0 \\ 1 & -4 & 0 & 0 \\ -2 & 0 & 6 & -3 \\ 0 & 0 & -3 & 2 \end{bmatrix}$$

Let us reduce A to the diagonal matrix. We know that $A = I_4 A I_4$.

$$\begin{bmatrix} 1 & 1 & -2 & 0 \\ 1 & -4 & 0 & 0 \\ -2 & 0 & 6 & -3 \\ 0 & 0 & -3 & 2 \end{bmatrix} = \begin{bmatrix} 1 & 0 & 0 & 0 \\ 0 & 1 & 0 & 0 \\ 0 & 0 & 1 & 0 \\ 0 & 0 & 0 & 1 \end{bmatrix} A \begin{bmatrix} 1 & 0 & 0 & 0 \\ 0 & 1 & 0 & 0 \\ 0 & 0 & 1 & 0 \\ 0 & 0 & 0 & 1 \end{bmatrix}$$

Operating $R_2 - R_1$, $R_3 + 2R_1$, we get

$$\begin{bmatrix} 1 & 1 & -2 & 0 \\ 0 & -5 & 2 & 0 \\ 0 & 2 & 2 & -3 \\ 0 & 0 & -3 & 2 \end{bmatrix} = \begin{bmatrix} 1 & 0 & 0 & 0 \\ -1 & 1 & 0 & 0 \\ 2 & 0 & 1 & 0 \\ 0 & 0 & 0 & 1 \end{bmatrix} A \begin{bmatrix} 1 & 0 & 0 & 0 \\ 0 & 1 & 0 & 0 \\ 0 & 0 & 1 & 0 \\ 0 & 0 & 0 & 1 \end{bmatrix}$$

Operating $C_2 - C_1$, $C_3 + 2C_1$, we get

$$\begin{bmatrix} 1 & 0 & 0 & 0 \\ 0 & -5 & 2 & 0 \\ 0 & 2 & 2 & -3 \\ 0 & 0 & -3 & 2 \end{bmatrix} = \begin{bmatrix} 1 & 0 & 0 & 0 \\ -1 & 1 & 0 & 0 \\ 2 & 0 & 1 & 0 \\ 0 & 0 & 0 & 1 \end{bmatrix} A \begin{bmatrix} 1 & -1 & 2 & 0 \\ 0 & 1 & 0 & 0 \\ 0 & 0 & 1 & 0 \\ 0 & 0 & 0 & 1 \end{bmatrix}$$

Operating $R_3 + (2/5)R_2$, we get

$$\begin{bmatrix} 1 & 0 & 0 & 0 \\ 0 & -5 & 2 & 0 \\ 0 & 0 & \frac{14}{5} & -3 \\ 0 & 0 & -3 & 2 \end{bmatrix} = \begin{bmatrix} 1 & 0 & 0 & 0 \\ -1 & 1 & 0 & 0 \\ \frac{8}{5} & \frac{2}{5} & 1 & 0 \\ 0 & 0 & 0 & 1 \end{bmatrix} A \begin{bmatrix} 1 & -1 & 2 & 0 \\ 0 & 1 & 0 & 0 \\ 0 & 0 & 1 & 0 \\ 0 & 0 & 0 & 1 \end{bmatrix}$$

Operating $C_3 + (2/5)\,C_2$, we get

$$\begin{bmatrix} 1 & 0 & 0 & 0 \\ 0 & -5 & 0 & 0 \\ 0 & 0 & \frac{14}{5} & -3 \\ 0 & 0 & -3 & 2 \end{bmatrix} = \begin{bmatrix} 1 & 0 & 0 & 0 \\ -1 & 1 & 0 & 0 \\ \frac{8}{5} & \frac{2}{5} & 1 & 0 \\ 0 & 0 & 0 & 1 \end{bmatrix} A \begin{bmatrix} 1 & -1 & \frac{8}{5} & 0 \\ 0 & 1 & \frac{2}{5} & 0 \\ 0 & 0 & 1 & 0 \\ 0 & 0 & 0 & 1 \end{bmatrix}$$

Operating $R_4 + \frac{15}{14} R_3$, we get

$$\begin{bmatrix} 1 & 0 & 0 & 0 \\ 0 & -5 & 0 & 0 \\ 0 & 0 & \frac{14}{5} & -3 \\ 0 & 0 & 0 & -\frac{17}{14} \end{bmatrix} = \begin{bmatrix} 1 & 0 & 0 & 0 \\ -1 & 1 & 0 & 0 \\ \frac{8}{5} & \frac{2}{5} & 1 & 0 \\ \frac{12}{7} & \frac{3}{7} & \frac{15}{14} & 1 \end{bmatrix} A \begin{bmatrix} 1 & -1 & \frac{8}{5} & 0 \\ 0 & 1 & \frac{2}{5} & 0 \\ 0 & 0 & 1 & 0 \\ 0 & 0 & 0 & 1 \end{bmatrix}$$

Operating $C_4 + \frac{15}{14} C_3$, we get

$$\begin{bmatrix} 1 & 0 & 0 & 0 \\ 0 & -5 & 0 & 0 \\ 0 & 0 & \frac{14}{5} & 0 \\ 0 & 0 & 0 & -\frac{17}{14} \end{bmatrix} = \begin{bmatrix} 1 & 0 & 0 & 0 \\ -1 & 1 & 0 & 0 \\ \frac{8}{5} & \frac{2}{5} & 1 & 0 \\ \frac{12}{7} & \frac{3}{7} & \frac{15}{14} & 1 \end{bmatrix} A \begin{bmatrix} 1 & -1 & \frac{8}{5} & \frac{12}{7} \\ 0 & 1 & \frac{2}{5} & \frac{3}{7} \\ 0 & 0 & 1 & \frac{15}{14} \\ 0 & 0 & 0 & 1 \end{bmatrix}$$

i.e., $\text{diag}\left(1, -5, \frac{14}{5}, -\frac{17}{14}\right) = P'AP$

$\therefore$ The canonical form of the given quadratic form is

$$Y'(P'AP)Y = Y' \text{ diag}\left(1, -5, \frac{14}{5}, -\frac{17}{14}\right) Y$$

$$= [y_1 y_2 y_3 y_4] \begin{bmatrix} 1 & 0 & 0 & 0 \\ 0 & -5 & 0 & 0 \\ 0 & 0 & \frac{14}{5} & 0 \\ 0 & 0 & 0 & -\frac{17}{14} \end{bmatrix} \begin{bmatrix} y_1 \\ y_2 \\ y_3 \\ y_4 \end{bmatrix}$$

$$= y_1^2 - 5y_2^2 + \frac{14}{5} y_3^2 - \frac{17}{14} y_4^2,$$ which is the sum of the squares.

Note. Here $\rho(A) = 4$, index = 2, signature = 2 – 2 = 0.

Example 3. *Find the matrix of the quadratic form* $q = 4x^2 - 2y^2 + z^2 - 2xy + 6zx$. (OU-Nov. 2006)

Sol. Compare the given expression with $ax^2 + by^2 + cz^2 + 2hxy + 2gzx + 2fyz$, we get

$$\begin{array}{l|l} a = 4 & 2h = -2 \Rightarrow h = -1 \\ b = -2 & 2g = 6 \Rightarrow g = 3 \\ c = 1 & 2f = 0 \Rightarrow f = 0 \end{array}$$

$$A = \begin{bmatrix} a & h & g \\ h & b & f \\ g & f & c \end{bmatrix} = \begin{bmatrix} 4 & -1 & 3 \\ -1 & -2 & 0 \\ 3 & 0 & 1 \end{bmatrix}.$$

1.33. REDUCTION TO CANONICAL FORM BY ORTHOGONAL TRANSFORMATION

Let X′AX be a given quadratic form. The modal matrix B of A is that matrix whose columns are characteristic vectors of A. If B represents the orthogonal matrix of A (the normalised modal matrix of A whose column vectors are pairwise orthogonal) then X = BY will reduce X′AX to Y′DY, where D = diag $(\lambda_1, \lambda_2, \ldots\ldots, \lambda_n)$ and $\lambda_1, \lambda_2, \ldots\ldots, \lambda_n$ are characteristic roots of A.

Note. This method works successfully if the characteristic vectors of A are linearly independent which are pairwise orthogonal.

Example 4. *Reduce* $8x^2 + 7y^2 + 3z^2 - 12xy + 4xz - 8yz$ *into canonical form by orthogonal transformation.* (OU-Dec. 2013)

Sol. The matrix of the quadratic form is

$$A = \begin{bmatrix} 8 & -6 & 2 \\ -6 & 7 & -4 \\ 2 & -4 & 3 \end{bmatrix}$$

The characteristic roots of A are given by $|A - \lambda I| = 0$

i.e.,
$$\begin{vmatrix} 8-\lambda & -6 & 2 \\ -6 & 7-\lambda & -4 \\ 2 & -4 & 3-\lambda \end{vmatrix} = 0 \quad \text{or} \quad \lambda(\lambda - 3)(\lambda - 15) = 0$$

$$\therefore \qquad \lambda = 0, 3, 15$$

Characteristic vector for $\lambda = 0$ is given by $[A - (0)I]X = O$

i.e.,
$$8x_1 - 6x_2 + 2x_3 = 0$$
$$-6x_1 + 7x_2 - 4x_3 = 0$$
$$2x_1 - 4x_2 + 3x_3 = 0$$

Solving first two, we get $\frac{x_1}{1} = \frac{x_2}{2} = \frac{x_3}{2}$ giving the eigen vector $X_1 = k_1(1, 2, 2)'$

When $\lambda = 3$, the corresponding characteristic vector is given by $[A - 3I]X = O$

i.e.,
$$5x_1 - 6x_2 + 2x_3 = 0$$
$$-6x_1 + 4x_2 - 4x_3 = 0$$
$$2x_1 - 4x_2 = 0$$

Solving any two equations, we get $X_2 = k_2(2, 1, -2)'$.

Similarly characteristic vector corresponding to

$$\lambda = 15 \text{ is } X_3 = k_3(2, -2, 1)'.$$

Now X_1, X_2, X_3 are pairwise orthogonal

i.e.,
$$X_1 \cdot X_2 = X_2 \cdot X_3 = X_3 \cdot X_1 = 0.$$

$\therefore$ The normalised modal matrix is

$$B = \left[\frac{X_1}{\|X_1\|}, \frac{X_2}{\|X_2\|}, \frac{X_3}{\|X_3\|}\right] = \begin{bmatrix} \frac{1}{3} & \frac{2}{3} & \frac{2}{3} \\ \frac{2}{3} & \frac{1}{3} & -\frac{2}{3} \\ \frac{2}{3} & \frac{-2}{3} & \frac{1}{3} \end{bmatrix}$$

Now B is orthogonal matrix and $|B| = 1$

i.e.,
$$B^{-1} = B^T \text{ and } B^{-1}AB = D = \text{diag}(0, 3, 15)$$

i.e.,
$$\begin{bmatrix} \frac{1}{3} & \frac{2}{3} & \frac{2}{3} \\ \frac{2}{3} & \frac{1}{3} & -\frac{2}{3} \\ \frac{2}{3} & -\frac{2}{3} & \frac{1}{3} \end{bmatrix} A \begin{bmatrix} \frac{1}{3} & \frac{2}{3} & \frac{2}{3} \\ \frac{2}{3} & \frac{1}{3} & \frac{-2}{3} \\ \frac{2}{3} & -\frac{2}{3} & \frac{1}{3} \end{bmatrix} = \begin{bmatrix} 0 & 0 & 0 \\ 0 & 3 & 0 \\ 0 & 0 & 15 \end{bmatrix}$$

$$X'AX = Y'(B^{-1}AB)Y = Y'DY$$

$$= [y_1\ y_2\ y_3] \begin{bmatrix} 0 & 0 & 0 \\ 0 & 3 & 0 \\ 0 & 0 & 15 \end{bmatrix} \begin{bmatrix} y_1 \\ y_2 \\ y_3 \end{bmatrix} = 0y_1^2 + 3y_2^2 + 15y_3^2$$

which is the required canonical form.

Note. Here the orthogonal transformation is X = BY, rank of the quadratic form = 2 ; index = 2, signature = 2. It is positive semi-definite.

Example 5. *Reduce* $6x_1^2 + 3x_2^2 + 3x_3^2 - 4x_1x_2 - 2x_2x_3 + 4x_3x_1$ *into canonical form.*

Sol. The matrix of the quadratic form is $A = \begin{bmatrix} 6 & -2 & 2 \\ -2 & 3 & -1 \\ 2 & -1 & 3 \end{bmatrix}$

The characteristic roots given by $| A - \lambda I | = 0$

i.e.,
$$\begin{vmatrix} 6-\lambda & -2 & 2 \\ -2 & 3-\lambda & -1 \\ 2 & -1 & 3-\lambda \end{vmatrix} = 0$$

or
$$\lambda^3 - 12\lambda^2 + 36\lambda - 32 = 0$$

which on solving gives $\lambda = 8, 2, 2$.

The vector corresponding to $\lambda = 8$ is given by $[A - 8I]X = O$

i.e.,
$$\begin{bmatrix} -2 & -2 & 2 \\ -2 & -5 & -1 \\ 2 & -1 & -5 \end{bmatrix} \begin{bmatrix} x_1 \\ x_2 \\ x_3 \end{bmatrix} = \begin{bmatrix} 0 \\ 0 \\ 0 \end{bmatrix}$$

or
$$-2x_1 - 2x_2 + 2x_3 = 0$$
$$-2x_1 - 5x_2 - x_3 = 0$$
$$2x_1 - x_2 - 5x_3 = 0$$

Solving any two of the equations, we get the vector as $k_1[2, -1, 1]'$.

The characteristic vector for $\lambda = 2$ is given by $[A - 2I]X = O$, which reduces to single equation

$$2x_1 - x_2 + x_3 = 0.$$

Putting $x_2 = 0$, we get $\frac{x_2}{1} = \frac{x_3}{1}$ or the vector is k_2 [0, 1, 1]′. Again by putting $x_2 = 0$, we get $\frac{x_1}{1} = \frac{x_3}{-2}$ or the vector is k_3 [1, 0 – 2]′.

Now, $X_1 = k_1 [2, -1, 1]'$; $X_2 = k_2 [0, 1, 1]'$ and $X_3 = k_3 [1, 0, -2]'$.

Here X_1, X_2, X_3 are not pairwise orthogonal

$\because$
$$X_1 . X_2 = 0 \ ; \quad X_2 . X_3 \neq 0 \ \text{ and } \ X_3 . X_1 = 0$$

To get X_3 orthogonal to X_2 assume a vector $[u, v, w]'$ orthogonal to X_2 also satisfying

$2x_1 - x_2 + x_3 = 0$ *i.e.,* $2u - v + w = 0$ and $0.u + 1.v + 1.w = 0$

Solving $[u, v, w]' = [1, 1, -1]' = X_3$ so that $X_1 . X_2 = X_2 . X_3 = X_3 . X_1 = 0$

$\therefore$ The normalised modal matrix is

$$B = \begin{bmatrix} \frac{1}{\sqrt{6}} & 0 & \frac{1}{\sqrt{3}} \\ -\frac{1}{\sqrt{6}} & \frac{1}{\sqrt{2}} & \frac{1}{\sqrt{3}} \\ \frac{1}{\sqrt{6}} & \frac{1}{\sqrt{2}} & -\frac{1}{\sqrt{3}} \end{bmatrix}$$

Now, B is orthogonal matrix and $|B| = 1$

i.e., $$B' = B^{-1} \text{ and } B^{-1}AB = D, \text{ where } D = \begin{bmatrix} 8 & 0 & 0 \\ 0 & 2 & 0 \\ 0 & 0 & 2 \end{bmatrix}$$

$\therefore$ $$X'AX = Y'(B^{-1}AB)Y = Y'DY$$

$$= [y_1 y_2 y_3] \begin{bmatrix} 8 & 0 & 0 \\ 0 & 2 & 0 \\ 0 & 0 & 2 \end{bmatrix} \begin{bmatrix} y_1 \\ y_2 \\ y_3 \end{bmatrix} = 8y_1^2 + 2y_2^2 + 2y_3^2$$

which is the required canonical form.

Note. Here rank of the quadratic form is 3, index = 3, signature = 3. It is positive definite.

Example 6. *Reduce the quadratic form* $3x_1^2 + 3x_2^2 + 3x_3^2 + 2x_1x_2 + 2x_1x_3 - 2x_2x_3$ *into a canonical form. Also give the transformation matrix.*

Sol. Here $$A = \begin{bmatrix} 3 & 1 & 1 \\ 1 & 3 & -1 \\ 1 & -1 & 3 \end{bmatrix}$$

The characteristic equation of A is $|A - \lambda I| = 0$

$$\Rightarrow \begin{vmatrix} 3-\lambda & 1 & 1 \\ 1 & 3-\lambda & -1 \\ 1 & 1 & 3-\lambda \end{vmatrix} = 0$$

$$\Rightarrow \lambda = 1, 4, 4$$

The corresponding eigen vectors are

$$X_1 = \begin{bmatrix} -1 \\ 1 \\ 1 \end{bmatrix}, X_2 = \begin{bmatrix} 1 \\ 0 \\ 1 \end{bmatrix}, X_3 = \begin{bmatrix} 0 \\ 1 \\ 1 \end{bmatrix}$$

$$\text{The modal matrix} = \begin{bmatrix} -1 & 1 & 0 \\ 1 & 0 & 1 \\ 1 & 1 & 1 \end{bmatrix}$$

The normalized modal matrix

$$P = \begin{bmatrix} -1/\sqrt{3} \\ 1/\sqrt{3} \\ 1/\sqrt{3} \end{bmatrix}$$

$$X_1^T X_2 = [-1 \ \ 1 \ \ 1] \begin{bmatrix} 1 \\ 0 \\ 1 \end{bmatrix} = [-1 + 0 + 1] = 0$$

$$X_2^T X_3 = [1 \ \ 0 \ \ 1] \begin{bmatrix} 0 \\ 1 \\ 1 \end{bmatrix} = [0 + 0 + 1] = 1 \neq 0$$

$$X_3^T X_1 = [0 \quad 1 \quad 1]\begin{bmatrix} -1 \\ 1 \\ 1 \end{bmatrix} = [0 + 1 + 1] = 2 \neq 0$$

Let $$X_3 = \begin{bmatrix} u \\ v \\ w \end{bmatrix}$$

Let $$X_2^T X_3 = 0$$

$$\Rightarrow \quad [1 \quad 0 \quad 1]\begin{bmatrix} u \\ v \\ w \end{bmatrix} = 0 \Rightarrow u + w = 0 \quad \Rightarrow u = -w \quad \text{and} \quad X_3^T X_1 = 0$$

$$\Rightarrow \quad [u \quad v \quad w]\begin{bmatrix} -1 \\ 1 \\ 1 \end{bmatrix} = 0 \Rightarrow -u + v + w = 0$$

$$\Rightarrow \quad w + v + w = 0 \Rightarrow v = -2w$$

$$\Rightarrow \quad \frac{v}{2} = -w$$

Hence, $$\frac{u}{1} = \frac{v}{2} = \frac{w}{-1}$$

$$X_3 = \begin{bmatrix} 1 \\ 2 \\ -1 \end{bmatrix}$$

$\therefore$ The modal matrix = $[X_1 \quad X_2 \quad X_3] = \begin{bmatrix} -1 & 1 & 1 \\ 1 & 0 & 2 \\ 1 & 1 & -1 \end{bmatrix}$

The normalised modal matrix

$$P = \begin{bmatrix} -1/\sqrt{3} & 1/\sqrt{2} & 1/\sqrt{6} \\ 1/\sqrt{3} & 0 & 2/\sqrt{6} \\ 1/\sqrt{3} & 1/\sqrt{2} & -1/\sqrt{6} \end{bmatrix} \text{ is an orthogonal matrix.}$$

Hence, $P^T = P^{-1}$

$$\therefore \quad P^T A P = \begin{bmatrix} 1 & 0 & 0 \\ 0 & 4 & 0 \\ 0 & 0 & 4 \end{bmatrix} = D \text{ (Prove)}$$

The required canonical form is

$$Q = Y^T D Y = [y_1 \quad y_2 \quad y_3]\begin{bmatrix} 1 & 0 & 0 \\ 0 & 4 & 0 \\ 0 & 0 & 4 \end{bmatrix}\begin{bmatrix} y_1 \\ y_2 \\ y_3 \end{bmatrix}$$

$$= y_1^2 + 4{y_2}^2 + 4y_3^2$$

Here P is the matrix of transformation.

EXERCISE 4

1. Write down the matrices of the following quadratic forms:

(i) $2x^2 + 3y^2 + 6xy$ (ii) $2x^2 + 5y^2 - 6z^2 - 2xy - yz + 8zx$

(iii) $x_1^2 + 2x_2^2 + 3x_3^2 + 4x_4^2 + 2x_1x_2 + 4x_1x_3 - 6x_1x_4 - 4x_2x_3 - 8x_2x_4 + 12x_3x_4$.

2. Write down the quadratic forms corresponding to the following matrices:

(i) $\begin{bmatrix} 2 & 4 & 5 \\ 4 & 3 & 1 \\ 5 & 1 & 1 \end{bmatrix}$ (ii) $\begin{bmatrix} 1 & 1 & -2 & 0 \\ 1 & -4 & 0 & 0 \\ -2 & 0 & 6 & -3 \\ 0 & 0 & -3 & 2 \end{bmatrix}$.

3. Reduce the following quadratic forms to canonical forms or to sum of squares by linear transformation. Write also the rank, index and signature.

(i) $2x^2 + 2y^2 + 3z^2 + 2xy - 4yz - 4xz$ (ii) $12x_1^2 + 4x_2^2 + 5x_3^2 - 4x_2x_3 + 6x_1x_3 - 6x_1x_2$

(iii) $2x^2 + 9y^2 + 6z^2 + 8xy + 8yz + 6zx$ (iv) $x^2 + 4y^2 + z^2 + 4xy + 6yz + 2zx$.

4. Reduce the following quadratic forms to canonical forms or to sum of squares by orthogonal transformation. Write also rank, index, signature.

(i) $3x^2 + 5y^2 + 3z^2 - 2xy - 2yz + 2zx$ (ii) $2x_1^2 + 2x_2^2 + 2x_3^2 - 2x_1x_2 + 2x_1x_3 - 2x_2x_3$

(iii) $3x^2 - 2y^2 - z^2 - 4xy + 8xz + 12yz$ (iv) $x^2 + 3y^2 + 3z^2 - 2yz$.

Answers

1. (i) $\begin{bmatrix} 2 & 3 \\ 3 & 3 \end{bmatrix}$ (ii) $\begin{bmatrix} 2 & -1 & 4 \\ -1 & 5 & -\frac{1}{2} \\ 4 & -\frac{1}{2} & -6 \end{bmatrix}$ (iii) $\begin{bmatrix} 1 & 1 & 2 & -3 \\ 1 & 2 & -2 & -4 \\ 2 & -2 & 3 & 6 \\ -3 & -4 & 6 & 4 \end{bmatrix}$

2. (i) $2x^2 + 3y^2 + z^2 + 8xy + 2yz + 10xz$ (ii) $x_1^2 - 4x_2^2 + 6x_3^2 + 2x_4^2 + 2x_1x_2 - 4x_1x_3 - 6x_3x_4$

3. (i) $2y_1^2 + \frac{3}{2}y_2^2 + \frac{1}{3}y_3^2$; Rank = 3, Index = 3, Sig. = 3

(ii) $12y_1^2 + \frac{13}{4}y_2^2 + \frac{49}{13}y_3^2$; Rank = 3, Index = 3, Sig. = 3

(iii) $2y_1^2 + y_2^2 - \frac{5}{2}y_3^2$; Rank = 3, Index = 2, Sig. = 1

(iv) $y_1^2 + 2y_2^2 - \frac{1}{2}y_3^2$; Rank = 3, Index = 2, Sig. = 1.

4. (i) $2y_1^2 + 3y_2^2 + 6y_3^2$; Rank = 3 , Index = 3, Sig. = 3

(ii) $4y_1^2 + y_2^2 + y_3^2$; Rank = 3, Index = 3, Sig. = 3

(iii) $3y_1^2 + 6y_2^2 - 9y_3^2$; Rank = 3, Index = 2, Sig. = 1

(iv) $y_1^2 + 2y_2^2 + 4y_3^2$; Rank = 3, Index = 3, Sig. = 3.

UNIT 2
Infinite Series

2.1. INTRODUCTION

Infinite series occur so frequently in all types of engineering problems that the necessity of studying their convergence or divergence is very important. Unless a series employed in an investigation is convergent, it may lead to absurd conclusions. Hence it is required that the students of engineering begin by acquiring an intelligent grasp of this topic.

2.2. SEQUENCE

A sequence is a function $f: \mathrm{N} \to \mathrm{S}$ whose domain is the set N of all natural numbers and the range may be any set S. A sequence is denoted by (a_n) or $\{a_n\}$ or $\langle a_n \rangle$, where a_n is known as nth term.

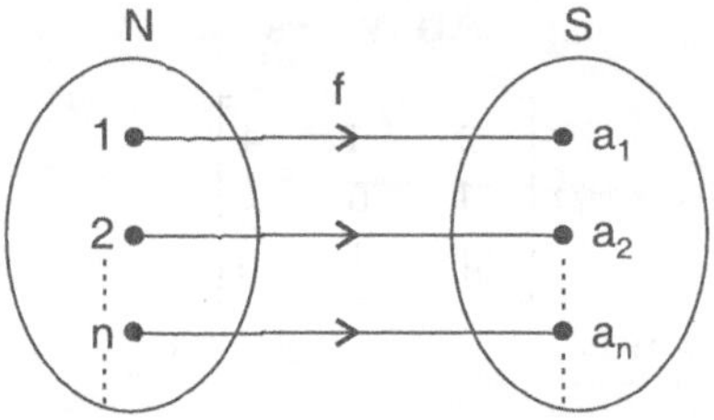

Examples:

1. $\{2, 4, 6, \ldots, (2n), \ldots\} = \{2n\}_{n=1}^{\infty}$, where $a_n = 2n$

2. $\left\{1, \frac{1}{3}, \frac{1}{5}, \ldots, \frac{1}{2n-1}, \ldots\right\} = \left\{\frac{1}{2n-1}\right\}_{n=1}^{\infty}$, where $a_n = \frac{1}{2n-1}$

3. $\{1, -1, 1, -1, \ldots, (-1)^{n-1}, \ldots\} = \{(-1)^{n-1}\}_{n=1}^{\infty}$, where $a_n = (-1)^{n-1}$

2.3. RANGE OF A SEQUENCE

The set of all **distinct** terms of a sequence is called its range.

Note. In a sequence $\{x_n\}$, since $n \in \mathrm{N}$ and N is an infinite set, **the number of terms of a sequence is always infinite.** The range of a sequence may be a finite set. *e.g.*, if $x_n = (-1)^n$ then $\{x_n\} = \{-1, 1, -1, 1, \ldots\ldots\}$

The range of sequence $\{x_n\} = \{-1, 1\}$ which is a finite set.

2.4. CONSTANT SEQUENCE

A sequence $\{x_n\}$ defined by $x_n = c \in R \quad \forall\ n \in N$ is called a constant sequence.

e.g., $\{x_n\} = \{c, c, c, ...\}$ is a constant sequence with range = $\{c\}$.

2.5. BOUNDED AND UNBOUNDED SEQUENCES

Bounded above sequence. A sequence $\{a_n\}$ is said to be bounded above if $\exists$ a real number K such that $a_n \le K \quad \forall\ n \in N$.

Bounded below sequence. A sequence $\{a_n\}$ is said to be bounded below if $\exists$ a real number K such that $a_n \ge k \quad \forall\ n \in N$.

Bounded sequence. A sequence $\{a_n\}$ is said to be bounded when it is bounded both above and below.

$\Rightarrow$ A sequence $\{a_n\}$ is bounded if $\exists$ two real numbers k and K ($k \le K$) such that

$$k \le a_n \le K \quad \forall\ n \in N.$$

Choosing M = maximum $\{|\ k\ |, |\ K\ |\}$, we can also define a sequence $\{a_n\}$ to be bounded if

$$|\ a_n\ | \le M \quad \forall\ \ n \in N.$$

Unbounded sequence. If $\exists$ no real number M such that $|\ a_n\ | \le M\ \forall\ n \in N$, then the sequence $\{a_n\}$ is said to be unbounded.

Examples: 1. The sequence $\{a_n\}$ defined by $a_n = \dfrac{1}{n}$.

Here $$\{a_n\} = \left\{\frac{1}{1}, \frac{1}{2}, \frac{1}{3}, ...\right\}$$

$\because$ $$0 < a_n \le 1 \quad \forall \quad n \in N$$

$\therefore$ $\{a_n\}$ is bounded above

2. The sequence $\{a_n\}$ defined by $a_n = 2^{n-1}$

Here $$\{a_n\} = \{1, 2, 2^2, 2^3, ...\}.$$

Since $a_n \ge 1, \quad \forall \quad n \in N$

$\therefore$ The sequence is bounded below.

2.6. CONVERGENT, DIVERGENT AND OSCILLATING SEQUENCES

Convergent sequence. A sequence $\{a_n\}$ is said to be convergent if $\underset{n\to\infty}{\text{Lt}}\ a_n$ is finite.

For example, consider the sequence $\dfrac{1}{2}, \dfrac{1}{2^2}, \dfrac{1}{2^3}, ..., \dfrac{1}{2^n}$ ......

Here $a_n = \dfrac{1}{2^n}$, $\underset{n\to\infty}{\text{Lt}}\ a_n = \underset{n\to\infty}{\text{Lt}}\ \dfrac{1}{2^n} = 0$, which is finite. $\left(\because\ 2^\infty = \infty, \dfrac{1}{\infty} = 0\right)$

$\Rightarrow$ The sequence $\{a_n\}$ is convergent.

Divergent sequence. A sequence $\{a_n\}$ is said to be divergent if $\underset{n\to\infty}{\text{Lt}}\ a_n$ is not finite, *i.e.,* if

$$\underset{n\to\infty}{\text{Lt}}\ a_n = +\infty \text{ or } -\infty.$$

For example:

(*i*) Consider the sequence $\{n^2\}$

Here $a_n = n^2$, $\underset{n\to\infty}{\text{Lt}}\ a_n = +\infty \Rightarrow$ The sequence $\{n^2\}$ is divergent.

(*ii*) Consider the sequence $\{-2^n\}$.

Here $$a_n = -2^n,\ \underset{n\to\infty}{\text{Lt}}\ a_n = \underset{n\to\infty}{\text{Lt}}\ (-2^n) = -\infty$$

$\Rightarrow$ The sequence $\{-2^n\}$ is divergent.

Oscillatory sequence. If a sequence $\{a_n\}$ neither converges to a finite number nor diverges to $+\infty$ or $-\infty$, it is called an oscillatory sequence. Oscillatory sequences are of two types :

(*i*) A bounded sequence which does not converge is said to **oscillate finitely.**

For example, consider the sequence $\{(-1)^n\}$.

Here $$a_n = (-1)^n$$

It is a bounded sequence. $$\underset{n\to\infty}{\text{Lt}}\ a_{2n} = \underset{n\to\infty}{\text{Lt}}\ (-1)^{2n} = 1$$

$$\underset{n\to\infty}{\text{Lt}}\ a_{2n+1} = \underset{n\to\infty}{\text{Lt}}\ (-1)^{2n+1} = -1.$$

Thus $\underset{n\to\infty}{\text{Lt}}\ a_n$ does not exist $\Rightarrow$ the sequence does not converge.

Hence this sequence oscillates finitely.

(*ii*) An unbounded sequence which does not diverge is said to **oscillate infinitely.**

For example, consider the sequence $\{(-1)^n\, n\}$.

Here $$a_n = (-1)^n\, n.$$

It is an unbounded sequence.

$$\underset{n\to\infty}{\text{Lt}}\ a_{2n} = \underset{n\to\infty}{\text{Lt}}\ (-1)^{2n}\,.\,2n = \underset{n\to\infty}{\text{Lt}}\ 2n = +\infty$$

$$\underset{n\to\infty}{\text{Lt}}\ a_{2n+1} = \underset{n\to\infty}{\text{Lt}}\ (-1)^{2n+1}\,(2n+1) = \underset{n\to\infty}{\text{Lt}}\ -(2n+1) = -\infty.$$

Thus the sequence does not diverge.

Hence this sequence oscillates infinitely.

Note. When we say $\underset{n\to\infty}{\text{Lt}}\ a_n = l$, it means $\underset{n\to\infty}{\text{Lt}}\ a_{2n} = \underset{n\to\infty}{\text{Lt}}\ a_{2n+1} = l$

Similarly, $\underset{n\to\infty}{\text{Lt}}\ a_n = +\infty$ means $\underset{n\to\infty}{\text{Lt}}\ a_{2n} = \underset{n\to\infty}{\text{Lt}}\ a_{2n+1} = +\infty.$

2.7. MONOTONIC SEQUENCES

(*i*) A sequence $\{a_n\}$ is said to be **monotonically increasing** if $a_{n+1} \ge a_n\ \forall\ n \in \text{N}$.

i.e., if $$a_1 \le a_2 \le a_3 \le \ldots\ldots \le a_n \le a_{n+1} \le \ldots\ldots$$

(*ii*) A sequence $\{a_n\}$ is said to be **monotonically decreasing** if $a_{n+1} \le a_n\ \forall\ n \in \text{N}$.

i.e., if $$a_1 \ge a_2 \ge a_3 \ge \ldots\ldots \ge a_n \ge a_{n+1} \ge \ldots\ldots$$

(*iii*) A sequence $\{a_n\}$ is said to be **monotonic** if it is either monotonically increasing or monotonically decreasing.

(*iv*) A sequence $\{a_n\}$ is said to be **strictly monotonically increasing** if

$$a_{n+1} > a_n\ \forall\ n \in \text{N}.$$

(*v*) A sequence $\{a_n\}$ is said to be **strictly monotonically decreasing** if

$$a_{n+1} < a_n\ \forall\ n \in \text{N}.$$

(*vi*) A sequence $\{a_n\}$ is said to be **strictly monotonic** if it is either strictly monotonically increasing or strictly monotonically decreasing.

2.8. LIMIT OF A SEQUENCE

A sequence $\{a_n\}$ is said to approach the limit l (say) when $n \to \infty$, if for each $\varepsilon > 0$, $\exists$ a +ve integer m (depending upon ε) such that $|a_n - l| < \varepsilon \quad \forall \quad n \geq m$.

In symbols, we write $\underset{n\to\infty}{\text{Lt}}\ a_n = l$.

Note. $|a_n - l| < \varepsilon \ \forall\ n \geq m \ \Rightarrow\ l - \varepsilon < a_n < l + \varepsilon$ for $n = m, m + 1, m + 2, \ldots\ldots$

2.9. EVERY CONVERGENT SEQUENCE IS BOUNDED

Let the sequence $\{a_n\}$ be convergent. Let it tend to the limit l.

Then given $\varepsilon > 0$, $\exists$ a +ve integer m, such that

$$|a_n - l| < \varepsilon \quad \forall \quad n \geq m$$

$$\Rightarrow \qquad l - \varepsilon < a_n < l + \varepsilon \quad \forall \quad n \geq m.$$

Let k and K be the least and the greatest of $a_1, a_2, a_3, \ldots\ldots, a_{m-1}, l - \varepsilon, l + \varepsilon$

Then $\qquad k \leq a_n \leq \text{K} \quad \forall \quad n \in \text{N},$

$\Rightarrow$ the sequence $\{a_n\}$ is bounded.

The converse is not always true *i.e.*, a sequence may be bounded, yet it may not be convergent. *e.g.*, consider $a_n = (-1)^n$, then the sequence $\{a_n\}$ is bounded but not convergent since it does not have a unique limiting point.

2.10. CONVERGENCE OF MONOTONIC SEQUENCES

Theorem I. *The necessary and sufficient condition for the convergence of a monotonic sequence is that it is bounded.*

A monotonic increasing sequence which is bounded above converges.

A monotonic decreasing sequence which is bounded below converges.

Theorem II. *If a monotonic increasing sequence is not bounded above, it diverges to* $+\infty$.

Theorem III. *If a monotonic decreasing sequence is not bounded below, it diverges to* $-\infty$.

Theorem IV. *If* $\{a_n\}$ *and* $\{b_n\}$ *are two convergent sequences, then sequence* $\{a_n + b_n\}$ *is also convergent.*

Or

If $\text{Lt}\ a_n = \text{A}$ *and* $\text{Lt}\ b_n = \text{B}$, *then* $\text{Lt}\ (a_n + b_n) = \text{A} + \text{B}$.

Theorem V. *If* $\{a_n\}$ *and* $\{b_n\}$ *are two convergent sequences such that* $\text{Lt}\ a_n = \text{A}$ *and* $\text{Lt}\ b_n = \text{B}$, *then*

(*i*) sequence $\{a_n b_n\}$ is also convergent and converges to AB.

(*ii*) sequence $\left\{\dfrac{a_n}{b_n}\right\}$ is also convergent and converges to $\dfrac{\text{A}}{\text{B}}$, $(\text{B} \neq 0)$.

Theorem VI. *The sequence* $\{|a_n|\}$ *converges to zero if and only if the sequence* $\{a_n\}$ *converges to zero.*

Theorem VII. *If a sequence* $\{a_n\}$ *converges to a and* $a_n \geq 0 \ \forall\ n$, *then* $a \geq 0$.

Theorem VIII. *If* $a_n \to a$, $b_n \to b$ *and* $a_n \leq b_n \ \forall\ n$, *then* $a \leq b$.

Theorem IX. *If* $a_n \to l$, $b_n \to l$, *and* $a_n \leq c_n \leq b_n$, $\forall\ n$, *then* $c_n \to l$. *(Squeeze Principle)*

2.11. INFINITE SERIES

If $\{u_n\}$ is a sequence of real numbers, then the expression $u_1 + u_2 + u_3 + ... + u_n ......$ [*i.e.*, the sum of the terms of the sequence, which are infinite in number] is called an infinite series.

The infinite series $u_1 + u_2 + ... + u_n + ...$ is usually denoted by $\sum_{n=1}^{\infty} u_n$ or more briefly, by Σu_n.

Example. $\Sigma n = 1 + 2 + 3 + ... + n + ... \infty$ is an infinite series.

2.12. SERIES OF POSITIVE TERMS

If all the terms of the series $\Sigma u_n = u_1 + u_2 + ... + u_n + ...$ are positive *i.e.*, if $u_n > 0, \forall n$, then the series Σu_n is called a series of positive terms.

2.13. ALTERNATING SERIES

A series in which the terms are alternate positive and negative is called an alternating series. Thus, the series $\Sigma (-1)^{n-1} u_n = u_1 - u_2 + u_3 - u_4 + ... + (-1)^{n-1} u_n + ...$, where $u_n > 0 \ \ \forall n$ is an alternating series.

2.14. PARTIAL SUMS

If $\Sigma u_n = u_1 + u_2 + u_3 + ... + u_n + ...$ is an infinite series, where the terms may be +ve or –ve, then

$S_n = u_1 + u_2 + ... + u_n$ is called the nth partial sum of Σu_n. Thus, the nth partial sum of an infinite series is the sum of its first n terms.

$S_1, S_2, S_3, ...$ are the first, second, third, ... partial sums of the series.

Since $n \in N$, $\{S_n\}$ is a sequence called the **sequence of partial sums** of the infinite series Σu_n.

$\therefore$ To every infinite series Σu_n, there corresponds a sequence $\{S_n\}$ of its partial sums.

2.15. CONVERGENT, DIVERGENT AND OSCILLATING SERIES

An infinite series Σu_n converges, diverges or oscillates (finitely or infinitely) according as the sequence $\{S_n\}$ of its partial sums converges, diverges or oscillates (finitely or infinitely).

(*i*) The series Σu_n converges (or is said to be convergent) if the sequence $\{S_n\}$ of its partial sums converges.

Thus, Σu_n is convergent if $\underset{n \to \infty}{\text{Lt}} S_n$ = finite.

(*ii*) The series Σu_n diverges (or is said to be divergent) if the sequence $\{S_n\}$ of its partial sums diverges.

Thus, Σu_n is divergent if $\underset{n \to \infty}{\text{Lt}} S_n = +\infty$ or $-\infty$.

(*iii*) The series Σu_n oscillates finitely if the sequence $\{S_n\}$ of its partial sums oscillates finitely.

Thus, Σu_n oscillates finitely if $\{S_n\}$ is bounded and neither converges nor diverges.

(*iv*) The series Σu_n oscillates infinitely if the sequence $\{S_n\}$ of its partial sums oscillates infinitely.

Thus, Σu_n oscillates infinitely if $\{S_n\}$ is unbounded and neither converges nor diverges.

ILLUSTRATIVE EXAMPLES

Example 1. *Discuss the convergence of the series*

$$\frac{1}{1\cdot 2}+\frac{1}{2\cdot 3}+\frac{1}{3\cdot 4}+......+\frac{1}{n(n+1)}+ ......\text{ to } \infty.$$

Sol. Here $u_n = \dfrac{1}{n(n+1)} = \dfrac{1}{n} - \dfrac{1}{n+1}$

Putting $n = 1, 2, 3, ..., n$

$$u_1 = \frac{1}{1}-\frac{1}{2}$$

$$u_2 = \frac{1}{2}-\frac{1}{3}$$

$$u_3 = \frac{1}{3}-\frac{1}{4}$$

....

....

$$u_n = \frac{1}{n}-\frac{1}{n+1}$$

Adding $S_n = 1 - \dfrac{1}{n+1}$

$$\underset{n\to\infty}{\text{Lt}}\ S_n = 1 - 0 = 1$$

$\Rightarrow$ $\{S_n\}$ converges to 1 $\Rightarrow$ Σu_n converges to 1.

Note. For another method, see Comparison Test.

Example 2. *Show that the series* $1^2 + 2^2 + 3^2 + ... + n^2 + ...$ *diverges to* $+\infty$.

Sol. $S_n = 1^2 + 2^2 + 3^2 + ... + n^2 = \dfrac{n(n+1)(2n+1)}{6}$

$$\underset{n\to\infty}{\text{Lt}}\ S_n = +\infty$$

$\Rightarrow$ $\{S_n\}$ diverges to $+\infty$

$\Rightarrow$ The given series diverges to $+\infty$.

Example 3. *Test the convergence of the series* $2 - 2 + 2 - 2 + 2 - ... \infty$.

Sol. Here $\Sigma u_n = 2 - 2 + 2 - 2 + 2 - ... \infty$

Then $S_n = 0$ if n is even

$= 2$ if n is odd.

Hence, the series is oscillatory.

Geometric Series Test:

Theorem 1. *The geometric series* $1 + x + x^2 + x^3 + ...$ *to* ∞

(*i*) *converges if* $-1 < x < 1$ *i.e.,* $|x| < 1$ (*ii*) *diverges if* $x \geq 1$

(*iii*) *oscillates finitely if* $x = -1$ (*iv*) *oscillates infinitely if* $x < -1$.

(OU-May 2009, June 2010, 2013)

Proof. (*i*) **When** $|x| < 1$

Since $|x| < 1$, $x^n \to 0$ as $n \to \infty$

$$S_n = 1 + x + x^2 + \ldots\ldots \text{ to } n \text{ terms} = \frac{1(1-x^n)}{1-x} = \frac{1}{1-x} - \frac{x^n}{1-x}$$

$$\underset{n\to\infty}{\text{Lt}}\ S_n = \frac{1}{1-x} \quad \Rightarrow \text{ the sequence } \{S_n\} \text{ is convergent}$$

$\Rightarrow$ the given series is convergent.

(*ii*) **When $x \geq 1$**

Sub-case I: When $x = 1$

$$S_n = 1 + 1 + 1 + \ldots\ldots \text{ to } n \text{ terms} = n$$

$$\underset{n\to\infty}{\text{Lt}}\ S_n = \infty \quad \Rightarrow \text{ the sequence } \{S_n\} \text{ diverges to } \infty.$$

$\Rightarrow$ the given series diverges to ∞.

Sub-case II: When $x > 1$, $x^n \to \infty$ as $n \to \infty$

$$S_n = 1 + x + x^2 + \ldots\ldots \text{ to } n \text{ terms} = \frac{1(x^n - 1)}{x - 1}$$

$$\underset{n\to\infty}{\text{Lt}}\ S_n = \infty \quad \Rightarrow \text{ the sequence } \{S_n\} \text{ diverges to } \infty$$

$\Rightarrow$ the given series diverges to ∞.

(*iii*) **When $x = -1$**

$$S_n = 1 - 1 + 1 - 1 + \ldots\ldots \text{ to } n \text{ terms}$$

$$= 1 \text{ or } 0 \text{ according as } n \text{ is odd or even.}$$

$$\Rightarrow \quad \underset{n\to\infty}{\text{Lt}}\ S_n = 1 \text{ or } 0 \quad \Rightarrow \text{ the sequence } \{S_n\} \text{ oscillates finitely.}$$

$\Rightarrow$ the given series oscillates finitely.

(*iv*) **When $x < -1$**

$$x < -1 \Rightarrow -x > 1$$

Let $r = -x$, then $r > 1$

$\therefore$ $r^n \to \infty$ as $n \to \infty$

$$S_n = 1 + x + x^2 + x^3 + \ldots\ldots \text{ to } n \text{ terms}$$

$$= \frac{1-x^n}{1-x} = \frac{1-(-r)^n}{1+r} \qquad [\because \ x = -r]$$

$$= \frac{1-r^n}{1+r} \text{ or } \frac{1+r^n}{1+r} \text{ according as } n \text{ is even or odd}$$

$$\underset{n\to\infty}{\text{Lt}}\ S_n = \frac{1-\infty}{1+r} \text{ or } \frac{1+\infty}{1+r} = -\infty \text{ or } +\infty$$

$\Rightarrow$ the sequence $\{S_n\}$ oscillates infinitely

$\Rightarrow$ the given series oscillates infinitely.

Note. If the given series is in the form of geometric series *i.e.*, $a + ar + ar^2 + ar^3 + \ldots..$ then geometric series test is applicable.

Problems on Geometric Series Test:

Example 1. *Test the nature of the series* $1 + \frac{1}{2} + \frac{1}{2^2} + \frac{1}{2^3} + \ldots \infty$.

Sol. The series is geometric series with common ratio $r = \frac{1}{2} < 1$.

Hence, the series is convergent.

2.16 GENERAL PROPERTIES OF A SERIES

1. The nature of an infinite series does not change:
   (*i*) by multiplication of all terms by a constant k.
   (*ii*) by addition or deletion of a finite number of terms.
2. If two series Σu_n and Σv_n are convergent, then $\Sigma(u_n + v_n)$ is also convergent.
3. If a series in which all the terms are positive is convergent, the series remains convergent even when some or all of its terms are negative.

Example 2. *Test for convergence the series* $2 + 3 + \frac{9}{2} + \frac{27}{4} + \cdots \infty$.

Sol. Leaving the first term, let the series be $3 + \frac{9}{2} + \frac{27}{4} + \cdots \infty$.

This is geometric series with common ratio $r = \frac{3}{2} = 1.5 > 1$.

Hence, the series is divergent.

Example 3. *Test for convergence* $\sum \frac{1}{2^{n+1}}$.

Sol. $$\sum_{1}^{\infty} \frac{1}{2^{n+1}} = \frac{1}{2^2} + \frac{1}{2^3} + \frac{1}{2^4} + \ldots$$

is a geometric series with common ratio $r = \frac{1}{2} = 0.5 < 1$.

Hence, the series is convergent.

EXERCISE 1

Examine the following series for convergence:

1. $1 + \frac{3}{4} + \frac{9}{16} + \frac{27}{64} + \ldots \infty$
2. $-2 + 1 + \frac{1}{3} + \frac{1}{3^2} + \frac{1}{3^3} + \ldots \infty$
3. $2 + 1 + \frac{4}{3} + \left(\frac{4}{3}\right)^2 + \left(\frac{4}{3}\right)^3 + \ldots \infty$
4. $\frac{5}{2} + \left(\frac{5}{2}\right)^2 + \left(\frac{5}{2}\right)^3 + \ldots \infty$
5. $\sum \frac{1}{3^n}$.

Answers

1. Convergent 2. Convergent 3. Divergent
4. Divergent 5. Convergent.

Necessary Condition for Convergence:

Theorem 2. *If a series Σu_n is convergent, then $\underset{n \to \infty}{Lt}\ u_n = 0$.* (OU-June 2014)

Proof. Let S_n denote the nth partial sum of the series Σu_n.

Then Σu_n is convergent $\Rightarrow$ $\{S_n\}$ is convergent

$\Rightarrow$ $\underset{n \to \infty}{\text{Lt}}\ S_n$ is finite and unique $= s$ (say). $\Rightarrow$ $\underset{n \to \infty}{\text{Lt}}\ S_{n-1} = s$

Now, $S_n - S_{n-1} = u_n$

$$\therefore \quad \underset{n \to \infty}{\text{Lt}}\ u_n = \underset{n \to \infty}{\text{Lt}}\ (S_n - S_{n-1}) = \underset{n \to \infty}{\text{Lt}}\ S_n - \underset{n \to \infty}{\text{Lt}}\ S_{n-1} = s - s = 0.$$

Hence, Σu_n is convergent $\Rightarrow$ $\underset{n \to \infty}{\text{Lt}}\ u_n = 0$.

The converse of the above theorem is not always true, *i.e.*, the nth term may tend to zero as $n \to \infty$ even if the series is not convergent.

For example, the series $1 + \frac{1}{2} + \frac{1}{3} + \ldots\ldots + \frac{1}{n} + \ldots\ldots$ diverges, though

$$\underset{n \to \infty}{\text{Lt}}\ u_n = \underset{n \to \infty}{\text{Lt}}\ \frac{1}{n} = 0.$$

Note 1. Σu_n is convergent $\Rightarrow$ $\underset{n \to \infty}{\text{Lt}}\ u_n = 0$.

Note 2. $\underset{n \to \infty}{\text{Lt}}\ u_n = 0$ $\Rightarrow$ Σu_n may or may not be convergent.

Note 3. $\underset{n \to \infty}{\text{Lt}}\ u_n \neq 0$ $\Rightarrow$ Σu_n is not convergent *i.e.*, Σu_n is divergent.

Note 4. In the following cases, we can apply necessary condition, that is if $\underset{n \to \infty}{\text{Lt}}\ u_n \neq 0$, then Σu_n is divergent.

Case 1: If $u_n = \dfrac{f(n)}{g(n)}$, where $f(n)$ and $g(n)$ are polynomials of same degree.

Case 2: If $u_n = \dfrac{f(n)}{g(n)}$, where $f(n)$ and $g(n)$ involves power terms with same base

Problems on Necessary Condition:

Example 1. *Test the series $\sum \dfrac{4n^3 + 2}{7n^3 + 2n}$ for convergence.* (OU-June 2013)

Sol. Here $$u_n = \frac{4n^3 + 2}{7n^3 + 2n}$$

$$\underset{n \to \infty}{\text{Lt}}\ u_n = \underset{n \to \infty}{\text{Lt}}\ \frac{4n^3 + 2}{7n^3 + 2n} = \underset{n \to \infty}{\text{Lt}}\ \frac{n^3\left(4 + \dfrac{2}{n^3}\right)}{n^3\left(7 + \dfrac{2}{n^2}\right)}$$

$$= \frac{4+\frac{2}{\infty}}{7+\frac{2}{\infty}} = \frac{4+0}{7+0} = \frac{4}{7} \neq 0 \qquad \left(\because \ \frac{x}{\infty} = 0\right)$$

$\therefore$ series is divergent.

Example 2. *Discuss the convergence of* $\sum \frac{(n^3+4)n}{(n^2+1)(n^2+4)}$. (OU-May 2012)

Sol. Consider $\quad u_n = \frac{(n^3+4)n}{(n^2+1)(n^2+4)}$

$$\underset{n\to\infty}{\text{Lt}}\ u_n = \underset{n\to\infty}{\text{Lt}} \frac{(n^3+4)n}{(n^2+1)(n^2+4)} = \underset{n\to\infty}{\text{Lt}} \frac{n^3\left(1+\frac{4}{n^3}\right)n}{n^2\left(1+\frac{1}{n^2}\right)n^2\left(1+\frac{4}{n^2}\right)}$$

$$= \frac{1+\frac{4}{\infty}}{\left(1+\frac{1}{\infty}\right)\left(1+\frac{4}{\infty}\right)} = \frac{1+0}{(1+0)(1+0)} = 1 \neq 0$$

$\therefore$ the given series is divergent.

Example 3. *Discuss the convergence* $\sum \frac{n+1}{2n+5}$. (OU-Jan. 2012)

Sol. Let $\quad u_n = \frac{n+1}{2n+5}$

$$\underset{n\to\infty}{\text{Lt}}\ u_n = \underset{n\to\infty}{\text{Lt}} \frac{n+1}{2n+5} = \underset{n\to\infty}{\text{Lt}} \frac{n\left(1+\frac{1}{n}\right)}{n\left(2+\frac{5}{n}\right)}$$

$$= \frac{1+\frac{1}{\infty}}{2+\frac{5}{\infty}} = \frac{1+0}{2+0} = \frac{1}{2} \neq 0$$

$\therefore \quad \Sigma u_n$ is divergent.

Example 4. *Discuss the convergence of the series* $\sum \frac{n^2+1}{n^2}$. (OU-June 2010)

Sol. Let $\quad u_n = \frac{n^2+1}{n^2}$

$$\underset{n\to\infty}{\text{Lt}}\ u_n = \underset{n\to\infty}{\text{Lt}} \frac{n^2+1}{n^2} = \underset{n\to\infty}{\text{Lt}} \frac{n^2\left(1+\frac{1}{n^2}\right)}{n^2}$$

$$= 1 + \frac{1}{\infty} = 1 + 0 = 1 \neq 0$$

Hence, the series is divergent.

Example 5. *Test the convergence of the series* $\sqrt{\frac{1}{4}} + \sqrt{\frac{2}{6}} + \sqrt{\frac{3}{8}} + \ldots + \sqrt{\frac{n}{2(n+1)}} + \ldots$.

Sol. Let $u_n = \sqrt{\frac{n}{2(n+1)}}$

$$\underset{n\to\infty}{\text{Lt}}\, u_n = \underset{n\to\infty}{\text{Lt}} \sqrt{\frac{n}{2(n+1)}} = \underset{n\to\infty}{\text{Lt}} \sqrt{\frac{n}{2n\left(1+\frac{1}{n}\right)}}$$

$$= \underset{n\to\infty}{\text{Lt}} \frac{1}{\sqrt{2\left(1+\frac{1}{n}\right)}} = \frac{1}{\sqrt{2\left(1+\frac{1}{\infty}\right)}} = \frac{1}{\sqrt{2(1+0)}} = \frac{1}{\sqrt{2}} \neq 0$$

$\therefore$ the series is divergent.

Example 6. *Examine for convergence of the series* $\frac{2^{n+1}+1}{2^n-3}$.

Sol. Here, $u_n = \frac{2^{n+1}+1}{2^n-3}$

$$\underset{n\to\infty}{\text{Lt}}\, u_n = \underset{n\to\infty}{\text{Lt}} \frac{2^{n+1}+1}{2^n-3} = \underset{n\to\infty}{\text{Lt}} \frac{2^n\cdot 2+1}{2^n-3}$$

$$= \underset{n\to\infty}{\text{Lt}} \frac{2^n\left(2+\frac{1}{2^n}\right)}{2^n\left(1-\frac{3}{2^n}\right)} = \frac{2+\frac{1}{2^\infty}}{1-\frac{3}{2^\infty}} \qquad (\because\ 2^\infty = \infty)$$

$$= \frac{2+\frac{1}{\infty}}{1-\frac{3}{\infty}} = \frac{2+0}{1-0} = 2 \neq 0$$

$\therefore$ Σu_n is divergent.

Example 7. *Test for convergence of the series* $\frac{1}{\sqrt{2}} + \frac{2}{\sqrt{5}} + \frac{4}{\sqrt{17}} + \ldots + \frac{2^n}{\sqrt{4^n+1}} + \ldots \infty$.

Sol. Here, nth term be $u_n = \frac{2^n}{\sqrt{4^n+1}}$ $\qquad (\because\ \sqrt{4^n} = \sqrt{(2^2)^n} = (2^{2n})^{\frac{1}{2}} = 2^n)$

$$\underset{n\to\infty}{\text{Lt}}\, u_n = \underset{n\to\infty}{\text{Lt}} \frac{2^n}{\sqrt{4^n+1}} = \underset{n\to\infty}{\text{Lt}} \frac{2^n}{\sqrt{4^n}\cdot\sqrt{1+\frac{1}{4^n}}}$$

$$= \underset{n\to\infty}{\text{Lt}} \frac{2^n}{2^n \cdot \sqrt{1+\frac{1}{4^n}}} = \frac{1}{\sqrt{1+\frac{1}{4^\infty}}}$$

$$= \frac{1}{\sqrt{1+\frac{1}{\infty}}} = \frac{1}{\sqrt{1+0}} = 1 \neq 0$$

$\therefore$ Σu_n is divergent.

EXERCISE 2

Test for convergence:

1. $\sum_{n=1}^{\infty} \frac{n}{n+1}$

2. $\sum_{n=1}^{\infty} \sqrt{\frac{n}{n+1}}$

3. $\frac{1}{2}+\frac{2}{3}+\frac{3}{4}+\frac{4}{5}+\ldots+\frac{n}{n+1}+\ldots\infty$

4. $\frac{1}{3}+\frac{3}{5}+\frac{8}{10}+\frac{15}{17}+\ldots+\frac{2^n-1}{2^n+1}+\ldots\infty$

5. $\sum \frac{n(n+1)(n+2)}{(3n+2)^3}$

6. $\sum \frac{3^{n-1}+1}{3^{n+1}-2}$.

Answers

1. Divergent
2. Divergent
3. Divergent
4. Divergent
5. Divergent
6. Divergent.

p-Series Test or Auxiliary Series Test or Harmonic Series Test:

Theorem 3. *An Important Test for Comparison.* $\sum \frac{1}{n^p}$. *[Hyper harmonic series or p-series]*

The series $\sum \frac{1}{n^p} = \frac{1}{1^p}+\frac{1}{2^p}+\frac{1}{3^p}+\ldots\ldots+\frac{1}{n^p}+\ldots\ldots$ *to* ∞ *converges if* $p > 1$ *and diverges if* $0 < p \leq 1$. (OU-Dec. 2007, 2008)

Proof. Case I: When p > 1

$$\frac{1}{1^p} = 1$$

$$\frac{1}{2^p}+\frac{1}{3^p} < \frac{1}{2^p}+\frac{1}{2^p} = \frac{2}{2^p} = \frac{1}{2^{p-1}} \qquad \left[\because \frac{1}{3^p} < \frac{1}{2^p}\right]$$

$$\frac{1}{4^p}+\frac{1}{5^p}+\frac{1}{6^p}+\frac{1}{7^p} < \frac{1}{4^p}+\frac{1}{4^p}+\frac{1}{4^p}+\frac{1}{4^p}$$

$$= \frac{4}{4^p} = \frac{1}{4^{p-1}} = \frac{1}{(2^{p-1})^2} \qquad \left[\because \frac{1}{5^p} < \frac{1}{4^p}, \frac{1}{6^p} < \frac{1}{4^p}, \text{etc.}\right]$$

Similarly, the sum of next eight terms

$$= \frac{1}{8^p} + \frac{1}{9^p} + \ldots\ldots + \frac{1}{15^p} < \frac{1}{8^p} + \frac{1}{8^p} + \ldots\ldots + \frac{1}{8^p} = \frac{8}{8^p}$$

$$= \frac{1}{8^{p-1}} = \frac{1}{(2^{p-1})^3} \text{ and so on.}$$

Now, $\sum \frac{1}{n^p} = \frac{1}{1^p} + \frac{1}{2^p} + \frac{1}{3^p} + \ldots\ldots + \frac{1}{n^p} + \ldots\ldots$

$$= \frac{1}{1^p} + \left(\frac{1}{2^p} + \frac{1}{3^p}\right) + \left(\frac{1}{4^p} + \frac{1}{5^p} + \frac{1}{6^p} + \frac{1}{7^p}\right) + \left(\frac{1}{8^p} + \frac{1}{9^p} + \ldots\ldots + \frac{1}{15^p}\right) + \ldots\ldots \quad \ldots(1)$$

$\therefore$ Each term of (1) after the first is less than the corresponding term in

$$1 + \frac{1}{2^{p-1}} + \frac{1}{(2^{p-1})^2} + \frac{1}{(2^{p-1})^3} + \ldots\ldots \quad \ldots(2)$$

But (2) is a G.P. whose common ratio $= \frac{1}{2^{p-1}} < 1$ $\qquad (\because \ p > 1)$

$\therefore$ (2) is convergent $\Rightarrow$ (1) is convergent.

Hence the given series is convergent.

Case II: When p = 1

$$\sum \frac{1}{n^p} = \sum \frac{1}{n} = 1 + \frac{1}{2} + \frac{1}{3} + \frac{1}{4} + \ldots\ldots$$

$$1 + \frac{1}{2} = 1 + \frac{1}{2}$$

$$\frac{1}{3} + \frac{1}{4} > \frac{1}{4} + \frac{1}{4} = \frac{2}{4} = \frac{1}{2}$$

$$\frac{1}{5} + \frac{1}{6} + \frac{1}{7} + \frac{1}{8} > \frac{1}{8} + \frac{1}{8} + \frac{1}{8} + \frac{1}{8} = \frac{4}{8} = \frac{1}{2} \quad \text{and so on.}$$

Now, $\sum \frac{1}{n} = 1 + \frac{1}{2} + \frac{1}{3} + \frac{1}{4} + \ldots\ldots + \frac{1}{n} + \ldots\ldots$

$$= 1 + \frac{1}{2} + \left(\frac{1}{3} + \frac{1}{4}\right) + \left(\frac{1}{5} + \frac{1}{6} + \frac{1}{7} + \frac{1}{8}\right) + \ldots\ldots \quad \ldots(1)$$

Each term of (1) after the second is greater than the corresponding term in

$$1 + \frac{1}{2} + \frac{1}{2} + \frac{1}{2} + \ldots\ldots \quad \ldots(2)$$

But after the second term (2) is a G.P. whose common ratio = 1.

$\therefore$ (2) is divergent. $\Rightarrow$ (1) is divergent.

Hence the given series is divergent.

Case III: When p < 1

$$p < 1 \ \Rightarrow \ n^p < n \ \Rightarrow \ \frac{1}{n^p} > \frac{1}{n} \quad \forall\, n$$

But the series $\sum \frac{1}{n}$ is divergent (Case II).

Hence $\sum \frac{1}{n^p}$ is also divergent.

Note. If the given series is in the form of $\sum \frac{1}{n^p}$, $p > 0$, then p-series test is applicable.

Problems on *p*-Series Test:

Example 1. *Test for convergence* $\sum \frac{1}{n}$.

Sol. Let $u_n = \frac{1}{n}$

Take $\Sigma u_n = \sum \frac{1}{n^p} = \sum \frac{1}{n'}$

Here $p = 1$, by *p*-series test, Σu_n is divergent.

Example 2. *Test for convergence* $\sum \frac{1}{n\sqrt{n}}$.

Sol. Take $\Sigma u_n = \sum \frac{1}{n^p} = \sum \frac{1}{n\sqrt{n}} = \sum \frac{1}{n^{3/2}}$

Here $p = \frac{3}{2} = 1.5 > 1$.

Hence, by *p*-series test, Σu_n is convergent.

Example 3. *Test for convergence* $\sum \frac{1}{\sqrt{n}}$.

Sol. Compare $\sum \frac{1}{\sqrt{n}}$ with $\sum \frac{1}{n^p}$, we get

$$p = \frac{1}{2} = 0.5 < 1$$

Hence, by *p*-series test, given series is divergent.

Example 4. *Test for convergence* Σn.

Sol. Here $u_n = n$, but do not take this as $\frac{1}{n^{-1}}$, where $p = -1$ because $p > 0$. In this case, *p*-series test is not applicable.

By the definition, the above series is divergent.

$\because$ $\Sigma u_n = \sum_{1}^{\infty} n = 1 + 2 + 3 + \cdots + n + \cdots \infty$

$$S_n = 1 + 2 + 3 + \ldots + n = \sum_{1}^{n} n = \frac{n(n+1)}{2}$$

$$\underset{n \to \infty}{\text{Lt}} S_n = \underset{n \to \infty}{\text{Lt}} \frac{n(n+1)}{2} = \infty$$

Hence, $\sum_{1}^{\infty} n$ is divergent.

Example 5. *Test the series* $\sum \left[\frac{1}{2^n} + \frac{1}{n}\right]$ *for convergence.* (OU-May 2015)

Sol. Take $\sum \left[\frac{1}{2^n} + \frac{1}{n}\right] = \sum \frac{1}{2^n} + \sum \frac{1}{n}$ ...(1)

Here $\sum \frac{1}{2^n}$ is a geometric series with common ratio $r = \frac{1}{2} < 1$.

$\therefore$ By geometric series test, $\sum \frac{1}{2^n}$ is convergent.

Compare $\sum \frac{1}{n}$ with $\sum \frac{1}{n^P}$, we get P = 1.

$\therefore$ By P-series test, $\sum \frac{1}{n}$ is divergent.

Since, convergent + divergent = divergent, we get $\sum \left[\frac{1}{2^n} + \frac{1}{n}\right]$ is divergent.

Theorem 4. Comparison Test:

Prove that if two positive term series Σu_n *and* Σv_n *be such that* $\underset{n\to\infty}{Lt} \frac{u_n}{v_n} = k =$ *non-zero finite quantity, then both* Σu_n *and* Σv_n *converge or diverge together.*

***Proof.** Since $\underset{n\to\infty}{\text{Lt}} \frac{u_n}{v_n} = k\ (\neq 0)$

By the definition of a limit, there exists a small positive number $\varepsilon > 0$, such that

$$\left|\frac{u_n}{v_n} - k\right| < \varepsilon \text{ for } n \geq m$$

$$\Rightarrow \qquad -\varepsilon < \frac{u_n}{v_n} - k < \varepsilon \text{ for } n \geq m$$

$$\Rightarrow \qquad k - \varepsilon < \frac{u_n}{v_n} < k + \varepsilon \text{ for } n \geq m$$

leaving the first m terms of both the series, we have

$$k - \varepsilon < \frac{u_n}{v_n} < k + \varepsilon \text{ for all } n \qquad \text{...(1)}$$

Case 1: Let Σv_n is convergent, then

$$\underset{n\to\infty}{\text{Lt}} (v_1 + v_2 + \ldots + v_n) = l = \text{a finite number} \qquad \text{...(2)}$$

From (1), $\frac{u_n}{v_n} < k + \varepsilon \Rightarrow u_n < (k + \varepsilon)v_n, \forall\, n$

$\therefore \underset{n\to\infty}{\text{Lt}} (u_1 + u_2 + \ldots + u_n) < (k + \varepsilon) \underset{n\to\infty}{\text{Lt}} (v_1 + v_2 + \ldots + v_n) = (k + \varepsilon)l$ [From (2)]

Hence, Σu_n is convergent.

*Proof is for student reference only.

Case 2: If Σv_n is divergent, then

$$\underset{n\to\infty}{\text{Lt}} (v_1 + v_2 + \ldots + v_n) \to \infty \qquad \ldots(3)$$

From (1), $$k - \varepsilon < \frac{u_n}{v_n} \Rightarrow u_n > (k - \varepsilon)v_n, \forall n$$

$$\therefore \quad \underset{n\to\infty}{\text{Lt}} (u_1 + u_2 + \ldots + u_n) > (k - \varepsilon) \underset{n\to\infty}{\text{Lt}} (v_1 + v_2 + \ldots + v_n) \to \infty \qquad \text{[From (3)]}$$

Hence, Σu_n is divergent.

Note. In the following cases comparison test can be applied.

Case 1: If $u_n = \dfrac{f(n)}{g(n)}$ and degree of $f(n)$ < degree of $g(n)$.

Example. $\sum \dfrac{2n^3 + 5}{4n^5 + 1}$.

Case 2: When u_n contains sin or tan terms and $\underset{\theta\to 0}{\text{Lt}} \dfrac{\sin\theta}{\theta} = 1$, $\underset{\theta\to 0}{\text{Lt}} \dfrac{\tan\theta}{\theta} = 1$ are applicable.

Example 1. $\sum \dfrac{1}{n} \sin\left(\dfrac{1}{n}\right)$.

Example 2. $\sum \dfrac{1}{\sqrt{n}} \tan\left(\dfrac{1}{n}\right)$.

Case 3: When two power terms with different base are involved.

Example. $\sum \dfrac{2^n + 1}{3^n - 2}$.

Case 4: When u_n involves square root terms or cube root terms in the form of $a - b$.

Example 1. $\sum (\sqrt{n^4 + 1} - n^2)$.

Example 2. $\sum (3\sqrt{n^3 + 1} - n)$.

ILLUSTRATIVE EXAMPLES

Problems on Comparison Test:

Example 1. *Test the convergence of the series* $\dfrac{1}{1\cdot 3\cdot 5} + \dfrac{2}{3\cdot 5\cdot 7} + \dfrac{3}{5\cdot 7\cdot 9} + \ldots$.

(OU-Dec. 2013)

Sol. First we need to find nth term u_n.

Here 1, 2, 3, ... are in A.P. and first term $a = 1$, difference $d = 1$.

In A.P., nth term is $a_n = a + (n - 1)d$

$$= 1 + (n - 1)1 = 1 + n - 1 = n$$

Also, 1, 3, 5, ... are in A.P. with $a = 1$, $d = 2$

$$b_n = 1 + (n - 1)2 = 1 + 2n - 2 = 2n - 1$$

3, 5, 7, ... in A.P. with $a = 3$, $d = 2$

$$c_n = 3 + (n - 1)2 = 3 + 2n - 2 = 2n + 1$$

5, 7, 9, ... in A.P. with $a = 5, d = 2$

$$d_n = 5 + (n - 1)2 = 5 + 2n - 2 = 2n + 3$$

Here, $$u_n = \frac{a_n}{b_n \cdot c_n \cdot d_n} = \frac{n}{(2n-1)(2n+1)(2n+3)}$$

$$v_n = \text{Ratio of highest degree terms}$$

$$= \frac{n}{n \cdot n \cdot n} = \frac{1}{n^2}$$

Compare with

$$\Sigma v_n = \sum \frac{1}{n^p} = \sum \frac{1}{n^2}$$

$\therefore$ $$p = 2 > 1,$$

By p-series test Σv_n is convergent.

Now, $$\underset{n\to\infty}{\text{Lt}} \frac{u_n}{v_n} = \underset{n\to\infty}{\text{Lt}} \frac{\frac{n}{(2n-1)(2n+1)(2n+3)}}{\frac{1}{n^2}}$$

$$= \underset{n\to\infty}{\text{Lt}} \frac{n^3}{n\left(2-\frac{1}{n}\right)n\left(2+\frac{1}{n}\right)n\left(2+\frac{3}{n}\right)}$$

$$= \frac{1}{\left(2-\frac{1}{\infty}\right)\left(2+\frac{1}{\infty}\right)\left(2+\frac{3}{\infty}\right)} = \frac{1}{(2-0)(2+0)(2+0)}$$

$$= \frac{1}{8} = \text{a non-zero finite value.}$$

By comparison test, Σu_n also convergent.

Example 2. *Test the convergence for the series* $\sum_{n=1}^{\infty} \frac{2n^3 + 5}{4n^5 + 1}$. (OU-June 2011)

Sol. Let $$u_n = \frac{2n^3 + 5}{4n^5 + 1}$$

$$v_n = \frac{n^3}{n^5} = \frac{1}{n^2}$$

Compare with

$$\Sigma v_n = \sum \frac{1}{n^p} = \sum \frac{1}{n^2}$$

$$p = 2 > 1$$

$\therefore$ By p-Test, Σv_n is convergent.

Now, $$\underset{n\to\infty}{\text{Lt}} \frac{u_n}{v_n} = \underset{n\to\infty}{\text{Lt}} \frac{2n^3+5}{4n^5+1} \times \frac{n^2}{1} = \underset{n\to\infty}{\text{Lt}} \frac{n^3\left(2+\frac{5}{n^3}\right)}{n^5\left(4+\frac{1}{n^5}\right)} \cdot n^2 = \frac{2+0}{4+0} = \frac{1}{2} \neq 0$$

By comparison test, Σu_n also convergent.

Example 3. *Test for convergence the series* $\frac{1}{1\cdot2\cdot3} + \frac{3}{2\cdot3\cdot4} + \frac{5}{3\cdot4\cdot5} + \ldots\infty$.

(OU-May 2009)

Sol. 1, 3, 5, ... are in A.P.

nth term is $a + (n-1)d = 1 + (n-1)2 = 2n - 1$

1, 2, 3, ... in A.P.

nth term is $= 1 + (n-1)1 = n$

2, 3, 4, ... in A.P.

nth term is $= 2 + (n-1)1 = n + 1$

3, 4, 5, ... in A.P.

nth term is $= 3 + (n-1)1 = n + 2$

$\therefore$ nth term of the series $$u_n = \frac{2n-1}{n(n+1)(n+2)}$$

$$v_n = \frac{n}{n\cdot n\cdot n} = \frac{1}{n^2}$$

Compare with $$\Sigma v_n = \sum \frac{1}{n^p} = \sum \frac{1}{n^2}$$

$$p = 2 > 1$$

By p-Test, Σv_n is convergent.

Consider $$\underset{n\to\infty}{\text{Lt}} \frac{u_n}{v_n} = \underset{n\to\infty}{\text{Lt}} \frac{2n-1}{n(n+1)(n+2)} \times \frac{n^2}{1}$$

$$= \underset{n\to\infty}{\text{Lt}} \frac{n\left(2-\frac{1}{n}\right)}{n\cdot n\left(1+\frac{1}{n}\right)n\cdot\left(1+\frac{2}{n}\right)} \times n^2 = \frac{2-0}{(1+0)(1+0)}$$

$= 2 =$ a non-zero finite value.

By comparison test, Σu_n also convergent.

Example 4. *Find whether the series* $\sum \frac{1}{n\sqrt{n^2-1}}$ *is convergent or divergent.*

(OU-Nov. 2006)

Sol. Let $$u_n = \frac{1}{n\sqrt{n^2-1}}$$

$$v_n = \frac{1}{n\sqrt{n^2}} = \frac{1}{n^2}$$

Since $\Sigma v_n = \sum \frac{1}{n^2}$ is of the form $\sum \frac{1}{n^p}$ with $p = 2 > 1$, by p-Test Σv_n is convergent.

Consider $$\underset{n\to\infty}{\text{Lt}} \frac{u_n}{v_n} = \underset{n\to\infty}{\text{Lt}} \frac{1}{n\sqrt{n^2-1}} \times \frac{n^2}{1}$$

$$= \underset{n\to\infty}{\text{Lt}} \frac{n^2}{n \cdot \sqrt{n^2}\sqrt{1-\frac{1}{n^2}}} = \frac{1}{\sqrt{1-0}} = 1 \neq 0$$

By comparison test, Σu_n also convergent.

Example 5. *Test for convergence of* $\sum \frac{(n+1)(n+2)}{n^3\sqrt{n}}$. (OU-Nov. 2006)

Sol. Here, $$u_n = \frac{(n+1)(n+2)}{n^3\sqrt{n}}$$

$$v_n = \frac{n \cdot n}{n^3\sqrt{n}} = \frac{1}{n\sqrt{n}} = \frac{1}{n^{3/2}}$$

Since $\Sigma v_n = \sum \frac{1}{n^{3/2}}$ is of the form $\sum \frac{1}{n^p}$ with $p = \frac{3}{2} = 1.5 > 1$, by p-Test Σv_n is convergent.

Now, $$\underset{n\to\infty}{\text{Lt}} \frac{u_n}{v_n} = \underset{n\to\infty}{\text{Lt}} \frac{(n+1)(n+2)}{n^3\sqrt{n}} \cdot \frac{n\sqrt{n}}{1}$$

$$= \underset{n\to\infty}{\text{Lt}} \frac{n\left(1+\frac{1}{n}\right)n\left(1+\frac{2}{n}\right)}{n^2} = (1+0)(1+0) = 1 \neq 0$$

By comparison test, Σu_n also convergent.

Note. Perform rationalisation when square roots are involved.

Example 6. *Test the series* $\sum \sqrt{n^4+1} - \sqrt{n^4-1}$ *for convergence.*

(OU-Dec. 2007, May 2009, June 2010)

Sol. Let $$u_n = \sqrt{n^4+1} - \sqrt{n^4-1}$$

Rationalising

$$u_n = (\sqrt{n^4+1} - \sqrt{n^4-1}) \times \frac{(\sqrt{n^4+1} + \sqrt{n^4-1})}{\sqrt{n^4+1} + \sqrt{n^4-1}}$$

$$\boxed{(a-b)(a+b) = a^2 - b^2}$$

$$u_n = \frac{(\sqrt{n^4+1})^2 - (\sqrt{n^4-1})^2}{\sqrt{n^4+1} + \sqrt{n^4-1}}$$

$$= \frac{n^4 + 1 - n^4 + 1}{\sqrt{n^4+1} + \sqrt{n^4-1}} = \frac{2}{\sqrt{n^4+1} + \sqrt{n^4-1}}$$

Here, $$v_n = \frac{1}{\sqrt{n^4}} = \frac{1}{n^2}$$

$$\Sigma v_n = \sum \frac{1}{n^p} = \sum \frac{1}{n^2}$$

Then $p = 2 > 1$, By p-Test, Σv_n is convergent.

$$\underset{n\to\infty}{\text{Lt}} \frac{u_n}{v_n} = \underset{n\to\infty}{\text{Lt}} \frac{2}{\sqrt{n^4+1} + \sqrt{n^4-1}} \times \frac{\sqrt{n^4}}{1}$$

$$= \underset{n\to\infty}{\text{Lt}} \frac{2\sqrt{n^4}}{\sqrt{n^4}\left(\sqrt{1+\frac{1}{n^4}} + \sqrt{1-\frac{1}{n^4}}\right)} = \frac{2}{\sqrt{1+0} + \sqrt{1-0}} = \frac{2}{2} = 1 \neq 0$$

By comparison test, Σu_n also convergent.

Example 7. *Test, the convergence or divergence of the series* $\sum (\sqrt{n^3+1} - \sqrt{n^3})$.

Sol. Let $u_n = (\sqrt{n^3+1} - \sqrt{n^3}) \times \frac{\sqrt{n^3+1} + \sqrt{n^3}}{\sqrt{n^3+1} + \sqrt{n^3}}$ (Rationalising)

$$= \frac{(\sqrt{n^3+1})^2 - (\sqrt{n^3})^2}{\sqrt{n^3+1} + \sqrt{n^3}} = \frac{n^3+1-n^3}{\sqrt{n^3+1} + \sqrt{n^3}}$$

$$u_n = \frac{1}{\sqrt{n^3+1} + \sqrt{n^3}}$$

Here, $$v_n = \frac{1}{\sqrt{n^3}} = \frac{1}{n^{3/2}}$$

$$\Sigma v_n = \sum \frac{1}{n^p} = \sum \frac{1}{n^{3/2}}$$

$$p = \frac{3}{2} = 1.5 > 1$$

By p-Test, Σv_n is convergent.

Consider $$\underset{n\to\infty}{\text{Lt}} \frac{u_n}{v_n} = \underset{n\to\infty}{\text{Lt}} \frac{1}{\sqrt{n^3+1} + \sqrt{n^3}} \times \frac{\sqrt{n^3}}{1}$$

$$= \underset{n\to\infty}{\text{Lt}} \frac{\sqrt{n^3}}{\sqrt{n^3}\left(\sqrt{1+\frac{1}{n^3}}+1\right)} = \frac{1}{\sqrt{1+0}+1} = \frac{1}{2} \neq 0$$

By comparison test, Σu_n also convergent.

Example 8. *Test for convergence* $\sum (\sqrt{n^2+1} - n)$.

Sol. Let $$u_n = (\sqrt{n^2+1} - n) \times \frac{(\sqrt{n^2+1}+n)}{\sqrt{n^2+1}+n} \quad \text{(Rationalising)}$$

$$= \frac{(\sqrt{n^2+1})^2 - n^2}{\sqrt{n^2+1}+n} = \frac{n^2+1-n^2}{\sqrt{n^2+1}+n} = \frac{1}{\sqrt{n^2+1}+n}$$

Here, $$v_n = \frac{1}{n}$$

$$\Sigma v_n = \sum \frac{1}{n^p} = \sum \frac{1}{n}$$

$\therefore$ $p = 1$, By p-Test Σv_n is divergent.

Now, we have $$\underset{n\to\infty}{\text{Lt}} \frac{u_n}{v_n} = \underset{n\to\infty}{\text{Lt}} \frac{1}{\sqrt{n^2+1}+n} \times \frac{n}{1}$$

$$= \underset{n\to\infty}{\text{Lt}} \frac{n}{n\left(\sqrt{1+\frac{1}{n^2}}+1\right)} = \frac{1}{\sqrt{1+0}+1} = \frac{1}{2} \neq 0$$

By comparison test, Σu_n also divergent.

Example 9. *Examine the convergence of the series* $\sum (\sqrt[3]{n^3+1} - n)$.

Sol. Here $$u_n = (n^3+1)^{1/3} - n$$

$$= \left[n^3\left(1+\frac{1}{n^3}\right)\right]^{1/3} - n = n\left(1+\frac{1}{n^3}\right)^{1/3} - n$$

Expanding by Binomial Theorem

$$(1 + X)^n = 1 + {}^nC_1X + {}^nC_2X^2 + {}^nC_3X^3 + \ldots$$

$$u_n = n\left(1 + \frac{1}{3}\cdot\frac{1}{n^3} + \frac{\frac{1}{3}\left(\frac{1}{3}-1\right)}{2!}\left(\frac{1}{n^3}\right)^2 + \ldots - 1\right)$$

$$= \frac{n}{n^3}\left[\frac{1}{3} - \frac{1}{9n^3} + \ldots\right] = \frac{1}{n^2}\left[\frac{1}{3} - \frac{1}{9n^3} + \ldots\right]$$

Take $$v_n = \frac{1}{n^2}$$

$$\Sigma v_n = \sum \frac{1}{n^p} = \sum \frac{1}{n^2}$$

$p = 2 > 1$, by p-Test Σv_n is convergent

$$\underset{n\to\infty}{\text{Lt}} \frac{u_n}{v_n} = \underset{n\to\infty}{\text{Lt}} \frac{\frac{1}{n^2}\left[\frac{1}{3} - \frac{1}{9n^3} + \ldots\right]}{\frac{1}{n^2}} = \frac{1}{3} - 0 + 0 - \ldots = \frac{1}{3} \neq 0$$

By comparison test, Σu_n also convergent.

Note. Rationalisation is effective only when square roots are involved whereas this method of Binomial Expansion is general.

Example 10. *Test the convergence of the series* $1 + \frac{1}{2^2} + \frac{2^2}{3^3} + \frac{3^3}{4^4} + \frac{4^4}{5^5} + \ldots$.

Sol. Leaving first term, we have $\frac{1^1}{2^2} + \frac{2^2}{3^3} + \frac{3^3}{4^4} + \ldots$

$$u_n = \frac{n^n}{(n+1)^{n+1}} = \frac{n^n}{(n+1)^n(n+1)}$$

($\because$ 1, 2, 3, ..., n, ... in A.P., 2, 3, 4, ..., $(n + 1)$, ... in A.P.)

Take
$$v_n = \frac{n^n}{n^n \cdot n} = \frac{1}{n}$$

$$\Sigma v_n = \sum \frac{1}{n^p} = \sum \frac{1}{n}$$

$p = 1$, By p-Test Σv_n is divergent.

$$\underset{n\to\infty}{\text{Lt}} \frac{u_n}{v_n} = \underset{n\to\infty}{\text{Lt}} \frac{n^n}{(n+1)^n(n+1)} \times \frac{n}{1}$$

$$= \underset{n\to\infty}{\text{Lt}} \frac{n^n \cdot n}{n^n\left(1+\frac{1}{n}\right)^n \cdot n\left(1+\frac{1}{n}\right)} = \frac{1}{\underset{n\to\infty}{\text{Lt}}\left(1+\frac{1}{n}\right)^n \cdot \underset{n\to\infty}{\text{Lt}}\left(1+\frac{1}{n}\right)}$$

$$= \frac{1}{e \cdot (1+0)} \qquad \left(\because \underset{n\to\infty}{\text{Lt}}\left(1+\frac{1}{n}\right)^n = e\right)$$

$$= \frac{1}{e} = \frac{1}{2.7183} \neq 0$$

By comparison test, Σu_n also divergent.

Example 11. *Test the convergence of the series:* $\sum_{n=1}^{\infty} \frac{1}{\sqrt{n} + \sqrt{n+1}}$.

Sol. We have $u_n = \frac{1}{\sqrt{n} + \sqrt{n+1}}$

Take $$v_n = \frac{1}{\sqrt{n}} = \frac{1}{n^{1/2}}$$

$$\Sigma v_n = \sum \frac{1}{n^p} = \sum \frac{1}{n^{1/2}}$$

$p = \frac{1}{2} = 0.5 < 1$, By p-Test Σv_n is divergent.

$$\underset{n\to\infty}{\text{Lt}} \frac{u_n}{v_n} = \underset{n\to\infty}{\text{Lt}} \frac{1}{\sqrt{n}+\sqrt{n+1}} \times \frac{\sqrt{n}}{1}$$

$$= \underset{n\to\infty}{\text{Lt}} \frac{\sqrt{n}}{\sqrt{n}\left(1+\sqrt{1+\frac{1}{n}}\right)} = \frac{1}{1+\sqrt{1+0}} = \frac{1}{2} \neq 0$$

By comparison test, Σu_n also divergent.

Example 12. *Discuss the convergence of* $\sum \frac{1}{\sqrt{n}} \sin \frac{1}{n}$.

Sol. Here, $$u_n = \frac{1}{\sqrt{n}} \sin \frac{1}{n}$$

Take $$v_n = \frac{1}{\sqrt{n}} \cdot \frac{1}{n} = \frac{1}{n\sqrt{n}} = \frac{1}{n^{3/2}}$$

$$\Sigma v_n = \sum \frac{1}{n^p} = \sum \frac{1}{n^{3/2}}$$

$p = \frac{3}{2} = 1.5 > 1$, By p-Test Σv_n is convergent

$$\underset{n\to\infty}{\text{Lt}} \frac{u_n}{v_n} = \underset{n\to\infty}{\text{Lt}} \frac{\frac{1}{\sqrt{n}} \sin \frac{1}{n}}{\frac{1}{n\sqrt{n}}}$$

$$= \underset{\frac{1}{n}\to 0}{\text{Lt}} \frac{\sin\left(\frac{1}{n}\right)}{\left(\frac{1}{n}\right)} = 1 \neq 0 \qquad \left(\because \underset{\theta\to 0}{\text{Lt}} \frac{\sin\theta}{\theta} = 1\right)$$

By comparison test, Σu_n also convergent.

Example 13. *Test for convergence* $\sum \frac{1}{n} \tan \frac{1}{n}$.

Sol. Here, $$u_n = \frac{1}{n} \tan \frac{1}{n}$$

Take $$v_n = \frac{1}{n^2}$$

$$\Sigma v_n = \sum \frac{1}{n^p} = \sum \frac{1}{n^2}$$

$p = 2 > 1$, By p-Test Σv_n is convergent

$$\underset{n\to\infty}{\text{Lt}} \frac{u_n}{v_n} = \underset{n\to\infty}{\text{Lt}} \frac{\frac{1}{n}\tan\frac{1}{n}}{\frac{1}{n^2}} = \underset{\frac{1}{n}\to 0}{\text{Lt}} \frac{\tan\left(\frac{1}{n}\right)}{\left(\frac{1}{n}\right)} = 1 \neq 0 \qquad \left(\because \underset{\theta\to 0}{\text{Lt}} \frac{\tan\theta}{\theta} = 1\right)$$

By comparison test, Σu_n also convergent.

Example 14. *Test for convergence of* $\sum \frac{1}{2^n + 3^n}$.

Sol. Let
$$u_n = \frac{1}{2^n + 3^n} = \frac{1}{3^n\left(1 + \frac{2^n}{3^n}\right)} = \frac{1}{3^n}\cdot\frac{1}{1 + \left(\frac{2}{3}\right)^n}$$

Take
$$v_n = \frac{1}{3^n}$$

Then $\Sigma v_n = \sum \left(\frac{1}{3}\right)^n$ is a geometric series with $r = \frac{1}{3} < 1$

$\therefore\ \Sigma v_n$ is convergent

$$\underset{n\to\infty}{\text{Lt}} \frac{u_n}{v_n} = \underset{n\to\infty}{\text{Lt}} \frac{\frac{1}{3^n}\cdot\frac{1}{1+\left(\frac{2}{3}\right)^n}}{\frac{1}{3^n}} = 1 \neq 0$$

$$\left(\because\ 0 < r < 1,\ \underset{n\to\infty}{\text{Lt}}\, r^n = 0. \text{ Hence } \underset{n\to\infty}{\text{Lt}} \left(\frac{2}{3}\right)^n = 0\right)$$

By comparison test, Σu_n also convergent.

Example 15. *Test for convergence* $\sum \left(\frac{2^n + 3}{3^n + 1}\right)^{\frac{1}{2}}$.

Sol.
$$u_n = \sqrt{\frac{2^n + 3}{3^n + 1}} = \sqrt{\frac{2^n\left(1 + \frac{3}{2^n}\right)}{3^n\left(1 + \frac{1}{3^n}\right)}} = \sqrt{\frac{2^n}{3^n}}\cdot\sqrt{\frac{1 + \frac{3}{2^n}}{1 + \frac{1}{3^n}}}$$

$$v_n = \sqrt{\frac{2^n}{3^n}} = \left(\sqrt{\frac{2}{3}}\right)^n$$

$$\Sigma v_n = \sum \left(\sqrt{\frac{2}{3}}\right)^n \text{ is a geometric series with } r = \sqrt{\frac{2}{3}} < 1$$

$\therefore$ Σv_n is convergent

$$\underset{n\to\infty}{\text{Lt}} \frac{u_n}{v_n} = \underset{n\to\infty}{\text{Lt}} \frac{\sqrt{\frac{2^n}{3^n}} \cdot \sqrt{\frac{1+\frac{3}{2^n}}{1+\frac{1}{3^n}}}}{\sqrt{\frac{2^n}{3^n}}} = \sqrt{\frac{1+\frac{3}{2^\infty}}{1+\frac{1}{3^\infty}}}$$

$$= \sqrt{\frac{1+\frac{3}{\infty}}{1+\frac{1}{\infty}}} = \sqrt{\frac{1+0}{1+0}} = 1 \neq 0$$

By comparison test Σu_n also convergent.

Example 16. *Examine the convergence of the series:*

(*i*) $\frac{3}{5} + \frac{4}{5^2} + \frac{3}{5^3} + \frac{4}{5^4} + \ldots\ldots \text{ to } \infty$ (*ii*) $1 + \frac{1}{4^{2/3}} + \frac{1}{9^{2/3}} + \frac{1}{16^{2/3}} + \ldots\ldots$.

Sol. (*i*) $\frac{3}{5} + \frac{4}{5^2} + \frac{3}{5^3} + \frac{4}{5^4} + \ldots\ldots \text{ to } \infty$

$$= \left(\frac{3}{5} + \frac{3}{5^3} + \ldots\ldots \text{ to } \infty\right) + \left(\frac{4}{5^2} + \frac{4}{5^4} + \ldots\ldots \text{ to } \infty\right) = \Sigma u_n + \Sigma v_n \text{ (say)}$$

Now Σu_n is a G.P. with common ratio $= \frac{1}{5^2}$ which is numerically less than 1,

$\therefore$ Σu_n is convergent.

Σv_n is also a G.P. with common ratio $= \frac{1}{5^2}$ which is numerically less than 1.

$\therefore$ Σv_n is convergent.

$\therefore$ The given series *viz*. $\Sigma(u_n + v_n)$ is also convergent.

(*ii*) $1 + \frac{1}{4^{2/3}} + \frac{1}{9^{2/3}} + \frac{1}{16^{2/3}} + \ldots\ldots \text{ to } \infty = 1 + \frac{1}{(2^2)^{2/3}} + \frac{1}{(3^2)^{2/3}} + \frac{1}{(4^2)^{2/3}} + \ldots\ldots \text{ to } \infty$

$$= \frac{1}{1^{4/3}} + \frac{1}{2^{4/3}} + \frac{1}{3^{4/3}} + \frac{1}{4^{4/3}} + \ldots\ldots \text{ to } \infty = \sum \frac{1}{n^{4/3}} = \sum \frac{1}{n^p} \text{ with } p = \frac{4}{3} > 1$$

$\therefore$ By *p*-series test, the given series is convergent.

Example 17. *Examine the convergence of the series:* $\frac{\sqrt{2}-1}{3^3-1} + \frac{\sqrt{3}-1}{4^3-1} + \frac{\sqrt{4}-1}{5^3-1} + \ldots\ldots$

Sol. Here
$$u_n = \frac{\sqrt{n+1}-1}{(n+2)^3-1} = \frac{\sqrt{n}\left(\sqrt{1+\frac{1}{n}} - \frac{1}{\sqrt{n}}\right)}{n^3\left[\left(1+\frac{2}{n}\right)^3 - \frac{1}{n^3}\right]} = \frac{\sqrt{1+\frac{1}{n}} - \frac{1}{\sqrt{n}}}{n^{5/2}\left[\left(1+\frac{2}{n}\right)^3 - \frac{1}{n^3}\right]}$$

Take $$v_n = \frac{1}{n^{5/2}}$$

$$\underset{n\to\infty}{\text{Lt}} \frac{u_n}{v_n} = \underset{n\to\infty}{\text{Lt}} \frac{\sqrt{1+\frac{1}{n}} - \frac{1}{\sqrt{n}}}{\left(1+\frac{2}{n}\right)^3 - \frac{1}{n^3}} = \frac{\sqrt{1+0} - 0}{(1+0)^3 - 0} = 1, \quad \text{which is finite and} \neq 0.$$

$\therefore$ Σu_n and Σv_n converge or diverge together.

Since $\Sigma v_n = \sum \frac{1}{n^{5/2}}$ is of the form $\sum \frac{1}{n^p}$ with $p = \frac{5}{2} > 1$.

Σv_n is convergent $\Rightarrow$ Σu_n is convergent.

Example 18. *Examine the convergence of the series:* $\sum \frac{\sqrt{n+1} - \sqrt{n}}{n^p}$.

Sol. Here
$$u_n = \frac{\sqrt{n+1} - \sqrt{n}}{n^p} = \frac{\sqrt{n+1} - \sqrt{n}}{n^p} \times \frac{\sqrt{n+1} + \sqrt{n}}{\sqrt{n+1} + \sqrt{n}} \quad \text{(Rationalising)}$$
$$= \frac{(n+1) - n}{n^p \cdot \sqrt{n}\left(\sqrt{1+\frac{1}{n}} + 1\right)} = \frac{1}{n^{p+1/2}\left(\sqrt{1+\frac{1}{n}} + 1\right)}$$

Take $$v_n = \frac{1}{n^{p+1/2}}$$

$$\underset{n\to\infty}{\text{Lt}} \frac{u_n}{v_n} = \underset{n\to\infty}{\text{Lt}} \frac{1}{\sqrt{1+\frac{1}{n}} + 1} = \frac{1}{1+1} = \frac{1}{2} \quad \text{which is finite and} \neq 0.$$

$\therefore$ Σu_n and Σv_n converge or diverge together.

Since $\Sigma v_n = \sum \frac{1}{n^{p+1/2}}$ is convergent if $p + \frac{1}{2} > 1$ and divergent if $p + \frac{1}{2} \leq 1$.

i.e., convergent if $p > \frac{1}{2}$ and divergent if $p \leq \frac{1}{2}$.

$\therefore$ Σu_n is convergent if $p > \frac{1}{2}$ and divergent if $p \leq \frac{1}{2}$.

EXERCISE 3

Test the convergence or divergence of the following series :

1. $\frac{1}{2} + \frac{1}{3^2} + \frac{1}{2^3} + \frac{1}{3^4} + \ldots\ldots \text{ to } \infty$

2. $\frac{1}{1.2} + \frac{1}{2.3} + \frac{1}{3.4} + \ldots\ldots \text{ to } \infty$

3. $\frac{1}{1.4} + \frac{1}{2.5} + \frac{1}{3.6} + \ldots\ldots \text{ to } \infty$

4. $\frac{1}{1.2} + \frac{2}{3.4} + \frac{3}{5.6} + \ldots\ldots \text{ to } \infty$

5. $\frac{1}{1.3} + \frac{2}{3.5} + \frac{3}{5.7} + \ldots\ldots \text{ to } \infty$

6. $\frac{1}{\sqrt{1.2}} + \frac{1}{\sqrt{2.3}} + \frac{1}{\sqrt{3.4}} + \ldots\ldots \text{ to } \infty$

7. $\frac{1}{1^p} + \frac{1}{3^p} + \frac{1}{5^p} + \ldots\ldots$ to ∞

8. $\frac{2}{1^p} + \frac{3}{2^p} + \frac{4}{3^p} + \ldots\ldots$ to ∞

9. $\sum_{n=1}^{\infty} \frac{n+1}{n(2n-1)}$

10. $\sum_{n=1}^{\infty} \frac{1}{n^p (n+1)^p}$

11. $\frac{2^p}{1^q} + \frac{3^p}{2^q} + \frac{4^p}{3^q} + \ldots\ldots$ to ∞

(*p* and *q* are positive numbers)

12. $\sum_{n=1}^{\infty} \frac{\sqrt{n}}{n^2+1}$

13. $\sum \frac{\sqrt{n^2-1}}{n^3+1}$

14. $\frac{1}{5} + \frac{\sqrt{2}}{7} + \frac{\sqrt{3}}{9} + \frac{\sqrt{4}}{11} + \ldots\ldots$ to ∞

15. $\frac{\sqrt{2}-\sqrt{1}}{1} + \frac{\sqrt{3}-\sqrt{2}}{2} + \frac{\sqrt{4}-\sqrt{3}}{3} + \ldots\ldots$

16. $\sum \left(\sqrt[3]{n+1} - \sqrt[3]{n}\right)$

17. $\sum \cot^{-1} n^2$

18. $\sum \left(\frac{1}{n} - \log \frac{n+1}{n}\right)$

19. $\sum_{n=2}^{\infty} \frac{1}{n \log n}$

20. $\sum \frac{1}{n^2+1}$. *(OU-Apr. 2016)*

Answers

1. Convergent
2. Convergent
3. Convergent
4. Divergent
5. Divergent
6. Divergent
7. Convergent for $p > 1$, divergent for $p \le 1$
8. Convergent for $p > 2$, divergent for $p \le 2$
9. Divergent
10. Convergent for $p > \frac{1}{2}$, divergent for $p \le \frac{1}{2}$
11. Convergent for $q > p + 1$, divergent for $q \le p + 1$
12. Convergent
13. Convergent
14. Divergent
15. Convergent
16. Divergent
17. Convergent
18. Convergent
19. Divergent
20. Convergent.

Theorem 5. D' Alembert's Ratio Test

Statement. *If Σu_n is a positive term series, and* $\underset{n \to \infty}{\text{Lt}} \frac{u_{n+1}}{u_n} = l$*, then*

(i) Σu_n is convergent if $l < 1$. *(ii) Σu_n is divergent if $l > 1$.*

Note. If $l = 1$, the test fails, *i.e.*, no conclusion can be drawn about the convergence or divergence of the series. The series may converge, it may diverge.

***Proof.** Since $\underset{n \to \infty}{\text{Lt}} \frac{u_{n+1}}{u_n} = l$,

∴ Given $\varepsilon > 0$, however small, there exists a positive integer *m* such that

$$\left| \frac{u_{n+1}}{u_n} - l \right| < \varepsilon \quad \forall\, n \ge m$$

$$\Rightarrow \qquad l - \varepsilon < \frac{u_{n+1}}{u_n} < l + \varepsilon \quad \forall\, n \ge m \qquad \ldots(i)$$

*Proof is for student reference only.

Case 1: When $l < 1$.

Choose $\varepsilon > 0$ such that $l < l + \varepsilon < 1$

Put $l + \varepsilon = r$, then $0 < r < 1$.

From (*i*), $\dfrac{u_{n+1}}{u_n} < r \ \forall\, n \geq m \ \Rightarrow\ u_{n+1} < ru_n \ \forall\, n \geq m$ $\qquad (\because\ u_n > 0)$

Putting $n = m, m + 1, m + 2, \ldots\ldots$, we get

$$u_{m+1} < ru_m$$
$$u_{m+2} < ru_{m+1} < r^2u_m$$
$$u_{m+3} < ru_{m+2} < r^3u_m \quad \text{and so on.}$$

Adding $u_{m+1} + u_{m+2} + u_{m+3} + \ldots\ldots < u_m\,(r + r^2 + r^3 + \ldots\ldots)$

$\Rightarrow$ each term of the given series Σu_n after leaving the first m terms (*i.e.*, a finite number of terms) is less than the corresponding term of a geometric series which is convergent ($\because$ its common ratio $r < 1$). Hence the given series is also convergent.

Case 2: When $l > 1$.

Choose $\varepsilon > 0$ such that $l - \varepsilon > 1$

Put $l - \varepsilon = \text{R}$, then $\text{R} > 1$.

From (*i*), $\dfrac{u_{n+1}}{u_n} > \text{R} \ \forall\, n \geq m \ \Rightarrow\ u_{n+1} > \text{R}u_n \ \forall\ n \geq m$ $\qquad (\because\ u_n > 0)$

Putting $n = m, m + 1, m + 2, \ldots\ldots$, we get

$$u_{m+1} > \text{R}u_m$$
$$u_{m+2} > \text{R}u_{m+1} > \text{R}^2u_m$$
$$u_{m+3} > \text{R}u_{m+2} > \text{R}^3u_m \quad \text{and so on.}$$

Adding $u_{m+1} + u_{m+2} + u_{m+3} + \ldots\ldots > u_m\,(\text{R} + \text{R}^2 + \text{R}^3 + \ldots\ldots)$

$\Rightarrow$ each term of the given series Σu_n after leaving the first m terms (*i.e.*, a finite number of terms) is greater than the corresponding term of a geometric series which is divergent ($\because$ its common ratio $\text{R} > 1$). Hence the given series is also divergent.

Practical Form of D' Alembert's Ratio Test:

In practice, Ratio Test is used in the following form :

If Σu_n is a positive term series, and $\underset{n\to\infty}{\text{Lt}} \dfrac{u_n}{u_{n+1}} = l$, then

(*i*) Σu_n is convergent if $l > 1$ $\qquad$ (*ii*) Σu_n is divergent if $l < 1$.

Note. In the following cases, D' Alembert's Ratio Test can be applied.

Case 1: When u_n contains factorial terms.

Example. $u_n = 1 + \dfrac{2^2}{2!} + \dfrac{3^2}{3!} + \dfrac{4^2}{4!} + \ldots\infty$.

Case 2: When u_n involves x^n term.

Example. $u_n = \dfrac{x^{2n}}{(n+1)\sqrt{n}}$.

Case 3: When u_n involves only one power term or same base terms.

Example 1. $u_n = \dfrac{3n-1}{2^n}$.

Example 2. $u_n = \frac{2^{n+1}-2}{2^n+1}x^{n-1}$.

Case 4: When u_n involves infinite numbers.

Example. $u_n = \frac{1\cdot 2\cdot 3\ldots n}{3\cdot 5\cdot 7\ldots(2n+1)}$.

ILLUSTRATIVE EXAMPLES

Example 1. *Discuss the convergence of the following series:*

(*i*) $1+\frac{2^p}{2!}+\frac{3^p}{3!}+\frac{4^p}{4!}+\ldots\ldots,(p>0)$ (*ii*) $\frac{1}{2}+\frac{1}{3}+\frac{1}{5}+\ldots\ldots+\frac{1}{2^{n-1}+1}+\ldots\ldots$

(*iii*) $\frac{1^2.2^2}{1!}+\frac{2^2.3^2}{2!}+\frac{3^2.4^2}{3!}+\frac{4^2.5^2}{4!}+\ldots\ldots$

Sol. (*i*) Here $u_n = \frac{n^p}{n!}$ $\left[\because\ 1=\frac{1^p}{1!}\right]$

$\therefore$ $u_{n+1} = \frac{(n+1)^p}{(n+1)!}$

$$\frac{u_n}{u_{n+1}} = \frac{n^p}{n!}\,\frac{(n+1)!}{(n+1)^p} = \frac{n^p.(n+1)\,n!}{n!\,(n+1)^p} = \frac{n^p}{(n+1)^{p-1}}$$

$$= \frac{n^p}{n^{p-1}\left(1+\frac{1}{n}\right)^{p-1}} = \frac{n}{\left(1+\frac{1}{n}\right)^{p-1}}$$

$$\underset{n\to\infty}{\text{Lt}}\frac{u_n}{u_{n+1}} = \underset{n\to\infty}{\text{Lt}}\frac{n}{\left(1+\frac{1}{n}\right)^{p-1}} = \infty > 1$$

$\therefore$ By D'Alembert's Ratio Test, Σu_n is convergent.

(*ii*) Here $u_n = \frac{1}{2^{n-1}+1}$ $\therefore$ $u_{n+1} = \frac{1}{2^n+1}$

$$\frac{u_n}{u_{n+1}} = \frac{2^n+1}{2^{n-1}+1} = \frac{2^n\left(1+\frac{1}{2^n}\right)}{2^{n-1}\left(1+\frac{1}{2^{n-1}}\right)} = 2.\frac{1+\frac{1}{2^n}}{1+\frac{1}{2^{n-1}}}$$

$$\underset{n\to\infty}{\text{Lt}}\frac{u_n}{u_{n+1}} = \underset{n\to\infty}{\text{Lt}}\ 2.\frac{1+\frac{1}{2^n}}{1+\frac{1}{2^{n-1}}} = 2 > 1$$

$\therefore$ By D'Alembert's Ratio Test, Σu_n is convergent.

(*iii*) Here $u_n = \frac{n^2(n+1)^2}{n!}$ $\therefore$ $u_{n+1} = \frac{(n+1)^2(n+2)^2}{(n+1)!}$

$$\frac{u_n}{u_{n+1}} = \frac{n^2 (n+1)^2}{n\,!} \cdot \frac{(n+1)!}{(n+1)^2 (n+2)^2} = \frac{n^2 \,.\, (n+1)\, n\,!}{n\,!\,(n+2)^2}$$

$$= \frac{n^3\left(1+\frac{1}{n}\right)}{n^2\left(1+\frac{2}{n}\right)^2} = n \,.\, \frac{1+\frac{1}{n}}{\left(1+\frac{2}{n}\right)^2}$$

$$\underset{n\to\infty}{\text{Lt}} \frac{u_n}{u_{n+1}} = \underset{n\to\infty}{\text{Lt}}\; n \,.\, \frac{1+\frac{1}{n}}{\left(1+\frac{2}{n}\right)^2} = \infty > 1$$

$\therefore$ By D'Alembert's Ratio Test, Σu_n is convergent.

Example 2. *Test the convergence of the following series:*

(*i*) $\sum \frac{n^3 + a}{2^n + a}$ (OU-May 2012) (*ii*) $\sum \frac{n\,!\,2^n}{n^n}$. (OU-Jan. 2012)

Sol. (*i*) Here $u_n = \frac{n^3 + a}{2^n + a} \quad \therefore \quad u_{n+1} = \frac{(n+1)^3 + a}{2^{n+1} + a}$

$$\frac{u_n}{u_{n+1}} = \frac{n^3 + a}{(n+1)^3 + a} \cdot \frac{2^{n+1} + a}{2^n + a}$$

$$= \frac{n^3\left(1+\frac{a}{n^3}\right)}{n^3\left[\left(1+\frac{1}{n}\right)^3 + \frac{a}{n^3}\right]} \cdot \frac{2^{n+1}\left(1+\frac{a}{2^{n+1}}\right)}{2^n\left(1+\frac{a}{2^n}\right)} = \frac{1+\frac{a}{n^3}}{\left(1+\frac{1}{n}\right)^3 + \frac{a}{n^3}} \cdot \frac{2\left(1+\frac{a}{2^{n+1}}\right)}{1+\frac{a}{2^n}}$$

$$\underset{n\to\infty}{\text{Lt}} \frac{u_n}{u_{n+1}} = \frac{1+0}{1+0} \,.\, 2 \,.\, \frac{1+0}{1+0} = 2 > 1$$

$\therefore$ By D'Alembert's Ratio Test, Σu_n is convergent.

(*ii*) Here $u_n = \frac{n\,!\,2^n}{n^n} \quad \therefore \quad u_{n+1} = \frac{(n+1)\,!\,2^{n+1}}{(n+1)^{n+1}}$

$$\frac{u_n}{u_{n+1}} = \frac{n\,!\,2^n}{(n+1)\,!\,2^{n+1}} \cdot \frac{(n+1)^{n+1}}{n^n} = \frac{1}{2(n+1)} \cdot \frac{(n+1)^{n+1}}{n^n}$$

$$= \frac{1}{2} \cdot \frac{(n+1)^n}{n^n} = \frac{1}{2} \cdot \left(\frac{n+1}{n}\right)^n = \frac{1}{2}\left(1+\frac{1}{n}\right)^n$$

$$\underset{n\to\infty}{\text{Lt}} \frac{u_n}{u_{n+1}} = \underset{n\to\infty}{\text{Lt}} \frac{1}{2}\left(1+\frac{1}{n}\right)^n = \frac{e}{2} \qquad (\because \; e = 2.7183)$$

Now, $2 < e < 3 \;\Rightarrow\; 1 < \frac{e}{2} < \frac{3}{2}$

$$\underset{n\to\infty}{\text{Lt}} \frac{u_n}{u_{n+1}} = \frac{e}{2} > 1 \;\Rightarrow\; \Sigma u_n \text{ is convergent.}$$

Example 3. *Discuss the convergence of the series:* $\sum \frac{n!}{n^n}$.

Sol. Here
$$u_n = \frac{n!}{n^n}$$
$$u_{n+1} = \frac{(n+1)!}{(n+1)^{n+1}} = \frac{(n+1)\,n!}{(n+1)^{n+1}} = \frac{n!}{(n+1)^n}$$
$$\frac{u_n}{u_{n+1}} = \frac{(n+1)^n}{n^n} = \left(\frac{n+1}{n}\right)^n = \left(1+\frac{1}{n}\right)^n$$
$$\underset{n\to\infty}{\text{Lt}} \frac{u_n}{u_{n+1}} = \underset{n\to\infty}{\text{Lt}} \left(1+\frac{1}{n}\right)^n = e > 1 \qquad (\because \; 2 < e < 3)$$

$\therefore$ By D'Alembert's Ratio Test, Σu_n is convergent.

Example 4. *Discuss the convergence of the series:* $\sum \frac{\sqrt{n}}{\sqrt{n^2+1}} x^n$.

Sol. Here
$$u_n = \sqrt{\frac{n}{n^2+1}}\, x^n$$
$\therefore$
$$u_{n+1} = \sqrt{\frac{n+1}{(n+1)^2+1}}\,.\, x^{n+1}$$
$$\frac{u_n}{u_{n+1}} = \sqrt{\frac{n}{n+1}\cdot\frac{n^2+2n+2}{n^2+1}}\,.\,\frac{1}{x} = \sqrt{\frac{1}{1+\frac{1}{n}}\cdot\frac{1+\frac{2}{n}+\frac{2}{n^2}}{1+\frac{1}{n^2}}}\,.\,\frac{1}{x}$$
$$\underset{n\to\infty}{\text{Lt}} \frac{u_n}{u_{n+1}} = \underset{n\to\infty}{\text{Lt}} \sqrt{\frac{1}{1+\frac{1}{n}}\cdot\frac{1+\frac{2}{n}+\frac{2}{n^2}}{1+\frac{1}{n^2}}}\,.\,\frac{1}{x} = \frac{1}{x}$$

$\therefore$ By D'Alembert's Ratio Test, Σu_n converges if $\frac{1}{x} > 1$ *i.e.,* $x < 1$ and diverges if $\frac{1}{x} < 1$ *i.e.,* $x > 1$

When $x = 1$, the Ratio Test fails.

When $x = 1$,
$$u_n = \sqrt{\frac{n}{n^2+1}} = \sqrt{\frac{n}{n^2\left(1+\frac{1}{n^2}\right)}} = \frac{1}{\sqrt{n}}\cdot\frac{1}{\sqrt{1+\frac{1}{n^2}}}$$

Take $v_n = \frac{1}{\sqrt{n}}$, $\underset{n\to\infty}{\text{Lt}} \frac{u_n}{v_n} = \underset{n\to\infty}{\text{Lt}} \frac{1}{\sqrt{1+\frac{1}{n^2}}} = 1$

which is finite and $\neq 0$.

$\therefore$ By Comparison Test, Σu_n and Σv_n converge or diverge together.

Since $\Sigma v_n = \sum \frac{1}{\sqrt{n}}$ is of the form $\sum \frac{1}{n^p}$ with $p = \frac{1}{2}$ (< 1),

Σv_n diverges $\Rightarrow$ Σu_n diverges.

Hence the given series Σu_n converges if $x < 1$ and diverges if $x \geq 1$.

Example 5. *Examine the convergence or divergence of the following series :*

$$\frac{1}{2\sqrt{1}} + \frac{x^2}{3\sqrt{2}} + \frac{x^4}{4\sqrt{3}} + \frac{x^6}{5\sqrt{4}} + \ldots\ldots \qquad \text{(OU-Apr. 2007, May 2008, Dec. 2008)}$$

Sol. Here $\quad u_n = \dfrac{x^{2n-2}}{(n+1)\sqrt{n}} \quad \therefore \quad u_{n+1} = \dfrac{x^{2n}}{(n+2)\sqrt{n+1}}$

$$\frac{u_n}{u_{n+1}} = \frac{(n+2)\sqrt{n+1}}{(n+1)\sqrt{n}} \cdot \frac{1}{x^2} = \frac{1+\dfrac{2}{n}}{1+\dfrac{1}{n}} \cdot \sqrt{1+\frac{1}{n}} \cdot \frac{1}{x^2}$$

$$\underset{n\to\infty}{\text{Lt}} \frac{u_n}{u_{n+1}} = \frac{1}{x^2}$$

By D'Alembert's Ratio Test, Σu_n converges if $\dfrac{1}{x^2} > 1$, *i.e.*, $x^2 < 1$

and diverges if $\dfrac{1}{x^2} < 1$ *i.e.*, $x^2 > 1$.

When $x^2 = 1$, $\quad u_n = \dfrac{1}{(n+1)\sqrt{n}} = \dfrac{1}{n^{3/2}\left(1+\dfrac{1}{n}\right)}$

Take $v_n = \dfrac{1}{n^{3/2}}$ By comparison test Σu_n is convergent $\qquad (\because \ p > 1)$

Hence Σu_n is convergent if $x^2 \leq 1$ and divergent if $x^2 > 1$.

Example 6. *Examine the convergence or divergence of the following series:*

$$1 + \frac{2}{5}x + \frac{6}{9}x^2 + \frac{14}{17}x^3 + \ldots\ldots + \frac{2^{n+1}-2}{2^{n+1}+1}x^n + \ldots\ldots \ (x > 0).$$

Sol. Here, leaving the first term, $\quad u_n = \dfrac{2^{n+1}-2}{2^{n+1}+1}x^n$

$$\therefore \qquad u_{n+1} = \frac{2^{n+2}-2}{2^{n+2}+1}x^{n+1}$$

$$\frac{u_n}{u_{n+1}} = \frac{2^{n+1}-2}{2^{n+1}+1} \cdot \frac{2^{n+2}+1}{2^{n+2}-2} \cdot \frac{1}{x} = \frac{2^{n+1}\left(1-\dfrac{2}{2^{n+1}}\right)}{2^{n+1}\left(1+\dfrac{1}{2^{n+1}}\right)} \cdot \frac{2^{n+2}\left(1+\dfrac{1}{2^{n+2}}\right)}{2^{n+2}\left(1-\dfrac{2}{2^{n+2}}\right)} \cdot \frac{1}{x}$$

$$\underset{n\to\infty}{\text{Lt}} \frac{u_n}{u_{n+1}} = \underset{n\to\infty}{\text{Lt}} \frac{1-\dfrac{1}{2^n}}{1+\dfrac{1}{2^{n+1}}} \cdot \frac{1+\dfrac{1}{2^{n+2}}}{1-\dfrac{1}{2^{n+1}}} \cdot \frac{1}{x} = \frac{1}{x}$$

$\therefore$ By D'Alembert's Ratio Test, Σu_n converges if $\dfrac{1}{x} > 1$, *i.e.*, $x < 1$

and diverges if $\dfrac{1}{x} < 1$ *i.e.*, $x > 1$ and Ratio Test fails if $x = 1$.

When $x = 1$, $$u_n = \frac{2^{n+1}-2}{2^{n+1}+1} = \frac{2^{n+1}\left(1-\frac{2}{2^{n+1}}\right)}{2^{n+1}\left(1+\frac{1}{2^{n+2}}\right)} = \frac{1-\frac{1}{2^n}}{1+\frac{1}{2^{n+1}}}$$

$\underset{n\to\infty}{\text{Lt}}\ u_n = 1 \neq 0 \Rightarrow \Sigma u_n$ does not converge. Being a series of +ve terms, it must diverge.

Hence Σu_n is convergent if $x < 1$ and divergent if $x \geq 1$.

Example 7. *Test for convergence the positive term series :*

$$1 + \frac{\alpha+1}{\beta+1} + \frac{(\alpha+1)(2\alpha+1)}{(\beta+1)(2\beta+1)} + \frac{(\alpha+1)(2\alpha+1)(3\alpha+1)}{(\beta+1)(2\beta+1)(3\beta+1)} + \ldots..$$

Sol. Leaving the first term, $u_n = \dfrac{(\alpha+1)(2\alpha+1)\ldots\ldots(n\alpha+1)}{(\beta+1)(2\beta+1)\ldots\ldots(n\beta+1)}$

$$\therefore \quad u_{n+1} = \frac{(\alpha+1)(2\alpha+1)\ldots\ldots(n\alpha+1)[(n+1)\alpha+1]}{(\beta+1)(2\beta+1)\ldots\ldots(n\beta+1)[(n+1)\beta+1]}$$

$$\frac{u_n}{u_{n+1}} = \frac{(n+1)\beta+1}{(n+1)\alpha+1} = \frac{\left(1+\frac{1}{n}\right)\beta+\frac{1}{n}}{\left(1+\frac{1}{n}\right)\alpha+\frac{1}{n}}$$

$$\underset{n\to\infty}{\text{Lt}} \frac{u_n}{u_{n+1}} = \underset{n\to\infty}{\text{Lt}} \frac{\left(1+\frac{1}{n}\right)\beta+\frac{1}{n}}{\left(1+\frac{1}{n}\right)\alpha+\frac{1}{n}} = \frac{\beta}{\alpha}$$

$\therefore$ By D'Alembert's Ratio Test, Σu_n converges if $\dfrac{\beta}{\alpha} > 1$ *i.e.,* $\beta > \alpha > 0$

and diverges if $\dfrac{\beta}{\alpha} < 1$ *i.e.,* $\beta < \alpha$ or $\alpha > \beta > 0$

When $\alpha = \beta$, the Ratio Test fails.

When $\alpha = \beta$, $u_n = 1 \quad \therefore \quad \underset{n\to\infty}{\text{Lt}}\ u_n = 1 \neq 0$

$\Rightarrow \Sigma u_n$ does not converge. Being a series of +ve terms, it must diverge.

Hence the given series is convergent if $\beta > \alpha > 0$ and divergent if $\alpha \geq \beta > 0$.

Example 8. *Test for convergence of the following series :* $\sum \dfrac{(n!)}{(2n)!} X^{2n}$.

(OU-Dec. 2013, June 2013)

Sol. Let $$u_n = \frac{(n!)^2}{(2n)!} X^{2n}$$

$$u_{n+1} = \frac{[(n+1)!]^2}{[2(n+1)]!} X^{2(n+1)} = \frac{[(n+1)n!]^2}{(2n+2)(2n)!} X^{2n+2}$$

$$\underset{n\to\infty}{\text{Lt}} \frac{u_n}{u_{n+1}} = \underset{n\to\infty}{\text{Lt}} \frac{(n!)^2}{(2n)!} \cdot \frac{(2n+2)(2n)!}{(n+1)^2(n!)^2} \cdot \frac{X^{2n}}{X^{2n}\cdot X^2}$$

$$= \underset{n\to\infty}{\text{Lt}} \frac{n\left(2+\frac{2}{n}\right)}{n^2\left(1+\frac{1}{n}\right)^2} \cdot \frac{1}{X^2} = \underset{n\to\infty}{\text{Lt}} \frac{1}{n} \cdot \frac{\left(2+\frac{2}{n}\right)}{\left(1+\frac{1}{n}\right)^2} \cdot \frac{1}{X^2}$$

$$= 0 < 1 \qquad \left(\because \underset{n\to\infty}{\text{Lt}} \frac{1}{n} = 0\right)$$

$\therefore$ By Ratio Test, Σu_n is divergent.

Example 9. *Test for convergence* $\sum_{1}^{\infty} \left[\frac{n^2}{2^n} + \frac{1}{n^2}\right]$. (OU-Jan. 2011)

Sol. Take $\sum \frac{n^2}{2^n}$

Let $u_n = \frac{n^2}{2^n}$

$$u_{n+1} = \frac{(n+1)^2}{2^{n+1}} = \frac{(n+1)^2}{2^n \cdot 2}$$

$$\underset{n\to\infty}{\text{Lt}} \frac{u_n}{u_{n+1}} = \underset{n\to\infty}{\text{Lt}} \frac{n^2}{2^n} \times \frac{2^n \cdot 2}{(n+1)^2} = \underset{n\to\infty}{\text{Lt}} \frac{2n^2}{n^2\left(1+\frac{1}{n}\right)^2}$$

$$= \frac{2}{(1+0)^2} = 2 > 1$$

$\therefore \quad \sum \frac{n^2}{2^n}$ is convergent ...(1)

By p-Test, $\sum \frac{1}{n^2}$ is convergent ...(2)

If Σu_n and Σv_n are convergent then $\Sigma(u_n + v_n)$ also convergent.

$\therefore$ From (1), (2), $\sum \left(\frac{n^2}{2^n} + \frac{1}{n^2}\right)$ is also convergent.

Example 10. *Test for convergence of* $\sum \frac{x^n}{n(n-1)(n-2)}$. (OU-Nov. 2006)

Sol. Let $u_n = \frac{x^n}{n(n-1)(n-2)}$

$$u_{n+1} = \frac{x^{n+1}}{(n+1)(n)(n-1)}$$

$$\underset{n\to\infty}{\text{Lt}} \frac{u_n}{u_{n+1}} = \underset{n\to\infty}{\text{Lt}} \frac{x^n}{n(n-1)(n-2)} \times \frac{(n+1)n(n-1)}{x^n \cdot x}$$

$$= \underset{n\to\infty}{\text{Lt}} \frac{n\left(1+\frac{1}{n}\right)n \cdot n\left(1-\frac{1}{n}\right)}{n \cdot n\left(1-\frac{1}{n}\right)n\left(1-\frac{2}{n}\right)x} = \frac{(1+0)(1-0)}{(1-0)(1-0)x} = \frac{1}{x}$$

If $\frac{1}{x} > 1 \;\Rightarrow\; 1 > x \;\Rightarrow\; x < 1$, then Σu_n is convergent

If $\frac{1}{x} < 1 \;\Rightarrow\; 1 < x \;\Rightarrow\; x > 1$, then Σu_n is divergent

If $\frac{1}{x} = 1 \;\Rightarrow\; x = 1$, then Ratio Test fails

when $x = 1$, $u_n = \frac{1}{n(n-1)(n-2)}$, then $v_n = \frac{1}{n^3}$.

By p-Test Σv_n is convergent

$$\underset{n\to\infty}{\text{Lt}} \frac{u_n}{v_n} = \underset{n\to\infty}{\text{Lt}} \frac{1}{n(n-1)(n-2)} \times \frac{n^3}{1} = 1 \neq 0$$

Σu_n also convergent.

$\therefore$ If $x \leq 1$, Σu_n is convergent.

If $x > 1$, Σu_n is divergent.

Example 11. *Discuss the convergence of the series:* $\frac{2}{1} + \frac{2 \cdot 5 \cdot 8}{1 \cdot 5 \cdot 9} + \frac{2 \cdot 5 \cdot 8 \cdot 11}{1 \cdot 5 \cdot 9 \cdot 13} + \ldots$ *to* ∞.

Sol. Here,
$$u_n = \frac{2 \cdot 5 \cdot 8 \cdot 11 \ldots (3n-1)}{1 \cdot 5 \cdot 9 \cdot 13 \ldots (4n-3)}$$

($\because$ 2, 5, 8, ... in A.P., $a_n = 2 + (n-1)3 = 3n - 1$
1, 5, 9, ... in A.P., $a_n = 1 + (n-1)4 = 4n - 3$)

$$u_{n+1} = \frac{2 \cdot 5 \cdot 8 \cdot 11 \ldots (3n-1)[3(n+1)-1]}{1 \cdot 5 \cdot 9 \cdot 13 \ldots (4n-3)[4(n+1)-3]}$$

$$\underset{n\to\infty}{\text{Lt}} \frac{u_n}{u_{n+1}} = \underset{n\to\infty}{\text{Lt}} \frac{4n+1}{3n+2} \qquad (\because \text{ After cancellation})$$

$$= \underset{n\to\infty}{\text{Lt}} \frac{n\left(4+\frac{1}{n}\right)}{n\left(3+\frac{2}{n}\right)} = \frac{4}{3} > 1$$

By Ratio Test, Σu_n is convergent.

Example 12. *Discuss the convergence of the series:* $\left(\frac{1}{3}\right)^2 + \left(\frac{1 \cdot 2}{3 \cdot 5}\right)^2 + \left(\frac{1 \cdot 2 \cdot 3}{3 \cdot 5 \cdot 7}\right)^2 + \left(\frac{1 \cdot 2 \cdot 3 \cdot 4}{3 \cdot 4 \cdot 7 \cdot 9}\right)^2 + \cdots$ *to* ∞.

Sol.
$$u_n = \left[\frac{1 \cdot 2 \cdot 3 \dots n}{3 \cdot 5 \cdot 7 \dots (2n+1)}\right]^2$$

$$u_{n+1} = \left[\frac{1 \cdot 2 \cdot 3 \dots n(n+1)}{3 \cdot 5 \cdot 7 \dots (2n+1)[2(n+1)+1]}\right]^2$$

$$\underset{n\to\infty}{\text{Lt}} \frac{u_n}{u_{n+1}} = \underset{n\to\infty}{\text{Lt}} \left(\frac{2n+3}{n+1}\right)^2 = \underset{n\to\infty}{\text{Lt}} \frac{n^2\left(2+\frac{3}{n}\right)^2}{n^2\left(1+\frac{1}{n}\right)^2} = 4 > 1$$

By Ratio Test, Σu_n is convergent.

Example 13. *Test the convergence of the series:* $\sum_{n=1}^{\infty} \frac{1}{X^n + X^{-n}}$. (OU-May 2008)

Sol. Let
$$u_n = \frac{1}{X^n + X^{-n}},\ u_{n+1} = \frac{1}{X^{n+1} + X^{-(n+1)}}$$

$$\underset{n\to\infty}{\text{Lt}} \frac{u_n}{u_{n+1}} = \underset{n\to\infty}{\text{Lt}} \frac{X^n \cdot X + X^{-n} \cdot X^{-1}}{X^n + X^{-n}} = \underset{n\to\infty}{\text{Lt}} \frac{X^n\left(X + \frac{X^{-2n}}{X}\right)}{X^n\,(1 + X^{-2n})} = X \text{ if } X \neq 0$$

if $X > 1$, Σu_n is convergent

if $X < 1$, Σu_n is divergent,

if $X = 1$, Ratio test fails.

When $X = 1$, $u_n = \frac{1}{2}$

$$\underset{n\to\infty}{\text{Lt}}\, u_n = \underset{n\to\infty}{\text{Lt}} \frac{1}{2} = \frac{1}{2} \neq 0$$

The series is divergent.

Hence, the given series is convergent for $X > 1$ and divergent for $X \leq 1$.

Example 14. *Discuss the convergence of the series:* $\sum \frac{(n+1)^4}{n^{n+1}} X^n,\ X > 0.$

(OU-June 2010)

Sol. Let
$$u_n = \frac{(n+1)^4}{n^{n+1}} X^n,\ u_{n+1} = \frac{(n+2)^4}{(n+1)^{n+2}} X^{n+1}$$

$$\underset{n\to\infty}{\text{Lt}} \frac{u_n}{u_{n+1}} = \underset{n\to\infty}{\text{Lt}} \frac{(n+1)^4}{n^{n+1}} \cdot \frac{(n+1)^{n+2}}{(n+2)^4} \cdot \frac{X^n}{X^{n+1}}$$

$$= \underset{n\to\infty}{\text{Lt}} \frac{n^4\left(1+\frac{1}{n}\right)^4}{n^{n+1}} \cdot \frac{n^{n+1}\left(1+\frac{1}{n}\right)^{n+1}(n+1)}{n^4\left(1+\frac{2}{n}\right)^4} \cdot \frac{1}{X} = \infty > 1.$$

$\therefore$ Σu_n is convergent.

EXERCISE 4

Discuss the convergence of the following series:

1. $1+\frac{2^2}{2!}+\frac{3^2}{3!}+\frac{4^2}{4!}+\ldots\ldots$ to ∞

2. $1+\frac{2!}{2^2}+\frac{3!}{3^3}+\frac{4!}{4^4}+\ldots\ldots$ to ∞

3. $\frac{1}{1+2}+\frac{2}{1+2^2}+\frac{3}{1+2^3}+\ldots\ldots$ to ∞

4. $\frac{2!}{3}+\frac{3!}{3^2}+\frac{4!}{3^3}+\ldots\ldots$ to ∞

5. $\sum \frac{2^{n-1}}{3^n+1}$

6. $\sum \frac{1}{n!}$

7. $\sum \frac{n^2(n+1)^2}{n!}$

8. $\sum \frac{x^n}{3^n \cdot n^2}, x>0$

9. $\sum \frac{x^n}{n}, x>0$

10. $\sum \frac{n}{n^2+1} x^n, x>0$

11. $\sum \sqrt{\frac{n+1}{n^3+1}} \cdot x^n, x>0$

12. $x+2x^2+3x^3+4x^4+\ldots\ldots$ to ∞

13. $1+\frac{x}{2}+\frac{x^2}{5}+\frac{x^3}{10}+\ldots\ldots+\frac{x^n}{n^2+1}+\ldots\ldots$ to ∞

14. $\frac{x}{1.3}+\frac{x^2}{3.5}+\frac{x^3}{5.7}+\ldots\ldots$ to ∞

15. $x+\frac{3}{5}x^2+\frac{8}{10}x^3+\frac{15}{17}x^4+\ldots\ldots+\frac{n^2-1}{n^2+1}x^n+\ldots\ldots$ to ∞

16. $\frac{x}{2\sqrt{3}}+\frac{x^2}{3\sqrt{4}}+\frac{x^3}{4\sqrt{5}}+\ldots\ldots$ to ∞

17. $\sum \frac{3^n-2}{3^n+1} \cdot x^{n-1}, x>0$

18. $\sum_{n=1}^{\infty} \frac{n^2}{3^n}$

19. $\sum_{n=1}^{\infty} \frac{x^n}{(2n)!}$.

Answers

1. Convergent
2. Convergent
3. Convergent
4. Divergent
5. Convergent
6. Convergent
7. Convergent
8. Convergent for $x \le 3$, divergent for $x > 3$
9. Convergent for $x < 1$, divergent for $x \ge 1$
10. Convergent for $x < 1$, divergent for $x \ge 1$
11. Convergent for $x < 1$, divergent for $x \ge 1$
12. Convergent for $x < 1$, divergent for $x \ge 1$
13. Convergent for $x \le 1$, divergent for $x > 1$
14. Convergent for $x \le 1$, divergent for $x > 1$
15. Convergent for $x < 1$, divergent for $x \ge 1$
16. Convergent for $x \le 1$, divergent for $x > 1$
17. Convergent for $x < 1$, divergent for $x \ge 1$
18. Convergent
19. Convergent.

Theorem 6. Cauchy's *n*th Root Test

Statement. *If* Σu_n *is a positive term series and* $\underset{n\to\infty}{Lt} (u_n)^{1/n} = l$, *then*

(*i*) Σu_n *is convergent if* $l < 1$ (*ii*) Σu_n *is divergent if* $l > 1$.

Note. If $l = 1$, the test fails *i.e.*, no conclusion can be drawn about the convergence or divergence of the series. The series may converge, it may diverge.

*__Proof.__ Since $\underset{n \to \infty}{\text{Lt}} (u_n)^{1/n} = l.$

$\therefore$ Given $\varepsilon > 0$, however small, there exists a +ve integer m such that

$$| (u_n)^{1/n} - l | < \varepsilon \qquad \forall \quad n \geq m$$

$$\Rightarrow \qquad l - \varepsilon < (u_n)^{1/n} < l + \varepsilon \qquad \forall \quad n \geq m$$

$$\Rightarrow \qquad (l - \varepsilon)^n < u_n < (l + \varepsilon)^n \qquad \forall \quad n \geq m \qquad ...(1)$$

(*i*) **When $l < 1$**

Choose $\varepsilon > 0$ such that $l < l + \varepsilon < 1$

Put $l + \varepsilon = r$, then $0 < r < 1$

From (1), $u_n < r^n \quad \forall \quad n \geq m$

Putting $n = m, m + 1, m + 2, ...$, we get $u_m < r^m, u_{m+1} < r^{m+1}, u_{m+2} < r^{m+2}$, ... and so on.

Adding $u_m + u_{m+1} + u_{m+2} + ... < r^m + r^{m+1} + r^{m+2} + ...$

$\Rightarrow$ each term of the given series Σu_n after leaving the first $(m - 1)$ terms, (*i.e.*, a finite number of terms) is less than the corresponding term of a geometric series which is convergent ($\because$ its common ratio $r < 1$). Hence the given series is also convergent.

(*ii*) **When $l > 1$**

Choose $\varepsilon > 0$ such that $l - \varepsilon > 1$

Put $l - \varepsilon = \text{R}$, then $\text{R} > 1$

From (1), $u_n > \text{R}^n \quad \forall \quad n \geq m$

Putting $n = m, m + 1, m + 2, ...$, we get $u_m > \text{R}^m, u_{m+1} > \text{R}^{m+1}, u_{m+2} > \text{R}^{m+2}$, ... and so on.

Adding $u_m + u_{m+1} + u_{m+2} + ... > \text{R}^m + \text{R}^{m+1} + \text{R}^{m+2} + ...$

$\Rightarrow$ each term of the series Σu_n after leaving the first $(m - 1)$ terms, (*i.e.*, a finite number of terms) is greater than the corresponding term of a geometric series which is divergent.

($\because$ its common ratio $\text{R} > 1$). Hence the given series is also divergent.

Note. If $u_n = (\ )^n$ or $(\ )^{n^2}$ or $(\ )^{-n}$ or $(\ )^{-n^2}$, then Cauchy's Root Test can be applied.

Formulae:

1. $\underset{n\to\infty}{\text{Lt}} \left(1 + \frac{1}{n}\right)^n = e$ **2.** $\underset{n\to\infty}{\text{Lt}} \left(1 + \frac{p}{n}\right)^n = e^p$

3. $\underset{n\to\infty}{\text{Lt}} \left(1 + \frac{1}{\sqrt{n}}\right)^{\sqrt{n}} = e.$

ILLUSTRATIVE EXAMPLES

Example 1. *Test the convergence of the following series:*

$$\sum \left(\frac{n}{n+1}\right)^{n^2} \quad or \quad \sum \left(1 + \frac{1}{n}\right)^{-n^2}.$$

(OU-Jan. 2011, 2012, June 2013, Apr. 2016)

*Proof is for student reference only.

Sol. Here $$u_n = \left(\frac{n}{n+1}\right)^{n^2}$$

$\therefore$ $$(u_n)^{1/n} = \left[\left(\frac{n}{n+1}\right)^{n^2}\right]^{1/n} = \left(\frac{n}{n+1}\right)^{n} = \left(\frac{n+1}{n}\right)^{-n} = \left[\left(1+\frac{1}{n}\right)^{n}\right]^{-1}$$

$$\underset{n\to\infty}{\text{Lt}} (u_n)^{1/n} = \underset{n\to\infty}{\text{Lt}} \left[\left(1+\frac{1}{n}\right)^{n}\right]^{-1} = e^{-1} = \frac{1}{e} < 1 \qquad (\because \; e = 2.7)$$

$\therefore$ By Cauchy's Root Test, the given series Σu_n is convergent.

Example 2. *Examine the convergence of the series:*

$$\left(\frac{2^2}{1^2}-\frac{2}{1}\right)^{-1} + \left(\frac{3^3}{2^3}-\frac{3}{2}\right)^{-2} + \left(\frac{4^4}{3^4}-\frac{4}{3}\right)^{-3} + \ldots\ldots$$

Sol. Here $$u_n = \left[\left(\frac{n+1}{n}\right)^{n+1} - \frac{n+1}{n}\right]^{-n}$$

$$(u_n)^{1/n} = \left[\left(\frac{n+1}{n}\right)^{n+1} - \frac{n+1}{n}\right]^{-1} = \left[\left(1+\frac{1}{n}\right)^{n+1} - \left(1+\frac{1}{n}\right)\right]^{-1}$$

$$= \left[\left(1+\frac{1}{n}\right)^{n} . \left(1+\frac{1}{n}\right) - \left(1+\frac{1}{n}\right)\right]^{-1}$$

$$\underset{n\to\infty}{\text{Lt}} (u_n)^{1/n} = \underset{n\to\infty}{\text{Lt}} \left[\left(1+\frac{1}{n}\right)^{n} \left(1+\frac{1}{n}\right) - \left(1+\frac{1}{n}\right)\right]^{-1}$$

$$= (e.1 - 1)^{-1} = \frac{1}{e-1} < 1 \qquad (\because \; e = 2.7)$$

$\therefore$ By Cauchy's Root Test, Σu_n is convergent.

Example 3. *Discuss the convergence of the series:* $\sum \left(1+\frac{1}{n^p}\right)^{n^{P+1}}$, *$p > 0$.*

(OU-Dec. 2013)

Sol. Here $$u_n = \left(1+\frac{1}{n^p}\right)^{n^{p+1}} = \left(1+\frac{1}{n^p}\right)^{n^p \cdot n}$$

$$\underset{n\to\infty}{\text{Lt}} (u_n)^{1/n} = \underset{n\to\infty}{\text{Lt}} \left[\left(1+\frac{1}{n^p}\right)^{n^p \cdot n}\right]^{\frac{1}{n}} = e = 2.7183 > 1$$

By Cauchy's Root Test, $\sum u_n$ is divergent.

Example 4. *Test the series:* $\sum \frac{[(n+1)x]^n}{n^{n+1}}$ *for convergence.*

(OU-Jan. 2011, Apr. 2007, May 2012)

Sol. Let $$u_n = \frac{[(n+1)x]^n}{n^{n+1}} = \left[\left(\frac{n+1}{n}\right)x\right]^n \cdot \frac{1}{n}$$

$$\underset{n\to\infty}{\text{Lt}} (u_n)^{1/n} = \underset{n\to\infty}{\text{Lt}} \left[\left[\left(\frac{n+1}{n}\right)x\right]^n \cdot \frac{1}{n}\right]^{\frac{1}{n}} = \underset{n\to\infty}{\text{Lt}} \left(1+\frac{1}{n}\right)x \cdot \underset{n\to\infty}{\text{Lt}} \frac{1}{n^{1/n}}$$

$$= x \cdot 1 = x \qquad \left(\because \underset{n\to\infty}{\text{Lt}} \frac{1}{n^{1/n}} = 1\right)$$

By Cauchy's Root Test

(*i*) if $x < 1$, Σu_n is convergent,

(*ii*) if $x > 1$, Σu_n is divergent,

(*iii*) if $x = 1$, test fails.

When $x = 1$, $u_n = \left(\frac{n+1}{n}\right)^n \cdot \frac{1}{n}$

Let $v_n = \frac{1}{n}$

$$\Sigma v_n = \sum \frac{1}{n^p} = \sum \frac{1}{n} \quad \Rightarrow \quad p = 1$$

By p-Test, Σv_n is divergent

Now, $$\underset{n\to\infty}{\text{Lt}} \frac{u_n}{v_n} = \underset{n\to\infty}{\text{Lt}} \frac{\left(\frac{n+1}{n}\right)^n \cdot \frac{1}{n}}{\frac{1}{n}} = \underset{n\to\infty}{\text{Lt}} \left(1+\frac{1}{n}\right)^n = e \neq 0$$

By comparison test, Σu_n also divergent.

$\therefore$ if $x < 1$, Σu_n is convergent and if $x \geq 1$, Σu_n is divergent.

Example 5. *Test for the convergence of the series:* $\sum \left(1+\frac{1}{\sqrt{n}}\right)^{-n^{3/2}}$.

Sol. Here $$u_n = \left(1+\frac{1}{\sqrt{n}}\right)^{-n^{3/2}} = \left(1+\frac{1}{\sqrt{n}}\right)^{-n\sqrt{n}}$$

$$\underset{n\to\infty}{\text{Lt}} (u_n)^{1/n} = \underset{n\to\infty}{\text{Lt}} \left[\left(1+\frac{1}{\sqrt{n}}\right)^{-n\sqrt{n}}\right]^{\frac{1}{n}}$$

$$= \frac{1}{\underset{n\to\infty}{\text{Lt}} \left(1+\frac{1}{\sqrt{n}}\right)^{\sqrt{n}}} = \frac{1}{e} = \frac{1}{2.7183} < 1$$

$\therefore$ By Root Test, Σu_n is convergent.

EXERCISE 5

Discuss the convergence of the following series:

1. $\sum \dfrac{1}{n^n}$
2. $\sum_{n=2}^{\infty} \dfrac{1}{(\log n)^n}$
3. $\sum \left(\dfrac{n+1}{3n}\right)^n$
4. $\sum \dfrac{(n-\log n)^n}{2^n \cdot n^n}$
5. $\sum \left(\dfrac{nx}{n+1}\right)^n$
6. $\sum 5^{-n-(-1)^n}$
7. $\sum \dfrac{(1+nx)^n}{n^n}$
8. $\dfrac{1}{2} + \dfrac{2}{3}x + \left(\dfrac{3}{4}\right)^2 x^2 + \left(\dfrac{4}{5}\right)^3 x^3 + \ldots\ldots$ to ∞.

Answers

1. Convergent
2. Convergent
3. Convergent
4. Convergent
5. Convergent for $x < 1$, divergent for $x \geq 1$
6. Convergent
7. Convergent for $x < 1$, divergent for $x \geq 1$
8. Convergent for $x < 1$, divergent for $x \geq 1$.

Theorem 7. Raabe's Test

Statement. *If Σu_n is a series of positive terms and* $\underset{n\to\infty}{Lt}\ n\left(\dfrac{u_n}{u_{n+1}} - 1\right) = l$, *then the series is convergent if $l > 1$ and divergent if $l < 1$.*

***Proof.** Let us compare the given series Σu_n with an auxiliary series $\Sigma v_n = \Sigma \dfrac{1}{n^p}$ which we know converges if $p > 1$ and diverges if $p \leq 1$.

Now,
$$\frac{v_n}{v_{n+1}} = \frac{\dfrac{1}{n^p}}{\dfrac{1}{(n+1)^p}} = \left(\frac{n+1}{n}\right)^p = \left(1+\frac{1}{n}\right)^p = 1 + \frac{p}{n} + \frac{p(p-1)}{2!}\cdot\frac{1}{n^2} + \ldots\ldots$$

Case I: *Let $\Sigma v_n = \Sigma \dfrac{1}{n^p}$ be convergent, so that $p > 1$.*

Then Σu_n will also converge if $\dfrac{u_n}{u_{n+1}} > \dfrac{v_n}{v_{n+1}}$

or if
$$\frac{u_n}{u_{n+1}} > 1 + \frac{p}{n} + \frac{p(p-1)}{2!}\cdot\frac{1}{n^2} + \ldots\ldots$$

or if
$$n\left(\frac{u_n}{u_{n+1}} - 1\right) > p + \frac{p(p-1)}{2!}\cdot\frac{1}{n} + \ldots\ldots$$

or if
$$\underset{n\to\infty}{\text{Lt}}\ n\left(\frac{u_n}{u_{n+1}} - 1\right) > p \quad \text{or} \quad \text{if } l > p$$

But p is itself greater than 1, $\therefore$ Σu_n is convergent if $l > 1$.

*Proof is for student reference only.

Case II: *Let Σv_n be divergent, so that $p \le 1$.*

Then Σu_n will also diverge if $\frac{u_n}{u_{n+1}} < \frac{v_n}{v_{n+1}}$

or if $$\frac{u_n}{u_{n+1}} < 1 + \frac{p}{n} + \frac{p(p-1)}{2!} \cdot \frac{1}{n^2} + \ldots\ldots$$

or if $$n\left(\frac{u_n}{u_{n+1}} - 1\right) < p + \frac{p(p-1)}{2!} \cdot \frac{1}{n} + \ldots\ldots$$

or if $$\underset{n\to\infty}{\text{Lt}}\ n\left(\frac{u_n}{u_{n+1}} - 1\right) < p \quad \text{or} \quad \text{if } l < p.$$

But p itself ≤ 1. Thus the given series Σu_n diverges if $l < 1$. This proves the result.

Note 1. If $\underset{n\to\infty}{\text{Lt}}\ n\left(\frac{u_n}{u_{n+1}} - 1\right) = 1$, then Raabe's test fails.

Note 2. Raabe's test is applicable if D'Alembert's Ratio test fails and in the ratio test, $\frac{u_n}{u_{n+1}}$ does not involve the number e.

ILLUSTRATIVE EXAMPLES

Problems on Raabe's Test:

Example 1. *Discuss the convergence of the series :* $\frac{1}{2} + \frac{1.3}{2.4} + \frac{1.3.5}{2.4.6} + \ldots\ldots$

Sol. Here $$u_n = \frac{1.3.5\ldots\ldots(2n-1)}{2.4.6.\ldots\ldots 2n}.$$

$$\therefore \quad u_{n+1} = \frac{1.3.5\ldots\ldots(2n-1)(2n+1)}{2.4.6.\ldots\ldots(2n)(2n+2)}$$

$$\therefore \quad \frac{u_n}{u_{n+1}} = \frac{1.3.5.\ldots\ldots(2n-1)}{2.4.6\ldots\ldots 2n} \times \frac{2.4.6.\ldots\ldots 2n(2n+2)}{1.3.5\ldots\ldots(2n-1)(2n+1)}$$

$$= \frac{2n+2}{2n+1} = \frac{1+\frac{1}{n}}{1+\frac{1}{2n}} \to 1 \text{ as } n \to \infty.$$

$\therefore$ D'Alembert's Ratio Test fails.

$$n\left[\frac{u_n}{u_{n+1}} - 1\right] = n\left[\frac{2n+2}{2n+1} - 1\right] = \frac{n}{2n+1} = \frac{1}{2+\frac{1}{n}}$$

$$\therefore \qquad \underset{n\to\infty}{\text{Lt}}\ n\left[\frac{u_n}{u_{n+1}} - 1\right] = \tfrac{1}{2} < 1.$$

$\therefore$ By Raabe's Test, Σu_n diverges.

Example 2. *Discuss the convergence of the series:* $\dfrac{1^2}{2^2} + \dfrac{1^2 . 3^2}{2^2 . 4^2} + \dfrac{1^2 . 3^2 . 5^2}{2^2 . 4^2 . 6^2} + \ldots\ldots$

Sol. Here
$$u_n = \frac{1^2 . 3^2 . 5^2 \ldots\ldots (2n-1)^2}{2^2 . 4^2 . 6^2 \ldots\ldots (2n)^2}$$

and
$$u_{n+1} = \frac{1^2 . 3^2 . 5^2 \ldots\ldots (2n-1)^2 (2n+1)^2}{2^2 . 4^2 . 6^2 \ldots\ldots (2n)^2 (2n+2)^2}$$

$$\therefore \qquad \frac{u_n}{u_{n+1}} = \frac{(2n+2)^2}{(2n+1)^2} = \frac{4n^2\left(1+\frac{1}{n}\right)^2}{4n^2\left(1+\frac{1}{2n}\right)^2} = \frac{\left(1+\frac{1}{n}\right)^2}{\left(1+\frac{1}{2n}\right)^2}$$

$$\therefore \qquad \underset{n\to\infty}{\text{Lt}}\ \frac{u_n}{u_{n+1}} = 1.$$

Hence the Ratio Test fails.

$$n\left(\frac{u_n}{u_{n+1}} - 1\right) = n\left[\frac{(2n+2)^2}{(2n+1)^2} - 1\right]$$

$$= n\left[\frac{4n^2+8n+4-(4n^2+4n+1)}{(2n+1)^2}\right] = n\,\frac{(4n+3)}{(2n+1)^2} = \frac{4n^2+3n}{(2n+1)^2}$$

$$= \frac{1+\dfrac{3}{4n}}{\left(1+\dfrac{1}{2n}\right)^2} \to 1 \text{ as } n \to \infty$$

$\therefore$ Raabe's Test also fails.

When D'Alembert Ratio Test fails, we can directly apply Gauss test. Now,

$$\frac{u_n}{u_{n+1}} = \frac{(2n+2)^2}{(2n+1)^2} = \frac{\left(1+\frac{1}{n}\right)^2}{\left(1+\frac{1}{2n}\right)^2} = \left(1+\frac{1}{n}\right)^2\left(1+\frac{1}{2n}\right)^{-2}$$

$$= \left(1+\frac{2}{n}+\frac{1}{n^2}\right)\left(1-\frac{2}{2n}+\frac{3}{4n^2}-\ldots\ldots\right) = 1+\frac{1}{n}+\frac{1}{n^2}\left(1-2+\frac{3}{4}\right)+\ldots\ldots$$

$$= 1+\frac{1}{n}-\frac{1}{4n^2}+\ldots\ldots = 1+\frac{1}{n}+O\left(\frac{1}{n^2}\right)$$

Comparing it with $\dfrac{u_n}{u_{n+1}} = 1+\dfrac{\lambda}{n}+O\left(\dfrac{1}{n^2}\right)$

We have $\lambda = 1$. Thus by Gauss test, the series Σu_n diverges.

Example 3. *Discuss the convergence of the series:*

$$\frac{x}{1}+\frac{1}{2}\cdot\frac{x^3}{3}+\frac{1.3}{2.4}\cdot\frac{x^5}{5}+\frac{1.3.5}{2.4.6}\cdot\frac{x^7}{7}+\ldots\ldots (x>0).$$

Sol. Neglecting the first term, we have $u_n = \dfrac{1.3.5\ldots\ldots(2n-1)}{2.4.6\ldots\ldots(2n)}\cdot\dfrac{x^{2n+1}}{2n+1}$

and
$$u_{n+1} = \frac{1.3.5\ldots\ldots(2n-1)(2n+1)}{2.4.6\ldots\ldots(2n)(2n+2)}\cdot\frac{x^{2n+3}}{2n+3}$$

$$\therefore \quad \frac{u_n}{u_{n+1}} = \frac{2n+2}{2n+1}\cdot\frac{2n+3}{2n+1}\cdot\frac{1}{x^2} = \frac{2n\left(1+\frac{1}{n}\right).2n\left(1+\frac{3}{2n}\right)}{2n\left(1+\frac{1}{2n}\right).2n\left(1+\frac{1}{2n}\right)}\cdot\frac{1}{x^2}$$

$$= \frac{\left(1+\frac{1}{n}\right)\left(1+\frac{3}{2n}\right)}{\left(1+\frac{1}{2n}\right)^2}\cdot\frac{1}{x^2}$$

$$\underset{n\to\infty}{\text{Lt}}\frac{u_n}{u_{n+1}} = \underset{n\to\infty}{\text{Lt}}\frac{\left(1+\frac{1}{n}\right)\left(1+\frac{3}{2n}\right)}{\left(1+\frac{1}{2n}\right)^2}\cdot\frac{1}{x^2} = \frac{1}{x^2}$$

$\therefore$ By Ratio Test, Σu_n is convergent if $\dfrac{1}{x^2} > 1$ *i.e.*, $x^2 < 1$ and divergent if $\dfrac{1}{x^2} < 1$ *i.e.*, $x^2 > 1$.

If $x^2 = 1$, then Ratio Test fails.

When $x^2 = 1$, we have $\dfrac{u_n}{u_{n+1}} = \dfrac{(2n+2)(2n+3)}{(2n+1)^2} = \dfrac{4n^2+10n+6}{4n^2+4n+1}$

$$\underset{n\to\infty}{\text{Lt}}\, n\left(\frac{u_n}{u_{n+1}}-1\right) = \underset{n\to\infty}{\text{Lt}}\, n\left(\frac{4n^2+10n+6}{4n^2+4n+1}-1\right)$$

$$= \underset{n\to\infty}{\text{Lt}}\frac{6n^2+5n}{4n^2+4n+1} = \underset{n\to\infty}{\text{Lt}}\frac{6+\frac{5}{n}}{4+\frac{4}{n}+\frac{1}{n^2}} = \frac{6}{4} = \frac{3}{2} > 1.$$

$\therefore$ By Raabe's Test, the series converges.

Hence Σu_n is convergent if $x^2 \le 1$ and divergent if $x^2 > 1$.

Example 4. *Discuss the convergence of the series:* $\sum \dfrac{4\cdot 7\cdot 10\ldots(3n+1)}{1\cdot 2\cdot 3\ldots n}X^n$.

(OU-Dec. 2007)

Sol. Let
$$u_n = \frac{4\cdot 7\cdot 10\ldots(3n+1)}{1\cdot 2\cdot 3\ldots n}X^n$$

$$u_{n+1} = \frac{4 \cdot 7 \cdot 10 \ldots (3n+1)[3(n+1)+1]}{1 \cdot 2 \cdot 3 \ldots n(n+1)} X^{n+1}$$

$$\underset{n\to\infty}{\text{Lt}} \frac{u_n}{u_{n+1}} = \underset{n\to\infty}{\text{Lt}} \frac{n+1}{3n+4} \cdot \frac{1}{X} \qquad (\because \text{ After cancellation})$$

$$= \underset{n\to\infty}{\text{Lt}} \frac{n\left(1+\frac{1}{n}\right)}{n\left(3+\frac{4}{n}\right)} \cdot \frac{1}{X} = \frac{1}{3X}$$

Thus by Ratio Test, the series converges for $\frac{1}{3X} > 1$, *i.e.*, for $X < \frac{1}{3}$ and diverges for $X > \frac{1}{3}$ and Ratio Test fails for $X = \frac{1}{3}$.

Hence, by Raabe's test

$$\underset{n\to\infty}{\text{Lt}}\, n\left(\frac{u_n}{u_{n+1}} - 1\right) = \underset{n\to\infty}{\text{Lt}}\, n\left(\frac{n+1}{3n+4} \cdot 3 - 1\right)$$

$$= \underset{n\to\infty}{\text{Lt}}\, n\left(\frac{3n+3-3n-4}{3n+4}\right) = \underset{n\to\infty}{\text{Lt}}\, n\left(\frac{-1}{3n+4}\right)$$

$$= \underset{n\to\infty}{\text{Lt}} \frac{-n}{n\left(3+\frac{4}{n}\right)} = -\frac{1}{3} < 1$$

Thus by Raabe's test, the series is divergent. Hence the given series is convergent for $X < \frac{1}{3}$ and divergent for $X \geq \frac{1}{3}$.

EXERCISE 6

Discuss the convergence of the following series:

1. $1 + \frac{2^2}{3^2} + \frac{2^2 . 4^2}{3^2 . 5^2} + \frac{2^2 . 4^2 . 6^2}{3^2 . 5^2 . 7^2} + \ldots\ldots \text{ to } \infty$

2. $\frac{1^2}{4^2} + \frac{1^2 . 5^2}{4^2 . 8^2} + \frac{1^2 . 5^2 . 9^2}{4^2 . 8^2 . 12^2} + \ldots\ldots \text{ to } \infty$

3. $1 + \frac{3}{7}x + \frac{3 . 6}{7 . 10}x^2 + \frac{3 . 6 . 9}{7 . 10 . 13}x^3 + \ldots\ldots \text{ to } \infty$

4. $1 + \frac{2}{1} . \frac{1}{2} + \frac{2 . 4}{1 . 3} . \frac{1}{3} + \frac{2 . 4 . 6}{1 . 3 . 5} . \frac{1}{4} + \ldots\ldots \text{ to } \infty$

5. $\frac{1^2}{2^2} + \frac{1^2 . 3^2}{2^2 . 4^2}x + \frac{1^2 . 3^2 . 5^2}{2^2 . 4^2 . 6^2}x^2 + \ldots\ldots \text{ to } \infty$

6. $x + \frac{2^2 x^2}{2!} + \frac{3^3 x^3}{3!} + \frac{4^4 x^4}{4!} + \frac{5^5 x^5}{5!} + \ldots\ldots \text{ to } \infty$

7. $1 + \frac{2x}{2!} + \frac{3^2 x^2}{3!} + \frac{4^3 x^3}{4!} + \ldots\ldots \text{ to } \infty$

8. $\frac{a+x}{1!} + \frac{(a+2x)^2}{2!} + \frac{(a+3x)^3}{3!} + \ldots\ldots \text{ to } \infty$

9. $1^p + \left(\frac{1}{2}\right)^p + \left(\frac{1 . 3}{2 . 4}\right)^p + \left(\frac{1 . 3 . 5}{2 . 4 . 6}\right)^p + \ldots\ldots \text{ to } \infty$

10. $\dfrac{a}{b}+\dfrac{a(a+1)}{b(b+1)}+\dfrac{a(a+1)(a+2)}{b(b+1)(b+2)}+\ldots\ldots$ to ∞

11. $\dfrac{a}{b}+\dfrac{a(a+d)}{b(b+d)}x+\dfrac{a(a+d)(a+2d)}{b(b+d)(b+2d)}x^2+\ldots\ldots$ to ∞ $(a>0, b>0, x>0)$

12. $\sum \dfrac{n!}{x(x+1)(x+2)\ldots\ldots(x+n-1)}$

13. $1+\dfrac{1}{2}x+\dfrac{1.3}{2.4}x^2+\dfrac{1.3.5}{2.4.6}x^3+\ldots\ldots$ to ∞ $(x>0)$

14. $\dfrac{x}{1.2}+\dfrac{x^2}{3.4}+\dfrac{x^3}{5.6}+\dfrac{x^4}{7.8}+\ldots\ldots$ to ∞ $(x>0)$

15. $\dfrac{x}{1.2}+\dfrac{x^2}{2.3}+\dfrac{x^3}{3.4}+\dfrac{x^4}{4.5}+\ldots\ldots$ to ∞

16. $\dfrac{1}{2}+\dfrac{2}{3}x+\left(\dfrac{3}{4}\right)^2x^2+\left(\dfrac{4}{5}\right)^3x^3+\ldots\ldots$ to ∞ $(x>0)$

17. $\sum \dfrac{1.3.5\ldots\ldots(2n-1)}{2.4.6\ldots\ldots(2n)}\cdot\dfrac{x^{2n}}{2n}$

18. $\sum \dfrac{(n!)^2}{(2n)!}x^{2n}$ $(x>0)$.

19. $1+\dfrac{(1!)^2}{2!}x^2+\dfrac{(2!)^2}{4!}x^4+\dfrac{(3!)^2}{6!}x^6+\ldots..$ to ∞ $(x>0)$

20. $x^2(\log 2)^q+x^3(\log 3)^q+x^4(\log 4)^q+\ldots\ldots.$ to ∞

Answers

1. Divergent
2. Convergent
3. Convergent for $x \le 1$, divergent for $x > 1$
4. Divergent
5. Convergent for $x < 1$, divergent for $x \ge 1$
6. Convergent for $x < \dfrac{1}{e}$, divergent for $x \ge \dfrac{1}{e}$
7. Convergent for $x \le \dfrac{1}{e}$, divergent for $x > \dfrac{1}{e}$
8. Convergent for $x < \dfrac{1}{e}$, divergent for $x \ge \dfrac{1}{e}$
9. Convergent for $p > 2$, divergent for $p \le 2$
10. Convergent for $b > a + 1$, divergent for $b \le a + 1$
11. Convergent for $x < 1$ or $x = 1$ and $b > a + d$, divergent for $x > 1$ or $x = 1$ and $b \le a + d$
12. Convergent for $x > 2$, divergent for $x \le 2$
13. Convergent for $x < 1$, divergent for $x \ge 1$
14. Convergent for $x \le 1$, divergent for $x > 1$
15. Convergent for $x \le 1$, divergent for $x > 1$
16. Convergent for $x < 1$, divergent for $x \ge 1$
17. Convergent for $x^2 \le 1$, divergent for $x^2 > 1$
18. Convergent for $x^2 < 4$, divergent for $x^2 \ge 4$
19. Convergent for $x^2 < 4$, divergent for $x^2 \ge 4$
20. Convergent for $x < 1$, divergent for $x \ge 1$.

Theorem 8. Leibnitz's Test on Alternating Series *(OU-Apr. 2016)*

Statement. *The alternating series:* $\Sigma(-1)^{n-1}u_n = u_1 - u_2 + u_3 - u_4 + \ldots\ldots$ $(u_n > 0\ \forall\ n)$ *converges if*

(i) $u_n \ge u_{n+1}$ $\forall$ *n and* *(ii)* $\underset{n\to\infty}{Lt}\, u_n = 0.$

***Proof.** Let S_n denote the nth partial sum of the series $\Sigma\,(-1)^{n-1} \cdot u_n$.

$$\begin{aligned} S_{2n} &= u_1 - u_2 + u_3 - u_4 + u_5 - \ldots\ldots - u_{2n-2} + u_{2n-1} - u_{2n} \\ &= u_1 - (u_2 - u_3) - (u_4 - u_5) \ldots\ldots - (u_{2n-2} - u_{2n-1}) - u_{2n} \\ &= u_1 - [(u_2 - u_3) + (u_4 - u_5) + \ldots\ldots + (u_{2n-2} - u_{2n-1}) + u_{2n}] \\ &< u_1 \qquad [\because\ u_n > u_{n+1} \text{ and } u_n > 0 \text{ for all } n] \end{aligned}$$

$\Rightarrow$ The sequence $\{S_{2n}\}$ is bounded above.

Also, $S_{2n+2} = S_{2n} + u_{2n+1} - u_{2n+2}$

---

*Proof is for student reference only.

$\Rightarrow \quad S_{2n+2} - S_{2n} = u_{2n+1} - u_{2n+2} > 0$ for all $n \Rightarrow S_{2n+2} > S_{2n}$

$\Rightarrow$ The sequence $\{S_{2n}\}$ is monotonically increasing.

Since every monotonically increasing sequence, which is bounded above converges, therefore, the sequence $\{S_{2n}\}$ converges. Let it converge to S, then $\underset{n\to\infty}{\text{Lt}}\ S_{2n} = S$

Now,
$$\underset{n\to\infty}{\text{Lt}}\ S_{2n+1} = \underset{n\to\infty}{\text{Lt}}\ (S_{2n} + u_{2n+1}) = \underset{n\to\infty}{\text{Lt}}\ S_{2n} + \underset{n\to\infty}{\text{Lt}}\ u_{2n+1}$$
$$= S + 0 \qquad \left[\because \underset{n\to\infty}{\text{Lt}}\ u_n = 0\right]$$
$$= S$$

$\therefore$ The sequences $\{S_{2n}\}$ and $\{S_{2n+1}\}$ converge to the same real number S.

$\therefore$ Given $\varepsilon > 0$ there exists positive integers m_1 and m_2 such that
$$| S_{2n} - S | < \varepsilon \ \forall \ 2n > m_1 \quad \text{and} \quad | S_{2n+1} - S | < \varepsilon \ \forall \ 2n + 1 > m_2$$

Let $m = \max. \{m_1, m_2\}$, then
$$| S_{2n} - S | < \varepsilon \ \forall \ n > m \quad \text{and} \quad | S_{2n+1} - S | < \varepsilon \ \forall \ n > m$$
$$\Rightarrow \quad | S_n - S | < \varepsilon \ \forall \ n > m$$

$\therefore$ The sequence $\{S_n\}$ converges to S.

Hence the given series is convergent.

Note 1. The alternating series is not convergent *i.e.*, oscillating series if any one of the above two conditions is not satisfied.

Note 2. To test convergence of an alternating series, we apply Leibnitz's test.

ILLUSTRATIVE EXAMPLES

Problems on Leibnitz's Test:

Example 1. *Examine the convergence of the series:* $1 - \frac{1}{2} + \frac{1}{3} - \frac{1}{4} + \ldots\ldots$

Sol. It is an alternating series.

(*i*) $u_n = \frac{1}{n}, \qquad u_{n+1} = \frac{1}{n+1}$

$\because \qquad \frac{1}{n} > \frac{1}{n+1} \ \forall \ n \quad \therefore \quad u_n > u_{n+1} \ \forall \ n$

(*ii*) $\underset{n\to\infty}{\text{Lt}}\ u_n = \underset{n\to\infty}{\text{Lt}}\ \frac{1}{n} = 0$

$\therefore$ Both the conditions of Leibnitz's Test are satisfied.

Hence the given series is convergent.

Example 2. *Examine the convergence of the series:*

(*a*) $2 - \frac{3}{2} + \frac{4}{3} - \frac{5}{4} + \ldots\ldots$

(*b*) $\frac{1}{2^3} - \frac{1}{3^3}(1 + 2) + \frac{1}{4^3}(1 + 2 + 3) - \frac{1}{5^3}(1 + 2 + 3 + 4) + \ldots\ldots$

Sol. (*a*) It is an alternating series

(*i*) $u_n = \dfrac{n+1}{n}, \qquad u_{n+1} = \dfrac{n+2}{n+1}$

$$u_n - u_{n+1} = \frac{n+1}{n} - \frac{n+2}{n+1} = \frac{(n+1)^2 - n(n+2)}{n(n+1)} = \frac{1}{n(n+1)} > 0 \quad \forall\, n$$

$\Rightarrow \qquad u_n > u_{n+1} \quad \forall\, n$

(*ii*) $\underset{n\to\infty}{\text{Lt}}\ u_n = \underset{n\to\infty}{\text{Lt}}\ \dfrac{n+1}{n} = \underset{n\to\infty}{\text{Lt}}\ \left(1+\dfrac{1}{n}\right) = 1 \neq 0.$

Since the second condition of Leibnitz's Test is not satisfied, the series is not convergent.

(*b*) It is an alternating series.

(*i*) $u_n = \dfrac{1}{(n+1)^3}\,[1+2+3+ \ldots\ldots + n] = \dfrac{1}{(n+1)^3}\cdot\dfrac{n(n+1)}{2} = \dfrac{1}{2}\cdot\dfrac{n}{(n+1)^2} \qquad \left(\because\ \sum n = \dfrac{n(n+1)}{2}\right)$

$$u_{n+1} = \frac{1}{2}\cdot\frac{n+1}{(n+2)^2}$$

$$u_n - u_{n+1} = \frac{1}{2}\left[\frac{n}{(n+1)^2} - \frac{n+1}{(n+2)^2}\right] = \frac{1}{2}\,\frac{n(n+2)^2 - (n+1)^3}{(n+1)^2(n+2)^2}$$

$$= \frac{1}{2}\cdot\frac{n^2+n-1}{(n+1)^2(n+2)^2} > 0 \quad \forall\, n$$

$\Rightarrow \qquad u_n > u_{n+1} \quad \forall\, n$

(*ii*) $\underset{n\to\infty}{\text{Lt}}\ u_n = \underset{n\to\infty}{\text{Lt}}\ \dfrac{n}{2(n+1)^2} = \underset{n\to\infty}{\text{Lt}}\ \dfrac{\dfrac{1}{n}}{2\left(1+\dfrac{1}{n}\right)^2} = 0.$

Since both the conditions of Leibnitz's Test are satisfied, the given series is convergent.

Example 3. *Test the convergence of the following series:* $\displaystyle\sum_{n=1}^{\infty} \frac{(-1)^{n-1}\,.\,n}{2n-1}$.

Sol. The given series is $\displaystyle\sum_{n=1}^{\infty} \frac{(-1)^{n-1}\,.\,n}{2n-1} = \frac{1}{1} - \frac{2}{3} + \frac{3}{5} - \frac{4}{7} + \ldots\ldots$

It is an alternating series

(*i*) $\qquad u_n = \dfrac{n}{2n-1}, \quad u_{n+1} = \dfrac{n+1}{2n+1}$

$$u_n - u_{n+1} = \frac{1}{4n^2-1} > 0 \quad \forall\ n \ \Rightarrow\ u_n > u_{n+1} \quad \forall\ n$$

(*ii*) $\underset{n\to\infty}{\text{Lt}}\ u_n = \underset{n\to\infty}{\text{Lt}}\ \dfrac{n}{2n-1} = \underset{n\to\infty}{\text{Lt}}\ \dfrac{1}{2-\dfrac{1}{n}} = \dfrac{1}{2} \neq 0.$

Here the second condition of Leibnitz's Test is not satisfied. Hence the given series is not convergent *i.e.,* the given series is an oscillating series.

Example 4. *Test the convergence of* $\sum \frac{(-1)^n}{n(n^2+1)}$. (OU-May 2012)

Sol. Let $\sum_1^{\infty} (-1)^n u_n = \sum_1^{\infty} \frac{(-1)^n}{n(n^2+1)}$ is an alternating series.

Take $$u_n = \frac{1}{n(n^2+1)}$$

$$u_{n+1} = \frac{1}{(n+1)[(n+1)^2+1]}$$

$$(i)\ u_n - u_{n+1} = \frac{1}{n(n^2+1)} - \frac{1}{(n+1)[(n+1)^2+1]}$$

$$= \frac{(n+1)[(n+1)^2+1] - n(n^2+1)}{n(n+1)(n^2+1)[(n+1)^2+1]} = \frac{(n+1)(n^2+2n+2) - n^3 - n}{n(n+1)(n^2+1)[(n+1)^2+1]}$$

$$= \frac{n^3+2n^2+2n+n^2+2n+2-n^3-n}{n(n+1)(n^2+1)[(n+1)^2+1]}$$

$$= \frac{3n^2+3n+2}{n(n+1)(n^2+1)[(n+1)^2+1]} > 0\ \forall\ n$$

$$\Rightarrow \quad u_n > u_{n+1},\ \forall\ n$$

$$(ii)\ \underset{n\to\infty}{\text{Lt}}\ u_n = \underset{n\to\infty}{\text{Lt}} \frac{1}{n(n^2+1)} = 0$$

Since both the conditions of Leibnitz's Test are satisfied, the given series is convergent.

Example 5. *Test for convergence* $\sum (-1)^n \frac{1}{2^{n+1}}$. (OU-Jan. 2012)

Sol. Let $$u_n = \frac{1}{2^{n+1}}$$

$$u_{n+1} = \frac{1}{2^{n+2}}$$

$$(i)\ u_n - u_{n+1} = \frac{1}{2^{n+1}} - \frac{1}{2^{n+1}\cdot 2} = \frac{1}{2^{n+1}}\left(1-\frac{1}{2}\right) = \frac{1}{2^{n+2}} > 0\ \forall\ n$$

$$(ii)\ \underset{n\to\infty}{\text{Lt}}\ u_n = \underset{n\to\infty}{\text{Lt}} \frac{1}{2^{n+1}} = 0$$

By Leibnitz's Test, the given series is convergent.

Example 6. *Discuss the convergence of the series:* $\sum_0^{\infty} \frac{(-1)^n}{n!}$. (OU-June 2011)

Sol. Let $$u_n = \frac{1}{n!},\ u_{n+1} = \frac{1}{(n+1)!} = \frac{1}{(n+1)n!}$$

$$(i)\ u_n - u_{n+1} = \frac{1}{n!} - \frac{1}{(n+1)n!} = \frac{1}{n!}\left[1 - \frac{1}{n+1}\right] = \frac{1}{n!}\cdot\frac{n}{n+1} = \frac{n}{(n+1)!} > 0,\ \forall\ n$$

$$(ii)\ \underset{n\to\infty}{\text{Lt}}\ u_n = \underset{n\to\infty}{\text{Lt}}\ \frac{1}{n!} = 0$$

By Leibnitz's Test, the given series is convergent.

Example 7. *Test for the convergence* $\sum_{1}^{\infty} (-1)^n \frac{n}{n^2+1}$. (OU-Jan. 2011)

Sol. Let $u_n = \frac{n}{n^2+1},\ u_{n+1} = \frac{n+1}{(n+1)^2+1}$

$$(i)\qquad u_n - u_{n+1} = \frac{n}{n^2+1} - \frac{n+1}{n^2+2n+2} = \frac{n(n^2+2n+2)-(n+1)(n^2+1)}{(n^2+1)(n^2+2n+2)}$$

$$= \frac{n^3+2n^2+2n-n^3-n-n^2-1}{(n^2+1)(n^2+2n+2)}$$

$$= \frac{n^2+n-1}{(n^2+1)(n^2+2n+2)} > 0,\ \forall\ n$$

$$\therefore \qquad u_n > u_{n+1},\ \forall\ n$$

$$(ii)\ \underset{n\to\infty}{\text{Lt}}\ u_n = \underset{n\to\infty}{\text{Lt}}\ \frac{n}{n^2+1} = \underset{n\to\infty}{\text{Lt}}\ \frac{n}{n^2\left(1+\frac{1}{n^2}\right)}$$

$$= \underset{n\to\infty}{\text{Lt}}\ \frac{1}{n}\cdot\frac{1}{1+\frac{1}{n^2}} = 0$$

By Leibnitz's Test, the given series is convergent.

Example 8. *Discuss the convergence of the series:* $\frac{x}{1+x} - \frac{x^2}{1+x^2} + \frac{x^3}{1+x^3} - \frac{x^4}{1+x^4} + \ldots$ *$(0 < x < 1)$.* (OU-Apr. 2007)

Sol. This is an alternating series.

Consider $$u_n = \frac{x^n}{1+x^n}$$

Now (*i*) $$u_n - u_{n+1} = \frac{x^n}{1+x^n} - \frac{x^{n+1}}{1+x^{n+1}} = \frac{x^n(1+x^{n+1})-(1+x^n)x^{n+1}}{(1+x^n)(1+x^{n+1})}$$

$$= \frac{x^n(1-x)}{(1+x^n)(1+x^{n+1})} > 0 \quad \text{for } 0 < x < 1$$

$\therefore \qquad u_n > u_{n+1}, \forall\, n$

(*ii*) $\displaystyle \underset{n\to\infty}{\text{Lt}}\, u_n = \underset{n\to\infty}{\text{Lt}} \frac{x^n}{1+x^n} = \frac{0}{1+0} = 0 \qquad \left(\because \text{ For } 0 < x < 1,\ \underset{n\to\infty}{\text{Lt}}\, x^n = 0\right)$

$\therefore$ By Leibnitz's Test, the series is convergent.

2.17. ABSOLUTE/CONDITIONAL CONVERGENCE OF A SERIES

Def. *If a convergent series whose terms are not all positive, remains convergent when all its terms are made positive, then it is called an* **absolutely convergent series**, *i.e.,*

An alternating series Σu_n is said to be absolutely convergent if $\Sigma\,|\,u_n\,|$ is a convergent series.

A series is said to be ***conditionally convergent*** *if it is convergent but does not converge absolutely i.e., An alternating series* $\sum u_n$ *is said to be conditionally convergent if* $\sum |u_n|$ *is a divergent series and* $\sum u_n$ *is a convergent series.* (OU-Dec. 2008)

Example. *Test whether the following series are absolutely convergent or conditionally convergent ?*

(*a*) $1 - \frac{1}{2^2} + \frac{1}{3^2} - \frac{1}{4^2} + \ldots\ldots$ (*b*) $\displaystyle\sum_{n=1}^{\infty} \frac{(-1)^{n-1}}{2n-1}$.

Sol. (*a*) The series is $1 - \frac{1}{2^2} + \frac{1}{3^2} - \frac{1}{4^2} + \ldots\ldots$

(*i*) In this alternating series, each term is less than the preceeding term numerically.

(*ii*) Moreover $u_n = \frac{1}{n^2}$ which $\to 0$ as $n \to \infty$.

Hence the series satisfies both the conditions of the test on alternating series and so the given series *converges*.

Again when all the term of the series are made positive, the series becomes

$$\sum |\,u_n\,| = 1 + \frac{1}{2^2} + \frac{1}{3^2} + \frac{1}{4^2} + \ldots\ldots = \sum \frac{1}{n^2}$$

which we know is a convergent series. ($\because$ Here $p = 2 > 1$)

Thus the given series converges absolutely.

(*b*) $\displaystyle\sum_{n=1}^{\infty} \frac{(-1)^{n-1}}{2n-1} = \Sigma\,(-1)^{n-1}\, u_n$ (say)

Putting $n = 1, 2, 3, \ldots\ldots$, the series becomes $1 - \frac{1}{3} + \frac{1}{5} - \frac{1}{7} + \ldots\ldots$

The series is clearly an alternating series.

(*i*) The terms go on decreasing numerically and

(*ii*) $\displaystyle \underset{n\to\infty}{\text{Lt}}\, u_n = \underset{n\to\infty}{\text{Lt}} \frac{1}{2n-1} = 0$

$\therefore$ By Leibnitz's Test, the series *converges*.

But when all terms are made positive, the series becomes,

$$\sum |u_n| = 1 + \frac{1}{3} + \frac{1}{5} + \frac{1}{7} + \ldots\ldots$$

Here $\quad u_n = \frac{1}{2n-1}$. Take $v_n = \frac{1}{n}$.

$$\therefore \quad \underset{n\to\infty}{\text{Lt}} \frac{u_n}{v_n} = \underset{n\to\infty}{\text{Lt}} \frac{n}{2n-1} = \underset{n\to\infty}{\text{Lt}} \left[\frac{1}{2-\frac{1}{n}}\right] = \frac{1}{2} = \text{finite} \neq 0$$

Hence by comparison test series Σu_n and Σv_n behave alike.

But $\Sigma v_n = \sum \frac{1}{n}$ is a divergent series ($\because$ Here $p = 1$), $\therefore$ Σu_n also diverges.

Hence the given series converges, and the series of absolute terms diverges, therefore, the given series converges conditionally.

EXERCISE 7

Examine the convergence of the following series :

1. $1 - \frac{1}{\sqrt{2}} + \frac{1}{\sqrt{3}} - \frac{1}{\sqrt{4}} + \ldots\ldots$ to ∞

2. $1 - \frac{1}{2^2} + \frac{1}{3^2} - \frac{1}{4^2} + \ldots\ldots$ to ∞

3. $1 - \frac{1}{2\sqrt{2}} + \frac{1}{3\sqrt{3}} - \frac{1}{4\sqrt{4}} + \ldots\ldots$ to ∞

4. $\frac{1}{a} - \frac{1}{a+b} + \frac{1}{a+2b} - \frac{1}{a+3b} + \ldots\ldots$ to ∞ $\quad (a > 0, b > 0)$

5. $\frac{1}{\log 2} - \frac{1}{\log 3} + \frac{1}{\log 4} - \frac{1}{\log 5} + \ldots\ldots$ to ∞

6. $\sum_{n=1}^{\infty} \frac{(-1)^{n-1}}{3n-2}$

7. $\sum_{n=1}^{\infty} \frac{n(-1)^{n-1}}{5^n}$

8. $\frac{1}{\sqrt{2}+1} - \frac{1}{\sqrt{3}+1} + \frac{1}{\sqrt{4}+1} - \frac{1}{\sqrt{5}+1} + \ldots\ldots$ to ∞

9. $\sum_{n=1}^{\infty} \frac{\cos n\pi}{n^2+1}$. $\quad$ **[Hint.** $\cos n\pi = (-1)^n$]

Answers

1. Convergent **2.** Convergent **3.** Convergent **4.** Convergent
5. Convergent **6.** Convergent **7.** Convergent **8.** Convergent.
9. Convergent.

ILLUSTRATIVE EXAMPLES

Example 1. *Prove that the series* $\frac{\sin x}{1^3} - \frac{\sin 2x}{2^3} + \frac{\sin 3x}{3^3} - \ldots\ldots$ *converges absolutely.*

(OU-June 2013)

Sol. The given series is $\sum_{n=1}^{\infty} u_n = \sum_{n=1}^{\infty} \frac{(-1)^{n-1} \sin nx}{n^3}$

Since $|u_n| = \frac{|\sin nx|}{n^3} \le \frac{1}{n^3} \ \forall\, n$ and $\sum \frac{1}{n^3}$ converges.

∴ By comparison test, the series $\sum |u_n|$ converges.

⇒ The given series converges absolutely.

Example 2. *Find the interval of convergence of the series* $x - \frac{x^2}{\sqrt{2}} + \frac{x^3}{\sqrt{3}} - \frac{x^4}{\sqrt{4}} + \ldots\ldots$

Sol. The given series is $\sum_{n=1}^{\infty} u_n = \sum_{n=1}^{\infty} \frac{(-1)^{n-1} x^n}{\sqrt{n}}$

Here $|u_n| = \frac{|x^n|}{\sqrt{n}} = \frac{|x|^n}{\sqrt{n}}$ and $|u_{n+1}| = \frac{|x|^{n+1}}{\sqrt{n+1}}$

∴ $$\frac{|u_n|}{|u_{n+1}|} = \sqrt{\frac{n+1}{n}} \cdot \frac{1}{|x|} = \sqrt{1+\frac{1}{n}} \cdot \frac{1}{|x|}$$

$$\lim_{n\to\infty} \frac{|u_n|}{|u_{n+1}|} = \frac{1}{|x|}$$

∴ By ratio test, the series $\sum |u_n|$ is convergent if $\frac{1}{|x|} > 1$ *i.e.,* if $|x| < 1$ *i.e.,* if $-1 < x < 1$ and divergent if $\frac{1}{|x|} < 1$ *i.e.,* if $|x| > 1$ *i.e.,* if $x > 1$ or $x < -1$.

Ratio test fails when $|x| = 1$, *i.e.,* when $x = 1$ or -1.

When $x = 1$, the series becomes $1 - \frac{1}{\sqrt{2}} + \frac{1}{\sqrt{3}} - \frac{1}{\sqrt{4}} + \ldots\ldots$

which is an alternating series and is convergent.

When $x = -1$, the series becomes $-\left(1 + \frac{1}{\sqrt{2}} + \frac{1}{\sqrt{3}} + \frac{1}{\sqrt{4}} + \ldots\ldots\right) = -\sum_{n=1}^{\infty} \frac{1}{\sqrt{n}}$

which is divergent by p-series test.

Hence the given series converges for $-1 < x \le 1$.

Example 3. *Test whether the series* $\sum \frac{(-1)^n}{n\sqrt{n}}$ *converges absolutely or not.*

(OU-Dec. 2013)

Sol. Let
$$\Sigma u_n = \sum \frac{(-1)^n}{n\sqrt{n}}$$
$$|u_n| = \frac{1}{n\sqrt{n}}$$

By p-Test, $\sum |u_n|$ is convergent.

Hence, by comparison test, $\sum |u_n|$ is convergent.

$\Rightarrow$ The given series converges absolutely.

Example 4. *Test the series* $\sum (-1)^{n-1} \frac{1}{n2^n}$ *for absolute convergence.*

(OU-June 2010)

Sol. Here
$$\Sigma u_n = \sum (-1)^{n-1} \cdot \frac{1}{n2^n}$$
$$|u_n| = \frac{1}{n2^n}, \ |u_{n+1}| = \frac{1}{(n+1)2^{n+1}}$$
$$\underset{n\to\infty}{\text{Lt}} \left| \frac{u_n}{u_{n+1}} \right| = \underset{n\to\infty}{\text{Lt}} \frac{1}{n2^n} \cdot \frac{(n+1)2^{n+1}}{1} = \underset{n\to\infty}{\text{Lt}} \frac{n+1}{n} \cdot \frac{2^n \cdot 2}{2^n}$$
$$= \underset{n\to\infty}{\text{Lt}} \left(1 + \frac{1}{n}\right) \cdot 2 = 2 > 1$$

$\therefore$ By Ratio Test, $\sum |u_n|$ is convergent. Hence, the given series is absolutely convergent.

Example 5. *Show that the series:* $\sum \frac{(-1)^{n-1}}{2^{n-1}}$ *is absolutely convergent.*

Sol. Let
$$\Sigma u_n = \sum \frac{(-1)^{n-1}}{2^{n-1}}$$
$$\therefore \quad \sum |u_n| = \sum \frac{1}{2^{n-1}}$$

This is a geometric series with common ratio $r = \frac{1}{2} < 1$

$\therefore$ $\sum |u_n|$ is convergent.

Hence, the series is absolutely convergent.

Example 6. *Examine the following series for absolute and conditional convergence*
$$\frac{1}{5\sqrt{2}} - \frac{1}{5\sqrt{3}} + \frac{1}{5\sqrt{4}} - \ldots + (-1)^n \frac{1}{5\sqrt{n}} + \ldots .$$

Sol. Here $\Sigma u_n = \sum (-1)^n \dfrac{1}{5\sqrt{n}}$

$$|u_n| = \frac{1}{5\sqrt{n}}$$

Take $v_n = \dfrac{1}{\sqrt{n}}$

$$\therefore \quad \underset{n\to\infty}{\text{Lt}} \frac{|u_n|}{v_n} = \underset{n\to\infty}{\text{Lt}} \frac{1}{5\sqrt{n}} \cdot \frac{\sqrt{n}}{1} = \frac{1}{5} \neq 0$$

By comparison test, $\Sigma\,|u_n|$ and Σv_n converge or diverge together.

Since $\Sigma v_n = \sum \dfrac{1}{n^{1/2}} = \sum \dfrac{1}{n^p}$ with $p = \dfrac{1}{2} < 1$

By p-Test, Σv_n is divergent

$\therefore \quad \sum |u_n|$ also divergent.

Now, we need to apply Leibnitz's test.

$$\sum f(n) = \sum (-1)^n \frac{1}{5\sqrt{n}} = \sum (-1)^n u_n$$

where $u_n = \dfrac{1}{5\sqrt{n}}$ and $u_{n+1} = \dfrac{1}{5\sqrt{n+1}}$

Since $\dfrac{1}{5\sqrt{n}} > \dfrac{1}{5\sqrt{n+1}}, \ \forall\, n$

$$u_n > u_{n+1}, \ \forall\, n$$

Also, $\underset{n\to\infty}{\text{Lt}}\ u_n = \underset{n\to\infty}{\text{Lt}} \dfrac{1}{5\sqrt{n}} = 0$

$\therefore$ By Leibnitz's test, the given series is convergent.
Hence the given series is conditionally convergent.

Example 7. *Test whether the series is absolutely convergent* $\sum \dfrac{\cos n\pi}{n^2 + 1}$.

Sol. $\sum \dfrac{\cos n\pi}{n^2+1} = \sum \dfrac{(-1)^n}{n^2+1}$ $\qquad [\because \ \cos n\pi = (-1)^n]$

Here $|u_n| = \dfrac{1}{n^2+1}$

Take $v_n = \dfrac{1}{n^2}$

$$\Sigma v_n = \sum \frac{1}{n^p} = \sum \frac{1}{n^2} \quad \Rightarrow \quad p = 2 > 1$$

By p-Test, Σv_n is convergent.

$$\underset{n\to\infty}{\text{Lt}} \frac{|u_n|}{v_n} = \underset{n\to\infty}{\text{Lt}} \frac{1}{n^2+1} \times \frac{n^2}{1} = \underset{n\to\infty}{\text{Lt}} \frac{n^2}{n^2\left(1+\frac{1}{n^2}\right)} = 1 \neq 0$$

By comparison test, $\sum |u_n|$ also convergent.

Hence the given series is absolutely convergent.

Example 8. *Test whether the following series is absolutely convergent/conditionally convergent* $\sum_{1}^{\infty} (-1)^{n+1}(\sqrt{n+1} - \sqrt{n})$.

Sol. Let
$$\Sigma u_n = \sum (-1)^{n+1} (\sqrt{n+1} - \sqrt{n})$$

$$|u_n| = \sqrt{n+1} - \sqrt{n} = (\sqrt{n+1} - \sqrt{n}) \times \frac{(\sqrt{n+1} + \sqrt{n})}{\sqrt{n+1} + \sqrt{n}}$$

$$= \frac{(n+1) - n}{\sqrt{n+1} + \sqrt{n}} = \frac{1}{\sqrt{n+1} + \sqrt{n}}$$

Take
$$v_n = \frac{1}{\sqrt{n}}$$

$$\Sigma v_n = \sum \frac{1}{n^p} = \sum \frac{1}{n^{1/2}} \Rightarrow p = \frac{1}{2} < 1$$

By p-Test, Σv_n is divergent.

Now,
$$\underset{n\to\infty}{\text{Lt}} \frac{|u_n|}{v_n} = \underset{n\to\infty}{\text{Lt}} \frac{1}{\sqrt{n+1} + \sqrt{n}} \times \frac{\sqrt{n}}{1}$$

$$= \underset{n\to\infty}{\text{Lt}} \frac{\sqrt{n}}{\sqrt{n}\left(\sqrt{1+\frac{1}{n}} + 1\right)} = 1 \neq 0$$

By comparison test, $\sum |u_n|$ also divergent. Now we have to apply Leibnitz's Test. The given series is

$$\sum f(n) = \sum (-1)^{n+1} (\sqrt{n+1} - \sqrt{n}) = \sum (-1)^{n+1} u_n$$

$$u_n = \sqrt{n+1} - \sqrt{n} = \frac{1}{\sqrt{n+1} + \sqrt{n}} \quad (\because \text{ After rationalising})$$

$$u_{n+1} = \frac{1}{\sqrt{n+2} + \sqrt{n+1}}$$

Clearly $u_n > u_{n+1}, \forall\ n$

Also,
$$\underset{n\to\infty}{\text{Lt}}\ u_n = \underset{n\to\infty}{\text{Lt}} \frac{1}{\sqrt{n+1} + \sqrt{n}} = 0$$

$\therefore$ By Leibnitz's test, the series is convergent.

Hence the given series is conditionally convergent.

EXERCISE 8

1. Prove that the series $\sum_{n=1}^{\infty} \frac{(-1)^{n-1}\cos^2 nx}{n\sqrt{n}}$ converges absolutely.

2. For what values of x are the following series convergent.

(i) $x - \frac{x^2}{2} + \frac{x^3}{3} - \frac{x^4}{4} + \ldots\ldots$ (ii) $1 + x + \frac{x^2}{2!} + \frac{x^3}{3!} + \ldots\ldots$

(iii) $x - \frac{x^3}{3} + \frac{x^5}{5} - \frac{x^7}{7} + \ldots\ldots$ (iv) $x - \frac{x^2}{2^2} + \frac{x^3}{3^2} - \frac{x^4}{4^2} + \ldots\ldots$

Answers

2. (i) $-1 < x \le 1$ (ii) all x (iii) $-1 < x \le 1$ (iv) $-1 < x \le 1$.

2.18 WORKING RULE FOR TESTING A SERIES FOR CONVERGENCE

Step 1: If the series is alternating series, apply the Leibnitz's rule.

Step 2: For a positive term series, first find u_n and if possible evaluate $\underset{n\to\infty}{\text{Lt}}\, u_n$.

If $\underset{n\to\infty}{\text{Lt}}\, u_n \neq 0$, the series is divergent.

If $\underset{n\to\infty}{\text{Lt}}\, u_n = 0$, compare Σu_n with $\sum \frac{1}{n^p}$ and apply comparison test.

If the comparison test is not applicable, apply the ratio test. If the ratio test fails, apply Raabe's test. If the Raabe's test fails, apply Cauchy's root test.

Step 3: For the power series, apply the ratio test. If the ratio test fails, test the series as in step 2 above.

UNIT 3
Differential Calculus

3.1. INTRODUCTION

Calculus is one of the most beautiful intellectual achievements of human being. The mathematical study of change, motion, growth or decay is calculus. One of the most important ideas of differential calculus is derivative which measures the rate of change of a given function. In engineering the concept of the derivative is very useful.

Definitions

Let f be a real valued function defined on a closed interval $[a, b]$ and $a < c < b$.

1. A function $f(x)$ is said to be **continuous** at $x = c$ if

   $\underset{x \to c}{\text{Lt}} f(x)$ exists, and $\underset{x \to c}{\text{Lt}} f(x) = f(c)$

2. A function $f(x)$ is said to be **right continuous** at $x = c$ if

   $\underset{x \to c^+}{\text{Lt}} f(x)$ exists, and $\underset{x \to c^+}{\text{Lt}} f(x) = f(c)$

3. A function $f(x)$ is said to be **left continuous** at $x = c$ if

   $\underset{x \to c^-}{\text{Lt}} f(x)$ exists, and $\underset{x \to c^-}{\text{Lt}} f(x) = f(c)$

4. A function $f(x)$ is said to be **differentiable** (or **derivable**) at $x = c$ if $\underset{x \to c}{\text{Lt}} \left(\frac{f(x) - f(c)}{x - c} \right)$ exists and finite. The limit is called the **derivative** of f at c and it is denoted by $f'(c)$.

Note. An existing result says that every differential function is continuous. If f' is derivable then the derivative of f' at c is denoted by either $f''(c)$ or $f^{(2)}(c)$. $f^{(2)}$ is called as second derivative of f. If $f^{(n-1)}$ is derivable, then the derivative of $f^{(n-1)}$ is denoted by $f^{(n)}$. This is called as the nth derivative of f.

Results

(*i*) Geometrically if $f(x)$ is continuous in $[a, b]$ then the graph of $f(x)$ is a continuous curve for all the points $x \in [a, b]$.

(*ii*) If $f(x)$ is continuous in $[a, b]$ then $f(x)$ is bounded and attains its bounds (that is, there exists $c, d \in [a, b]$ such that $f(c) = M$ and $f(d) = m$, where M, m are the greatest and least values of $f(x)$ that attains between $f(a)$ and $f(b)$.

(*iii*) Suppose $f(a)$ and $f(b)$ have opposite signs then there exists $c \in (a, b)$ such that $f(c) = 0$. In Fig. 3.1(*b*), $f(a)$ is positive and $f(b)$ is negative.

(*iv*) A function $f(x)$ is said to be **derivable** in (a, b) if the curve $f(x)$ has a unique tangent at each point $c \in (a, b)$.

(*v*) Geometrically if $f(x)$ is derivable in (a, b) then the curve $f(x)$ has a unique tangent at each $c \in (a, b)$.

(*vi*) If $f(x)$ is derivable at $x = c$, then $f(x)$ is continuous at $x = c$.

(*vii*) If $f'(c) > 0$, then $f(x)$ is increasing at c observe Fig. 3.1 (*c*) and if $f'(c) < 0$, then $f(x)$ is decreasing at c observe Fig. 3.1 (*d*).

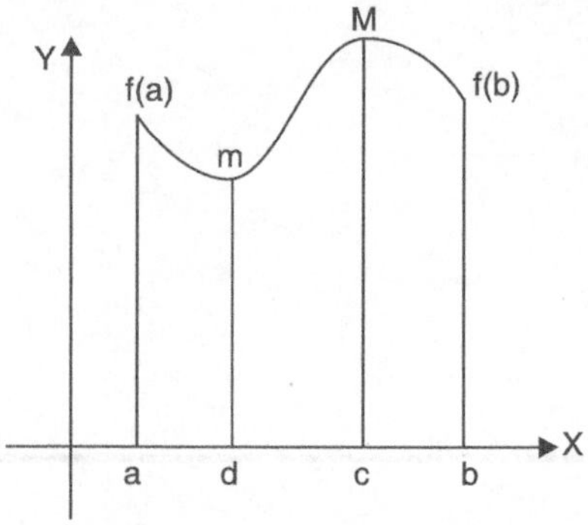

Fig. 3.1(*a*)

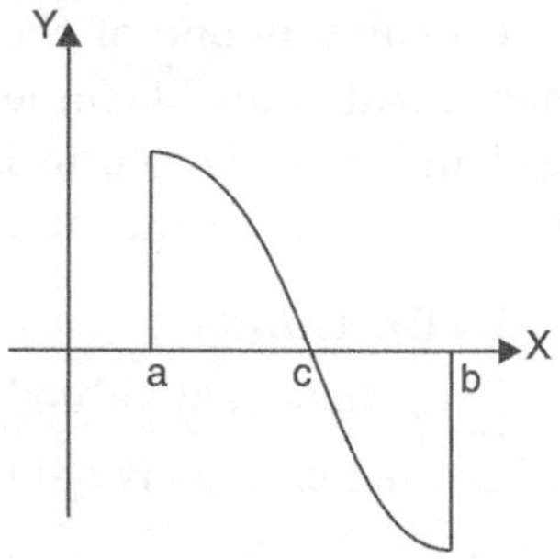

Fig. 3.1(*b*)

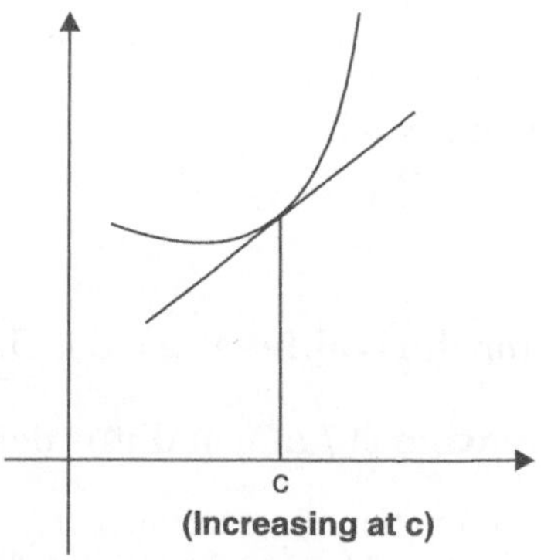

Fig. 3.1(*c*)

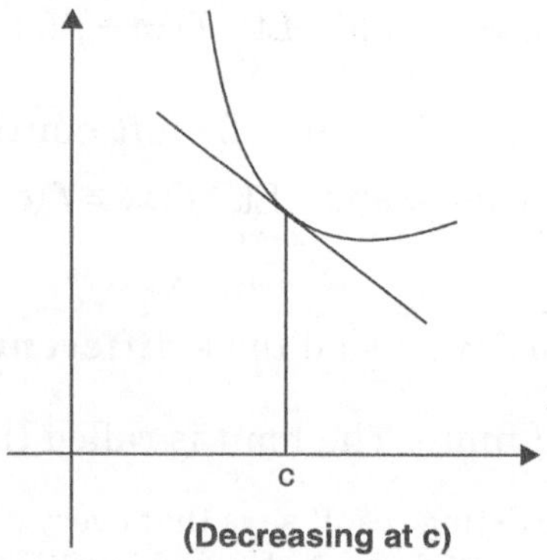

Fig. 3.1(*d*)

3.2. MEAN VALUE THEOREMS

Continuous and differentiable functions have many interesting properties. Some of them are studied through *Rolle's Theorem, Lagrange's Theorem and Cauchy's Theorem.* Two theorems *Taylor's* and *Maclaurin's,* which are generalizations of the mean value theorem, are useful in approximating transcendental functions.

Mean value theorems, especially Taylor's theorem play an important role in differentiation. The values of a function and its successive derivatives at a point help us in finding the value of the function in the neighbourhood of that point using Taylor's theorem. That is, Taylor's theorem provides expansion of $f(a + h)$ in ascending powers of h and the derivatives of f at a.

3.2.1. Rolle's Theorem (OU-Apr. 2007, Dec. 2008, 2013, May 2009)

Statement. Let $f(x)$ be a function that satisfy

(*i*) it is continuous in $[a, b]$; (*ii*) it is derivable in (a, b); and (*iii*) $f(a) = f(b)$, then there exists at least one point c in (a, b) such that $f'(c) = 0$.

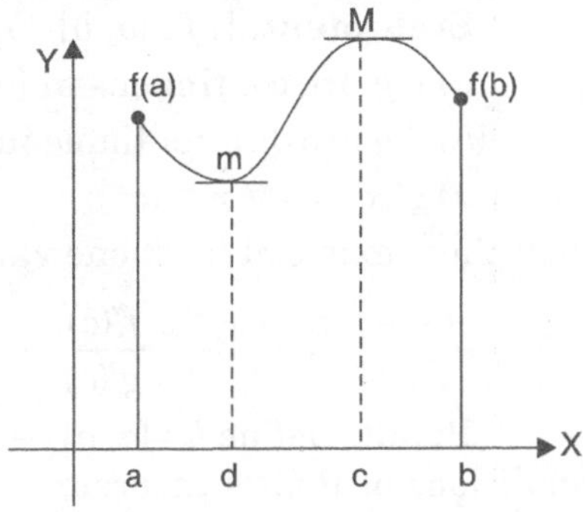

We use this Rolle's theorem in the proof of the following:

Proof. Since $f(x)$ is continuous in $[a, b]$, it is also bounded in $[a, b]$.

Let m = greatest lower bound and M = least upper bound.

Case I: If $M = m$, then $f(x)$ is constant *i.e.*, $f(x) = c \Rightarrow f'(x) = 0, \forall x \in (a, b)$.

Hence, the given statement is true.

Case II: If $M \neq m$, then take $f(c) = M$, $c \in (a, b)$

Since M is lub, $f(c + h) \leq M$ where h is very small positive value.

$$\Rightarrow \quad f(c+h) - f(c) \leq 0 \qquad [\because \; M = f(c)]$$

If $h > 0$ then $\underset{h \to 0}{\text{Lt}} \dfrac{f(c+h) - f(c)}{h} \leq 0 \;\Rightarrow\; f'(c) \leq 0$

If $h < 0$ then $\underset{h \to 0}{\text{Lt}} \dfrac{f(c+h) - f(c)}{h} \geq 0 \;\Rightarrow\; f'(c) \geq 0$

From above, we get $f'(c) = 0$ where $c \in (a, b)$.

$\therefore$ The given statement is true *i.e.*, there exists a point $c \in (a, b)$ such that $f'(c) = 0$. Similarly, if $f(d) = m$ where $d \in (a, b)$, we get $f'(d) = 0$.

3.2.2. Lagrange's Mean Value Theorem (OU-Apr. 2007, June 2013)

Statement. Let $f(x)$ be the function such that

(*i*) f is continuous in $[a, b]$ and

(*ii*) f is differentiable in (a, b)

then there exists at least one value $c \in (a, b)$ such that

$$f'(c) = \frac{f(b) - f(a)}{b - a}$$

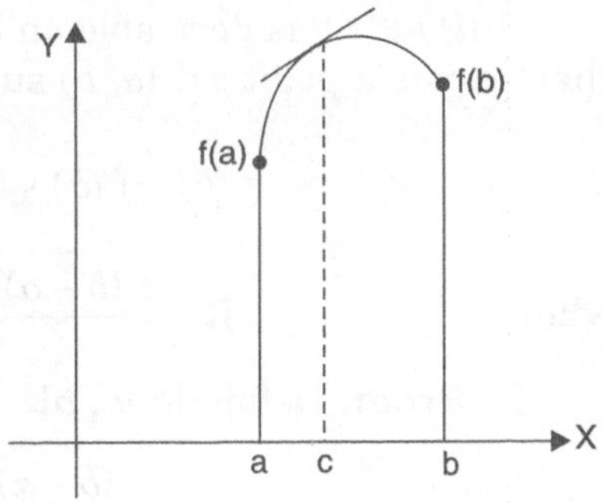

Proof. Define $g(x) = f(x) + Ax$ in $[a, b]$ ...(1)

where A is a constant to be determined and satisfies all the conditions of Rolle's theorem

Hence, $g(a) = g(b)$

$$\Rightarrow \quad f(a) + Aa = f(b) + Ab \;\Rightarrow\; A(a - b) = f(b) - f(a)$$

$$\Rightarrow \quad A = \frac{f(b) - f(a)}{-(b-a)} \qquad \text{...(2)}$$

Substitute the value of A in (1), then $g(x)$ satisfies all the conditions of Rolle's theorem. Hence $\exists \; c \in (a, b)$ such that $g'(c) = 0$

$$\Rightarrow \quad f'(c) + A = 0 \qquad (\because \; g(x) = f(x) + Ax \Rightarrow g'(x) = f'(x) + A)$$

$$\Rightarrow \quad f'(c) = -A = \frac{f(b) - f(a)}{b - a} \qquad | \text{ Using (2)}$$

3.2.3. Cauchy's Mean Value Theorem

(OU-June 2010, May 2009, 2008, Dec. 2008)

Statement. If $f : [a, b] \to \mathrm{R}$ and $g : [a, b] \to \mathrm{R}$ such that

(*i*) f, g are continuous in $[a, b]$,

(*ii*) f, g are differentiable in (a, b), and

(*iii*) $g'(x) \neq 0 \ \forall \ x \in (a, b)$

then there exists at least one value $c \in (a, b)$ such that

$$\frac{f'(c)}{g'(c)} = \frac{f(b) - f(a)}{g(b) - g(a)}$$

Proof. Define $h : [a, b] \to \mathrm{R}$ such that $h(x) = f(x) + \mathrm{A}g(x)$ where $\mathrm{A} \in \mathrm{R}$ and satisfies all the conditions of Rolle's theorem.

$$\therefore \qquad h(a) = h(b) \Rightarrow f(a) + \mathrm{A}g(a) = f(b) + \mathrm{A}g(b)$$

$$\Rightarrow \qquad \mathrm{A}[g(a) - g(b)] = f(b) - f(a)$$

$$\Rightarrow \qquad \mathrm{A} = -\left[\frac{f(b) - f(a)}{g(b) - g(a)}\right] \qquad \ldots(1)$$

where $\qquad g(b) - g(a) \neq 0$

By Rolle's theorem, we have

$$h'(c) = 0 \quad \Rightarrow \quad f'(c) + \mathrm{A}g'(c) = 0 \quad \Rightarrow \quad \mathrm{A} = -\frac{f'(c)}{g'(c)} \qquad \ldots(2)$$

From (1) and (2), we get

$$\frac{f'(c)}{g'(c)} = \frac{f(b) - f(a)}{g(b) - g(a)}.$$

3.2.4. Taylor's Theorem (Generalized Mean Value Theorem)

Statement. If $f : [a, b] \to \mathrm{R}$ satisfy (*i*) $f^{(n-1)}$ is continuous on $[a, b]$ and

(*ii*) $f^{(n-1)}$ is derivable on (a, b) (or) $f^{(n)}$ exists on (a, b) and p is a positive integer, then there exists a point $c \in (a, b)$ such that

$$f(b) = f(a) + \frac{(b-a)}{1!} f'(a) + \frac{(b-a)^2}{2!} f''(a) + \ldots + \frac{(b-a)^{n-1}}{(n-1)!} f^{(n-1)}(a) + \mathrm{R}_n$$

where $\qquad \mathrm{R}_n = \dfrac{(b-a)^p (b-c)^{x-p} f^{(x)}(c)}{(n-1)!\, p}.$

***Proof.** Define $\mathrm{F}: [a, b] \to \mathrm{R}$ by

$$\mathrm{F}(x) = f(b) - f(x) - \frac{(b-x)}{1!} f'(x) - \frac{(b-x)^2}{2!} f''(x) - \ldots - \frac{(b-x)^{n-1}}{(n-1)!} f^{(n-1)}(x) - \mathrm{A}\left[\frac{b-x}{b-a}\right]^p$$

where A is a real number chosen such that $\mathrm{F}(a) = 0$.

$\mathrm{F}(a) = 0$

$$\Rightarrow \quad 0 = f(b) - f(a) - \frac{(b-a)}{1!} f'(a) - \frac{(b-a)^2}{2!} f''(a) - \ldots - \frac{(b-a)^{n-1}}{(n-1)!} f^{(n-1)}(a) - \mathrm{A}\left[\frac{b-a}{b-a}\right]^p$$

$$\Rightarrow \quad f(b) = f(a) + \frac{(b-a)}{1!} f'(a) + \ldots + \frac{(b-a)^{(n-1)}}{(n-1)!} f^{(n-1)}(a) + \mathrm{A} \qquad \ldots(1)$$

*Proof is for student reference only.

$f^{(n-1)}$ is continuous on $[a, b]$ $\Rightarrow$ $f, f' \ldots f^{(n-1)}$ are continuous on $[a, b]$

$f^{(n-1)}$ is derivable on (a, b) $\Rightarrow$ $f, f', \ldots f^{(n-1)}$ are derivable on (a, b).

Since $(b-x), (b-x)^2, \ldots, (b-x)^{n-1}, (b-x)^n$ are polynomial functions, we conclude that they are continuous and derivable on R. In particular they are continuous on $[a, b]$ and derivable on (a, b).

Hence, F is continuous on $[a, b]$ and derivable on (a, b).

Also, $$F(b) = f(b) - f(b) - \frac{(b-b)}{1!} f'(b) \ldots \frac{(b-b)^{n-1}}{(n-1)!} f^{(n-1)}(b) - A\left[\frac{b-b}{b-a}\right]^p = 0$$

Since $F(a) = 0$ by the definition of F, we have that $F(a) = F(b)$

$F : [a, b] \to R$ satisfies all the conditions of Rolle's theorem.

Hence there exists $c \in (a, b)$ such that $F'(c) = 0$.

For $x \in (a, b)$, $F'(x) = -f'(x) - \{-f'(x) + (b-x) f''(x)\}$

$$-\left\{-(b-x) f''(x) + \frac{(b-x)^2}{2!} f'''(x)\right\} \ldots$$

$$-\left\{-\frac{(b-x)^{n-2}}{(n-2)!} f^{(n-1)}(x) + \frac{(b-x)^{n-1}}{(n-1)!} f^n(x)\right\} + \frac{Ap}{(b-a)^p}(b-x)^{p-1}$$

That is, $$F'(x) = \frac{-(b-x)^{n-1}}{(n-1)!} f^{(n)}(x) + \frac{Ap}{(b-a)^p}(b-x)^{p-1}$$

For $c \in (a, b)$, $F'(c) = 0$

$$\frac{-(b-c)^{n-1}}{(n-1)!} f^{(n)}(c) + \frac{Ap}{(b-a)^p}(b-c)^{p-1} = 0$$

$$\Rightarrow \quad A = \frac{(b-a)^p (b-c)^{n-p} f^{(n)}(c)}{(n-1)!\, p} \quad \ldots(2)$$

From equations (1) and (2), we have that

$$f(b) = f(a) + \frac{(b-a)}{1!} f'(a) + \frac{(b-a)^2}{2!} f''(a) + \ldots + \frac{(b-a)^{n-1}}{(n-1)!} f^{(n-1)}(a)$$
$$+ \frac{(b-a)^p (b-c)^{n-p} f^{(n)}(c)}{p(n-1)!} \quad \ldots(A)$$

The proof is complete.

Note. There are different forms of remainder value R_n. Each of these remainder values may have its own significance. We learn the following forms:

(*i*) SCHLÖMILCH-ROCHE'S form of remainder:

$$R_n = \frac{(b-a)^p (b-c)^{n-p} f^{(n)}(c)}{p(n-1)!}$$

(*ii*) LAGRANGE'S form of remainder:

(Put $p = n$ in SCHLÖMILCH-ROCHE'S form of remainder)

$$R_n = \frac{(b-a)^n f^{(n)}(c)}{n!}$$

(*iii*) CAUCHY'S form of remainder

(Put $p = 1$ in SCHLÖMILCH-ROCHES form of remainder)

$$R_n = \frac{(b-a)(b-c)^{n-1} f^{(n)}(c)}{(n-1)!}$$

Note. Suppose the remainder after nth term tends to 0 as $n \to \infty$. Then

$$f(x) = f(a) + (x-a) f'(a) + \frac{(x-a)^2}{2!} f''(a) + \ldots \frac{(x-a)^n}{n!} f^{(n)}(a) + \ldots$$

is called Taylor's series expansion of $f(x)$ about $x = a$, assuming that $f(x)$ has successive derivatives of all orders for $x \in [a, b]$.

3.2.5. Another Form of Taylor's Theorem

If f: $[a, a+h] \to R$ satisfy

(*i*) $f', f'', \ldots, f^{(n-1)}$ are continuous on $[a, a+h]$;

(*ii*) $f^{(n-1)}$ is derivable on $(a, a+h)$ and p is a positive integer, then there exists a real number $0 < \theta < 1$ such that

$$f(a+h) = f(a) + \frac{h}{1!} f'(a) + \frac{h^2}{2!} f''(a) + \ldots \frac{h^{n-1}}{(n-1)!} f^{(n-1)}(a) + R_n \qquad \ldots(B)$$

where $R_n = \dfrac{h^n (1-\theta)^{n-p} f^{(n)}(a+\theta h)}{p(n-1)!}$ (This is known as Schlömilch-Roches form of remainder).

This theorem is called as Taylor's Theorem with Schlömilch-Roches's form of remainder.

If $p = n$, then $R_n = \dfrac{h^n f^{(n)}(a+\theta h)}{n!}$ (In this case R_n is known as Lagrange's form of remainder).

In this case the theorem is called as Taylor's theorem with Lagrange's form of remainder.

If $p = 1$, then $R_n = \dfrac{h^n (1-\theta)^{n-1} f^{(n)}(a+\theta h)}{(n-1)!}$ (In this case R_n is known as Cauchy's form of remainder).

In this case, the theorem is called as Taylor's theorem with Cauchy's form of remainder.

3.2.6. Maclaurin's Theorem

If f: $[0, x] \to R$ is satisfy

(*i*) $f', f^{(2)}, \ldots, f^{(n-1)}$ will exist and continuous on $[0, x]$;

(*ii*) $f^{(n-1)}$ is derivable on the open interval $(0, x)$ and p is a positive integer, then there exists a real number $\theta \in (0, 1)$ such that

$$f(x) = f(0) + xf'(0) + \frac{x^2}{2!} f''(0) + \ldots + \frac{x^{n-1}}{(n-1)!} f^{(n-1)}(0) + \frac{x^n (1-\theta)^{n-p}}{p(n-1)!} f^{(n)}(\theta x) \qquad \ldots(C)$$

Here $R_n = \dfrac{x^n(1-\theta)^{n-p}}{p(n-1)!} f^{(n)}(\theta x)$ is called as Schlömilch-Rocheses form of remainder.

Proof. If we put $a = 0$, $b = x$ and $c = \theta x$ where $0 < \theta < 1$ in Taylor's theorem then, we get Meclaurin's theorem. So, there is no basic difference between Taylor's and Maclaurin's Theorem.

Note 1. (*i*) If $p = n$ then $R_n = \dfrac{x^n}{n!} f^{(n)}(\theta x)$.

In this case, R_n is called as Lagrange's form of remainder.

(*ii*) If $p = 1$ then $R_n = \dfrac{x^n(1-\theta)^{n-1} f^{(n)}(\theta x)}{(n-1)!}$.

In this case, R_n is called as Cauchy's form of remainder.

Note 2. The expressions of $f(b)$ in (A) (of 3.2.4), $f(a + h)$ in (B) (of 3.2.5) and $f(x)$ in C (of 3.2.6) are finite sums. However as $n \to \infty$, the expressions (A), (B), (C) become infinite series. These are convergent if the remainder after n terms, $R_n \to 0$ as $n \to \infty$.

3.2.7. Maclaurin's Series Expansion

If $f(x)$ has derivatives of all orders for $x \in [0, b]$, and if the remainder after nth term is in Lagrange's form of remainder $R_n = \dfrac{x^n}{n!} f^{(n)}(\theta x)$ tends to 0, where $0 < \theta < 1$ as $n \to \infty$, then

$f(x) = f(0) + xf'(0) + \dfrac{x^2}{2!} f''(0) + \ldots + \dfrac{x^n}{n!} f^{(n)}(0) + \ldots$ is called as Maclaurin's series expansion of $f(x)$.

Note. In most of the elementary cases, the Taylor's series expansion of $f(x)$ about $x = a$ and the Maclaurin's series expansion of $f(x)$ (which is just the Taylor's series expansion of $f(x)$ about $x = 0$) convergent to $f(x)$ in the interval, where $f(x)$ has successive derivatives of all orders, with the condition that $R_n \to 0$ as $n \to \infty$.

ILLUSTRATIVE EXAMPLES

Example 1. *Verify Rolle's theorem for* $f(x) = x^2 - 2x - 3$ *in* $(1, -3)$.

Sol. Given that $f(x) = x^2 - 2x - 3, x \in (1, -3)$

Also, $f(1) = 1 - 2 - 3 = -4$

and $f(-3) = 9 + 6 - 3 = 12$

Since $f(1) \neq f(-3)$, the condition of Rolle's theorem is not satisfied.

Example 2. *Verify Rolle's theorem for* $f(x) = x^{2/3} - 2x^{1/3}$ *in the interval* $(0, 8)$.

Sol. Given that $f(x) = x^{2/3} - 2x^{1/3}, x \in (0, 8)$.

Also, $f(0) = 0 - 2 \cdot 0 = 0$

and $f(8) = 8^{2/3} - 2(8)^{1/3} = 4 - 4 = 0$.

Now, $f(0) = f(8)$.

Therefore, f satisfies all the conditions of Rolle's, theorem.

Thus, there exists a number $c \in (0, 8)$ such that $f'(c) = 0$.

So $f'(c) = \frac{2}{3}c^{-1/3} - \frac{2}{3}c^{-2/3} = 0$

$\Rightarrow$ $c^{-1/3} = c^{-2/3}$ or $c(c-1) = 0$ or $c = 1$ or 0.

As $c = 0 \notin (0, 8)$, we take $c = 1$

Hence, we verified the theorem.

Example 3. *Using Rolle's theorem. Show that $g(x) = 8x^3 - 6x^2 - 2x + 1$ has a zero between 0 and 1.* [JNTU 2008 (Set 4)]

Sol. Given that $g(x) = 8x^3 - 6x^2 - 2x + 1,\ x \in [0, 1]$

Now $g(x)$ is continuous in $[0, 1]$ and differentiable in $(0, 1)$.

Also, $g(0) = 1, g(1) = 1$.

So g satisfies the conditions of Rolle's theorem.

Therefore, there exists a number $c \in (0, 1)$ such that $g'(c) = 0$.

$$g'(x) = 24x^2 - 12x - 2$$

Now, $g'(c) = 0 \Rightarrow 24c^2 - 12c - 2 = 0 \Rightarrow c = \frac{3 \pm \sqrt{21}}{12}$.

But $c = \frac{3 \pm \sqrt{21}}{12} = 0.6318 \in (0, 1)$.

Example 4. *Verify Rolle's theorem for $f(x) = 9x^3 - 4x$ in the interval $\left[-\frac{2}{3}, \frac{2}{3}\right]$.* (JNTU 2010)

Sol. Given function is $f(x) = 9x^3 - 4x$.

This $f(x)$ is continuous in $\left[-\frac{2}{3}, \frac{2}{3}\right]$ and differentiable in $\left(-\frac{2}{3}, \frac{2}{3}\right)$

$$f\left(-\frac{2}{3}\right) = -\frac{8}{3} + \frac{8}{3} = 0,$$

$$f\left(\frac{2}{3}\right) = \frac{8}{3} - \frac{8}{3} = 0.$$

Thus, $f\left(-\frac{2}{3}\right) = f\left(\frac{2}{3}\right)$

Therefore, f satisfies all the conditions of Rolle's theorem. So there exists a number $c \in \left(-\frac{2}{3}, \frac{2}{3}\right)$ such that $f'(c) = 0$.

Now, $f'(x) = 27x^2 - 4$

and $f'(c) = 0 \Rightarrow 27c^2 - 4 = 0$

$\Rightarrow$ $c^2 = \frac{4}{27}$.

$\Rightarrow$ $c = \pm\frac{2}{3\sqrt{3}} = \pm\frac{2\sqrt{3}}{9} = \pm 0.385$

So the value of c is 0.385 and – 0.385. Both the values lies in $\left(-\frac{2}{3}, \frac{2}{3}\right)$.

$\therefore$ Rolle's theorem is verified.

Example 5. *Verify whether Rolle's theorem can be applied to the following functions in the intervals cited.*

(i) $f(x) = \tan x$ *in* $[0, \pi]$ *(ii)* $f(x) = \dfrac{1}{x}$ *in* $[-1, 1]$

(iii) $f(x) = x^2$ *in* $[1, 3]$.

Sol. (*i*) Given $f(x) = \tan x$ in $[0, \pi]$. Since $f\left(\dfrac{\pi}{2}\right) = \tan \dfrac{\pi}{2} = \infty$ where $\dfrac{\pi}{2} \in [0, \pi]$

$f(x)$ is discontinuous in $[0, \pi]$

$\therefore$ Rolle's theorem cannot be applied to $f(x) = \tan x$ in $[0, \pi]$.

(*ii*) Given $f(x) = \dfrac{1}{x}$ in $[-1, 1]$.

Since $f(0) = \dfrac{1}{0} = \infty$ where $0 \in [-1, 1]$,

$f(x)$ is not continuous in $[-1, 1]$.

Hence, Rolle's theorem is not applicable.

(*iii*) Clearly $f(x) = x^2$ is continuous in $[1, 3]$ and differentiable in $(1, 3)$. $(\because f'(x) = 2x)$

But $f(1) = 1^2 = 1$ and $f(3) = 3^2 = 9$

$\therefore$ $f(1) \neq f(3)$

Hence, Rolle's theorem is not applicable.

Example 6. *Verify Rolle's theorem for the function* $f(x) = \dfrac{\sin x}{e^x}$ *in* $[0, \pi]$.

Sol. Given $f(x) = \dfrac{\sin x}{e^x}$ in $[0, \pi]$

(*i*) Since $\sin x$ and e^x are continuous functions in $[0, \pi]$, $f(x) = \dfrac{\sin x}{e^x}$ also continuous in $[0, \pi]$

(*ii*) Since $\sin x$ and e^x are differentiable in $(0, \pi)$, $f(x) = \dfrac{\sin x}{e^x}$ also differentiable in $(0, \pi)$.

(*iii*) $f(0) = \dfrac{\sin 0}{e^0} = \dfrac{0}{1} = 0$

$$f(\pi) = \frac{\sin \pi}{e^\pi} = \frac{0}{e^\pi} = 0$$

$\therefore$ $f(0) = f(\pi)$

Hence, Rolle's theorem can be applied.

$$f'(x) = \frac{e^x \cos x - \sin x \cdot e^x}{(e^x)^2} \qquad \left(\because \frac{d^{u/v}}{dx} = \frac{vu' - uv'}{v^2}\right)$$

$$= \frac{\cos x - \sin x}{e^x}$$

By Rolle's theorem, $f'(c) = 0 \Rightarrow \dfrac{\cos c - \sin c}{e^c} = 0$

$$\Rightarrow \quad \cos c - \sin c = 0 \Rightarrow \sin c = \cos c \Rightarrow \frac{\sin c}{\cos c} = 1$$

$$\Rightarrow \tan c = 1 = \tan \frac{\pi}{4} \Rightarrow c = \frac{\pi}{4} \in (0, \pi)$$

$\therefore$ Rolle's theorem is verified.

Example 7. *Verify Rolle's theorem for the function* $log \left[\dfrac{x^2 + ab}{x(a + b)}\right]$ *in* $[a, b], a > 0, b > 0.$

Sol. Given $f(x) = \log \left[\dfrac{x^2 + ab}{x(a + b)}\right]$

$$\Rightarrow \qquad f(x) = \log (x^2 + ab) - \log x - \log (a + b)$$

(*i*) Since $a > 0, b > 0$, clearly $f(x)$ is continuous in $[a, b]$.

(*ii*) $f'(x) = \dfrac{2x}{x^2 + ab} - \dfrac{1}{x} = \dfrac{x^2 - ab}{x(x^2 + ab)}$ exists $\forall\, x \in (a, b)$

Hence $f(x)$ is differentiable in (a, b)

(*iii*) $f(a) = \log \left[\dfrac{a^2 + ab}{a(a + b)}\right] = \log 1 = 0$

$$f(b) = \log \left[\frac{b^2 + ab}{b(a + b)}\right] = \log 1 = 0$$

$$\therefore \qquad f(a) = f(b)$$

Hence, by Rolle's theorem,

$\exists\, c \in (a, b)$ such that $f'(c) = 0$

$$\Rightarrow \qquad \frac{c^2 - ab}{c(c^2 + ab)} = 0$$

$$\Rightarrow \qquad c^2 - ab = 0 \quad \Rightarrow \quad c^2 = ab$$

$$\Rightarrow \qquad c = \pm\sqrt{ab}$$

Take $c = \sqrt{ab} \in (a, b)$ $\qquad$ ($\because$ $\sqrt{ab}$ is a geometric mean of a, b)

Hence, Rolle's theorem is verified.

Example 8. *Verify Rolle's theorem for* $f(x) = |\, x \,|$ *in* $[-1, 1]$.

Sol. Take $x = 0 \in (-1, 1)$. Then

$$\text{LHD} = f'(0-) = \underset{h \to 0^-}{\text{Lt}} \frac{f(0 + h) - f(0)}{h}$$

$$= \underset{h \to 0^-}{\text{Lt}} \frac{|\, h \,| - 0}{h} = \underset{h \to 0^-}{\text{Lt}} \frac{-h}{h} = -1 \qquad (\because\ h < 0)$$

$$\text{RHD} = f'(0+) = \underset{h \to 0+}{\text{Lt}} \frac{f(0 + h) - f(0)}{h}$$

$$= \underset{h \to 0+}{\text{Lt}} \frac{|h| - 0}{h} = \underset{h \to 0+}{\text{Lt}} \frac{h}{h} = 1 \qquad (\because\ h > 0)$$

Since, LHD $\neq$ RHD at $x = 0$,

$f(x)$ is not differentiable at $x = 0 \in (-1, 1)$.

Hence, Rolle's theorem cannot be applied.

Example 9. *Verify Rolle's theorem for $f(x) = (x - a)^m (x - b)^n$ where m, n are positive integers in $[a, b]$.*

Sol. Clearly, $f(x)$ is continuous in $[a, b]$ and differentiable in (a, b).

Also, $f(a) = f(b) = 0$

$$f'(x) = m(x-a)^{m-1}(x-b)^n + n(x-a)^m (x-b)^{n-1}$$

$(\because \quad (uv)' = u'v + uv')$

By Rolle's theorem, we have

$$f'(c) = 0$$

$$\Rightarrow \quad m(c-a)^{m-1}(c-b)^n + n(c-a)^m(c-b)^{n-1} = 0$$

$$\Rightarrow \quad (c-a)^{m-1}(c-b)^{n-1}[m(c-b) + n(c-a)] = 0$$

$$\Rightarrow \quad mc - mb + nc - na = 0$$

$$\Rightarrow \quad c = \frac{mb + na}{m + n} \in (a, b)$$

Hence, Rolle's theorem is verified.

Example 10. *Verify Lagrange's mean value theorem for $f(x) = (x - 1)(x - 2)(x - 3)$ in $[0, 4]$.*

(OU-May 2012, 2009)

Sol. Let $a = 0$, $b = 4$ and

$$f(a) = f(0) = (0-1)(0-2)(0-3) = -6,$$

$$f(b) = f(4) = (4-1)(4-2)(4-3) = 6$$

$$f(x) = (x-1)(x-2)(x-3) = x^3 - 6x^2 + 11x - 6$$

$$\Rightarrow \quad f'(x) = 3x^2 - 12x + 11$$

$$\Rightarrow \quad f'(c) = 3c^2 - 12c + 11$$

Clearly, $f(x)$ is continuous in $[0, 4]$ and differentiable in $(0, 4)$.

By Lagrange's mean value theorem, we have

$$f'(c) = \frac{f(b) - f(a)}{b - a}$$

$$\Rightarrow \quad 3c^2 - 12c + 11 = \frac{6+6}{4-0} = \frac{12}{4} = 3$$

$$\Rightarrow \quad 3c^2 - 12c + 8 = 0$$

$$c = \frac{12 \pm \sqrt{144 - 96}}{6} = \frac{6 \pm 2\sqrt{3}}{3} \in (0, 4)$$

$\therefore$ Lagrange's mean value theorem is verified.

Example 11. *Verify Lagrange's mean value theorem for* $f(x) = x^3 - 3x - 1$ *in* $\left(-\frac{11}{7}, \frac{13}{7}\right)$.

Sol. Here $a = -\frac{11}{7}$, $b = \frac{13}{7}$

$$f(a) = f\left(-\frac{11}{7}\right) = \left(-\frac{11}{7}\right)^3 - 3\left(-\frac{11}{7}\right) - 1$$

$$= -\frac{1331}{343} + \frac{33}{7} - 1 = \frac{-1331 + 1617 - 343}{343} = \frac{-57}{343}$$

$$f(b) = f\left(\frac{13}{7}\right) = \left(\frac{13}{7}\right)^3 - 3\left(\frac{13}{7}\right) - 1 = \frac{2197}{343} - \frac{39}{7} - 1$$

$$= \frac{2197 - 1911 - 343}{343} = \frac{-57}{343}$$

Given $\quad f(x) = x^3 - 3x - 1$

$\Rightarrow \quad f'(x) = 3x^2 - 3$

$\Rightarrow \quad f'(c) = 3c^2 - 3$

Clearly $f(x)$ is continuous in $\left[-\frac{11}{7}, \frac{13}{7}\right]$ and differentiable in $\left(-\frac{11}{7}, \frac{13}{7}\right)$.

By Lagrange's mean value theorem, we have

$$f'(c) = \frac{f(b) - f(a)}{b - a}$$

$$\Rightarrow \quad 3c^2 - 3 = \frac{\frac{-57}{343} + \frac{57}{343}}{\frac{13}{7} + \frac{11}{7}} = 0$$

$$\Rightarrow \quad 3c^2 - 3 = 0 \Rightarrow c^2 = 1$$

$$\Rightarrow \quad c = \pm 1 \in \left(-\frac{11}{7}, \frac{13}{7}\right)$$

Hence, Lagrange's mean value theorem is verified.

Example 12. *If* $f(x) = \sin^{-1} x$ *and* $0 < a < 1$, *then prove that*

$$\frac{b-a}{\sqrt{1-a^2}} < \sin^{-1} b - \sin^{-1} a < \frac{b-a}{\sqrt{1-b^2}}.$$

Sol. Given that $f(x) = \sin^{-1} x$, $0 < a < b < 1$.

Now, $f'(x) = \frac{1}{\sqrt{1-x^2}}$ which exists in (0, 1).

By using generalized mean value theorem for the function $f(x)$ in $[a, b]$ with Lagrange's form of remainder, we get

$$f(b) = f(a) + \frac{(b-a)}{1!} f'(c) \text{ for some } c \in (a, b)$$

$$\Rightarrow \quad \frac{f(b) - f(a)}{b - a} = f'(c);\ a < c < b$$

$$\Rightarrow \qquad \frac{\sin^{-1} b - \sin^{-1} a}{b - a} = \frac{1}{\sqrt{1 - c^2}} \qquad ...(1)$$

Since $a < c < b \Rightarrow a^2 < c^2 < b^2 \Rightarrow - a^2 > - c^2 > - b^2 \Rightarrow 1 - a^2 > 1 - c^2 > 1 - b^2$

$$\Rightarrow \qquad \sqrt{1 - a^2} > \sqrt{1 - c^2} > \sqrt{1 - b^2}$$

$$\Rightarrow \qquad \frac{1}{\sqrt{1 - a^2}} < \frac{1}{\sqrt{1 - c^2}} < \frac{1}{\sqrt{1 - b^2}}$$

Now from (1), we get that

$$\frac{1}{\sqrt{1 - a^2}} < \frac{\sin^{-1} b - \sin^{-1} a}{b - a} < \frac{1}{\sqrt{1 - b^2}}$$

$$\Rightarrow \qquad \frac{b - a}{\sqrt{1 - a^2}} < \sin^{-1} b - \sin^{-1} a < \frac{b - a}{\sqrt{1 - b^2}}.$$

Example 13. *Prove using mean value theorem* $|\sin u - \sin v| \leq |u - v|$.

[JNTU 2008 (Set 3)]

Sol. Suppose $f(x) = \sin x, x \in [u, v]$

Using mean value theorem for $f(x)$ in $[u, v]$ and neglecting higher powers, we get that $f(v) = f(u) + (v - u) f'(c)$ for some $c \in [u, v]$.

$$\Rightarrow \qquad \sin v - \sin u = (v - u) \cos c$$

$$\Rightarrow \qquad |\sin u - \sin v| = |u - v| \cos c \qquad | \text{ Since } |\cos c| \leq 1$$

$$\Rightarrow \qquad |\sin u - \sin v| \leq |u - v|.$$

Example 14. *Using Lagrange mean value theorem, show that* $|\cos b - \cos a| \leq |b - a|$.

(OU-May 2008)

Sol. Let $f(x) = \cos x \Rightarrow f'(x) = -\sin x$

By Lagrange's mean value theorem, we have

$$f'(c) = \frac{f(b) - f(a)}{b - a}$$

$$\Rightarrow \qquad -\sin c = \frac{\cos b - \cos a}{b - a}$$

$$\Rightarrow \qquad \sin c = \frac{\cos a - \cos b}{b - a}$$

Since $\qquad |\sin c| \leq 1 \Rightarrow \left| \dfrac{\cos a - \cos b}{b - a} \right| \leq 1$

$$\Rightarrow \qquad |\cos b - \cos a| \leq |b - a|.$$

Example 15. *State whether Lagrange's mean value theorem can be applied to the function* $x^{1/3}$ *in* $[-1, 1]$. *Justify your answer.*

Sol. Let $f(x) = x^{1/3}$

$$\Rightarrow \qquad f'(x) = \frac{1}{3} x^{\frac{1}{3} - 1} = \frac{1}{3} x^{\frac{-2}{3}} = \frac{1}{3x^{2/3}}$$

Clearly, $f(x)$ is continuous in $[-1, 1]$ but $f'(0) = \frac{1}{0} = \infty$ where $0 \in (-1, 1)$.

Hence $f(x)$ is not differentiable in $(-1, 1)$.

$\therefore$ Lagrange's mean value theorem cannot be applied.

Example 16. *Find c of Cauchy's mean value theorem for the functions $f(x) = \sin x$, $g(x) = \cos x$ in (a, b).* (OU-Nov. 2006)

Sol. Since $\sin x$ and $\cos x$ are continuous and differentiable, $f(x)$ and $g(x)$ are continuous in $[a, b]$ and differentiable in (a, b).

$f'(x) = \cos x$ and $g'(x) = -\sin x$

By Cauchy's mean value theorem, we have

$$\frac{f'(c)}{g'(c)} = \frac{f(b)-f(a)}{g(b)-g(a)}$$

$$\Rightarrow \quad \frac{\cos c}{-\sin c} = \frac{\sin b - \sin a}{\cos b - \cos a} = \frac{2\cos\left(\frac{b+a}{2}\right)\sin\left(\frac{b-a}{2}\right)}{-2\sin\left(\frac{b+a}{2}\right)\sin\left(\frac{b-a}{2}\right)}$$

$$\Rightarrow \quad -\cot c = -\cot\left(\frac{b+a}{2}\right)$$

$$\Rightarrow \quad c = \frac{a+b}{2} \in (a, b) \qquad \left(\because \frac{a+b}{2} \text{ is an arithmetic mean of } a, b\right)$$

$\therefore$ Cauchy's mean value theorem is verified.

Example 17. *Verify Cauchy's mean value theorem for $f(x) = x^2$, $g(x) = x^3$ in $[1, 2]$.*

Sol. Given $f(x) = x^2$, $g(x) = x^3$ in $[1, 2]$

$$\Rightarrow \quad f'(x) = 2x,\ g'(x) = 3x^2$$

Clearly f, g are continuous in $[1, 2]$ and differentiable in $(1, 2)$.

By Cauchy's mean value theorem, we have

$$\frac{f'(c)}{g'(c)} = \frac{f(2)-f(1)}{g(2)-g(1)}$$

$$\Rightarrow \quad \frac{2c}{3c^2} = \frac{4-1}{8-1}$$

$$\Rightarrow \quad \frac{2}{3c} = \frac{3}{7}$$

$$\Rightarrow \quad 14 = 9c$$

$$\Rightarrow \quad c = \frac{14}{9} \in (1, 2)$$

Hence, Cauchy's mean value theorem is verified.

Example 18. *Find c of Cauchy's mean value theorem on $[a, b]$ for*

$f(x) = e^x$ and $g(x) = e^{-x}$ $(a, b > 0)$.

Sol. Given $\qquad f(x) = e^x,\ g(x) = e^{-x}$

$\Rightarrow \qquad f'(x) = e^x,\ g'(x) = -e^{-x}$

Since e^x is continuous and differentiable, f, g are continuous in $[a, b]$ and differentiable in (a, b).

By Cauchy's mean value theorem

$$\frac{f'(c)}{g'(c)} = \frac{f(b)-f(a)}{g(b)-g(b)}$$

$$\Rightarrow \qquad \frac{e^c}{-e^{-c}} = \frac{e^b - e^a}{e^{-b} - e^{-a}}$$

$$\Rightarrow \qquad -e^{2c} = \frac{e^b - e^a}{\dfrac{1}{e^b} - \dfrac{1}{e^a}} = \frac{-(e^a - e^b)}{\dfrac{(e^a - e^b)}{e^b \cdot e^a}}$$

$$\Rightarrow \qquad e^{2c} = e^{a+b}$$

$$\Rightarrow \qquad 2c = a + b$$

$$\Rightarrow \qquad c = \frac{a+b}{2} \in (a, b) \qquad \left(\because \frac{a+b}{2} \text{ is A.M. of } a, b\right)$$

Example 19. *Verify Cauchy's mean value theorem for the functions f(x) and f′(x) in (1, e) given f(x) = log x.*

Sol. $f(x) = \log x$

$$\Rightarrow \qquad f'(x) = \frac{1}{x} \Rightarrow f'(c) = \frac{1}{c}$$

Given $f(x) = \log x,\ g(x) = f'(x) = \dfrac{1}{x}$ (say)

$$\Rightarrow \qquad g'(x) = -\frac{1}{x^2} \Rightarrow g'(c) = -\frac{1}{c^2}$$

Clearly, f, g are continuous in $[1, e]$ and differentiable in $(1, e)$.

By Cauchy's mean value theorem

$$\frac{f'(c)}{g'(c)} = \frac{f(e)-f(1)}{g(e)-g(1)}$$

$$\Rightarrow \qquad \frac{\dfrac{1}{c}}{-\dfrac{1}{c^2}} = \frac{\log e - \log 1}{\dfrac{1}{e} - \dfrac{1}{1}}$$

$$\Rightarrow \qquad -c = \frac{1}{\left(\dfrac{1-e}{e}\right)} = \frac{e}{-(e-1)}$$

$$\Rightarrow \qquad c = \frac{e}{e-1} = \frac{2.7183}{1.7183} \in (1, e)$$

Hence, Cauchy's mean value theorem is verified.

Example 20. *Consider the natural logarithm log x and find its Taylor series with centre $x_0 = 1$.*

Sol. Write $f(x) = \ln(x)$. So $f(1) = \ln 1 = 0$.

$$f'(x) = \frac{1}{x}, \quad \text{and} \quad f'(1) = +1.$$

$$f''(x) = \frac{-1}{x^2}, \quad \text{and} \quad f''(1) = -1 = (-1)^3(1!).$$

$$f'''(x) = \frac{+2}{x^3}, \quad \text{and} \quad f'''(1) = +2 = (-1)^4(2!).$$

$$f^{iv}(x) = \frac{-2.3}{x^4}, \quad \text{and} \quad f^{iv}(1) = -2.3 = (-1)^5(3!)$$

$$f^{v}(1) = +2 \cdot 3 \cdot 4 = 4! = (-1)^6(4!).$$

Thus, we get that $f^{(n)}(1) = (-1)^{n+1}(n-1)!$

Check that this works for all $n = 1, 2, ...$

This yields the following Taylor series

$$f(x) = f(1) + (x-1)f'(1) + \frac{(x-1)^2}{2!}f''(1) + \ldots$$

$$\Rightarrow \quad \log x = \sum_{n=1}^{\infty} \frac{(-1)^{n+1}(n-1)!}{n!}(x-1)^n$$

$$= \sum_{n=1}^{\infty} \frac{(-1)^{n+1}}{n}(x-1)^n$$

$$= (x-1) - \frac{(x-1)^2}{2} + \frac{(x-1)^3}{3} - \ldots$$

Example 21. *Find the Taylor series with center $x_0 = 0$ for $f(x) = \cos(2x)$.*

Sol. First, we compute the derivatives at $x_0 = 0$.

Write $f(x) = \cos(2x)$. Now $f(0) = 1$.

$$f'(x) = -2\sin(2x) \quad \Rightarrow \quad f'(0) = 0$$

$$f''(x) = -2^2\cos(2x) \quad \Rightarrow \quad f''(0) = -2^2$$

$$f'''(x) = +2^3\sin(2x) \quad \Rightarrow \quad f'''(0) = 0$$

$$f^{iv}(x) = 2^4\cos(2x) \quad \Rightarrow \quad f^{iv}(0) = 2^4$$

....

....

Here only the even order derivatives are non-zero [since cos (2x) is an even function]. We can model this for the general term by using the counter $2n$ instead of n (we use $2n + 1$, if we want to catch odd terms).

Since the power of the term 2 equals the order of the derivative, the general term has the form $f^{(2n)}(0) = \pm 2^{2n}$.

What about the alternating sign?

If $n = 0$, we want to create a positive term, so let's use $(-1)^n$.

For $n = 1$, this produces the desired negative term (associated with the second derivative).

Thus $f^{(2n)}(0) = (-1)^n 2^{2n}$

This yields the following Taylor series expansion.

$$f(x) = f(0) + xf'(0) + \frac{x^2}{2!} f''(0) + \ldots$$

$$\Rightarrow \qquad \cos(2x) = \sum_{n=0}^{\infty} \frac{(-1)^n 2^{2n}}{(2n)!} x^{2n} = 1 - \frac{4}{2!}x^2 + \frac{16}{4!}x^4 - \frac{64}{6!}x^6 + \ldots$$

Example 22. *Find the Taylor series for which* $f(x) = \dfrac{2}{x-5}$ *with center* $x_0 = 2$.

Sol. (Another method) $f(x) = \dfrac{2}{x-5} = \dfrac{2}{(x-2)-3} = \dfrac{2}{3}\left[\dfrac{1}{\dfrac{(x-2)}{3} - 1}\right] = (-1)\dfrac{2}{3}\left[\dfrac{1}{1 - \dfrac{(x-2)}{3}}\right]$

Since $x_0 = 2 \Rightarrow |x-2| < 1 \Rightarrow \left|\dfrac{x-2}{3}\right| < 1$, By Binomial Theorem, we get

$$f(x) = -\frac{2}{3}\left[\sum_{n=0}^{\infty}\left(\frac{x-2}{3}\right)^n\right] = -\frac{2}{3}\sum_{n=0}^{\infty}\frac{(x-2)^n}{3^n} = -2\sum_{n=0}^{\infty}\frac{(x-2)^n}{3^{n+1}}$$

Note. The Taylor series will represent the function as long as $\left|\dfrac{x-2}{3}\right| < 1 \iff |x-2| < 3$

Thus the series can be obtained for $-1 < x < 5$.

Example 23. *Find the first three terms of the Taylor series at* $x = 1$ *for* $f(x) = \dfrac{x}{1+x}$.

Sol. We compute the first two derivatives of $f(x)$.

$$f'(x) = \frac{1}{(1+x)^2} \quad \Rightarrow \quad f'(1) = \frac{1}{4}$$

$$f''(x) = \frac{-2}{(1+x)^3} \quad \Rightarrow \quad f''(1) = \frac{-2}{2^3} = \frac{-1}{4}$$

Thus, we get the Taylor expansion as

$$f(x) = f(1) + (x-1)f'(1) + \frac{(x-1)^2}{2!} f''(1) + \ldots$$

$$\Rightarrow \qquad \frac{x}{1+x} = \frac{1}{2} + \frac{x-1}{4} - \frac{1}{4}\frac{(x-1)^2}{2!} + \ldots = \frac{1}{2} + \frac{x-1}{4} - \frac{(x-1)^2}{8} + \ldots$$

Example 24. *Determine the Taylor polynomial of degree 2 for* $f(x) = e^{\sin x}$ *and expand it about the point* π.

Sol. Write $\quad f(x) = e^{\sin x}$

$$f'(x) = e^{\sin x}\cos x$$

$$f''(x) = e^{\sin x}(\cos^2 x - \sin x)$$

and $$f(\pi) = 1, f'(\pi) = -1, f''(\pi) = 1$$

Thus, the Taylor expansion is given by

$$f(x) = f(\pi) + (x-\pi)f'(\pi) + \frac{(x-\pi)^2}{2!}f''(\pi)$$

$$\Rightarrow \qquad e^{\sin x} = 1 - (x-\pi) + \frac{(x-\pi)^2}{2}.$$

Example 25. *Find the Taylor series (with center $x_0 = -1$) of the polynomial $p(x) = x^2 - 3x + 5$.*

Sol. We consider the derivatives of $p(x)$ and then substitute $x_0 = -1$

$$p(-1) = 1 + 3 + 5 = 9.$$

$$p'(x) = 2x - 3 \quad \text{and} \quad p'(-1) = -2 - 3 = -5.$$

$$p''(x) = 2 \quad \text{and} \quad p''(-1) = 2.$$

All higher derivatives are 0, and so the Taylor series terminates.

Using the formula, we obtain that

$$p(x) = p(-1) + (x+1)p'(-1) + \frac{(x+1)^2}{2!}p''(-1)$$

$$\Rightarrow \quad x^2 - 3x + 5 = 9 - 5(x+1) + \frac{2}{2}(x+1)^2 = (x+1)^2 - 5(x+1) + 9.$$

Example 26. *Use Taylor series to express the polynomial $2x^3 + 7x^2 + x - 6$ in powers of $x - 2$.*

Sol. Write $f(x) = 2x^3 + 7x^2 + x - 6$

$$f(x) = f(2 + x - 2)$$

$$= f(2+h) \text{ where } h = x - 2$$

$$= f(2) + hf'(2) + \frac{h^2}{2!}f''(2) + \ldots \qquad \ldots(1)$$

$$f(x) = 2x^3 + 7x^2 + x - 6 \quad \Rightarrow \quad f(2) = 16 + 28 + 2 - 6 = 40$$

$$f'(x) = 6x^2 + 14x + 1 \quad \Rightarrow \quad f'(2) = 24 + 28 + 1 = 53$$

$$f''(x) = 12x + 14 \quad \Rightarrow \quad f''(2) = 24 + 14 = 38$$

$$f'''(x) = 12 \quad \Rightarrow \quad f'''(2) = 12$$

We substitute (2) in (1), to get ...(2)

$$2x^3 + 7x^2 + x - 6 = 40 + 53h + 38\frac{h^2}{2!} + 12\frac{h^3}{3!} + \ldots$$

$$= 40 + 53(x-2) + 19(x-2)^2 + 2(x-2)^3 + \ldots \text{ (Note that } h = x - 2)$$

Example 27. *Verify Taylor's theorem for $f(x) = (1-x)^{5/2}$ with Lagrange's form of remainder up to 2 terms in the interval [0, 1].* (JNTU 2003, 2007 S, 2008 S)

Sol. Consider $f(x) = (1-x)^{5/2}$ in [0, 1].

Since $f(x)$, $f'(x)$ are continuous in [0, 1] and

$f''(x)$ is differentiable in (0, 1), we have that

$f(x)$ satisfies all the conditions of the hypothesis of the Taylor's theorem.

We consider the Taylor's theorem with Lagrange's form of remainder upto 2 terms.

$$f(x) = f(0) + xf'(0) + \frac{x^2}{2!}f''(\theta x) \text{ for some } \theta \text{ with } 0 < \theta < 1 \quad \text{...(1)}$$

Here $n = p = 2$, $a = 0$ and $b = 1$
So $f(x) = (1 - x)^{5/2} \Rightarrow f(0) = 1$

$$f'(x) = \frac{-5}{2}(1 - x)^{3/2} \Rightarrow f'(0) = -5/2$$

$$f''(x) = \frac{15}{4}(1 - x)^{1/2} \Rightarrow f''(\theta x) = \frac{15}{4}(1 - \theta x)^{1/2}$$

$$f''(\theta) = \frac{15}{4}(1 - \theta)^{1/2} \text{ and } f(1) = 0$$

On substituting the above values in equation (1), we get that

$$0 = 1 + 1\left(\frac{-5}{2}\right) + \frac{1^2}{2!}\frac{15}{4}(1 - \theta)^{1/2}$$

$$\Rightarrow \qquad \theta = \frac{9}{25} = 0.36.$$

Therefore, θ lies between 0 and 1 and hence the verification is complete.

Example 28. *Expand $\log_e x$ in powers of $(x - 1)$ and hence evaluate $\log_e 1.1$ correct to 4 decimal places.*

Sol. Write $f(x) = \log_e(x)$. So $f(1) = 0$.

Since, $f'(x) = \frac{1}{x}$, we have that $f'(1) = +1$.

Since, $f''(x) = \frac{-1}{x^2}$, we have that $f''(1) = -1$.

$$f'''(x) = \frac{+2}{x^3} \text{ and so } f'''(1) = +2.$$

$$f^{iv}(x) = \frac{-2 \cdot 3}{x^4} \text{ and so } f^{iv}(1) = -6, \text{ and so on.}$$

By substituting these values in Taylor's series, we get that

$$f(x) = f(1) + (x - 1)f'(1) + \frac{(x-1)^2}{2!}f''(1) + \ldots$$

$$\Rightarrow \qquad \log_e x = (x - 1) - \frac{(x-1)^2}{2} + \frac{(x-1)^3}{3} - \frac{(x-1)^4}{4} + \ldots$$

By putting $x = 1.1$, we get that

$$\log_e (1.1) = (1.1 - 1) - \frac{(1.1-1)^2}{2} + \frac{(1.1-1)^3}{3} - \frac{(1.1-4)^4}{4} + \ldots$$
$$= 0.1 - 0.005 + 0.0003 - 0.00002 + \ldots = 0.0953.$$

Example 29. *Obtain the Maclaurin's series expansion for the following:*

(i) e^x *(ii)* $\cos x$

(iii) $\sin x$ *(iv)* $\cosh x$

(v) $\sinh x$ *(vi)* $\log_e (1 + x)$.

Sol. The Maclaurin's series expansion is

$$f(x) = f(0) + xf'(0) + \frac{x^2}{2!} f''(0) + \ldots \qquad \ldots(1)$$

(i) Let $f(x) = e^x$, then

$$f'(x) = f''(x) = f'''(x) = \ldots = e^x$$

Therefore, $f(0) = f'(0) = f''(0) = \ldots = f^n(0) = e^0 = 1$

Hence the Maclaurin's series expansion of e^x is given by from equation (1)

$$e^x = 1 + x + \frac{x^2}{2!} + \frac{x^3}{3!} + \ldots$$

(ii) Let $f(x) = \cos x$, then

$$f'(x) = -\sin x;\ f''(x) = -\cos x;\ f'''(x) = \sin x;\ f^{iv}(x) = \cos x;\ \ldots$$
$$f(0) = 1;\ f'(0) = 0;\ f''(0) = -1;\ f'''(0) = 0;\ f^{iv}(0) = 1;\ \ldots$$

Hence the Maclaurin's series expansion $f(x) = \cos x$ is given by from equation (1)

$$\cos x = 1 - \frac{x^2}{2!} + \frac{x^4}{4!} - \frac{x^6}{6!} + \ldots = \sum_{n=0}^{\infty} (-1)^n \frac{x^{2n}}{(2n)!}$$

(iii) Let $f(x) = \sin x$, then

$$f'(x) = \cos x;\ f''(x) = -\sin x;\ f'''(x) = -\cos x;\ f^{iv}(x) = \sin x;\ \ldots$$

Hence, $f(0) = 0,\ f'(0) = 1,\ f''(0) = 0;\ f'''(0) = -1,\ f^{iv}(0) = 0,\ \ldots$

Hence the Maclaurin's series expansion of $f(x) = \sin x$ is given by from equation (1)

$$\sin x = x - \frac{x^3}{3!} + \frac{x^5}{5!} - \frac{x^7}{7!} + \ldots = \sum_{n=0}^{\infty} \frac{(-1)^n x^{2n+1}}{(2n+1)!}$$

(iv) Let $f(x) = \cosh x$, then

$$f'(x) = \sinh x;\ f''(x) = \cosh x;\ f'''(x) = \sinh x,\ \ldots$$
$$f(0) = 1,\ f'(0) = 0,\ f''(0) = 1,\ f'''(0) = 0,\ \ldots$$

Hence from equation (1), we have that

$$\cosh x = 1 + \frac{x^2}{2!} + \frac{x^4}{4!} + \frac{x^6}{6!} + \ldots = \sum_{n=0}^{\infty} \frac{x^{2n}}{(2n)!}$$

(v) Let $f(x) = \sinh x$, then

$$f'(x) = \cosh x;\ f''(x) = \sinh x;\ f'''(x) = \cosh x,\ \ldots$$

$$f(0) = 0, f'(0) = 1, f''(0) = 0, f'''(0) = 1, \ldots$$

Hence from equation (1), we have that

$$\sinh x = x + \frac{x^3}{3!} + \frac{x^5}{5!} + \ldots = \sum_{n=0}^{\infty} \frac{x^{2n+1}}{(2n+1)!}$$

(*vi*) Let $f(x) = \log_e (1 + x)$. Then $f(0) = 0$

$$f'(x) = \frac{1}{1+x} \quad \Rightarrow \quad f'(0) = 1$$

$$f''(x) = \frac{-1}{(1+x)^2} \quad \Rightarrow \quad f''(0) = -1$$

$$f'''(x) = \frac{(-1)^2 2!}{(1+x)^3} \quad \Rightarrow \quad f'''(0) = 2!$$

$$f^{iv}(x) = \frac{(-1)^3\, 3!}{(1+x)^4} \quad \Rightarrow \quad f^{iv}(0) = (-1)3!$$

In general, $\quad f^{(0)} = (-1)^{n-1}\,(n-1)!$

Hence from equation (1), we have that

$$\log_e (1 + x) = 0 + x(+1) + \frac{x^2}{2!}(-1) + \frac{x^3}{3!}(2!) + \frac{x^4}{4!}(-1)3! \ldots$$

$$= x - \frac{x^2}{2!}(1!) + \frac{x^3}{3!}(2!) - \frac{x^4}{4!}(3!) + \ldots$$

$$= \sum_{n-1}^{\infty} \frac{(-1)^{n-1}(n-1)!}{n!} x^n$$

Hence, $\quad \log_e (1 + x) = x - \dfrac{x^2}{2} + \dfrac{x^3}{3} - \dfrac{x^4}{4} + \ldots$

Example 30. *Verify Maclaurin's theorem for $f(x) = (1 - x)^{5/2}$ with Lagrange's form of remainder upto 3 terms with $x = 1$.*

Sol. Maclaurin's theorem for $f(x)$ is

$$f(x) = f(0) + \frac{x}{1!} f'(0) + \frac{x^2}{2!} f''(0) + \frac{x^3}{3!} f'''(0) + \ldots \qquad \ldots(1)$$

$$f(x) = (1 - x)^{5/2} \quad \Rightarrow \quad f(0) = 1$$

$$f'(x) = \frac{5}{2}(1 - x)^{3/2}(-1) \quad \Rightarrow \quad f'(0) = -5/2$$

$$f''(x) = \frac{5}{2} \bullet \frac{3}{2}(1 - x)^{1/2}(-1)^2 \quad \Rightarrow \quad f''(0) = \frac{15}{4}$$

$$f'''(x) = \frac{5}{2} \cdot \frac{3}{2} \cdot \frac{1}{2}(1 - x)^{-1/2}(-1)^3 \quad \Rightarrow \quad f'''(\theta x) = \frac{-15}{8}(1 - \theta)^{-1/2}$$

At $x = 1 \;\Rightarrow\; f(x) = f(1) = 0$

On substituting $x = 1$ and the above values in equation (1), we get that

$$0 = 1 + \frac{1}{1!}\left(\frac{-5}{2}\right) + \frac{15}{4}\cdot\frac{1}{2} - \frac{1}{6}\cdot\frac{15}{8}(1-\theta)^{-1/2}$$

$$\Rightarrow \quad 1 - \frac{5}{2} + \frac{15}{8} - \frac{5}{16}(1-\theta)^{-1/2} = 0$$

$$\Rightarrow \quad \frac{5}{16}\cdot\frac{1}{(1-\theta)^{1/2}} = \frac{8 - 20 + 15}{8}$$

$$\Rightarrow \quad 3(1-\theta)^{1/2} = \frac{5}{2}$$

$$\Rightarrow \quad (1-\theta)^{1/2} = \frac{5}{6}$$

$$\Rightarrow \quad 1 - \theta = \frac{25}{36}$$

$$\Rightarrow \quad \theta = 1 - \frac{25}{36} = \frac{11}{36} \in (0, 1)$$

Hence Maclaurin's theorem is verified.

Example 31. *Expand $\tan^{-1} x$ in powers of $(x - 1)$ up to the term containing fourth degree.*

Sol. Taylor's series expansion in powers of $(x - 1)$ is given by

$$f(x) = f(1) + \frac{(x-1)}{1!} f'(1) + \frac{(x-1)^2}{2!} f''(1) + \frac{(x-1)^3}{3!} f'''(1) + \ldots \qquad \ldots(1)$$

$$f(x) = \tan^{-1} x \qquad \Rightarrow \quad f(1) = \pi/4$$

Find derivatives $f'(x)$, $f''(x)$, $f'''(x)$, $f^{(iv)}(x)$, we get the following:

$$f'(x) = \frac{1}{1+x^2} \quad \text{and} \quad f'(1) = \frac{1}{2}$$

$$f''(x) = -2\cdot\frac{x}{(1+x^2)^2} \quad \text{and} \quad f''(1) = -2\cdot\frac{1}{2^2} = -\frac{1}{2}$$

$$f'''(x) = -2\left[\frac{1-x^2}{(1+x^2)^4}\right] \quad \text{and} \quad f'''(1) = 0$$

$$f^{(iv)}(x) = \frac{4x(5-3x^2)}{(1+x^2)^5} \quad \text{and} \quad f^{(iv)}(1) = \frac{1}{4}$$

Substituting the values in (1), we get that

$$\tan^{-1}(x) = \frac{\pi}{4} + \frac{1}{2}(x-1) - \frac{1}{4}(x-1)^2 + \frac{1}{96}(x-1)^4.$$

Example 32. *Find the Taylor's series expansion of $f(x) = 2^x$ about $x = 0$.*

(OU-Dec. 2013)

Sol. We have

$$f(x) = f(0) + xf'(0) + \frac{x^2}{2!} f''(0) + \frac{x^3}{3!} f'''(0) + \ldots \qquad \ldots(1)$$

Given $\quad f(x) = 2^x \quad \Rightarrow \quad f(0) = 2^0 = 1$

$$f'(x) = 2^x \log 2 \quad \Rightarrow \quad f'(0) = \log 2 \qquad \left(\because \; \frac{da^x}{dx} = a^x \log a\right)$$

$$f''(x) = 2^x (\log 2)^2 \quad \Rightarrow \quad f''(0) = (\log 2)^2$$

$$f'''(x) = 2^x (\log 2)^3 \quad \Rightarrow \quad f'''(0) = (\log 2)^3 \text{ and so on.}$$

Substituting in (1), we get

$$2^x = 1 + x \log 2 + \frac{x^2}{2!}(\log 2)^2 + \frac{x^3}{3!}(\log 2)^3 + \ldots$$

$$= 1 + \frac{(x \log 2)^1}{1!} + \frac{(x \log 2)^2}{2!} + \frac{(x \log 2)^3}{3!} + \ldots.$$

Example 33. *Find the Taylor series of $f(x) = \sin x$ about $x = \frac{\pi}{4}$.* (OU-June 2013)

Sol. We have

$$f(x) = f\left(\frac{\pi}{4}\right) + \left(x - \frac{\pi}{4}\right) f'\left(\frac{\pi}{4}\right) + \frac{\left(x - \frac{\pi}{4}\right)^2}{2!} f''\left(\frac{\pi}{4}\right) + \frac{\left(x - \frac{\pi}{4}\right)^3}{3!} f'''\left(\frac{\pi}{4}\right) + \ldots \quad \ldots(1)$$

Given $\quad f(x) = \sin x \quad \Rightarrow \quad f\left(\frac{\pi}{4}\right) = \sin \frac{\pi}{4} = \frac{1}{\sqrt{2}}$

$$f'(x) = \cos x \quad \Rightarrow \quad f'\left(\frac{\pi}{4}\right) = \cos \frac{\pi}{4} = \frac{1}{\sqrt{2}}$$

$$f''(x) = -\sin x \quad \Rightarrow \quad f''\left(\frac{\pi}{4}\right) = -\sin \frac{\pi}{4} = -\frac{1}{\sqrt{2}}$$

$$f'''(x) = -\cos x \quad \Rightarrow \quad f'''\left(\frac{\pi}{4}\right) = -\cos \frac{\pi}{4} = -\frac{1}{\sqrt{2}} \text{ and so on.}$$

Substitute in (1)

$$\sin x = \frac{1}{\sqrt{2}} + \left(x - \frac{\pi}{4}\right)\left(\frac{1}{\sqrt{2}}\right) + \frac{\left(x - \frac{\pi}{4}\right)^2}{2!}\left(-\frac{1}{\sqrt{2}}\right) + \ldots$$

$$= \frac{1}{\sqrt{2}}\left[1 + \left(x - \frac{\pi}{4}\right) - \frac{\left(x - \frac{\pi}{4}\right)^2}{2!} - \ldots\right].$$

Example 34. *Expand $f(x) = \tan x$ about $x = \frac{\pi}{4}$ upto x^4.* (OU-May 2012)

Sol. Given $f(x) = \tan x$

$$\Rightarrow \quad f\left(\frac{\pi}{4}\right) = \tan \frac{\pi}{4} = 1$$

$$f'(x) = \sec^2 x \quad \Rightarrow \quad f'\left(\frac{\pi}{4}\right) = \sec^2 \frac{\pi}{4} = (\sqrt{2})^2 = 2$$

$$f''(x) = 2 \sec x \cdot \sec x \tan x = 2 \sec^2 x \tan x$$
$$= 2(1 + \tan^2 x) \tan x = 2 \tan x + 2 \tan^3 x$$

$$\Rightarrow \quad f''\left(\frac{\pi}{4}\right) = 2 + 2 = 4$$

$$f'''(x) = 2 \sec^2 x + 2 \cdot 3 \tan^2 x \cdot \sec^2 x$$
$$= 2(1 + \tan^2 x) + 6 \tan^2 x (1 + \tan^2 x)$$
$$= 2 + 8 \tan^2 x + 6 \tan^4 x$$

$$\Rightarrow \quad f'''\left(\frac{\pi}{4}\right) = 2 + 8 + 6 = 16$$

$$f^{iv} = 0 + 16 \tan x \cdot \sec^2 x + 24 \tan^3 x \cdot \sec^2 x$$
$$= 16 \tan x (1 + \tan^2 x) + 24 \tan^3 x (1 + \tan^2 x)$$
$$= 16 \tan x + 40 \tan^3 x + 24 \tan^5 x$$

$$\Rightarrow \quad f^{iv}\left(\frac{\pi}{4}\right) = 16 + 40 + 24 = 80 \text{ and so on.}$$

$\therefore$ Substitute in

$$f(x) = f\left(\frac{\pi}{4}\right) + \left(x - \frac{\pi}{4}\right) f'\left(\frac{\pi}{4}\right) + \ldots$$

$$\tan x = 1 + \left(x - \frac{\pi}{4}\right)(2) + \frac{\left(x - \frac{\pi}{4}\right)^2}{2!}(4) + \frac{\left(x - \frac{\pi}{4}\right)^3}{3!}(16) + \frac{\left(x - \frac{\pi}{4}\right)^4}{4!}(80) + \ldots.$$

Example 35. *Find the series expansion of $\sin^{-1} x$ in powers of x.* (OU-Apr. 2007)

Sol. Maclaurin's series expansion is

$$f(x) = f(0) + x f'(0) + \frac{x^2}{2!} f''(0) + \frac{x^3}{3!} f'''(0) + \ldots \qquad \ldots(1)$$

Let $f(x) = \sin^{-1} x \;\Rightarrow\; f(0) = \sin^{-1} 0 = 0$

$$f'(x) = \frac{1}{\sqrt{1 - x^2}}$$

$$\Rightarrow \quad f'(0) = \frac{1}{\sqrt{1 - 0}} = 1$$

$$f''(x) = -\frac{1}{2(1 - x^2)\sqrt{1 - x^2}}(-2x)$$

$$\Rightarrow \quad f''(0) = 0$$

$$f''(x) = \frac{x}{(1 - x^2)^{3/2}}$$

$$\Rightarrow \quad f'''(x) = \frac{(1 - x^2)^{3/2} \cdot 1 - x \cdot \frac{3}{2}(1 - x^2)^{\frac{3}{2} - 1}(-2x)}{(1 - x^2)^3}$$

$$= \frac{(1 - x^2)^{3/2} + 3x^2 \sqrt{(1 - x^2)}}{(1 - x^2)^3}$$

$\Rightarrow$ $f'''(0) = 1$ and so on.

Substitute in (1), we get

$$\sin^{-1} x = 0 + x(1) + \frac{x^2}{2!}(0) + \frac{x^3}{3!}(1) + \ldots = x + \frac{x^3}{3!} + \ldots.$$

EXERCISE 1

1. Verify Rolle's theorem for $f(x) = 2x^3 + x^2 - 4x - 2$ in $[-\sqrt{3}, \sqrt{3}]$.
2. Verify Lagrange's mean value theorem for $f(x) = \log_e x$ in $[1, e]$.
3. If $a < b$, prove that $\frac{b-a}{1+b^2} < \tan^{-1} b - \tan^{-1} a < \frac{b-a}{1+a^2}$ using Lagrange's mean value theorem.

   Deduce the following:

   (*i*) $\frac{\pi}{4} + \frac{3}{35} < \tan^{-1}\frac{4}{3} < \frac{\pi}{4} + \frac{1}{6}$

   (*ii*) $\frac{5\pi + 4}{20} < \tan^{-1} 2 < \frac{\pi + 2}{4}$
4. Show that for any $x > 0$, $1 + x < e^x < 1 + xe^x$.
5. Prove that $\frac{\pi}{3} - \frac{1}{5\sqrt{3}} > \cos^{-1}\frac{3}{5} > \frac{\pi}{3} - \frac{1}{8}$ using Lagrange's mean value theorem.
6. Find c of Cauchy's mean value theorem for $f(x) = \sqrt{x}$ and $g(x) = \frac{1}{\sqrt{x}}$ in $[a, b]$ where $0 < a < b$.
7. Discuss the applicability of Cauchy's mean value theorem for the functions $f(x) = \frac{1}{x^2}$, $g(x) = \frac{1}{x}$ on $[a, b]$.
8. Using Taylor's Theorem, express the following: (*i*) e^x (*ii*) $\tan^{-1} x$ in powers of $(x - 1)$ up to 4 terms.
9. Expand $\sin x$ in powers of $(x - \pi/2)$. Hence find the value of sin 91° (correct to 4 decimal places).
10. Prove that: $\log \sin x = \log \sin a + (x - a) \cot a - \frac{1}{2}(x - a)^2 \operatorname{cosec}^2 a + \ldots$
11. Find the Taylor's series expansion for $\log \cos x$ about $\pi/3$
12. Using Maclaurin's series, expand the following functions:

    (*a*) $\sec x$ (*b*) $e^{x \sin x}$

    (*c*) $\frac{x}{\sin x}$ (*d*) $e^x \cos x$.

Answers

6. $\sqrt{ab}$

8. (*i*) $e\left\{1 + \frac{(x-1)}{1!} + \frac{(x-1)^2}{2!} + \frac{(x-1)^3}{3!} + \ldots\right\}$

(ii) $\frac{\pi}{4}+\frac{1}{2}(x-1)-\frac{1}{4}(x-1)^2+\frac{1}{12}(x-1)^3$

9. $1-\frac{(x-\pi/2)^2}{2!}+\frac{(x-\pi/2)^4}{4!}\ldots$; 0.9998

11. $\log(0.5)-\sqrt{3}(x-\pi/3)-2(x-\pi/3)^2-\frac{4\sqrt{3}}{3}(x-\pi/3)^3+\ldots$

12. (a) $1+\frac{x^2}{2!}+\frac{5x^4}{4!}+\ldots$ (b) $1+x+\frac{x^2}{2}-\frac{x^4}{8}+\ldots$

(c) $1+\frac{x^2}{6}+\frac{x^4}{40}+\ldots$ (d) $1+x-\frac{2x^3}{3!}+\frac{2^2x^4}{4!}+\ldots$

3.3. CURVATURE *(OU-May 2008)*

The shape of a plane curve C is characterized by the degree of bentness or curvedness. A straight line has no bending while a circle has constant bending.

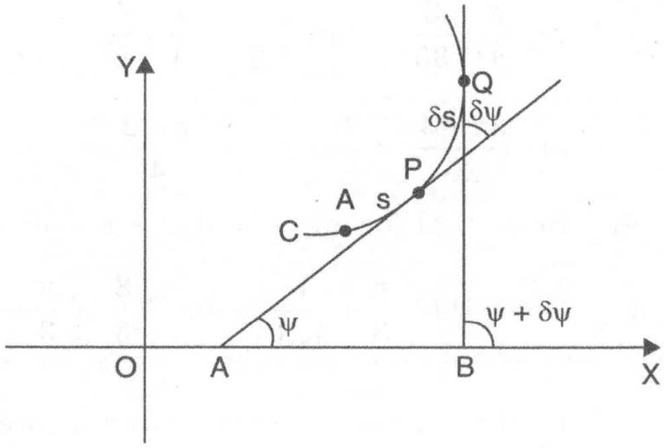

Definition: Curvature of a curve is a measure of rate of change of curvedness.

Here AP = arc length = s units

AQ = $s + \delta s$ units

Hence, PQ = δs units

The total curvature of the arc PQ is defined to be the angle $\delta\psi$. The average curvature of the arc PQ is $\frac{\delta\psi}{\delta s}$. Curvature of a curve C at a point P is denoted by K and is given by

$$K = \underset{Q\to P}{\text{Lt}}\frac{\delta\psi}{\delta s} = \underset{\delta s\to 0}{\text{Lt}}\frac{\delta\psi}{\delta s} = \frac{d\psi}{ds}$$

Note 1. The curvature at any point of the circle is equal to the reciprocal of its radius.

Note 2. The curvature of the circle decreases as the radius increases.

3.4. RADIUS OF CURVATURE *(OU-May 2008)*

The reciprocal of the curvature at any point of a curve is called radius of curvature at that point and is denoted by ρ. Thus $\rho = \frac{1}{K}$.

3.4.1. Radius of Curvature: Cartesian Form (x, y)

1. If the equation of the curve is $y = f(x)$, then $\rho = \frac{(1+y_1^2)^{3/2}}{y_2}$ where $y_1 = \frac{dy}{dx}$ and $y_2 = \frac{d^2y}{dx^2}$.

2. If $\frac{dy}{dx} = \infty$ at any point then $\rho = \dfrac{\left[1+\left(\frac{dx}{dy}\right)^2\right]^{3/2}}{\frac{d^2x}{dy^2}}$.

Example 1. *Find the radius of curvature of* $\sqrt{x} + \sqrt{y} = \sqrt{a}$ *at* $\left(\frac{a}{4}, \frac{a}{4}\right)$.

(OU-Nov. 2006, Jan. 2011, May 2012)

Sol. Radius of curvature in cartesian form is

$$\rho = \frac{\left[1+\left(\frac{dy}{dx}\right)^2\right]^{3/2}}{\frac{d^2y}{dx^2}} \qquad \ldots(1)$$

Given $\sqrt{x} + \sqrt{y} = \sqrt{a}$

Differentiating w.r.t. x on both sides

$$\frac{1}{2\sqrt{x}} + \frac{1}{2\sqrt{y}}\frac{dy}{dx} = 0 \Rightarrow \frac{dy}{dx} = \frac{-\sqrt{y}}{\sqrt{x}} \qquad \ldots(2)$$

At $\left(\frac{a}{4}, \frac{a}{4}\right)$

$$\frac{dy}{dx} = -\frac{\sqrt{a/4}}{\sqrt{a/4}} = -1$$

Differentiating (2) w.r.t. x on both sides

$$\frac{d^2y}{dx^2} = -\left[\frac{\sqrt{x}\cdot\frac{1}{2\sqrt{y}}\frac{dy}{dx} - \sqrt{y}\cdot\frac{1}{2\sqrt{x}}}{x}\right]$$

Put $x = \frac{a}{4}, y = \frac{a}{4}, \frac{dy}{dx} = -1$, we get

$$\frac{d^2y}{dx^2} = -\left[\frac{\frac{\sqrt{a/4}}{2\sqrt{a/4}}(-1) - \frac{\sqrt{a/4}}{2\sqrt{a/4}}}{\frac{a}{4}}\right] = -\left[\frac{-\frac{1}{2}-\frac{1}{2}}{\frac{a}{4}}\right] = \frac{1}{\frac{a}{4}} = \frac{4}{a}$$

Substitute in (1), we get

$$\rho = \frac{[1+(-1)^2]^{3/2}}{\frac{4}{a}} = \frac{2^{3/2}a}{4} = \frac{2\sqrt{2}a}{4} = \frac{a}{\sqrt{2}}.$$

Example 2. *Find the radius of curvature of* $x^2 + y^2 = 16$ *at any point on this curve.*

(OU-May 2012)

Sol. Given $x^2 + y^2 = 16$

Differentiating w.r.t. x on both sides

$$2x + 2y\frac{dy}{dx} = 0 \Rightarrow \frac{dy}{dx} = -\frac{x}{y} \qquad \ldots(1)$$

Again differentiating w.r.t. x

$$\frac{d^2y}{dx^2} = -\left[\frac{y \cdot 1 - x\dfrac{dy}{dx}}{y^2}\right] = -\left[\frac{y - x\left(-\dfrac{x}{y}\right)}{y^2}\right] = -\left[\frac{y^2 + x^2}{y^3}\right]$$

Substitute in $$\rho = \frac{\left[1 + \left(\dfrac{dy}{dx}\right)^2\right]^{3/2}}{\dfrac{d^2y}{dx^2}} = \frac{\left[1 + \left(-\dfrac{x}{y}\right)^2\right]^{3/2}}{-\left[\dfrac{y^2 + x^2}{y^3}\right]} = -\frac{(y^2 + x^2)^{3/2}}{y^3} \times \frac{y^3}{y^2 + x^2}$$

$$= -\sqrt{y^2 + x^2} = -4$$

Since $\rho > 0$, we have $\rho = 4$.

Example 3. *Find the radius of curvature for $y^2 = a^2(a - x)/x$ at $(a, 0)$.*

(OU-June 2011)

Sol. Given $$y^2 = \frac{a^2(a - x)}{x} \Rightarrow xy^2 = a^2(a - x)$$

$$\Rightarrow \quad xy^2 = a^3 - a^2x$$

Differentiating w.r.t. x

$$x \cdot 2y\frac{dy}{dx} + y^2 \cdot 1 = -a^2$$

$$\frac{dy}{dx} = \frac{-a^2 - y^2}{2xy}$$

At $(a, 0)$, we get $$\frac{dy}{dx} = \infty$$

Then $$\frac{dx}{dy} = -\frac{2xy}{a^2 + y^2} \qquad \text{...(2)}$$

At $(a, 0)$, $$\frac{dx}{dy} = 0$$

Differentiating (2) w.r.t. y

$$\frac{d^2x}{dy^2} = -2\left[\frac{(a^2 + y^2)\left(x \cdot 1 + y\dfrac{dx}{dy}\right) - xy(2y)}{(a^2 + y^2)^2}\right]$$

At $(a, 0)$ $$\frac{d^2x}{dy^2} = -2\left[\frac{(a^2 + 0)(a + 0) - 0}{(a^2 + 0)^2}\right] = -2\left[\frac{a^3}{a^4}\right] = -\frac{2}{a}$$

$$\therefore \quad \rho = \frac{\left[1 + \left(\dfrac{dx}{dy}\right)^2\right]^{3/2}}{\dfrac{d^2x}{dy^2}} = \frac{(1 + 0)^{3/2}}{-\dfrac{2}{a}} = -\frac{a}{2}.$$

Since $\rho > 0$, $\rho = \dfrac{a}{2}$.

Example 4. *Show that the radius of curvature for the rectangular parabola* $xy = c^2$ *is* $\frac{(x^2+y^2)^{3/2}}{2c^2}$. (OU-Dec. 2007)

Sol. Given $xy = c^2$

$$\Rightarrow \qquad y = \frac{c^2}{x}$$

Differentiating w.r.t. x

$$\frac{dy}{dx} = c^2\left(-\frac{1}{x^2}\right)$$

Again differentiating w.r.t. x

$$\frac{d^2y}{dx^2} = c^2\left(\frac{2}{x^3}\right)$$

Substitute in $$\rho = \frac{\left[1+\left(\frac{dy}{dx}\right)^2\right]^{3/2}}{\frac{d^2y}{dx^2}}$$

$$\rho = \frac{\left[1+\frac{c^4}{x^4}\right]^{3/2}}{\frac{2c^2}{x^3}} = \frac{(x^4+c^4)^{3/2}}{x^6} \times \frac{x^3}{2c^2}$$

$$= \frac{\left[(x^2)\left(x^2+\frac{c^4}{x^2}\right)\right]^{3/2}}{2x^3c^2} = \frac{x^3(x^2+y^2)^{3/2}}{2x^3c^2}$$

$$= \frac{(x^2+y^2)^{3/2}}{2c^2}. \qquad \left(\because\ y = \frac{c^2}{x} \Rightarrow y^2 = \frac{c^4}{x^2}\right)$$

Radius of Curvature at the Origin:

I. Newton's Method: To find radius of curvature at the origin we should apply Newton's method when x-axis or y-axis is a tangent at the origin.

Statement 1. If the curve passes through the origin and x-axis is the tangent at the origin, then $\rho = \underset{\substack{x\to 0\\ y\to 0}}{\text{Lt}}\left(\frac{x^2}{2y}\right)$.

Statement 2. If the curve passes through the origin and y-axis is the tangent at the origin, then $\rho = \underset{\substack{x\to 0\\ y\to 0}}{\text{Lt}}\left(\frac{y^2}{2x}\right)$.

Note. The tangents at the origin are obtained by equating the lowest degree terms of the equation to zero.

Example 1. *The tangents at the origin of the curve* $x^3 + y^3 = 3xy$ *are* $3xy = 0$

$$\Rightarrow \qquad x = 0,\ y = 0.$$

Example 2. *The tangents at the origin of the curve*

$$2x^4 + 2y^4 + 4x^2y + xy - y^2 + 2x = 0 \text{ is } 2x = 0 \Rightarrow x = 0.$$

Example 3. *Find the radius of curvature of the curve* $x^3 + y^3 + 4x^2y + 2xy^2 + 3x^2 + 3xy + 2y^2 + 3x = 0$ *at the origin.*

Sol. Since the curve passes through the origin, the tangents at the origin are

$$3x = 0 \Rightarrow x = 0 \text{ i.e., } y\text{-axis}$$

By Newton's method, $$\rho = \underset{\substack{x\to 0\\ y\to 0}}{\text{Lt}} \left(\frac{y^2}{2x}\right) \qquad \ldots(1)$$

Dividing the given equation by $2x$, we get

$$\frac{x^2}{2} + y\left(\frac{y^2}{2x}\right) + 2xy + y^2 + \frac{3}{2}x + \frac{3}{2}y + 2\left(\frac{y^2}{2x}\right) + \frac{3}{2} = 0$$

Applying $\underset{\substack{x\to 0\\ y\to 0}}{\text{Lt}}$ on both sides, we get

$$0 + 0(\rho) + 0 + 0 + 0 + 0 + 2\rho + \frac{3}{2} = 0$$

$$\Rightarrow \qquad 2\rho = -\frac{3}{2}$$

$$\Rightarrow \qquad \rho = -\frac{3}{4}$$

Since $\rho > 0$, we have $\rho = \frac{3}{4}$.

II. Method of Expansion: If x-axis or y-axis is not the tangent at the origin, we should apply the method of expansion.

Statement. To find p and q, substitute

$$y = px + \frac{qx^2}{2!} + \ldots$$

in the given equation and equate the coefficients of like powers of x on both sides. Then

$$\rho = \frac{(1 + p^2)^{3/2}}{q}.$$

Example 4. *Find the radius of curvature of the curve* $y^2 = x^2\left(\frac{a + x}{a - x}\right)$ *at the origin.*

Sol. Given $\qquad y^2(a - x) = x^2(a + x)$

Since it passes through the origin, the tangents at the origin are obtained by equating the lowest degree terms to zero.

$$\Rightarrow \qquad a(y^2 - x^2) = 0 \Rightarrow y = \pm x$$

Applying method of expansion:

Substitute $y = px + \frac{qx^2}{2!} + \ldots$ in the given equation, we get,

$$\left(px + \frac{qx^2}{2!} + \ldots\right)^2 (a - x) = x^2(a + x)$$

Equating coefficients of x^2, $ap^2 = a \Rightarrow p = \pm 1$

Equating coefficients of x^3, $-p^2 + apq = 1$ ...(1)

If $p = 1$, from (1) $q = \frac{2}{a}$

$$\therefore \qquad \rho = \frac{(1+p^2)^{3/2}}{q} = \frac{(1+1)^{3/2}}{\frac{2}{a}} = \sqrt{2}a$$

If $p = -1$, from (1) $q = -\frac{2}{a}$ and $\rho = -\sqrt{2}a$.

Example 5. *Find the radius of curvature at origin of the curve $y - x = x^2 + 2xy + y^2$.*

(OU-Apr. 2007, Jan. 2012)

Sol. Since the curve passes through the origin, the tangents at the origin are $y - x = 0$ $\Rightarrow y = x$ which is not either x-axis or y-axis.

Hence, apply method of expansion.

Substitute $y = px + \frac{qx^2}{2!} + \ldots$ in the given equation.

$$\left[px + \frac{qx^2}{2!} + \ldots\right] - x = x^2 + 2x\left[px + \frac{qx^2}{2!} + \ldots\right] + \left[px + \frac{qx^2}{2!} + \ldots\right]^2$$

Equating coefficients of x, $p - 1 = 0 \Rightarrow p = 1$

Equating coefficients of x^2, $\frac{q}{2} = 1 + 2p + p^2$

$$\Rightarrow \qquad \frac{q}{2} = 1 + 2 + 1 \qquad (\because\ p = 1)$$

$$\Rightarrow \qquad q = 8$$

$$\therefore \qquad \rho = \frac{(1+p^2)^{3/2}}{q} = \frac{(1+1)^{3/2}}{8} = \frac{2\sqrt{2}}{8} = \frac{1}{2\sqrt{2}}.$$

3.4.2. Radius of Curvature: Polar Form (r, θ)

$$\rho = \frac{(r^2 + r_1^2)^{3/2}}{r^2 + 2r_1^2 - rr_2}, \text{ where } r_1 = \frac{dr}{d\theta}, r_2 = \frac{d^2r}{d\theta^2}.$$

Example 1. *Find the radius of curvature at any point of the cardioid $r = a(1 - \cos\theta)$.*

Sol. Given $r = a(1 - \cos\theta)$

Differentiate w.r.t. 'θ'

$$r_1 = \frac{dr}{d\theta} = a(0 + \sin\theta) = a\sin\theta$$

Again differentiate w.r.t. θ

$$r_2 = \frac{d^2r}{d\theta^2} = a\cos\theta$$

We have $$\rho = \frac{(r^2 + r_1^2)^{3/2}}{r^2 + 2r_1^2 - rr_2} \qquad \ldots(1)$$

Numerator $$(r^2 + r_1^2)^{3/2} = [a^2(1 - \cos\theta)^2 + a^2\sin^2\theta]^{3/2}$$
$$= (a^2)^{3/2}\,[1 + \cos^2\theta - 2\cos\theta + \sin^2\theta]^{3/2}$$
$$= a^3[2 - 2\cos\theta]^{3/2} = a^3(2)^{3/2}\,(1 - \cos\theta)^{3/2}$$
$$= a^3 2\sqrt{2}\left(2\sin^2\frac{\theta}{2}\right)^{3/2} = 2\sqrt{2}a^3\cdot 2\sqrt{2}\sin^3\frac{\theta}{2} = 8a^3\sin^3\frac{\theta}{2}$$

Denominator $$= r^2 + 2r_1^2 - rr_2$$
$$= a^2(1 - \cos\theta)^2 + 2(a\sin\theta)^2 - a(1 - \cos\theta)\,a\cos\theta$$
$$= a^2[1 + \cos^2\theta - 2\cos\theta + 2\sin^2\theta - \cos\theta + \cos^2\theta]$$
$$= a^2[3 - 3\cos\theta] = 3a^2(1 - \cos\theta) = 3a^2\left(2\sin^2\frac{\theta}{2}\right) = 6a^2\sin^2\frac{\theta}{2}$$

From (1), we get

$$\rho = \frac{8a^3\sin^3\frac{\theta}{2}}{6a^2\sin^2\frac{\theta}{2}} = \frac{4}{3}a\sin\frac{\theta}{2}.$$

Example 2. *Find the radius of curvature at any point on the curve $r^n = a^n\cos n\theta$.*

Sol. Given $r^n = a^n\cos n\theta$

$\Rightarrow$ $\log_e(r^n) = \log_e(a^n\cos n\theta)$ $\Rightarrow$ $n\log r = \log(a^n) + \log(\cos n\theta)$

Differentiating w.r.t. 'θ', we get

$$n\cdot\frac{1}{r}\frac{dr}{d\theta} = 0 + \frac{1}{\cos n\theta}(-\sin n\theta)n \quad\Rightarrow\quad r_1 = \frac{dr}{d\theta} = -r\tan n\theta$$

Again differentiating w.r.t. 'θ'

$$r_2 = \frac{d^2r}{d\theta^2} = -\left[r\cdot\sec^2 n\theta\cdot n + \tan n\theta\cdot\frac{dr}{d\theta}\right]$$
$$= -[nr\sec^2 n\theta + \tan n\theta(-r\tan n\theta)] = -r[n\cdot\sec^2 n\theta - \tan^2 n\theta]$$

We have $$\rho = \frac{(r^2 + r_1^2)^{3/2}}{r^2 + 2r_1^2 - rr_2} = \frac{(r^2 + r^2\tan^2 n\theta)^{3/2}}{r^2 + 2(r^2\tan^2 n\theta) - r[-r(n\sec^2 n\theta - \tan^2 n\theta)]}$$
$$= \frac{(r^2)^{3/2}(1 + \tan^2 n\theta)^{3/2}}{r^2[1 + 2\tan^2 n\theta + n\sec^2 n\theta - \tan^2 n\theta]}$$
$$= \frac{r^3(\sec^2 n\theta)^{3/2}}{r^2[1 + \tan^2 n\theta + n\sec^2 n\theta]} = \frac{r\sec^3 n\theta}{\sec^2 n\theta + n\sec^2 n\theta}$$
$$= \frac{r\sec^3 n\theta}{(1 + n)\sec^2 n\theta} = \frac{r\sec n\theta}{1 + n}.$$

Example 3. *Find the radius of curvature of the curve $r = a \sin\theta + b\cos\theta$ at $\theta = \frac{\pi}{2}$.*

(OU-Dec. 2013)

Sol. Given $r = a\sin\theta + b\cos\theta$

$$r_1 = \frac{dr}{d\theta} = a\cos\theta - b\sin\theta$$

$$r_2 = \frac{d^2r}{d\theta^2} = -a\sin\theta - b\cos\theta$$

At $\theta = \frac{\pi}{2}$, we have

$$r = a\sin\frac{\pi}{2} + b\cos\frac{\pi}{2} = a + 0 = a$$

$$r_1 = a\cos\frac{\pi}{2} - b\sin\frac{\pi}{2} = 0 - b = -b$$

$$r_2 = -a\sin\frac{\pi}{2} - b\cos\frac{\pi}{2} = -a - 0 = -a$$

We have
$$\rho = \frac{(r^2 + r_1^2)^{3/2}}{r^2 + 2r_1^2 - rr_2} = \frac{(a^2+b^2)^{3/2}}{a^2 + 2b^2 - a(-a)}$$

$$= \frac{(a^2+b^2)\sqrt{a^2+b^2}}{2(a^2+b^2)} = \frac{\sqrt{a^2+b^2}}{2}.$$

3.4.3. Radius of Curvature: Parametric form

$$x = f(t),\ y = g(t)$$

or
$$x = f(\theta),\ y = g(\theta)$$

$$\rho = \frac{\left[1 + \left(\frac{dy}{dx}\right)^2\right]^{3/2}}{\frac{d^2y}{dx^2}},\ \text{where } \frac{dy}{dx} = \frac{dy/dt}{dx/dt} \text{ or } \frac{dy/d\theta}{dx/d\theta}$$

Or

$$\rho = \frac{(\dot{x}^2 + \dot{y}^2)^{3/2}}{\dot{x}\ddot{y} - \ddot{x}\dot{y}},\ \text{where } \dot{x} = \frac{dx}{dt},\ \dot{y} = \frac{dy}{dt},\ \ddot{x} = \frac{d^2x}{dt^2},\ \ddot{y} = \frac{d^2y}{dt^2}.$$

Example 1. *Find the curvature and radius of curvature of the curve $x = a(t - \sin t)$, $y = a(1 - \cos t)$ at $t = \pi$.*

(OU-Dec. 2008)

Sol. Given $x = a(t - \sin t)$, $y = a(1 - \cos t)$

$$\Rightarrow \quad \frac{dx}{dt} = a(1-\cos t),\ \frac{dy}{dt} = a\sin t$$

$$y_1 = \frac{dy}{dx} = \frac{dy/dt}{dx/dt} = \frac{a\sin t}{a(1-\cos t)} = \frac{2\sin\frac{t}{2}\cos\frac{t}{2}}{2\sin^2\frac{t}{2}} = \cot\left(\frac{t}{2}\right) \quad \ldots(1)$$

At $t = \pi$, $\frac{dy}{dx} = \cot\left(\frac{\pi}{2}\right) = 0$

Differentiating (1) w.r.t. x

$$y_2 = \frac{d^2y}{dx^2} = -\operatorname{cosec}^2\left(\frac{t}{2}\right)\cdot\frac{1}{2}\cdot\frac{dt}{dx} = \frac{-\frac{1}{2}\operatorname{cosec}^2\frac{t}{2}}{\frac{dx}{dt}}$$

$$= \frac{-\frac{1}{2}\operatorname{cosec}^2\frac{t}{2}}{a(1-\cos t)} = -\frac{1}{2a}\frac{\operatorname{cosec}^2\left(\frac{t}{2}\right)}{2\sin^2\left(\frac{t}{2}\right)}$$

$$y_2 = -\frac{1}{4a}\operatorname{cosec}^4\left(\frac{t}{2}\right)$$

At $t = \pi$

$$y_2 = -\frac{1}{4a}\operatorname{cosec}^4\left(\frac{\pi}{2}\right) = -\frac{1}{4a}$$

$$\therefore \qquad \rho = \frac{(1+y_1^2)^{3/2}}{y_2} = \frac{(1+0)^{3/2}}{-\frac{1}{4a}} = -4a$$

Since $\rho > 0$, we have $\rho = 4a$.

Example 2. *Find the radius of curvature $y^2 = 4ax$ at $(at^2, 2at)$.* (OU-Jan. 2012)

Sol. Let $x = at^2$, $y = 2at$

$$\Rightarrow \qquad \frac{dx}{dt} = 2at, \ \frac{dy}{dt} = 2a$$

$$y_1 = \frac{dy}{dx} = \frac{dy/dt}{dx/dt} = \frac{2a}{2at} = \frac{1}{t}$$

Differentiating w.r.t. 'x'

$$y_2 = \frac{d^2y}{dx^2} = -\frac{1}{t^2}\frac{dt}{dx} = -\frac{1/t^2}{dx/dt} = -\frac{1/t^2}{2at} = -\frac{1}{2at^3}$$

$$\therefore \qquad \rho = \frac{(1+y_1^2)^{3/2}}{y_2} = \frac{\left(1+\frac{1}{t^2}\right)^{3/2}}{-\frac{1}{2at^3}} = -\frac{(t^2+1)^{3/2}}{t^3}\cdot 2at^3$$

$$\rho = -(t^2+1)^{3/2}\cdot 2a$$

Since $\rho > 0$, we have $\rho = 2a(1 + t^2)^{3/2}$.

Example 3. *Find the radius of curvature for the cycloid $x = a(t + \sin t)$, $y = a(1 - \cos t)$.* (OU-May 2009)

Sol. Given $x = a(t + \sin t)$, $y = a(1 - \cos t)$

$$\Rightarrow \qquad \frac{dx}{dt} = a(1+\cos t), \ \frac{dy}{dt} = a\sin t$$

$$y_1 = \frac{dy}{dx} = \frac{dy/dt}{dx/dt} = \frac{a \sin t}{a(1+\cos t)} = \frac{2 \sin \frac{t}{2} \cos \frac{t}{2}}{2 \cos^2 \frac{t}{2}} = \tan \frac{t}{2}$$

Now, differentiating w.r.t. 'x'

$$y_2 = \frac{d^2y}{dx^2} = \sec^2\left(\frac{t}{2}\right) \cdot \frac{1}{2} \cdot \frac{dt}{dx} = \frac{\frac{1}{2}\sec^2\left(\frac{t}{2}\right)}{\frac{dx}{dt}}$$

$$= \frac{\frac{1}{2}\sec^2\left(\frac{t}{2}\right)}{a(1+\cos t)} = \frac{1}{2a} \frac{\sec^2\left(\frac{t}{2}\right)}{2\cos^2\left(\frac{t}{2}\right)} = \frac{1}{4a}\sec^4\left(\frac{t}{2}\right)$$

$$\therefore \quad \rho = \frac{(1+y_1^2)^{3/2}}{y_2} = \frac{\left(1+\tan^2\left(\frac{t}{2}\right)\right)^{3/2}}{\frac{1}{4a}\sec^4\left(\frac{t}{2}\right)} = \frac{4a \cdot \left(\sec^2\left(\frac{t}{2}\right)\right)^{3/2}}{\sec^4\left(\frac{t}{2}\right)}$$

$$= \frac{4a \cdot \sec^3\left(\frac{t}{2}\right)}{\sec^4\left(\frac{t}{2}\right)} = 4a \cdot \cos\left(\frac{t}{2}\right).$$

Example 4. *Find the radius of curvature at any point of the astroid* $x^{2/3} + y^{2/3} = a^{2/3}$.

Sol. To make the solution easy, take parametric equations

$$x = a\cos^3\theta,\ y = a\sin^3\theta$$

$$\Rightarrow \quad \frac{dx}{d\theta} = 3a\cos^2\theta\,(-\sin\theta) = -3a\sin\theta\cos^2\theta$$

and $$\frac{dy}{d\theta} = 3a\sin^2\theta \cdot \cos\theta = 3a\cos\theta\sin^2\theta$$

$$y_1 = \frac{dy}{dx} = \frac{dy/d\theta}{dx/d\theta} = \frac{3a\cos\theta\sin^2\theta}{-3a\sin\theta\cos^2\theta} = -\tan\theta$$

Differentiating w.r.t. 'x'

$$y_2 = \frac{d^2y}{dx^2} = -\sec^2\theta \cdot \frac{d\theta}{dx} = \frac{-\sec^2\theta}{\frac{dx}{d\theta}} = \frac{-\sec^2\theta}{-3a\sin\theta\cos^2\theta}$$

$$= \frac{1}{3a}\operatorname{cosec}\theta\sec^4\theta$$

$$\therefore \quad \rho = \frac{(1+y_1^2)^{3/2}}{y_2} = \frac{(1+\tan^2\theta)^{3/2}}{\frac{1}{3a}\operatorname{cosec}\theta\sec^4\theta}$$

$$= 3a\,\frac{(\sec^2\theta)^{3/2}}{\operatorname{cosec}\theta\sec^4\theta} = 3a\sin\theta\cos\theta.$$

3.5. CENTRE OF CURVATURE

The centre of curvature C(x, y) at any point P on the curve is the point which lies on the positive direction of the normal at P and is at a distance equal to radius of curvature from it.

Here positive direction means the direction of the tangent in the x increasing direction on the curve. The positive direction of the normal is obtained by rotating the positive direction of the tangent through 90° in anticlockwise direction.

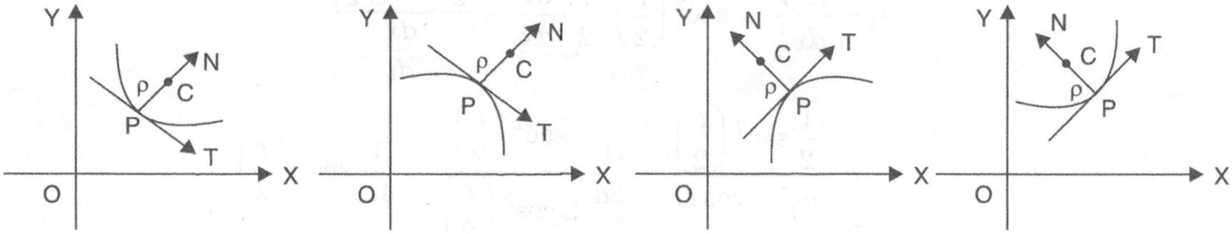

Formula for centre of curvature: If (X, Y) are the coordinates of the centre of curvature at any point P(x, y) on the curve $y = f(x)$ then

$$X = x - \left[\frac{y_1(1+y_1^2)}{y_2}\right] \quad \ldots(1)$$

$$Y = y + \left[\frac{1+y_1^2}{y_2}\right] \quad \ldots(2)$$

where $y_1 = \dfrac{dy}{dx}$ and $y_2 = \dfrac{d^2y}{dx^2}$.

3.6. CIRCLE OF CURVATURE

The circle of curvature at any point of a curve is the circle with centre at the centre of curvature at P and radius equal to the radius of curvature at that point. If (X, Y) be the centre of curvature and ρ be the radius of curvature, then the equation of the circle of curvature at P(x, y) is given by $(x - X)^2 + (y - Y)^2 = \rho^2$.

3.7. EVOLUTE

The locus of centre of curvature is called the evolute and the curve itself is called involute of evolute.

To find locus, eliminate x, y from above equations (1) and (2) and obtain a relation between X and Y which gives evolute.

Example 1. *Find the centre of curvature at the point* $\left(\frac{a}{4}, \frac{a}{4}\right)$ *of the curve* $\sqrt{x} + \sqrt{y} = \sqrt{a}$. *Find also the equation of the circle of curvature at that point.*

Sol. Given $\sqrt{x} + \sqrt{y} = \sqrt{a}$

Differentiating w.r.t. 'x'

$$\frac{1}{2\sqrt{x}} + \frac{1}{2\sqrt{y}}\frac{dy}{dx} = 0 \;\Rightarrow\; \frac{dy}{dx} = -\left[\frac{\sqrt{y}}{\sqrt{x}}\right]$$

at $\left(\frac{a}{4}, \frac{a}{4}\right), y_1 = \frac{dy}{dx} = -1$

Again differentiating w.r.t. x

$$\frac{d^2y}{dx^2} = -\left[\frac{\sqrt{x}\cdot\frac{1}{2\sqrt{y}}\frac{dy}{dx} - \sqrt{y}\cdot\frac{1}{2\sqrt{x}}}{x}\right]$$

At $\left(\frac{a}{4}, \frac{a}{4}\right)$

$$y_2 = \frac{d^2y}{dx^2} = -\left[\frac{-\frac{\sqrt{a/4}}{2\sqrt{a/4}}(-1) - \frac{\sqrt{a/4}}{2\sqrt{a/4}}}{a/4}\right] = -\left[\frac{-\frac{1}{2}-\frac{1}{2}}{a/4}\right]$$

$$= -\left[\frac{-1}{a/4}\right] = \frac{4}{a}$$

$$\therefore \quad \rho = \frac{(1+y_1^2)^{3/2}}{y_2} = \frac{(1+1)^{3/2}}{4/a} = \frac{2\sqrt{2}a}{4} = \frac{a}{\sqrt{2}}$$

$$X = x - \left[\frac{y_1(1+y_1^2)}{y_2}\right] = \frac{a}{4} - \left[\frac{(-1)(1+1)}{4/a}\right] = \frac{a}{4} + \frac{2a}{4} = \frac{3a}{4}$$

$$Y = y + \left[\frac{1+y_1^2}{y_2}\right] = \frac{a}{4} + \left[\frac{1+1}{4/a}\right] = \frac{a}{4} + \frac{2a}{4} = \frac{3a}{4}$$

$\therefore$ Centre of curvature = (X, Y) = $\left(\frac{3a}{4}, \frac{3a}{4}\right)$

$\therefore$ Equation of circle of curvature is

$$(x - X)^2 + (y - Y)^2 = \rho^2 \Rightarrow \left(x - \frac{3a}{4}\right)^2 + \left(y - \frac{3a}{4}\right)^2 = \frac{a^2}{2}.$$

Example 2. *Show that the equation of evolute of the parabola $y^2 = 4ax$ is $27ay^2 = 4(x - 2a)^3$.* (OU-Apr. 2007, May 2008, June 2013, Dec. 2013)

Sol. Given $y^2 = 4ax$

Differentiating w.r.t. x

$$2y\frac{dy}{dx} = 4a \Rightarrow y_1 = \frac{dy}{dx} = \frac{2a}{y}$$

and

$$\frac{d^2y}{dx^2} = 2a\left(-\frac{1}{y^2}\right)\left(\frac{dy}{dx}\right) = -\frac{2a}{y^2}\cdot\frac{2a}{y} = -\frac{4a^2}{y^3} = y_2$$

$$X = x - \left[\frac{y_1(1+y_1^2)}{y_2}\right] = x - \left[\frac{\frac{2a}{y}\left(1+\frac{4a^2}{y^2}\right)}{-\frac{4a^2}{y^3}}\right]$$

$$= x + \frac{y^2}{2a}\left[\frac{y^2 + 4a^2}{y^2}\right] = x + \frac{y^2 + 4a^2}{2a} = x + \frac{y^2}{2a} + 2a$$

But $\qquad y^2 = 4ax$

Hence, $\qquad X = x + \frac{4ax}{2a} + 2a = x + 2x + 2a = 3x + 2a$

$$\Rightarrow \qquad 3x = X - 2a \Rightarrow x = \frac{X - 2a}{3} \qquad \ldots(1)$$

$$Y = y + \left[\frac{1 + y_1^2}{y_2}\right] = -\frac{2x\sqrt{x}}{\sqrt{a}} \text{ on substituting of } y_1 \text{ and } y_2$$

$$\Rightarrow \qquad Y^2 = \frac{4x^3}{a} = \frac{4}{a}\left(\frac{X - 2a}{3}\right)^3 = \frac{4(X - 2a)^3}{27a}$$

$$\Rightarrow \qquad 27aY^2 = 4(X - 2a)^3$$

Hence, the locus of (X, Y) is

$$27ay^2 = 4(x - 2a)^3$$

This is an evolute of the curve.

Example 3. *Find the evolute of the parabola $x^2 = 4ay$.* (OU-June 2011)

Sol. Given $\qquad x^2 = 4ay \Rightarrow 2x = 4a\frac{dy}{dx}$

$$\Rightarrow \qquad y_1 = \frac{dy}{dx} = \frac{x}{2a} \text{ and } \frac{d^2y}{dx^2} = \frac{1}{2a}$$

$$X = x - \left[\frac{y_1(1 + y_1^2)}{y_2}\right] = x - \left[\frac{\frac{x}{2a}\left(1 + \frac{x^2}{4a^2}\right)}{\frac{1}{2a}}\right]$$

$$= x - x - \frac{x^3}{4a^2} \Rightarrow X = -\frac{x^3}{4a^2} \Rightarrow X^2 = \frac{x^6}{16a^4}$$

$$Y = y + \left[\frac{1 + y_1^2}{y_2}\right] = y + \left[\frac{1 + \frac{x^2}{4a^2}}{\frac{1}{2a}}\right] = \frac{x^2}{4a} + 2a\left[1 + \frac{x^2}{4a^2}\right]$$

$$= \frac{x^2}{4a} + 2a + \frac{2x^2}{4a} = \frac{3x^2}{4a} + 2a$$

$$Y - 2a = \frac{3x^2}{4a} \Rightarrow (Y - 2a)^3 = \frac{27x^6}{64a^3}$$

$$\Rightarrow \qquad 4(Y - 2a)^3 = 27aX^2 \qquad \left(\because X^2 = \frac{x^6}{16a^4}\right)$$

Locus of (X, Y) is $4(y - 2a)^3 = 27ax^2$.

This is an evolute.

Example 4. *Show that the evolute of the cycloid $x = a(t + \sin t)$, $y = a(1 - \cos t)$ is the cycloid $x = a(t - \sin t)$, $y = 2a + a(1 + \cos t)$.*

Sol. Given $x = a(t + \sin t)$, $y = a - a \cos t$

$$\Rightarrow \qquad \frac{dx}{dt} = a(1 + \cos t), \quad \frac{dy}{dt} = a \sin t$$

$$y_1 = \frac{dy}{dx} = \frac{dy/dt}{dx/dt} = \frac{a \sin t}{a(1+\cos t)} = \frac{2 \sin \frac{t}{2} \cos \frac{t}{2}}{2 \cos^2 \left(\frac{t}{2}\right)} = \tan\left(\frac{t}{2}\right)$$

$$y_2 = \frac{d^2y}{dx^2} = \sec^2\left(\frac{t}{2}\right) \cdot \frac{1}{2} \cdot \frac{dt}{dx} = \frac{\frac{1}{2}\sec^2\left(\frac{t}{2}\right)}{\frac{dx}{dt}}$$

$$= \frac{\frac{1}{2}\sec^2\left(\frac{t}{2}\right)}{a(1+\cos t)} = \frac{1}{2a} \cdot \frac{\sec^2\left(\frac{t}{2}\right)}{2\cos^2\left(\frac{t}{2}\right)} = \frac{1}{4a}\sec^4\left(\frac{t}{2}\right)$$

$$X = x - \left[\frac{y_1(1+y_1^2)}{y_2}\right] = a(t + \sin t) - \left[\frac{\tan \frac{t}{2}\left(1 + \tan^2 \frac{t}{2}\right)}{\frac{1}{4a}\sec^4\left(\frac{t}{2}\right)}\right]$$

$$= a\left[t + \sin t - 4 \tan \frac{t}{2} \cdot \frac{\sec^2 \frac{t}{2}}{\sec^4 \frac{t}{2}}\right] = a\left[t + \sin t - 4 \tan \frac{t}{2} \cdot \cos^2 \frac{t}{2}\right]$$

$$= a\left[t + \sin t - 2 \cdot 2 \sin \frac{t}{2} \cos \frac{t}{2}\right]$$

$$= a[t + \sin t - 2 \sin t] = a[t - \sin t]$$

$$Y = y + \left[\frac{1+y_1^2}{y_2}\right] = a(1 - \cos t) + \left[\frac{1 + \tan^2 \frac{t}{2}}{\frac{1}{4a}\sec^4\left(\frac{t}{2}\right)}\right]$$

$$= a\left[1 - \cos t + 4\frac{\sec^2 \frac{t}{2}}{\sec^4\left(\frac{t}{2}\right)}\right] = a\left[1 - \cos t + 4\cos^2 \frac{t}{2}\right]$$

$$= a\left[1 - \cos t + 4\left(\frac{1+\cos t}{2}\right)\right] = a[1 - \cos t + 2 + 2 \cos t]$$

$$= a[3 + \cos t]$$

Hence the locus of (X, Y) is

$x = a(t - \sin t)$, $y = a(3 + \cos t)$ is also a cycloid.

Example 5. *Find the evolute of the astroid $x = a\cos^3\theta$, $y = a\sin^3\theta$.* (OU-May 2009)

Sol. Given $x = a\cos^3\theta, y = a\sin^3\theta$

$$y_1 = \frac{dy}{dx} = \frac{dy/d\theta}{dx/d\theta} = \frac{3a\sin^2\theta\cos\theta}{-3a\cos^2\theta\sin\theta} = -\tan\theta$$

$$y_2 = \frac{d^2y}{dx^2} = -\sec^2\theta\cdot\frac{d\theta}{dx} = \frac{1}{3a\cos^4\theta\sin\theta}$$

$$X = x - \left[\frac{y_1(1+y_1^2)}{y_2}\right] = a\cos^3\theta + 3a\sin^2\theta\cos\theta$$

$$Y = y + \left[\frac{1+y_1^2}{y_2}\right] = a\sin^3\theta + 3a\cos^2\theta\sin\theta$$

To find the evolute, eliminate θ.

Consider $(X+Y)^{2/3} + (X-Y)^{2/3} = a^{2/3}[(\cos\theta+\sin\theta)^2 + (\cos\theta-\sin\theta)^2] = 2a^{2/3}$

∴ The locus of (X, Y) is $(x+y)^{2/3} + (x-y)^{2/3} = 2a^{2/3}$

which is an evolute.

Example 6. *Find the evolute of $xy = C^2$.* (OU-Jan. 2011)

Sol. Given $xy = C^2 \Rightarrow y = \frac{C^2}{x}$

$$y_1 = \frac{dy}{dx} = -\frac{C^2}{x^2} = -\frac{y}{x} \text{ and } y_2 = \frac{d^2y}{dx^2} = \frac{2C^2}{x^3} = \frac{2y}{x^2}$$

$$X = x - \left[\frac{y_1(1+y_1^2)}{y_2}\right] = \frac{3x^2+y^2}{2x} \quad \text{...(1) on substituting } y_1 \text{ and } y_2$$

$$Y = y + \left[\frac{1+y_1^2}{y_2}\right] = \frac{3y^2+x^2}{2y} \quad \text{...(2)}$$

To get an evolute, we have to eliminate x and y from (1) and (2)

Take
$$X+Y = \frac{3x^2+y^2}{2x} + \frac{3y^2+x^2}{2y} = \frac{(x+y)^3}{2C^2} \quad (\because\ xy = C^2)$$

$$\Rightarrow \quad (X+Y)^{2/3} = \frac{(x+y)^2}{(2C^2)^{2/3}} \quad \text{...(3)}$$

$$X-Y = \frac{3x^2+y^2}{2x} - \frac{3y^2+x^2}{2y} = -\frac{(x-y)^3}{2C^2}$$

$$\Rightarrow \quad (X-Y)^{2/3} = \frac{(x-y)^2}{(2C^2)^{2/3}} \quad \text{...(4)}$$

$$(3)\text{–}(4) \Rightarrow (X+Y)^{2/3} - (X-Y)^{2/3} = \frac{1}{(2C^2)^{2/3}}[(x+y)^2 - (x-y)^2]$$

$$= \frac{1}{2^{2/3}\cdot C^{4/3}}\cdot 4xy = (4C)^{2/3} \quad (\because\ xy = C^2)$$

∴ Locus of (X, Y) is $(x+y)^{2/3} - (x-y)^{2/3} = (4C)^{2/3}$

This is an evolute.

Example 7. *Find the evolute of the ellipse* $\dfrac{x^2}{a^2}+\dfrac{y^2}{b^2}=1.$

Sol. Let $x = a\cos\theta$, $y = b\sin\theta$ satisfies the given equation.

$$\therefore \qquad \frac{dx}{d\theta} = -a\sin\theta,\ \frac{dy}{d\theta} = b\cos\theta$$

$$y_1 = \frac{dy}{dx} = \frac{dy/d\theta}{dx/d\theta} = \frac{b\cos\theta}{-a\sin\theta} = -\frac{b}{a}\cot\theta$$

$$y_2 = \frac{d^2y}{dx^2} = -\frac{b}{a}(-\operatorname{cosec}^2\theta)\frac{d\theta}{dx} = -\frac{b}{a^2}\operatorname{cosec}^3\theta$$

Substituting y_1, y_2 and $x = a\cos\theta$, $y = b\sin\theta$ in

$$X = x - \left[\frac{y_1(1+y_1^2)}{y_2}\right] \text{ and } Y = y + \frac{1+y_1^2}{y_2}, \text{ we get}$$

$$X = \left(\frac{a^2-b^2}{a}\right)\cos^3\theta,\ Y = -\left(\frac{a^2-b^2}{b}\right)\sin^3\theta$$

$$\Rightarrow \qquad (aX)^{2/3} = (a^2-b^2)^{2/3}\cos^2\theta \text{ and}$$

$$(bY)^{2/3} = (a^2-b^2)^{2/3}\sin^2\theta$$

$$\therefore \qquad (aX)^{2/3} + (bY)^{2/3} = (a^2-b^2)^{2/3}$$

Locus of (X, Y) is $(ax)^{2/3} + (by)^{2/3} = (a^2-b^2)^{2/3}$ is a required evolute.

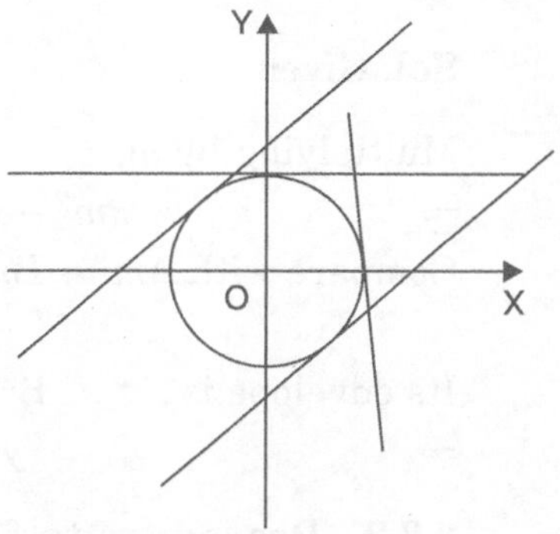

3.8. ENVELOPE

A curve which touches each member of a given family of curves is called envelope of that family.

Example. Here circle $x^2 + y^2 = a^2$ is an envelope of family of straight lines $x\cos\alpha + y\sin\alpha = a$.

3.8.1. Envelope of $f(x, y, \alpha) = 0$ which is quadratic in α

If the given equation is $A\alpha^2 + B\alpha + C = 0$ where α is a parameter then envelope is $B^2 - 4AC = 0$ *i.e.*, discriminant = 0.

Example 1. *Find the envelope of the family of curves* $y = (x - P)^2$, *where P is the parameter.*

(OU-June 2013)

Sol. Given $y = x^2 + P^2 - 2Px$

$$\Rightarrow \qquad P^2 - 2x\cdot P + (x^2 - y) = 0$$

Compare with $AP^2 + BP + C = 0$, then $A = 1$, $B = -2x$, $C = x^2 - y$

Its envelope is $\qquad B^2 - 4AC = 0$

$$\Rightarrow \qquad (-2x)^2 - 4(1)(x^2 - y) = 0$$

$$\Rightarrow \qquad 4x^2 - 4x^2 + 4y = 0$$

$$\Rightarrow \qquad y = 0 \text{ i.e., } x\text{-axis is an envelope.}$$

Example 2. *Find the envelope of the family of straight lines* $y = mx + \sqrt{a^2m^2 + b^2}$ *, m being a parameter.* (OU-Nov. 2006)

Sol. Given $y - mx = \sqrt{a^2m^2 + b^2}$

Squaring on both sides

$$y^2 + m^2x^2 - 2mxy = a^2m^2 + b^2$$

$$\Rightarrow \quad (x^2 - a^2)m^2 - (2xy)m + (y^2 - b^2) = 0$$

Compare with $Am^2 + Bm + C = 0$, we get

$$A = x^2 - a^2, B = -2xy, C = y^2 - b^2$$

Its envelope is $\quad B^2 - 4AC = 0$

$$\Rightarrow \quad (-2xy)^2 - 4(x^2 - a^2)(y^2 - b^2) = 0$$

$$\Rightarrow \quad 4x^2y^2 - 4x^2y^2 + 4x^2b^2 + 4a^2y^2 - 4a^2b^2 = 0$$

Dividing by $4a^2b^2$, we get

$$\frac{x^2}{a^2} + \frac{y^2}{b^2} = 1 \text{ is an envelope.}$$

Example 3. *Find the envelope of the family of straight lines* $y = mx + \frac{a}{m}$*, where m is a parameter.*

Sol. Given $\quad y = mx + \frac{a}{m}$

Multiplying by m, $\quad my = m^2x + a$

$$\Rightarrow \quad xm^2 - ym + a = 0$$

Compare with $Am^2 + Bm + C = 0$, we get

$$A = x, B = -y, C = a$$

Its envelope is $\quad B^2 - 4AC = 0$

$$\Rightarrow \quad y^2 - 4ax = 0.$$

3.8.2. Procedure to find Envelope of the Equation with one Parameter α and is not Quadratic in α

Let $\quad f(x, y, \alpha) = 0 \quad$ ...(1)

where α is a parameter.

Differentiating (1) w.r.t. α, we get $\frac{df}{d\alpha} = 0$ ...(2)

Eliminate α from (1) and (2) to get envelope of the given family of curves.

Example 4. *Find the envelope of the family of curves* $x \tan \alpha + y \sec \alpha = 5$*,* α *is a parameter.* (OU-Dec. 2013)

Sol. Given $x \tan \alpha + y \sec \alpha = 5$ ...(1)

Differentiating w.r.t. α, we get

$$x \sec^2 \alpha + y \sec \alpha \tan \alpha = 0$$

$$\Rightarrow \quad x \sec \alpha + y \tan \alpha = 0 \quad \text{...(2)}$$

$(1)^2 - (2)^2$

$$\Rightarrow \quad x^2 \tan^2 \alpha + y^2 \sec^2 \alpha + 2xy \tan \alpha \sec \alpha = 25$$
$$x^2 \sec^2 \alpha + y^2 \tan^2 \alpha + 2xy \tan \alpha \sec \alpha = 0$$
$$\underline{\quad - \qquad - \qquad - \qquad - \quad}$$
$$-x^2(\sec^2 \alpha - \tan^2 \alpha) + y^2(\sec^2 \alpha - \tan^2 \alpha) = 25$$
$$-x^2 + y^2 = 25 \qquad (\because \sec^2 \alpha - \tan^2 \alpha = 1)$$
$$x^2 - y^2 + 25 = 0$$

is a required envelope.

Example 5. *Find the envelope of the family of straight lines $x \cos \alpha + y \sin \alpha = p$ where α is a parameter.* (OU-May 2012)

Sol. Given $x \cos \alpha + y \sin \alpha = p$ ...(1)

Differentiating w.r.t. α, we get

$$-x \sin \alpha + y \cos \alpha = 0 \qquad ...(2)$$

$(1)^2 + (2)^2$

$$\Rightarrow \quad x^2 \cos^2 \alpha + y^2 \sin^2 \alpha + 2xy \sin \alpha \cos \alpha = p^2$$
$$x^2 \sin^2 \alpha + y^2 \cos^2 \alpha - 2xy \sin \alpha \cos \alpha = 0$$
$$x^2 + y^2 = p^2 \qquad (\because \sin^2 \alpha + \cos^2 \alpha = 1)$$

is a required envelope.

Example 6. *Find the envelope of $\frac{x}{a} \cos \alpha + \frac{y}{b} \sin \alpha = 1$, α is a parameter.*

(OU-May 2009, June 2010)

Sol. Given $\frac{x}{a} \cos \alpha + \frac{y}{b} \sin \alpha = 1$ ...(1)

Differentiating w.r.t. α, we get $-\frac{x}{a} \sin \alpha + \frac{y}{b} \cos \alpha = 0$ ...(2)

$(1)^2 + (2)^2 \quad \Rightarrow \quad \frac{x^2}{a^2} + \frac{y^2}{b^2} = 1$ is an envelope of the given family of straight lines.

Example 7. *Find the envelope of the family of straight lines $x \cos \alpha + y \sin \alpha = a \sin \alpha \cos \alpha$, where a is a constant and α is parameter.*

Sol. Given $x \cos \alpha + y \sin \alpha = a \sin \alpha \cos \alpha$

Dividing by $\sin \alpha \cos \alpha$, we get

$$x \operatorname{cosec} \alpha + y \sec \alpha = a \qquad ...(1)$$

Differentiating w.r.t. α, we get

$$-x \operatorname{cosec} \alpha \cot \alpha + y \sec \alpha \tan \alpha = 0 \qquad ...(2)$$

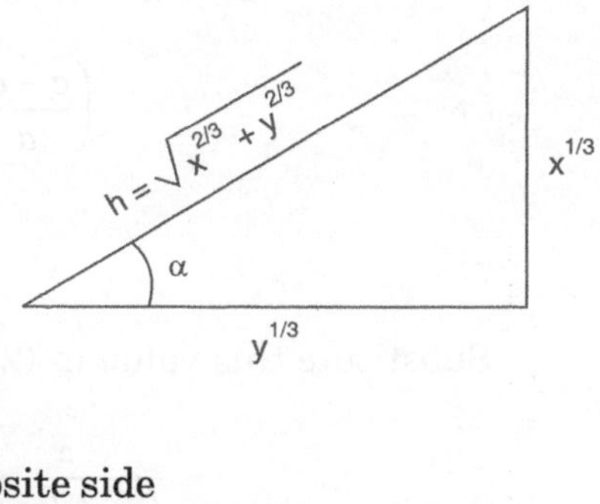

$$\Rightarrow \quad y \sec \alpha \tan \alpha = x \operatorname{cosec} \alpha \cot \alpha$$

$$\Rightarrow \quad \frac{\sec \alpha \tan \alpha}{\operatorname{cosec} \alpha \cot \alpha} = \frac{x}{y}$$

$$\Rightarrow \quad \tan^3 \alpha = \frac{x}{y} \quad \Rightarrow \quad \tan \alpha = \frac{x^{1/3}}{y^{1/3}} = \frac{\text{opposite side}}{\text{adjacent side}}$$

$$\operatorname{cosec} \alpha = \frac{\text{hypotenuse}}{\text{opposite side}} = \frac{h}{x^{1/3}}$$

$$\sec\alpha = \frac{\text{hypotenuse}}{\text{adjacent side}} = \frac{h}{y^{1/3}}$$

Substituting cosec α, sec α values in (1), we get

$$x\left(\frac{h}{x^{1/3}}\right) + y\left(\frac{h}{y^{1/3}}\right) = a \quad \Rightarrow \quad h[x^{2/3} + y^{2/3}] = a$$

$$\Rightarrow \quad (x^{2/3} + y^{2/3})^{3/2} = a \qquad (\because \; h = \sqrt{x^{2/3} + y^{2/3}})$$

$$\Rightarrow \quad x^{2/3} + y^{2/3} = a^{2/3}$$ is a required envelope.

3.8.3. Envelope of Two Parameter Family of Curves

Let $f(x, y, a, b) = 0$ where a, b are parameters connected by the relation $g(a, b) = 0$, be the given family of curves.

To find envelope, write one parameter in terms of other parameter, say $b = \phi(a)$ or $a = \psi(b)$ so that the given family reduces to one parameter and hence follow the above method 3.8.2.

Example 8. *Find the envelope of the family of straight lines* $\frac{x}{a} + \frac{y}{b} = 1$ *where* $a + b = c$, *c is a constant.* (OU-Dec. 2008, June 2011, Jan. 2012)

Sol. Given
$$\frac{x}{a} + \frac{y}{b} = 1 \qquad \ldots(1)$$

where a, b are two parameters.

Since, $a + b = c \Rightarrow b = c - a$

Substituting in (1), we get

$$\frac{x}{a} + \frac{y}{c-a} = 1 \qquad \ldots(2)$$

where a is one parameter.

Differentiating (2) w.r.t. 'a', we get

$$-\frac{x}{a^2} - \frac{y}{(c-a)^2}(-1) = 0 \quad \Rightarrow \quad \frac{x}{a^2} = \frac{y}{(c-a)^2}$$

$$\Rightarrow \quad \left(\frac{c-a}{a}\right)^2 = \frac{y}{x} \quad \Rightarrow \quad \frac{c}{a} - 1 = \frac{\sqrt{y}}{\sqrt{x}}$$

$$\Rightarrow \quad \frac{c}{a} = \frac{\sqrt{x} + \sqrt{y}}{\sqrt{x}} \quad \Rightarrow \quad a = \frac{c\sqrt{x}}{\sqrt{x} + \sqrt{y}}$$

Substitute this value in (2), we get

$$\frac{x}{\left(\frac{c\sqrt{x}}{\sqrt{x}+\sqrt{y}}\right)} + \frac{y}{c - \frac{c\sqrt{x}}{\sqrt{x}+\sqrt{y}}} = 1$$

$$\Rightarrow \qquad \frac{\sqrt{x}(\sqrt{x}+\sqrt{y})}{c} + \frac{(\sqrt{x}+\sqrt{y})y}{c(\sqrt{x}+\sqrt{y}-\sqrt{x})} = 1$$

$$\Rightarrow \qquad \sqrt{x}(\sqrt{x}+\sqrt{y}) + \sqrt{y}(\sqrt{x}+\sqrt{y}) = c$$

$$\Rightarrow \qquad (\sqrt{x}+\sqrt{y})^2 = c$$

$$\Rightarrow \qquad \sqrt{x}+\sqrt{y} = \sqrt{c}$$

is a required envelope.

Example 9. *Find the envelope of the straight lines* $\frac{x}{a}+\frac{y}{b} = 1$ *where a, b are parameters connected by* $ab = c^2$. (OU-Jan. 2011)

Sol. Given $$\frac{x}{a}+\frac{y}{b} = 1 \qquad \text{...(1)}$$

Also given $$ab = c^2 \;\Rightarrow\; b = \frac{c^2}{a} \qquad \text{...(2)}$$

Substitute (2) in (1), we get $$\frac{x}{a}+\frac{ya}{c^2} = 1 \qquad \text{...(3)}$$

Differentiating (3) w.r.t. 'a', we get

$$-\frac{x}{a^2}+\frac{y}{c^2} = 0$$

$$\Rightarrow \qquad \frac{x}{a^2} = \frac{y}{c^2} \;\Rightarrow\; a = \frac{c\sqrt{x}}{\sqrt{y}} \qquad \text{...(4)}$$

Substitute (4) in (3), we get

$$\frac{x\sqrt{y}}{c\sqrt{x}} + \frac{c\sqrt{x}\cdot y}{c^2\sqrt{y}} = 1 \;\Rightarrow\; \sqrt{x}\sqrt{y}+\sqrt{x}\sqrt{y} = c$$

$$\Rightarrow \qquad 2\sqrt{xy} = c \;\Rightarrow\; 4xy = c^2$$

is a required envelope.

EXERCISE 2

1. Find the radius of curvature at any point on the curve $y = c\cosh\frac{x}{c}$.
2. Find the radius of curvature at any point on the curve $x = a(\cos t + t\sin t)$, $y = a(\sin t - t\cos t)$.
3. Find the radius of curvature at the point $\left(\frac{3a}{2}, \frac{3a}{2}\right)$ of the curve $x^3 + y^3 = 3axy$.
4. Find the radius of curvature of the curve $xy^2 = a^3 - x^3$ at the point $(a, 0)$.

5. Find radius of curvature at (0, 0) for the curve $x^3 + y^3 - 2x^2 + by = 0$.

6. Find the evolute of the hyperbola $\frac{x^2}{a^2} - \frac{y^2}{b^2} = 1$.

7. Show that the evolute of the curve $x = a\left[\cos t + \log \tan \frac{t}{2}\right]$, $y = a \sin t$ is the catenary $y = a \cosh\left(\frac{x}{a}\right)$.

8. Find the equation of the circle of curvature of the curve $x + y = ax^2 + by^2 + cx^3$ at the origin.

9. Find the envelope of the family of curves $\frac{a^2}{x}\cos\theta - \frac{b^2}{y}\sin\theta = c$ and θ being the parameter.

10. Find the envelope of the family of lines $y = mx + \sqrt{1+m^2}$, where m is the parameter.

11. Find the envelope of the ellipses $\frac{x^2}{a^2} + \frac{y^2}{b^2} = 1$ for which $a + b = c$.

Answers

1. $c \cosh^2\left(\frac{x}{c}\right)$ 2. at 3. $\frac{3a\sqrt{2}}{16}$

4. $\frac{3a}{2}$ 5. $\frac{3}{2}$ 6. $x^{2/3} - y^{2/3} = (2a)^{2/3}$

8. $(a + b)(x^2 + y^2) = 2(x + y)$ 9. $\frac{a^4}{x^2} + \frac{b^4}{y^2} = c^2$ 10. $x^2 + y^2 = 1$

11. $x^{2/3} + y^{2/3} = c^{2/3}$.

Asymptotes of a Curve

An asymptote of a curve is a line such that the distance between the curve and the line approaches zero as they tend to $+\infty$ and $-\infty$.

There are three types of asymptotes:

Type I:

Horizontal asymptote. An asymptote which is paralle to x-axis is called a horizontal asymptote. This is denoted by $y = k$ or $y = -k$.

Type I:

Vertical asymptote. An asymptote which is parallel to y-axis is called a vertical asymptote. This is denoted by $x = k$ or $x = -k$.

Type III:

Oblique asymptote. An asymptote which is neither parallel to x-axis nor to y-axis is called an oblique asymptote. This is denoted by $y = mx + c$.

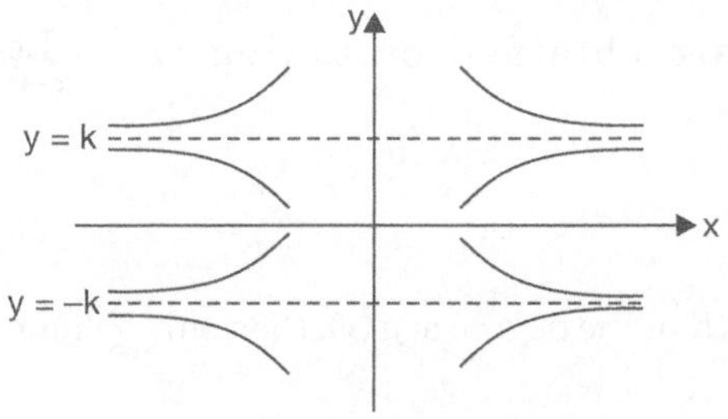

Horizontal asymptotes

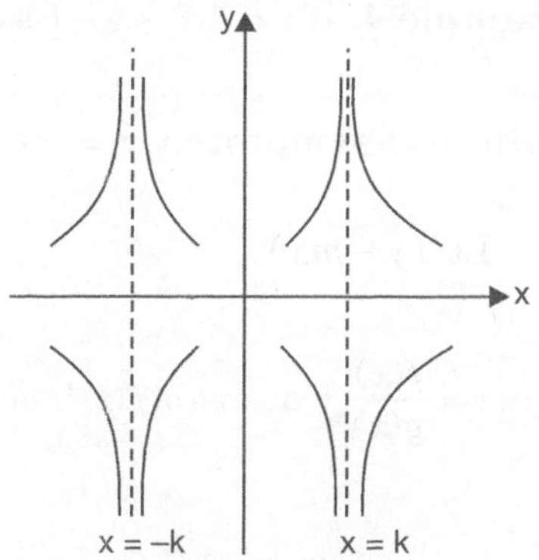

Vertical asymptotes

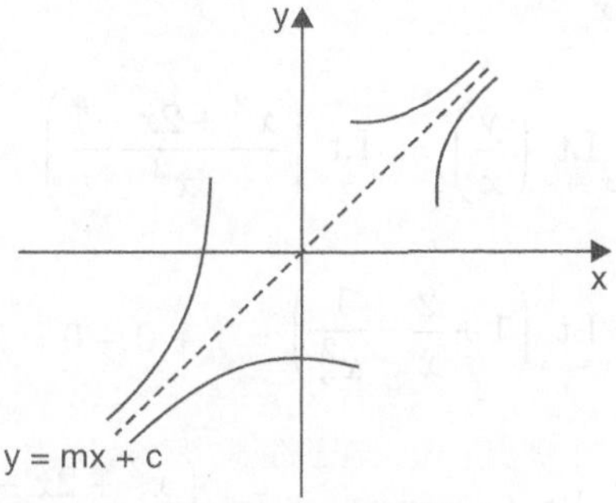

Oblique asymptote

Working rule. Let $y = \dfrac{f(x)}{g(x)}$ be the given curve. Then

(*i*) Vertical asymptotes are obtained by equating denominator $g(x)$ to zero.

Example 1. If $y = \dfrac{x^2 - 3x}{x - 1}$, then $x = 1$ is a V.A.

Example 2. If $y = \dfrac{x^2 + 1}{x^2 - 1}$, then $x = 1, -1$ are V.A.

Example 3. If $y = \dfrac{x^2 + 2x - 1}{x(x - 2)}$, then $x = 0, 2$ are V.A.

Example 4. If $y = (x^2 - x - 6)(x - 7)$, then there is no V.A.

(*ii*) Horizontal asymptotes are obtained by taking y = ratio of highest and same degree terms.

Example 1. If $y = \dfrac{2x^2 + 1}{3x^2 - 1}$, then $y = \dfrac{2}{3}$ is a H.A.

Example 2. If $y = \dfrac{x^2 + 1}{(x + 1)(x + 2)}$, then $y = 1$ is a H.A.

Example 3. If $y = \dfrac{x^2 - 3x}{x - 1}$, then there is no H.A.

Example 4. If $y = (x^2 - x - 6)(x - 7)$, then there is no H.A.

(*iii*) Oblique asymptotes $y = mx + c$ are obtained by taking $m = \underset{x\to\infty}{\text{Lt}}\left(\dfrac{y}{x}\right)$ and $c = \underset{x\to\infty}{\text{Lt}}\,(y - mx)$.

Note. In $y = \dfrac{f(x)}{g(x)}$ if degree of $f(x)$ is one more than the degree of $g(x)$, then only oblique asymptotes exist.

Example 1. Let $y = \dfrac{x^2 + 2x - 1}{x}$ be the given curve.

To get O.A. take
$$m = \underset{x\to\infty}{\text{Lt}}\left(\frac{y}{x}\right) = \underset{x\to\infty}{\text{Lt}}\left(\frac{x^2 + 2x - 1}{x^2}\right)$$
$$= \underset{x\to\infty}{\text{Lt}}\left(1 + \frac{2}{x} - \frac{1}{x^2}\right) = 1 + 0 - 0 = 1$$

and
$$c = \underset{x\to\infty}{\text{Lt}}\,(y - mx) = \underset{x\to\infty}{\text{Lt}}\left(\frac{x^2 + 2x - 1}{x} - x\right)$$
$$= \underset{x\to\infty}{\text{Lt}}\left(\frac{2x - 1}{x}\right) = \underset{x\to\infty}{\text{Lt}}\left(2 - \frac{1}{x}\right) = 2 - 0 = 2$$

So $y = mx + c = x + 2$ is an oblique asymptote.

Example 2. $y = \dfrac{x^2 + 1}{x^2 - 1}$ has no O.A.

Example 3. $y = (x^2 - x - 6)(x - 7)$ has no O.A.

ILLUSTRATIVE EXAMPLES

Example 1. *Find all asymptotes of the curve* $y = x + \dfrac{1}{x}$. (OU-June 2014)

Sol. Given $y = \dfrac{x^2 + 1}{x}$

(*i*) Denominator = 0 $\Rightarrow$ $x = 0$ is a vertical asymptote.

(*ii*) No horizontal asymptote because in $y = \dfrac{f(x)}{g(x)}$, $f(x)$ and $g(x)$ are not having same higher degree terms.

(*iii*) Take $m = \underset{x\to\infty}{\text{Lt}} \left(\dfrac{y}{x}\right) = \underset{x\to\infty}{\text{Lt}} \left(\dfrac{x^2+1}{x^2}\right) = \underset{x\to\infty}{\text{Lt}} \left(1+\dfrac{1}{x^2}\right) = 1 + 0 = 1$

and $c = \underset{x\to\infty}{\text{Lt}} (y - mx) = \underset{x\to\infty}{\text{Lt}} \left(\dfrac{x^2+1}{x} - x\right) = \underset{x\to\infty}{\text{Lt}} \left(\dfrac{1}{x}\right) = 0$

$\therefore$ $y = mx + c = 1 \cdot x + 0 = x$ is an oblique asymptote.

Example 2. *Find all asymptotes of* $x^2y - y - x = 0$.

Sol. Given $x^2y - y - x = 0 \Rightarrow y(x^2 - 1) = x \Rightarrow y = \dfrac{x}{x^2-1}$ which is of the form $y = \dfrac{f(x)}{g(x)}$.

(*i*) Denominator = 0 $\Rightarrow$ $x^2 - 1 = 0$ $\Rightarrow$ $x = 1$ and -1 are vertical asymptotes.

(*ii*) No horizontal asymptote because $f(x)$ and $g(x)$ does not have same higher degree terms.

(*iii*) No oblique asymptote because degree of $f(x)$ is not one more than the degree of $g(x)$.

3.9. CURVE TRACING : Cartesian Form

I. Symmetry

Apply the following rules to find whether the given curve is symmetrical about any line or not.

(*i*) *Symmetry about the x-axis.*

If the equation of a curve remains unaltered when y is changed to $-y$ (*i.e.*, if only **even** powers of y occur in the equation), then the curve is symmetrical about x-axis.

For example, the parabola $y^2 = 4ax$ is symmetrical about the x-axis.

(*ii*) *Symmetry about the y-axis.*

If only **even** powers of x occur in the equation of the curve, the curve is symmetrical about the y-axis.

For example, the curve $x^2 = 4y$ is symmetrical about the y-axis.

(*iii*) *Symmetry in opposite quadrants.*

If the equation of the curve remains unaltered when **both** x and y are changed to $-x$ and $-y$ respectively, then the curve is symmetrical in opposite quadrants.

For example, the curve $xy = c^2$ is symmetrical in opposite quadrants.

(*iv*) *Symmetry about the line* $y = x$.

If the equation of the curve is unchanged when x and y are interchanged (*i.e.*, x is changed to y and y to x), the curve is symmetrical about the line $y = x$.

For example, the curve $x^3 + y^3 = 3axy$ is symmetrical about the line $y = x$.

II. Origin

(*i*) Find whether the curve passes through the origin or not. It will pass through the origin if the equation of the curve is free from the constant term.

(*ii*) If the curve passes through the origin, find the equations of the tangents at the origin by equating to zero the lowest degree terms in the equation of the curve.

(*iii*) If there are two tangents at the origin, then origin is a *double point.*

If the two tangents are real and different, then origin is a *node*.

If the two tangents are real and coincident, then origin is a *cusp*.

If the two tangents are imaginary, then origin is a *conjugate point or isolated point.*

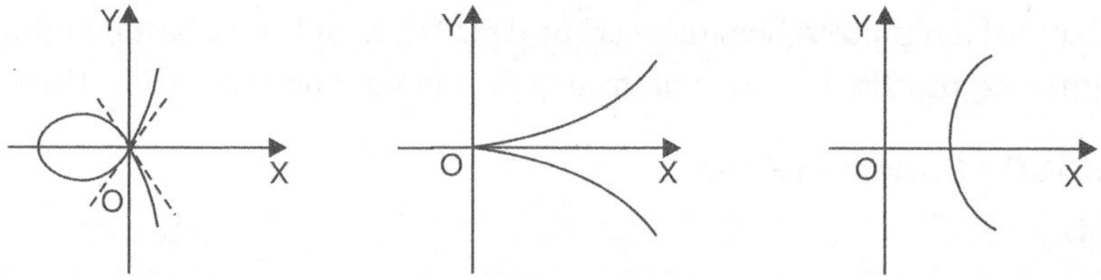

Node Cusp Conjugate point

III. Axes Intersection

Find the points where the curve cuts the axes of co-ordinate. Find also the tangents at these points if necessary and the position of the curve relative to these tangents.

If $y = x$ or $y = -x$ is a line of symmetry, find the points of intersection of the curve and the line and also the tangents at that point.

IV. Asymptotes

Find all the asymptotes (parallel to the axes and oblique) of the curve if any.

Also find the points of intersection of the curve and its asymptotes.

V. Region

Find those regions of the plane where the curve does not exist. Such a region is obtained on solving the given equation for one variable in terms of the other variable and finding out the set of values of one variable which make the other imaginary.

VI. Tangents

If the curve passes through the origin then the equations of the tangent at the origin are obtained by equating to zero, the lowest degree term in the equation of the curve. Find the points where the tangents is parallel to the x-axis (*i.e.,* the points where $dy/dx = 0$) or perpendicular to the x-axis (*i.e.,* the points where $dy/dx = \infty$).

VII. Curve Rising and Falling

(*i*) If $\dfrac{dy}{dx} > 0, \forall\, x \in [a, b]$, the curve is rising in $[a, b]$.

(*ii*) If $\dfrac{dy}{dx} < 0, \forall\, x \in [a, b]$, the curve is falling in $[a, b]$.

(*iii*) If $\dfrac{dy}{dx} = 0$ at any point P, then P is a stationary point.

VIII. Concavity Upwards and Downwards

(*i*) If $\frac{d^2y}{dx^2} > 0, \forall x \in [a, b]$, then the curve is concave upwards in $[a, b]$.

(*ii*) If $\frac{d^2y}{dx^2} < 0, \forall x \in [a, b]$, then the curve is concave downwards in $[a, b]$.

(*iii*) If $\frac{d^2y}{dx^2} = 0$ at a point P then P is point of inflexion.

IX. Table of points

x	$-\infty$ .... -1 0 1 ...... ∞
y	Find corresponding as values of x

Illustration. Consider the curve $y^2(2a - x) = x^3$.

Here $$y = \pm x\sqrt{\frac{x}{2a - x}}$$

When x is +ve and $> 2a$, y is imaginary (Since $2a - x$ is –ve) and so no part of the curve lies to the right of the line $x = 2a$.

Again when $x < 0, y$ is again imaginary, so the curve does not lie to the left of the line $x = 0$. That is y-axis.

For limitations of the curve, we sometimes change from the cartesian to polar or parametric form.

Note 1. In the examples, inconvenient steps may be omitted without any disadvantage.

Note 2. To obtain asymptotes parallel to x-axis, equate to zero the real linear factors of the co-efficient of highest power of x in the equation of the curve. If this co-efficient is a constant or its factors are not real, then there is no asymptote parallel to x-axis.

To obtain asymptotes parallel to y-axis, equate to zero the real linear factors of the co-efficient of the highest power of y in the equation of the curve.

ILLUSTRATIVE EXAMPLES

Example 1. *Trace the curve* $y^2 (a + x) = x^2 (3a - x)$. [JNTU 2006 S (Set 4)]

Sol. The equation of the curve is

$$y^2 (a + x) = x^2 (3a - x) \quad ...(1)$$

1. Symmetry. Since the equation (1) contains only even powers of y.

Therefore the curve is symmetrical about x-axis.

The curve is *not* symmetrical about y-axis.

2. Origin. (*i*) The curve passes through the origin.

(*ii*) Equating to zero, the lowest degree terms, the tangents at the origin are given by

$ay^2 - 3ax^2 = 0$ or $y = \pm \sqrt{3}\,x$.

The two tangents are real and different.

Therefore origin is a node.

(*iii*) From (1), we have $y^2 = x^2 \dfrac{(3a - x)}{a + x}$.

Since $\dfrac{3a - x}{a + x} < \dfrac{3a}{a} = 3$ for small +ve x,

Therefore, y is less than $\sqrt{3}x$.

Hence the curve lies below the tangent $y = \sqrt{3}x$ for small +ve values of x.

3. Axes Intersection. The curve meets x-axis, where putting $y = 0$, we have

$$x^2 (3a - x) = 0$$

Therefore, $x = 0$ or $3a$

That is, in the points (0, 0) and A $(3a, 0)$ curve meets x-axis.

Similarly, it meets y-axis in the origin only.

Shifting the origin to $(3a, 0)$, the equation (1) transforms to

$$y^2 (4a + x) = (3a + x)^2 (- x),$$

Equating to zero the lowest degree terms, the tangent at the new origin is given by $- 9a^2x = 0$ or $x = 0$. Thus the tangent at A $(3a, 0)$ is || to y-axis.

4. Asymptotes. Equating to zero the denominator of y^2, the asymptote || to y-axis is $x + a = 0$.

There is no other asymptote of the curve.

5. Special points. Differentiating (1) with respect to x, we get

$$2y \frac{dy}{dx} (a + x) + y^2 = 6ax - 3x^2$$

or

$$2y \frac{dy}{dx} (a + x) = 6ax - 3x^2 - y^2 = 6ax - 3x^2 - \frac{x^2 (3a - x)}{a + x}$$

$$= \frac{6a^2x - 3ax^2 + 6ax^2 - 3x^3 - 3ax^2 + x^3}{a + x}$$

$$= \frac{6a^2x - 2x^3}{a + x} = \frac{2x (3a^2 - x^2)}{a + x}$$

Therefore,

$$\frac{dy}{dx} = \frac{x(3a^2 - x^2)}{y(a + x)^2} = \frac{x (3a^2 - x^2)}{(a + x)^2} \cdot \frac{\sqrt{a + x}}{x\sqrt{3a - x}}$$

$$= \frac{(3a^2 - x^2)}{(a + x)^{3/2} (3a - x)^{1/2}}$$

(*i*) Now $\dfrac{dy}{dx} = 0$ when $3a^2 - x^2 = 0$ or $x = \pm\sqrt{3}a$.

The tangent is || to x-axis at the points where $x = \pm\sqrt{3}a$.

(*ii*) Again $\frac{dy}{dx} \to \infty$, when $a + x \to 0$ or $3a - x \to 0$

When $x \to -a, y \to \infty$, showing that $x = -a$ is an asymptote. When $x = 3a, y = 0$. Thus the tangent is || to y-axis at $(3a, 0)$.

(*iii*) Again when $x = a, y = a$ and when $x = 2a$,

$$y = \frac{2a}{\sqrt{3}} = 1.2a \text{ nearly.}$$

6. Region. Solving for y and considering the +ve values (Since the curve is symmetrical about the x-axis), we have

$$y = x\sqrt{\frac{3a - x}{a + x}} \qquad ...(2)$$

(*i*) When $x = 0, y = 0$. When x is +ve and small, y is real. As x goes on increasing, y also increases up to the point for which $x = \sqrt{3}\,a$ and then decreases and finally becomes zero at $x = 3a$.

(*ii*) When $x > 3a$, y is imaginary. Hence no part of the curve lies beyond the line $x = 3a$.

(*iii*) When x is –ve and numerically small, from (2), y is real.

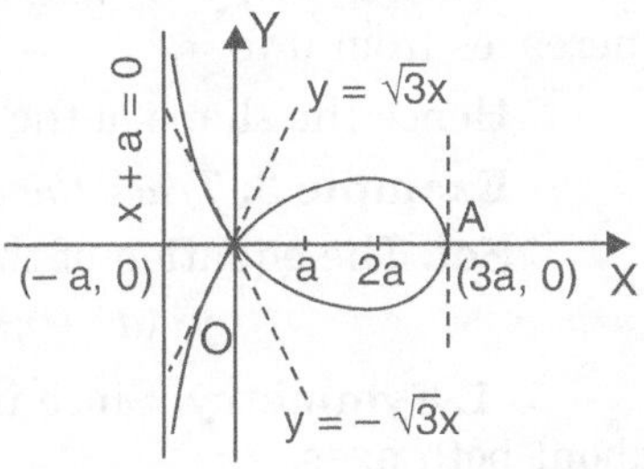

Also for values of x under consideration,

$$3a - x > 3\,(a + x).$$

Therefore, y is numerically $> -\sqrt{3}x$.

Therefore, the curve lies above the tangent in the second quadrant.

(*iv*) When $x < -a$, from (2), y is imaginary. Therefore the curve does not lie beyond $x = -a$.

Again when x is –ve and numerically increases from 0 to a, y is –ve and decreases from 0 to $-\infty$. Hence the shape of the curve is as shown in the figure.

Example 2. *Trace the curve:* $y^2(2a - x) = x^3$. [JNTU 2009 (Set I)]

Sol. The equation of the curve is $y^2\,(2a - x) = x^3$ ...(1)

1. Symmetry. Since (1) contains only even powers of y, the curve is symmetrical about x-axis.

2. Origin. (*i*) The curve passes through the origin.

(*ii*) The tangents at the origin are given by $y^2 = 0$. Since the two tangents are real and coincident, therefore origin is a cusp.

3. Axes Intersection. The curve meets x-axis and y-axis at the origin only.

4. Asymptotes. Equating to zero the co-efficient of y^2, the highest degree term in y, the asymptote || to y-axis is $x = 2a$. There is no other asymptote of the curve.

5. Special points. From (1),

$$y = \frac{x^{3/2}}{\sqrt{2a - x}} = x^{3/2}\,(2a - x)^{-1/2} \qquad ...(2)$$

Therefore,

$$\frac{dy}{dx} = \frac{3}{2}x^{1/2}\,(2a - x)^{-1/2} - \frac{1}{2}\,(2a - x)^{-3/2} \cdot (-1)\,.\,x^{3/2}$$

$$= \frac{1}{2}\,(2a - x)^{-3/2} \cdot x^{1/2}\,[3(2a - x) + x] = \frac{\sqrt{x}\,(3a - x)}{(2a - x)^{3/2}}$$

Therefore, $\dfrac{dy}{dx} = 0$ when $\sqrt{x}\ (3a - x) = 0$ or $x = 0, 3a$.

Rejecting $x = 3a$. Since when $x = 3a$, from (2), y is imaginary.

Therefore, the tangent at $x = 0$ is || to x-axis.

Again $\dfrac{dy}{dx} \to \infty$ when $x \to 2a$. From (2), when $x \to 2a$, $y \to \infty$.

Thus $x = 2a$ is an asymptote.

6. Region. From (1), $y = x\sqrt{\dfrac{x}{2a - x}}$ | Taking +ve root

When $x < 0$ or when $x > 2a$, y is imaginary.

Thus the curve does not lie to the left of y-axis and to the right of the line $x = 2a$.

When $x = 0$, $y = 0$. As x increases from 0 to $2a$, y also increases from 0 to ∞.

Hence the shape of the curve is as shown in the figure.

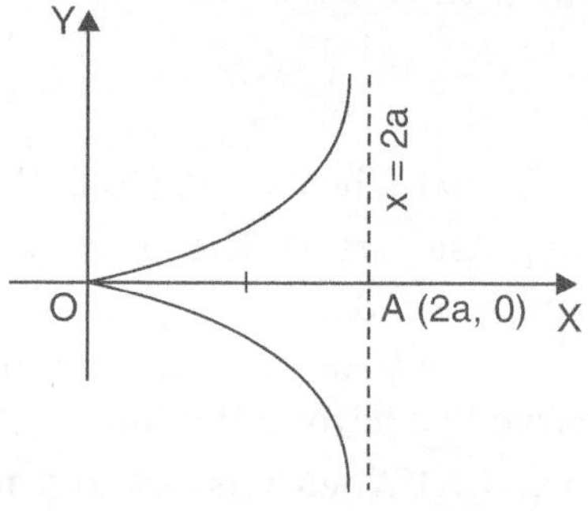

Example 3. *Trace the curve* $y^2 (a^2 + x^2) = x^2 (a^2 - x^2)$.

Sol. The equation of the curve is

$$y^2 (a^2 + x^2) = x^2 (a^2 - x^2) \quad \text{...(1)}$$

1. Symmetry. Since in (1), powers of both x and y are even, the curve is symmetrical about both axes.

2. Origin. The curve passes through the origin and the tangents at the origin are $y^2 = x^2$ or $y = \pm x$, which are real and distinct showing that origin is a node.

3. Axes Intersection. The curve meets x-axis at A $(a, 0)$ and A$'(-a, 0)$. Transferring the origin to $(a, 0)$ the equation (1) transforms to

$$\text{Y}^2 [a^2 + (\text{X} + a)^2] = (\text{X} + a)^2 [a^2 - (\text{X} + a)^2]$$

or
$$\text{Y}^2[2a^2 + 2a\text{X} + \text{X}^2] = (\text{X} + a)^2 [-2a\text{X} - \text{X}^2]$$

Equating to zero the lowest degree terms, the tangent at the new origin is X = 0, that is, new y-axis. Thus the tangent at A $(a, 0)$ is parallel to y-axis. Similarly, the tangent at A$'$ $(-a, 0)$ is also parallel to y-axis (by symmetry).

4. Asymptotes. The curve has no asymptotes.

5. Region. (*i*) Solving (1) for y, we have $y = \pm\, x\sqrt{\dfrac{a^2 - x^2}{a^2 + x^2}}$

Now y is real if and only if $a^2 - x^2 \ge 0$ or if $x^2 \le a^2$ or $|x| \le a$.

(*ii*) For the branch in the first quadrant.

$$y = x\sqrt{\dfrac{a^2 - x^2}{a^2 + x^2}}$$

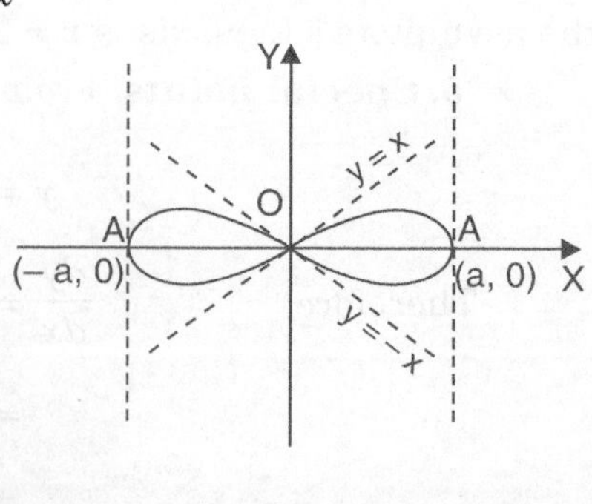

When $x = 0, y = 0$. As x increases, y also increases and goes on increasing until

$$x = a\sqrt{\sqrt{2}-1}, \text{ when } \frac{dy}{dx} = 0$$

i.e., the tangent is parallel to x-axis. As x increases from $a\sqrt{\sqrt{2}-1}$ to a, y decreases and ultimately becomes zero when $x = a$.

The shape of the curve is as shown in the figure.

Example 4. *Trace the curve* $y^2 = (x-2)(x-3)^2$. [JNTU 2006 (Set 2)]

Sol. Given that the equation of the curve is $y^2 = (x-2)(x-3)^2$

1. Symmetry. Since the equation contains even powers of y, we have that the curve is symmetric about x-axis and the curve is not symmetric about y-axis.

2. Origin. The curve does not pass through origin.

3. Axes Intersection. The curve meets y-axis when $x = 0 \Rightarrow y^2 = -18$, that is, y is imaginary and so there is no y-intercept.

Next putting $y = 0$, we have $x = 2$ and $= 3$

So the curve meets the x-axis at A (2, 0), B (3, 0).

4. Tangents. At A (2, 0), the line $x = 2$ is tangent, and at B (3, 0) there are two tangents $y = \pm(3-2)^{1/2}(x-3)$. Thus B is a node.

5. Region. We have $y = \pm\sqrt{x-2}\,(x-4)$

Thus for $x < 2$, y is imaginary. So the curve exists only for $x \geq 2$.

6. Special points. Differentiating with respect to x, we get that

$$2y\frac{dy}{dx} = (x-3)^2 + 2(x-2)(x-3)$$

$$2y\frac{dy}{dx} = x^2 - 6x + 9 + 2[x^2 - 5x + 6]$$

$$2y\frac{dy}{dx} = 3x^2 - 16x + 21$$

$$\frac{dy}{dx} = \frac{3x^2 - 16x + 21}{2\sqrt{x-2}\,(x-4)}$$

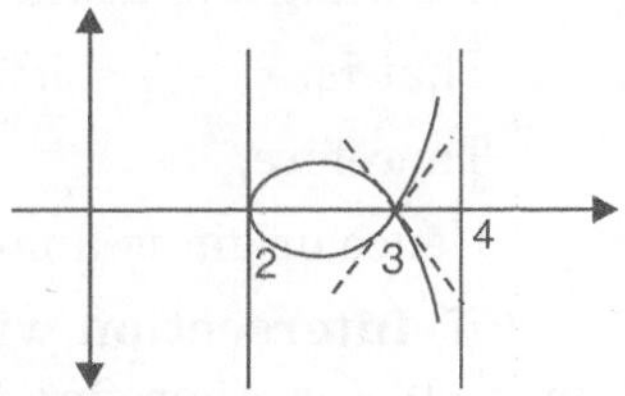

$$\frac{dy}{dx} = 0 \Rightarrow (x-3)(3x-7) = 0$$

$$\Rightarrow \quad x = 3, x = 7/3.$$

Tangent is || to x-axis at $x = 3$ and $x = 7/3 = 2.33$

$$\frac{dy}{dx} = \infty \Rightarrow (x-4)\sqrt{x-2} = 0$$

$$\Rightarrow \quad x = 4, x = 2$$

Lines $x = 2$ and $x = 4$ are asymptotes.

The shape of the curve is given in the figure.

Example 5. *Trace the curve* $y = (x-1)(x-2)$. [JNTU 2010 (Set 3)]

Sol. (*i*) **Symmetry.** The given curve is symmetric about y-axis

(*ii*) **Origin.** The given curve does not pass through origin as

$y = x^2 - 3x + 2 \Rightarrow$ when $x = 0, y = 2$

(*iii*) **Points of intersection.** If $y = 0$, then $x = 1, 2$

If $x = 0$, then $y = 2$

The points of intersection with x-axis are (1, 0), (2, 0) and the points of intersection with y-axis is (0, 2)

$$y = (x - 1)(x - 2) \Rightarrow \frac{dy}{dx} = (x - 1) + (x - 2) = 2x - 3$$

At (1, 0), $\left.\frac{dy}{dx}\right|_{x=1} = -1$, slope is –ve, $\theta > 90$

At (2, 0), $\left.\frac{dy}{dx}\right|_{x=2} = 1$, slope is +ve, $\theta < 90$

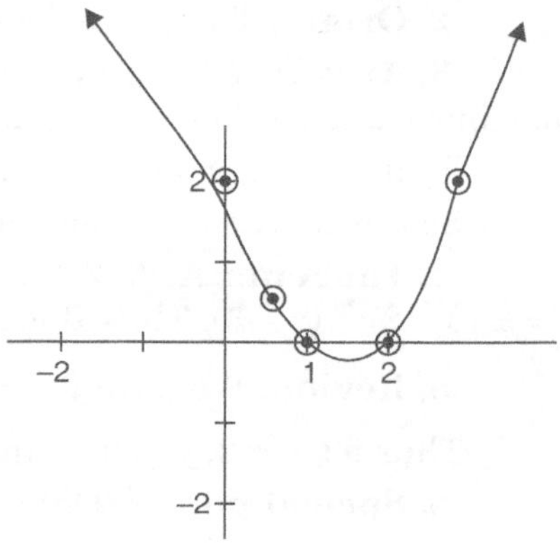

Next, if $0 < x < 1$, then $y > 0$

If $1 < x < 2$, then $y < 0$

If $x > 2$, then $y > 0$ and goes to ∞

If $x < 0$, then $y > 0$

If $x \to -\infty$, then $y \to \infty$.

This shows that the curve extends infinitely in the second quadrant.

(*iv*) **Asymptotes.** The curve has no asymptotes.

Example 6. *Trace the curve* $a^2y^2 = x^2 (a^2 - x^2)$.

Sol. (*i*) **Symmetry.** The given curve is symmetric about both the axises x and y.

(*ii*) **Origin.** The given curve passes through origin.

The tangents at the origin are $a^2y^2 - a^2x^2 = 0$

That is, $\quad a^2(y^2 - x^2) = 0 \quad$ or $\quad y^2 = x^2$

Therefore, $\quad y = \pm x$

Hence origin is a node.

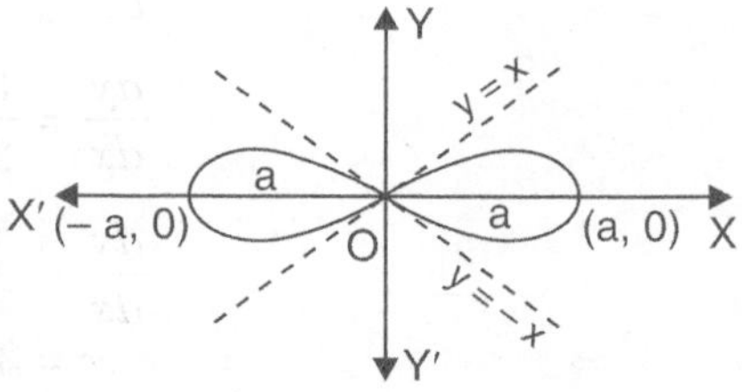

(*iii*) **Intersection with co-ordinate axes.** If $y = 0$, then $x = 0, \pm a$. Therefore the curve cuts the x-axis at the points (0, 0), $(a, 0)$ and $(-a, 0)$.

The given curve cuts the y-axis only at the origin.

(*iv*) **Asymptotes.** There are no asymptotes.

(*v*) **Region.** When $x > a$ or when $x < -a$, we have that y is imaginary.

$$y^2 = \frac{x^2(a^2 - x^2)}{a^2} \quad \text{or} \quad y = \frac{x}{a}\sqrt{a^2 - x^2}.$$

Therefore, the curve does not exist for $x > a$ and $x < -a$.

Hence the curve is as shown in the figure.

Example 7. *Trace the curve* $x^2 = \dfrac{y+1}{y-1}$.

Sol. The given curve is $x^2 = \dfrac{y+1}{y-1}$

(*i*) **Symmetry.** Since the power of x is even, the curve is symmetrical about y-axis.

(*ii*) **Origin.** The given curve does not pass through the origin.

(*iii*) **Asymptotes.** The highest power of x is 2 and its coefficient is $y - 1$.

Therefore, the asymptotes parallel to x-axis are given by $y - 1 = 0$ (That is, $y = 1$).

Also the given curve can be written as $y = \dfrac{x^2+1}{x^2-1}$ or $(x^2 - 1)y = (x^2 + 1)$. The highest power of y is 1 and its coefficient is $x^2 - 1$

Therefore, the asymptotes parallel to y-axis are given by $x^2 - 1 = 0$, that is, $x = \pm 1$.

Hence the asymptotes are $x = 1$, $x = -1$ and $y = 1$.

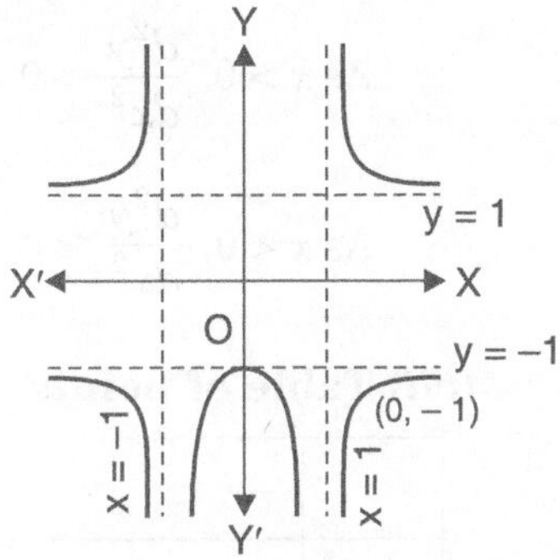

(*iv*) If $x = 0$, then $y = -1$. If $y = 0$ then x^2 is negative.

(means, x is imaginary)

Therefore, the given curve will not intersect the x-axis, but it will cross the y-axis at $(0, -1)$.

(*v*) **Region.** If the value of y lies between -1 and 1, the value of x^2 is negative.

Therefore, the curve does not lie between the lines $y = -1$ and $y = 1$.

The curve is as shown in the figure.

Example 8. *Trace the curve* $y = x^3 - 12x - 16$. (OU-Dec. 2013)

Sol. (*i*) Symmetry. The curve is not symmetric about any axis or any line.

(*ii*) Origin. Since the given equation is not satisfied by (0, 0), the curve does not passes through the origin.

(*iii*) Asymptotes. Since the curve is not in the form of $y = \dfrac{f(x)}{g(x)}$, the curve has no symptotes.

(*iv*) Points of intersection.

At $x = 0$, $\quad y = -16 \Rightarrow (0, -16)$

At $y = 0$, $\quad x^3 - 12x - 16 = 0 \Rightarrow (x + 2)^2(x - 4) = 0$

$\Rightarrow x = -2, 4 \quad \Rightarrow (-2, 0), (4, 0)$

$\therefore$ The curve touches x-axis at $(-2, 0)$ and $(4, 0)$ and y-axis at $(0, -16)$.

(*v*) Rising and falling.

$$\frac{dy}{dx} = 3x^2 - 12 = 3(x^2 - 4) = 3(x - 2)(x + 2)$$

Since $\frac{dy}{dx} > 0$ in $(-\infty, -2]$ and $[2, \infty)$, the curve is rising in $(-\infty, -2]$ and $[2, \infty)$.

Since $\frac{dy}{dx} < 0$ in $(-2, 2)$, the curve is falling in $(-2, 2)$.

(*vi*) Concavity upwards and downwards.

$$\frac{d^2y}{dx^2} = 6x$$

As $x > 0$, $\frac{d^2y}{dx^2} > 0 \Rightarrow$ The curve is concave upwards in $(0, \infty)$.

As $x < 0$, $\frac{d^2y}{dx^2} < 0 \Rightarrow$ The curve is concave downwards in $(-\infty, 0)$.

(*vii*) Table of points

x	$-\infty$	...	-3	-2	-1	1	2	4	...	∞
y	$-\infty$	...	-7	0	-5	-27	-32	0	...	∞

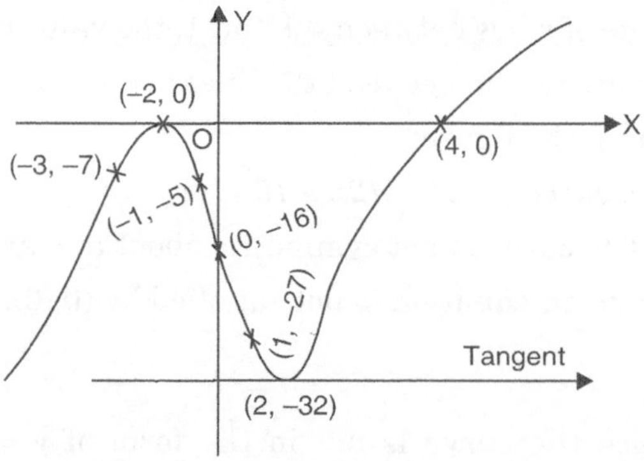

Example 9. *Trace the curve* $y = x^3$. (OU-June 2013)

Sol. (*i*) Symmetry. The curve is not symmetrical about any axis, but if we change x to $-x$ and y to $-y$, the equation does not change. Hence if (x, y) is a point on the curve then $(-x, -y)$ is also a point on the curve.

(*ii*) Origin. Since the equation is satisfied by (0, 0) the curve passes through the origin.

(*iii*) Tangents. Equating the lowest degree term to zero *i.e.,* $y = 0$ is a tangent to the curve at origin.

(*iv*) Asymptotes. Since the equation is not in the form of $y = \frac{f(x)}{g(x)}$, there are no asymptotes.

(*v*) **Table of points**

x	$-\infty$	...	-4	-1	0	1	4	...	∞
y	$-\infty$	...	-64	-1	0	1	64	...	∞

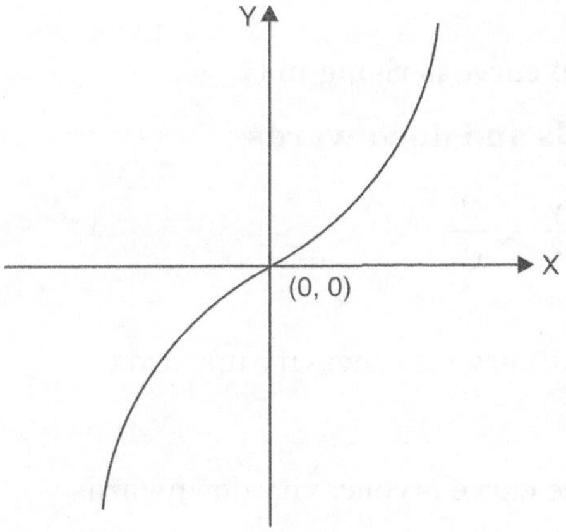

Example 10. *Sketch the graph of the curve* $x^2 - xy + 1 = 0$. (OU-May 2009, June 2010)

Sol. I. Symmetry. The curve is not symmetrical about any axis, but the equation does not change by changing x to $-x$ and y to $-y$. Thus if (x, y) is a point on the curve then $(-x, -y)$ also a point on the curve.

II. Origin. The curve does not passes through the origin.

III. Asymptotes. Given $x^2 - xy + 1 = 0 \Rightarrow y = \dfrac{x^2+1}{x} = x + \dfrac{1}{x}$

(*i*) To get vertical asymptotes, take denominator = 0 *i.e.*, $x = 0$ is a vertical asymptote.

(*ii*) To get horizontal asymptotes, take y = ratio of same and highest degree terms. Here, there is no horizontal asymptote.

(*iii*) In $y = \dfrac{f(x)}{g(x)}$ if degree of $f(x)$ – degree of $g(x) = 1$ then oblique asymptotes exist. To find oblique asymptote put $y = mx + C$ in the given equation. Thus $x^2 - x(mx + C) + 1 = 0$

equating x^2 coefficients, $1 - m = 0 \Rightarrow m = 1$

equating x coefficients, $-C = 0 \Rightarrow C = 0$

$\therefore \quad y = mx + C = x$ is an oblique asymptote.

IV. Rising and falling.

$$\frac{dy}{dx} = 1 - \frac{1}{x^2} = 0 \Rightarrow \frac{1}{x^2} = 1 \Rightarrow x^2 = 1 \Rightarrow x = \pm 1$$

If $x < -1$, $\frac{dy}{dx} > 0$ then curve is rising in $(-\infty, -1)$

If $-1 < x < 0$, $\frac{dy}{dx} < 0$ then curve is falling in $(-1, 0)$

If $0 < x < 1$, $\frac{dy}{dx} < 0$ then curve is falling in $(0, 1)$

If $x > 1$, $\frac{dy}{dx} > 0$ then curve is rising in $(1, \infty)$.

V. Concavity upwards and downwards

$$\frac{d^2y}{dx^2} = \frac{2}{x^3}$$

As $x > 0$, $\frac{d^2y}{dx^2} > 0$ the curve is concavity upwards

As $x < 0$, $\frac{d^2y}{dx^2} < 0$ the curve is concavity downwards

VI. Table of points

x	– 2	– 1	0	1	2	...	∞
y	– 2.5	– 2	∞	2	2.5	...	∞

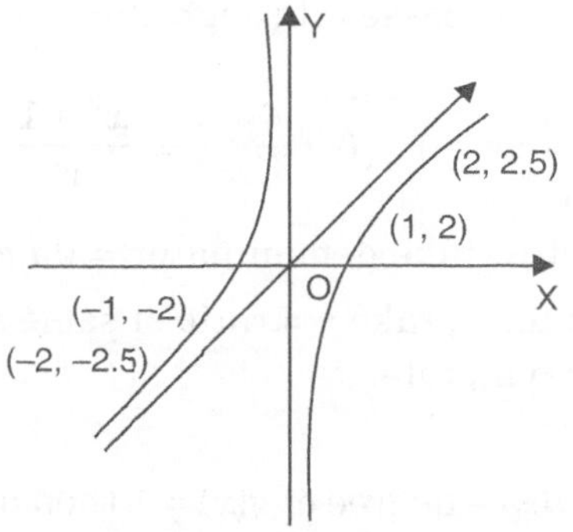

Example 11. *Trace the curve* $x^2y^2 = a^2(y^2 - x^2)$. (OU-Dec. 2007)

Sol. I. Symmetry. Since powers of x and y are even, curve is symmetrical about x and y-axis.

II. Origin. The curve passes through origin. Tangents at origin are given by $y^2 - x^2 = 0 \Rightarrow y = \pm x$. Since the tangents are real and distinct, origin is a node.

III. Asymptotes. Given $y = \pm \frac{ax}{\sqrt{a^2 - x^2}}$

As $x \to \pm a$, we get $x \to \pm \infty$

$\therefore \quad x = \pm a$ are vertical asymptotes.

IV. Points of intersection.

At $x = 0, y = 0 \Rightarrow (0, 0)$

At $y = 0, x = 0 \Rightarrow (0, 0)$

The curve meets the axes only at the origin.

V. Region. When $x > a$ or $x < -a$, y is imaginary.

$\therefore$ The curve lies between the lines $x = -a$ and $x = a$.

VI. Table of points

x	$-a$	0	$a/2$	a
y	$-\infty$	0	$a/\sqrt{3}$	∞

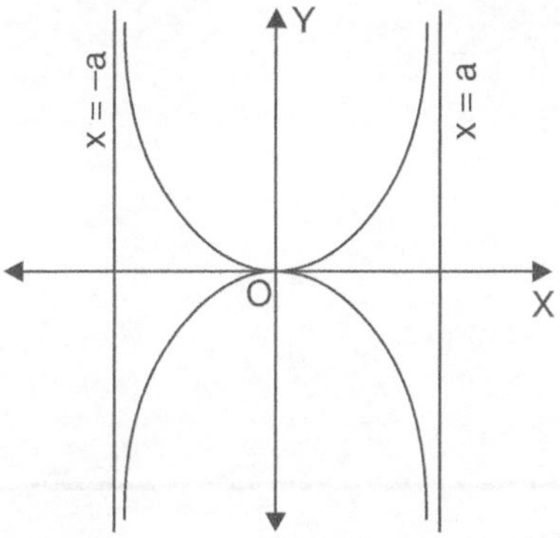

EXERCISE 3

Trace the following curves:

1. $y = c \cosh \dfrac{x}{c}$

2. $y^2(a - x) = x^2(a + x)$

3. $y^2 = x^3$

4. $x^3 + y^3 = 3axy, a > 0$

5. $y = \dfrac{8a^3}{x^2 + 4a^2}, (a > 0)$

6. $\left(\dfrac{x}{a}\right)^{\frac{2}{3}} + \left(\dfrac{y}{b}\right)^{\frac{2}{3}} = 1$

7. $y = x + \dfrac{1}{x}$

8. $y^2 = (x - a)^3, a > 0$

Answers

1.
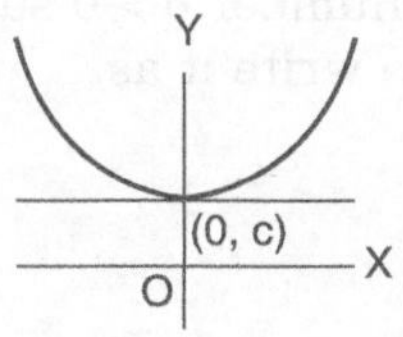

2.
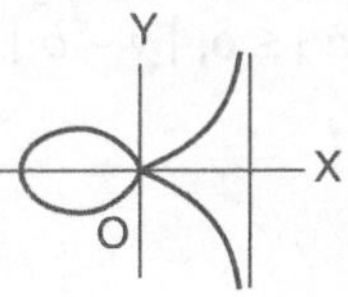

3.
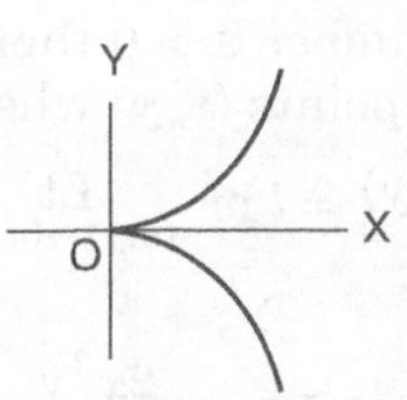

4.
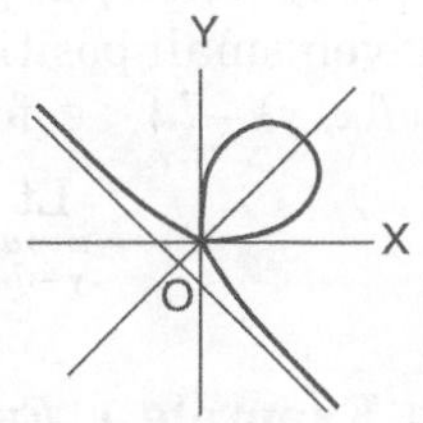

5.
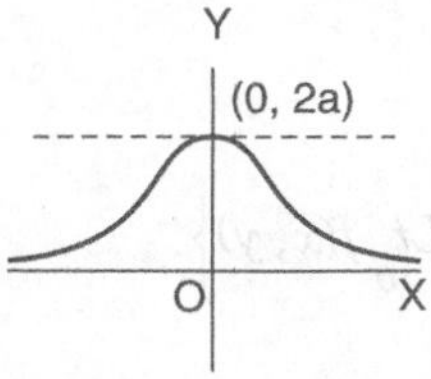

6.
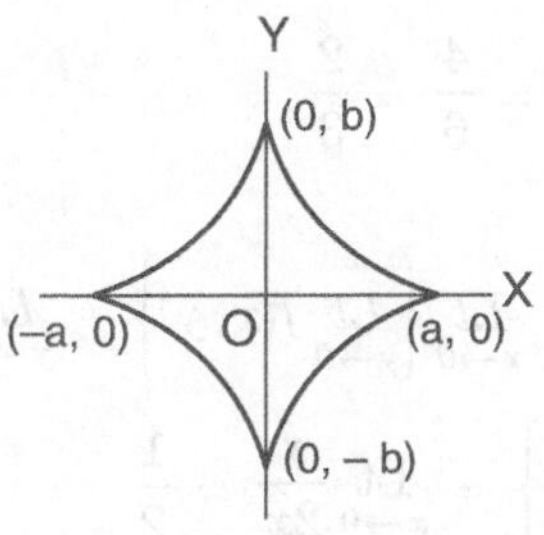

7.
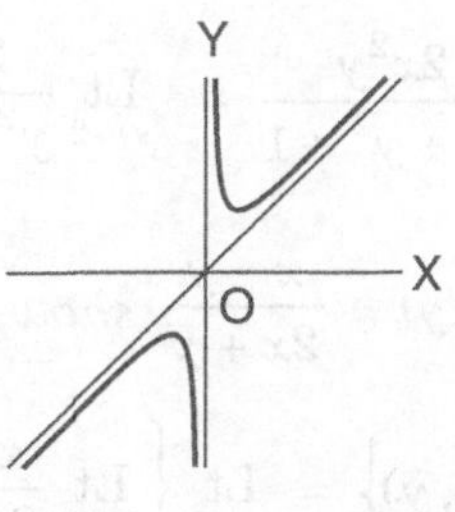

8.
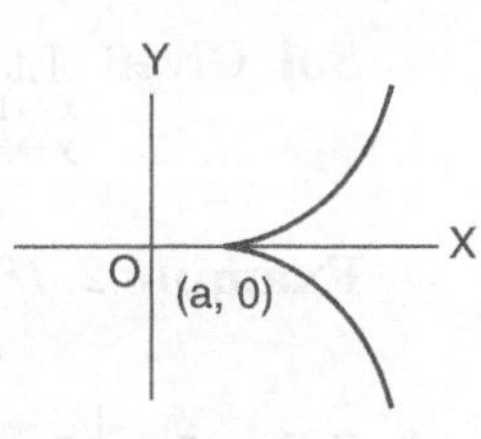

UNIT 4

Functions of Several Variables

4.1. FUNCTIONS OF TWO VARIABLES

If three variables x, y, z are so related that the value of z depends upon the values of x and y, then z is called a function of two variables x and y, and this is denoted by $z = f(x, y)$.

z is called the dependent variable while x and y are called independent variables.

For example, the area of a triangle is determined when its base and altitude are known. Thus, area of a triangle is a function of two variables, base and altitude.

In a similar way, a function of more than two variables can be defined.

Geometrical Interpretation :

Let $z = f(x, y)$ be a function of two independent variables x and y defined for all pairs of values of x and y which belong to an area A of the xy-plane. Then to each point (x, y) of this area corresponds a value of z given by the relation $z = f(x, y)$. Representing all these values (x, y, z) by points in space, we get a surface.

Hence the function $z = f(x, y)$ represents a surface.

4.2. LIMIT

A function $f(x, y)$ is said to tend to the limit l as (x, y) tends to (a, b) if corresponding to any given small positive number $\in > 0$ there exists another small positive number $\delta > 0$ such that $| f(x, y) - l | < \in$ for all points (x, y) whenever $| x - a | \le \delta$, $| y - b | \le \delta$. We write it as

$$\underset{\substack{x \to a \\ y \to b}}{\text{Lt}} f(x, y) = l \text{ or } \underset{(x,y)\to(a,b)}{\text{Lt}} f(x, y) = l.$$

Example 1. *Evaluate* $\underset{\substack{x \to 1 \\ y \to 2}}{Lt} \dfrac{2x^2 y}{x^2 + y^2 + 1}$.

Sol. Given $\underset{\substack{x \to 1 \\ y \to 2}}{\text{Lt}} \dfrac{2x^2 y}{x^2 + y^2 + 1} = \underset{y \to 2}{\text{Lt}} \dfrac{2y}{y^2 + 2} = \dfrac{4}{6} = \dfrac{2}{3}$.

Example 2. *If* $f(x, y) = \dfrac{x - y}{2x + y}$ *show that* $\underset{x \to 0}{Lt} \left\{ \underset{y \to 0}{Lt} f(x, y) \right\} \neq \underset{y \to 0}{Lt} \left\{ \underset{x \to 0}{Lt} f(x, y) \right\}$.

Sol. $\underset{x \to 0}{\text{Lt}} \left\{ \underset{y \to 0}{\text{Lt}} f(x, y) \right\} = \underset{x \to 0}{\text{Lt}} \left\{ \underset{y \to 0}{\text{Lt}} \dfrac{x - y}{2x + y} \right\} = \underset{x \to 0}{\text{Lt}} \dfrac{x}{2x} = \dfrac{1}{2}$

Now, $\underset{y \to 0}{\text{Lt}} \left\{ \underset{x \to 0}{\text{Lt}} f(x, y) \right\} = \underset{y \to 0}{\text{Lt}} \left\{ \underset{x \to 0}{\text{Lt}} \dfrac{x - y}{2x + y} \right\}$

$$= \underset{y \to 0}{\text{Lt}} \frac{-y}{y} = -1$$

Hence the result follows.

Example 3. *Show that* $\underset{(x, y) \to (0,0)}{Lt} \dfrac{x^2 - x\sqrt{y}}{x^2 + y}$ *does not exist.* (OU-Dec. 2013)

Sol. I. $\underset{y \to 0}{\text{Lt}} \left\{ \underset{x \to 0}{\text{Lt}} \dfrac{x^2 - x\sqrt{y}}{x^2 + y} \right\} = \underset{y \to 0}{\text{Lt}} \left(\dfrac{0 - 0}{y} \right) = \underset{y \to 0}{\text{Lt}} (0) = 0$

II. $\underset{x \to 0}{\text{Lt}} \left\{ \underset{y \to 0}{\text{Lt}} \dfrac{x^2 - x\sqrt{y}}{x^2 + y} \right\} = \underset{x \to 0}{\text{Lt}} \left\{ \dfrac{x^2}{x^2} \right\} = \underset{x \to 0}{\text{Lt}} (1) = 1$

Since, $\underset{y \to 0}{\text{Lt}} \left\{ \underset{x \to 0}{\text{Lt}} \dfrac{x^2 - x\sqrt{y}}{x^2 + y} \right\} \neq \underset{x \to 0}{\text{Lt}} \left\{ \underset{y \to 0}{\text{Lt}} \dfrac{x^2 - x\sqrt{y}}{x^2 + y} \right\}$

$\underset{(x, y) \to (0,0)}{\text{Lt}} \dfrac{x^2 - x\sqrt{y}}{x^2 + y}$ does not exist.

Example 4. *Determine* $\underset{(x, y) \to (1,\, 1)}{Lt} \dfrac{x(y-1)}{y(x-1)}$. (OU-June 2011, May 2012)

Sol. I. $\underset{y \to 1}{\text{Lt}} \left\{ \underset{x \to 1}{\text{Lt}} \dfrac{x(y-1)}{y(x-1)} \right\} = \underset{y \to 1}{\text{Lt}} \left\{ \dfrac{y-1}{y(0)} \right\} = \infty$

II. $\underset{x \to 1}{\text{Lt}} \left\{ \underset{y \to 1}{\text{Lt}} \dfrac{x(y-1)}{y(x-1)} \right\} = \underset{x \to 1}{\text{Lt}} \dfrac{0}{x-1} = \underset{x \to 1}{\text{Lt}}\ 0 = 0$

Since, $\underset{y \to 1}{\text{Lt}} \left\{ \underset{x \to 1}{\text{Lt}} \dfrac{x(y-1)}{y(x-1)} \right\} \neq \underset{x \to 1}{\text{Lt}} \left\{ \underset{y \to 1}{\text{Lt}} \dfrac{x(y-1)}{y(x-1)} \right\}$

$\underset{(x, y) \to (1,1)}{\text{Lt}} \dfrac{x(y-1)}{y(x-1)}$ does not exist.

Example 5. *Determine* $\underset{(x, y) \to (0,0)}{Lt} \left[\dfrac{xy}{x^2 + y^2} \right]$. (OU-Jan. 2011, 2012, May 2008)

Sol. I. $\underset{y \to 0}{\text{Lt}} \left\{ \underset{x \to 0}{\text{Lt}} \dfrac{xy}{x^2 + y^2} \right\} = \underset{y \to 0}{\text{Lt}} (0) = 0$

II. $\underset{x \to 0}{\text{Lt}} \left\{ \underset{y \to 0}{\text{Lt}} \dfrac{xy}{x^2 + y^2} \right\} = \underset{x \to 0}{\text{Lt}} (0) = 0$

III. $\underset{x \to 0}{\text{Lt}} \left\{ \underset{y \to mx}{\text{Lt}} \dfrac{xy}{x^2 + y^2} \right\} = \underset{x \to 0}{\text{Lt}} \left\{ \dfrac{mx^2}{x^2 + m^2x^2} \right\} = \underset{x \to 0}{\text{Lt}} \dfrac{mx^2}{x^2(1+m^2)} = \dfrac{m}{1+m^2}$

It takes different values for different m. Hence, limit does not exist.

Example 6. *Determine* $\underset{(x,y)\to(0,0)}{Lt} \dfrac{x^2+y^2}{x-y}$, *if it exists.* (OU-June 2010)

Sol. I. $\underset{y\to 0}{\text{Lt}}\left\{\underset{x\to 0}{\text{Lt}} \dfrac{x^2+y^2}{x-y}\right\} = \underset{y\to 0}{\text{Lt}}\left\{\dfrac{y^2}{-y}\right\} = \underset{y\to 0}{\text{Lt}}\,(-y) = 0$

II. $\underset{x\to 0}{\text{Lt}}\left\{\underset{y\to 0}{\text{Lt}} \dfrac{x^2+y^2}{x-y}\right\} = \underset{x\to 0}{\text{Lt}}\left\{\dfrac{x^2}{x}\right\} = \underset{x\to 0}{\text{Lt}}\,(x) = 0$

III. $\underset{x\to 0}{\text{Lt}}\left\{\underset{y\to mx}{\text{Lt}} \dfrac{x^2+y^2}{x-y}\right\} = \underset{x\to 0}{\text{Lt}}\left\{\dfrac{x^2+m^2x^2}{x-mx}\right\} = \underset{x\to 0}{\text{Lt}} \dfrac{x^2(1+m^2)}{x(1-m)} = \underset{x\to 0}{\text{Lt}} \dfrac{x(1+m^2)}{1-m} = 0$

Hence, $\underset{(x,y)\to(0,0)}{\text{Lt}} \dfrac{x^2+y^2}{x-y} = 0.$

Example 7. *Show that* $\underset{(x,y)\to(0,1)}{Lt} \tan^{-1}\left(\dfrac{y}{x}\right)$ *does not exist.*

Sol. We have

$$\underset{(x,y)\to(0,1)}{\text{Lt}} \tan^{-1}\left(\frac{y}{x}\right) = \tan^{-1}(\pm\infty) = \pm\frac{\pi}{2}$$

If we approach from left along the line $y = 1$, we obtain the limit as $-\dfrac{\pi}{2}$ and if we approach from right along the line $y = 1$, we obtain the limit as $\dfrac{\pi}{2}$. Since the limit is not unique, the limit does not exist.

Example 8. *Show that* $\underset{(x,y)\to(2,1)}{Lt} (3x+4y) = 10.$ (OU-Dec. 2008)

Sol. $\underset{y\to 1}{\text{Lt}}\left\{\underset{x\to 2}{\text{Lt}}\,(3x+4y)\right\} = \underset{y\to 1}{\text{Lt}}\,\{6+4y\} = 6+4 = 10$

Also, $\underset{x\to 2}{\text{Lt}}\left\{\underset{y\to 1}{\text{Lt}}\,(3x+4y)\right\} = \underset{x\to 2}{\text{Lt}}\,\{3x+4\} = 6+4 = 10$

Hence, $\underset{(x,y)\to(2,1)}{\text{Lt}}\,(3x+4y) = 10.$

Example 9. *Does* $\underset{(x,y)\to(0,0)}{Lt} \dfrac{x+\sqrt{y}}{x^2+y^2}$ *exists?*

Sol. Choose the path $y = mx^2$

$$\underset{(x,y)\to(0,0)}{\text{Lt}} \frac{x+\sqrt{y}}{x^2+y^2} = \underset{x\to 0}{\text{Lt}}\left\{\underset{y\to mx^2}{\text{Lt}} \frac{x+\sqrt{y}}{x^2+y^2}\right\}$$

$$= \underset{x\to 0}{\text{Lt}}\left\{\frac{x+x\sqrt{m}}{x^2+m^2x^4}\right\} = \underset{x\to 0}{\text{Lt}} \frac{x(1+\sqrt{m})}{x^2(1+m^2x^2)}$$

$$= \underset{x \to 0}{\text{Lt}} \frac{1+\sqrt{m}}{x(1+m^2x^2)} = \infty$$

Hence, the limit does not exist.

Example 10. *Show that* $\underset{(x, y) \to (0,0)}{Lt} \dfrac{x^3 y}{x^6 + y^2}$ *does not exist.* (OU-June 2014)

Sol. Choose the path $y = mx^3$

$$\underset{(x,y)\to(0,0)}{\text{Lt}} \frac{x^3y}{x^6+y^2} = \underset{x\to 0}{\text{Lt}} \left\{ \underset{y\to mx^3}{\text{Lt}} \frac{x^3y}{x^6+y^2} \right\} = \underset{x\to 0}{\text{Lt}} \left\{ \frac{mx^6}{x^6+m^2x^6} \right\}$$

$$= \underset{x\to 0}{\text{Lt}} \left\{ \frac{m}{1+m^2} \right\} = \frac{m}{1+m^2}$$

It takes different values for different m.
Hence, limit does not exist.

4.3. CONTINUITY

A function $f(x, y)$ is said to be continuous at a point (a, b) if, for any arbitrarily chosen positive number ε, however small, we can find a corresponding number δ such that $| f(x, y) - f(a, b) | < \varepsilon$ for every point (x, y) within the circle with its centre at (a, b) and radius δ.

i.e., whenever $0 < (x - a)^2 + (y - b)^2 < \delta^2$.

Alternatively, $f(x, y)$ is said to be continuous at (a, b) if $\lim\limits_{\substack{x \to a \\ y \to b}} f(x, y) = f(a, b)$ irrespective of the path along which $x \to a, y \to b$.

It should not be assumed that the path along which the point (x, y) tends to (a, b) is immaterial, because $\lim\limits_{x \to a} \{ \lim\limits_{y \to b} f(x, y)\}$ is not always equal to $\lim\limits_{y \to b} \{ \lim\limits_{x \to a} f(x, y)\}$.

However, *usually*, the limit is the same irrespective of the path of approach.

In what follows, we shall assume that the functions considered are continuous and their partial differential co-efficients as defined in the next section, exist.

Example 1. *Investigate the continuity of the function $f(x, y)$ at (0, 0) where*

$$f(x, y) = \begin{cases} \dfrac{xy}{x^2 + y^2}, & (x, y) \neq (0, 0) \\ 0, & (x, y) = (0, 0) \end{cases}.$$ (OU-Apr. 2007)

Sol. Let us consider the limit of the function for testing the continuity along the line $y = mx$.

$$\underset{\substack{y\to mx \\ x\to 0}}{\text{Lt}} f(x, y) = \underset{\substack{y\to mx \\ x\to 0}}{\text{Lt}} \frac{xy}{x^2+y^2} = \underset{x\to 0}{\text{Lt}} \frac{mx^2}{x^2+m^2x^2}$$

$$= \underset{x\to 0}{\text{Lt}} \frac{mx^2}{x^2(1+m^2)} = \frac{m}{1+m^2}$$

which is not unique *i.e.*, it will be different for the different m.

$\therefore$ $\underset{\substack{x\to 0 \\ y\to 0}}{\text{Lt}} f(x, y)$ does not exist.

Example 2. *Examine for continuity at origin of the function defined by*

$$f(x, y) = \frac{x^2}{\sqrt{x^2 + y^2}} \text{ for } x \neq 0, y \neq 0$$
$$= 0 \text{ for } x = 0, y = 0.$$

Sol. Given $f(0, 0) = 0$

I. $\underset{\substack{x \to 0 \\ y \to 0}}{\text{Lt}} f(x, y) = \underset{y \to 0}{\text{Lt}} \left\{ \underset{x \to 0}{\text{Lt}} \frac{x^2}{\sqrt{x^2 + y^2}} \right\} = \underset{y \to 0}{\text{Lt}} \frac{0}{\sqrt{y^2}} = \underset{y \to 0}{\text{Lt}}\ 0 = 0$

II. $\underset{\substack{y \to 0 \\ x \to 0}}{\text{Lt}} f(x, y) = \underset{x \to 0}{\text{Lt}} \left\{ \underset{y \to 0}{\text{Lt}} \frac{x^2}{\sqrt{x^2 + y^2}} \right\} = \underset{x \to 0}{\text{Lt}} \left(\frac{x^2}{x} \right) = \underset{x \to 0}{\text{Lt}}\ x = 0$

III. $\underset{\substack{y \to mx \\ x \to 0}}{\text{Lt}} f(x, y) = \underset{x \to 0}{\text{Lt}} \left\{ \underset{y \to mx}{\text{Lt}} \frac{x^2}{\sqrt{x^2 + y^2}} \right\} = \underset{x \to 0}{\text{Lt}} \left\{ \frac{x^2}{\sqrt{x^2 + m^2 x^2}} \right\}$

$$= \underset{x \to 0}{\text{Lt}} \left\{ \frac{x^2}{x\sqrt{1 + m^2}} \right\} = \underset{x \to 0}{\text{Lt}} \frac{x}{\sqrt{1 + m^2}} = 0$$

Hence the function $f(x, y)$ is continuous at the origin.

Example 3. *Show that the function* $f(x, y) = \begin{cases} \dfrac{x^2 + y^2}{x - y}, & (x, y) \neq (0, 0) \\ 0, & (x, y) = (0, 0) \end{cases}$ *is continuous at (0, 0).*

(OU-June 2013)

Sol. Consider

I. $\underset{\substack{y \to 0 \\ x \to 0}}{\text{Lt}} f(x, y) = \underset{\substack{y \to mx \\ x \to 0}}{\text{Lt}} f(x, y) = \underset{x \to 0}{\text{Lt}} \left\{ \underset{y \to mx}{\text{Lt}} \frac{x^2 + y^2}{x - y} \right\}$

$$= \underset{x \to 0}{\text{Lt}} \left\{ \frac{x^2 + m^2 x^2}{x - mx} \right\} = \underset{x \to 0}{\text{Lt}} \frac{x^2(1 + m^2)}{x(1 - m)} = \underset{x \to 0}{\text{Lt}} \frac{x(1 + m^2)}{1 - m} = 0$$

II. $\underset{\substack{x \to 0 \\ y \to 0}}{\text{Lt}} f(x, y) = \underset{y \to 0}{\text{Lt}} \left\{ \underset{x \to 0}{\text{Lt}} \frac{x^2 + y^2}{x - y} \right\} = \underset{y \to 0}{\text{Lt}} \left(\frac{y^2}{-y} \right) = \underset{y \to 0}{\text{Lt}} (-y) = 0$

III. $\underset{\substack{y \to 0 \\ x \to 0}}{\text{Lt}} f(x, y) = \underset{x \to 0}{\text{Lt}} \left\{ \underset{y \to 0}{\text{Lt}} \frac{x^2 + y^2}{x - y} \right\} = \underset{x \to 0}{\text{Lt}} \left\{ \frac{x^2}{x} \right\} = \underset{x \to 0}{\text{Lt}} (x) = 0$

IV. $\underset{\substack{y \to mx^2 \\ x \to 0}}{\text{Lt}} f(x, y) = \underset{x \to 0}{\text{Lt}} \left\{ \underset{y \to mx^2}{\text{Lt}} \frac{x^2 + y^2}{x - y} \right\} = \underset{x \to 0}{\text{Lt}} \frac{x^2 + m^2 x^4}{x - mx^2} = \underset{x \to 0}{\text{Lt}} \frac{x^2(1 + m^2 x^2)}{x(1 - mx)}$

$$= \underset{x \to 0}{\text{Lt}} \frac{x(1+m^2x^2)}{1-mx} = 0$$

Since $f(0, 0) = 0$,

$f(x, y)$ is continuous at $(0, 0)$.

4.4. PARTIAL DERIVATIVES OF FIRST ORDER

Let $z = f(x, y)$ be a function of two independent variables x and y. If y is kept constant and x alone is allowed to vary, then z becomes a function of x only. The derivative of z, with respect to x, treating y as constant, is called partial derivative of z w.r.t. x and is denoted by

$$\frac{\partial z}{\partial x} \quad \text{or} \quad \frac{\partial f}{\partial x} \quad \text{or} \quad f_x.$$

Thus,
$$\frac{\partial z}{\partial x} = \lim_{h \to 0} \frac{f(x+h, y) - f(x, y)}{h}$$

Similarly, the derivative of z, with respect to y, treating x as constant, is called partial derivative of z w.r.t. y and is denoted by $\frac{\partial z}{\partial y}$ or $\frac{\partial f}{\partial y}$ or f_y

Thus,
$$\frac{\partial z}{\partial y} = \lim_{k \to 0} \frac{f(x, y+k) - f(x, y)}{k}$$

$\frac{\partial z}{\partial x}$ and $\frac{\partial z}{\partial y}$ are called **first order partial derivatives of z.**

[**In general,** if z is a function of two or more independent variables, then the **partial derivative of z** w.r.t. **any one** of the independent variables is the **ordinary derivative of z** w.r.t. that variable, **treating all other variables as constant.**]

Geometrical Interpretation :

Let $z = f(x, y)$ be a function of two variables x and y. Then by Art. 4.1, it represents a surface S. If $y = k$, a constant, then $y = k$ represents a plane parallel to the zx-plane.

$\therefore$ $z = f(x, y)$ and $y = k$ represent a plane curve C which is the section of S by $y = k$.

$\frac{\partial z}{\partial x}$ represents the slope of tangent to C at (x, k, z).

Thus, $\frac{\partial z}{\partial x}$ gives the slope of the tangent drawn to the curve of intersection of the surface $z = f(x, y)$ and a plane parallel to zx-plane.

Similarly, $\frac{\partial z}{\partial y}$ gives the slope of the tangent drawn to the curve of intersection of the surface $z = f(x, y)$ and a plane parallel to yz-plane.

4.5. PARTIAL DERIVATIVES OF HIGHER ORDER

Since the first order partial derivatives $\frac{\partial z}{\partial x}$ and $\frac{\partial z}{\partial y}$ are themselves functions of x and y, they can be further differentiated partially w.r.t. x as well as y. These are called second order partial derivatives of z. The usual notations for these second order partial derivatives are :

$$\frac{\partial}{\partial x}\left(\frac{\partial z}{\partial x}\right)=\frac{\partial^2 z}{\partial x^2} \quad \text{or} \quad f_{xx}\,; \frac{\partial}{\partial y}\left(\frac{\partial z}{\partial y}\right)=\frac{\partial^2 z}{\partial y^2} \quad \text{or} \quad f_{yy}$$

$$\frac{\partial}{\partial x}\left(\frac{\partial z}{\partial y}\right)=\frac{\partial^2 z}{\partial x \partial y} \quad \text{or} \quad f_{xy}\,; \frac{\partial}{\partial y}\left(\frac{\partial z}{\partial x}\right)=\frac{\partial^2 z}{\partial y \partial x} \quad \text{or} \quad f_{yx}$$

In general, $$\frac{\partial^2 z}{\partial x \partial y}=\frac{\partial^2 z}{\partial y \partial x} \quad \text{or} \quad f_{xy}=f_{yx}.$$

Note 1. If $z = f(x)$, a function of single independent variable x, we get $\frac{dz}{dx}$.

If $z = f(x, y)$, a function of two independent variables x and y, we get $\frac{\partial z}{\partial x}$ and $\frac{\partial z}{\partial y}$.

Similarly, for a function of more than two independent variables $x_1, x_2, \ldots\ldots, x_n$, we get $\frac{\partial z}{\partial x_1}, \frac{\partial z}{\partial x_2}, \ldots\ldots, \frac{\partial z}{\partial x_n}$.

Note 2. (*i*) If $z = u + v$, where $u = f(x, y)$, $v = \phi(x, y)$, then z is a function of x and y.

$$\therefore \qquad \frac{\partial z}{\partial x}=\frac{\partial u}{\partial x}+\frac{\partial v}{\partial x}\,; \quad \frac{\partial z}{\partial y}=\frac{\partial u}{\partial y}+\frac{\partial v}{\partial y}.$$

(*ii*) If $z = uv$, where $u = f(x, y)$, $v = \phi(x, y)$, then

$$\frac{\partial z}{\partial x}=\frac{\partial}{\partial x}(uv)=u\frac{\partial v}{\partial x}+v\frac{\partial u}{\partial x}\,; \quad \frac{\partial z}{\partial y}=\frac{\partial}{\partial y}(uv)=u\frac{\partial v}{\partial y}+v\frac{\partial u}{\partial y}$$

(*iii*) If $z = \frac{u}{v}$, where $u = f(x, y)$, $v = \phi(x, y)$, then

$$\frac{\partial z}{\partial x}=\frac{\partial}{\partial x}\left(\frac{u}{v}\right)=\frac{v\frac{\partial u}{\partial x}-u\frac{\partial v}{\partial x}}{v^2}\,; \quad \frac{\partial z}{\partial y}=\frac{\partial}{\partial y}\left(\frac{u}{v}\right)=\frac{v\frac{\partial u}{\partial y}-u\frac{\partial v}{\partial y}}{v^2}$$

(*iv*) If $z = f(u)$, where $u = \phi(x, y)$, then $\frac{\partial z}{\partial x}=\frac{dz}{du}\cdot\frac{\partial u}{\partial x}\,; \quad \frac{\partial z}{\partial y}=\frac{dz}{du}\cdot\frac{\partial u}{\partial y}.$

ILLUSTRATIVE EXAMPLES

Example 1. *Find the first order partial derivatives of the following :*

(*i*) $u = \tan^{-1}\frac{x^2+y^2}{x+y}$ (*ii*) $u = \cos^{-1}\left(\frac{x}{y}\right)$.

Sol. (*i*) $$u = \tan^{-1}\frac{x^2+y^2}{x+y}$$

$$\frac{\partial u}{\partial x}=\frac{1}{1+\left(\frac{x^2+y^2}{x+y}\right)^2}\cdot\frac{\partial}{\partial x}\left(\frac{x^2+y^2}{x+y}\right)$$

$$=\frac{(x+y)^2}{(x+y)^2+(x^2+y^2)^2}\cdot\frac{(x+y)\frac{\partial}{\partial x}(x^2+y^2)-(x^2+y^2)\frac{\partial}{\partial x}(x+y)}{(x+y)^2}$$

$$=\frac{(x+y)\cdot 2x-(x^2+y^2)\cdot 1}{(x+y)^2+(x^2+y^2)^2}$$

$$\Rightarrow \qquad \frac{\partial u}{\partial x} = \frac{x^2 + 2xy - y^2}{(x+y)^2 + (x^2+y^2)^2} \qquad \ldots(1)$$

Since u remains the same if we interchange x and y, u is symmetrical w.r.t. x and y. Interchanging x and y in (1), we have

Similarly, $$\frac{\partial u}{\partial y} = \frac{y^2 + 2xy - x^2}{(x+y)^2 + (x^2+y^2)^2}$$

(ii) $$u = \cos^{-1}\left(\frac{x}{y}\right)$$

$$\frac{\partial u}{\partial x} = \frac{-1}{\sqrt{1-\left(\frac{x}{y}\right)^2}} \cdot \frac{\partial}{\partial x}\left(\frac{x}{y}\right) = \frac{-y}{\sqrt{y^2-x^2}} \cdot \frac{1}{y} = \frac{-1}{\sqrt{y^2-x^2}}$$

$$\frac{\partial u}{\partial y} = \frac{-1}{\sqrt{1-\left(\frac{x}{y}\right)^2}} \cdot \frac{\partial}{\partial y}\left(\frac{x}{y}\right) = \frac{-y}{\sqrt{y^2-x^2}}\left(-\frac{x}{y^2}\right) = \frac{x}{y\sqrt{y^2-x^2}}.$$

Example 2. *If* $z(x+y) = x^2 + y^2$, *show that* $\left(\frac{\partial z}{\partial x} - \frac{\partial z}{\partial y}\right)^2 = 4\left(1 - \frac{\partial z}{\partial x} - \frac{\partial z}{\partial y}\right)$.

Sol. $$z = \frac{x^2+y^2}{x+y} \qquad [z \text{ is symmetrical w.r.t. } x \text{ and } y]$$

$$\frac{\partial z}{\partial x} = \frac{(x+y)\frac{\partial}{\partial x}(x^2+y^2) - (x^2+y^2)\frac{\partial}{\partial x}(x+y)}{(x+y)^2}$$

$$= \frac{(x+y)\cdot 2x - (x^2+y^2)\cdot 1}{(x+y)^2} = \frac{x^2+2xy-y^2}{(x+y)^2}$$

Similarly, $$\frac{\partial z}{\partial y} = \frac{y^2+2xy-x^2}{(x+y)^2}$$

Now, $$\left(\frac{\partial z}{\partial x} - \frac{\partial z}{\partial y}\right)^2 = \left[\frac{2x^2-2y^2}{(x+y)^2}\right]^2 = \frac{4(x+y)^2(x-y)^2}{(x+y)^4} = \frac{4(x-y)^2}{(x+y)^2}$$

$$4\left(1 - \frac{\partial z}{\partial x} - \frac{\partial z}{\partial y}\right) = 4\left[1 - \frac{x^2+2xy-y^2}{(x+y)^2} - \frac{y^2+2xy-x^2}{(x+y)^2}\right]$$

$$= 4\left[\frac{x^2+2xy+y^2-x^2-2xy+y^2-y^2-2xy+x^2}{(x+y)^2}\right]$$

$$= \frac{4(x^2-2xy+y^2)}{(x+y)^2} = \frac{4(x-y)^2}{(x+y)^2}$$

$$\therefore \qquad \left(\frac{\partial z}{\partial x} - \frac{\partial z}{\partial y}\right)^2 = 4\left(1 - \frac{\partial z}{\partial x} - \frac{\partial z}{\partial y}\right).$$

Example 3. *Prove that if* $f(x, y) = \frac{1}{\sqrt{y}} . e^{-\frac{(x-a)^2}{4y}}$, *then* $f_{xy} = f_{yx}$.

Sol. $$f(x, y) = \frac{1}{\sqrt{y}} . e^{-\frac{(x-a)^2}{4y}} = y^{-\frac{1}{2}} e^{-\frac{(x-a)^2}{4y}}$$

$$f_x = \frac{\partial f}{\partial x} = y^{-\frac{1}{2}} . e^{-\frac{(x-a)^2}{4y}} . \frac{\partial}{\partial x}\left[-\frac{(x-a)^2}{4y}\right]$$

$$= y^{-\frac{1}{2}} . e^{-\frac{(x-a)^2}{4y}} . \left[-\frac{2(x-a)}{4y}\right] = -\frac{1}{2} y^{-\frac{3}{2}} (x-a) e^{-\frac{(x-a)^2}{4y}}$$

$$f_y = \frac{\partial f}{\partial y} = -\frac{1}{2} y^{-\frac{3}{2}} . e^{-\frac{(x-a)^2}{4y}} + y^{-\frac{1}{2}} . e^{-\frac{(x-a)^2}{4y}} . \frac{\partial}{\partial y}\left[-\frac{(x-a)^2}{4y}\right]$$

$$= e^{-\frac{(x-a)^2}{4y}} \left[-\frac{1}{2} y^{-\frac{3}{2}} + y^{-\frac{1}{2}} . \frac{(x-a)^2}{4y^2}\right] = \frac{1}{4} y^{-\frac{3}{2}} e^{-\frac{(x-a)^2}{4y}} [-2 + y^{-1}(x-a)^2]$$

$$f_{xy} = \frac{\partial}{\partial x}\left(\frac{\partial f}{\partial y}\right)$$

$$= \frac{1}{4} y^{-\frac{3}{2}} \left\{ e^{-\frac{(x-a)^2}{4y}} . \frac{\partial}{\partial x}\left[-\frac{(x-a)^2}{4y}\right] . [-2 + y^{-1}(x-a)^2] + e^{-\frac{(x-a)^2}{4y}} . 2y^{-1}(x-a) \right\}$$

$$= \frac{1}{4} y^{-\frac{3}{2}} . e^{-\frac{(x-a)^2}{4y}} \left\{ -\frac{2(x-a)}{4y} [-2 + y^{-1}(x-a)^2] + 2y^{-1}(x-a) \right\}$$

$$= \frac{1}{4} y^{-\frac{3}{2}} . e^{-\frac{(x-a)^2}{4y}} . \frac{x-a}{y} \left\{ -\frac{1}{2} [-2 + y^{-1}(x-a)^2] + 2 \right\}$$

$$= \frac{1}{4} y^{-\frac{5}{2}} (x-a) e^{-\frac{(x-a)^2}{4y}} \left[3 - \frac{(x-a)^2}{2y}\right]$$

$$f_{yx} = \frac{\partial}{\partial y}\left(\frac{\partial f}{\partial x}\right) = -\frac{1}{2}(x-a)\left[-\frac{3}{2} y^{-\frac{5}{2}} . e^{-\frac{(x-a)^2}{4y}} + y^{-\frac{3}{2}} . e^{-\frac{(x-a)^2}{4y}} . \frac{(x-a)^2}{4y^2}\right]$$

$$= -\frac{1}{4}(x-a) y^{-\frac{5}{2}} . e^{-\frac{(x-a)^2}{4y}} \left[-3 + \frac{(x-a)^2}{2y}\right]$$

$$= \frac{1}{4} y^{-\frac{5}{2}} (x-a) e^{-\frac{(x-a)^2}{4y}} \left[3 - \frac{(x-a)^2}{2y}\right]$$

$\therefore \quad f_{xy} = f_{yx}$.

Example 4. *If* $u = x^y$, *show that* $\frac{\partial^3 u}{\partial x^2 \partial y} = \frac{\partial^3 u}{\partial x \partial y \partial x}$.

Sol. $$u = x^y$$

$$\frac{\partial u}{\partial y} = x^y \log x$$

$$\frac{\partial^2 u}{\partial x \partial y} = \frac{\partial}{\partial x}\left(\frac{\partial u}{\partial y}\right) = yx^{y-1} \log x + x^y \cdot \frac{1}{x} = x^{y-1}(y \log x + 1)$$

$$\frac{\partial^3 u}{\partial x^2 \partial y} = \frac{\partial}{\partial x}\left(\frac{\partial^2 u}{\partial x \partial y}\right) = \frac{\partial}{\partial x}[x^{y-1}(y \log x + 1)] \quad ...(1)$$

$$\frac{\partial u}{\partial x} = yx^{y-1}$$

$$\frac{\partial^2 u}{\partial y \partial x} = \frac{\partial}{\partial y}\left(\frac{\partial u}{\partial x}\right) = x^{y-1} + yx^{y-1} \log x = x^{y-1}(y \log x + 1)$$

$$\frac{\partial^3 u}{\partial x \partial y \partial x} = \frac{\partial}{\partial x}\left(\frac{\partial^2 u}{\partial y \partial x}\right) = \frac{\partial}{\partial x}[x^{y-1}(y \log x + 1)] \quad ...(2)$$

From (1) and (2), $\frac{\partial^3 u}{\partial x^2 \partial y} = \frac{\partial^3 u}{\partial x \partial y \partial x}$.

Example 5. *If $u = (1 - 2xy + y^2)^{-1/2}$, prove that*

$$\frac{\partial}{\partial x}\left\{(1 - x^2)\frac{\partial u}{\partial x}\right\} + \frac{\partial}{\partial y}\left\{y^2 \frac{\partial u}{\partial y}\right\} = 0,$$

Sol. $u = (1 - 2xy + y^2)^{-1/2} = V^{-1/2}$, where $V = 1 - 2xy + y^2$

$$\frac{\partial u}{\partial x} = -\frac{1}{2} V^{-3/2} \cdot \frac{\partial V}{\partial x} = -\frac{1}{2} V^{-3/2}(-2y) = yV^{-3/2}$$

$$\frac{\partial^2 u}{\partial x^2} = y \cdot \frac{\partial}{\partial x}(V^{-3/2}) = y \cdot \left(-\frac{3}{2}\right) V^{-5/2} \cdot \frac{\partial V}{\partial x} = -\frac{3}{2} yV^{-5/2}(-2y) = 3y^2 V^{-5/2}$$

$$\therefore \quad \frac{\partial}{\partial x}\left\{(1 - x^2)\frac{\partial u}{\partial x}\right\} = (1 - x^2) \cdot \frac{\partial^2 u}{\partial x^2} + \frac{\partial u}{\partial x} \cdot \frac{\partial}{\partial x}(1 - x^2)$$

$$= (1 - x^2) \cdot 3y^2 V^{-5/2} + yV^{-3/2}(-2x) = yV^{-3/2}[3yV^{-1}(1 - x^2) - 2x] \quad ...(1)$$

Also, $$\frac{\partial u}{\partial y} = -\frac{1}{2} V^{-3/2} \frac{\partial V}{\partial y} = -\frac{1}{2} V^{-3/2} \cdot (-2x + 2y) = V^{-3/2} \cdot (x - y)$$

$$\frac{\partial^2 u}{\partial y^2} = V^{-3/2} \cdot \frac{\partial}{\partial y}(x - y) + (x - y) \cdot \frac{y}{\partial y}(V^{-3/2})$$

$$= V^{-3/2} \cdot (-1) + (x - y) \cdot \left(-\frac{3}{2} V^{-5/2}\right) \cdot \frac{\partial V}{\partial y}$$

$$= -V^{-3/2} - \frac{3}{2}(x - y) V^{-5/2} \cdot (-2x + 2y) = -V^{-3/2} + 3(x - y)^2 V^{-5/2}$$

$$\therefore \quad \frac{\partial}{\partial y}\left\{y^2 \frac{\partial u}{\partial y}\right\} = y^2 \cdot \frac{\partial^2 u}{\partial y^2} + \frac{\partial u}{\partial y} \cdot \frac{\partial}{\partial y}(y^2)$$

$$= y^2[-V^{-3/2} + 3(x - y)^2 V^{-5/2}] + V^{-3/2}(x - y) \cdot 2y$$

$$= y\, V^{-3/2}\, [-y + 3y(x-y)^2\, V^{-1} + 2(x-y)]$$

$$= y\, V^{-3/2}\, [3y\,(x-y)^2\, V^{-1} + (2x - 3y)] \qquad \ldots(2)$$

Adding (1) and (2), we have

$$\frac{\partial}{\partial x}\left\{(1-x^2)\frac{\partial u}{\partial x}\right\} + \frac{\partial}{\partial y}\left\{y^2\frac{\partial u}{\partial y}\right\} = yV^{-3/2}\,[3yV^{-1}(1-x^2) - 2x + 3y\,(x-y)^2\,V^{-1} + 2x - 3y]$$

$$= y\, V^{-3/2}[3yV^{-1}\,(1 - x^2 + x^2 - 2xy + y^2) - 3y]$$

$$= yV^{-3/2}[3yV^{-1}(1 - 2xy + y^2) - 3y]$$

$$= yV^{-3/2}\,[3y - 3y] \qquad | \because \quad V = 1 - 2xy + y^2$$

$$= 0.$$

Example 6. *If $u = \log\,(x^3 + y^3 + z^3 - 3xyz)$, show that*

(*i*) $\left(\dfrac{\partial}{\partial x} + \dfrac{\partial}{\partial y} + \dfrac{\partial}{\partial z}\right)^2 u = -\dfrac{9}{(x+y+z)^2}$

(*ii*) $\dfrac{\partial^2 u}{\partial x^2} + \dfrac{\partial^2 u}{\partial y^2} + \dfrac{\partial^2 u}{\partial z^2} + 2\dfrac{\partial^2 u}{\partial y \partial z} + 2\dfrac{\partial^2 u}{\partial z \partial x} + 2\dfrac{\partial^2 u}{\partial x \partial y} = \dfrac{-9}{(x+y+z)^2}$.

Sol. (*i*) $\qquad u = \log\,(x^3 + y^3 + z^3 - 3xyz)$

$$\frac{\partial u}{\partial x} = \frac{3x^2 - 3yz}{x^3 + y^3 + z^3 - 3xyz}\,; \quad \frac{\partial u}{\partial y} = \frac{3y^2 - 3zx}{x^3 + y^3 + z^3 - 3xyz}\,;$$

$$\frac{\partial u}{\partial z} = \frac{3z^2 - 3xy}{x^3 + y^3 + z^3 - 3xyz}$$

Adding, $\qquad \dfrac{\partial u}{\partial x} + \dfrac{\partial u}{\partial y} + \dfrac{\partial u}{\partial z} = \dfrac{3(x^2 + y^2 + z^2 - xy - yz - zx)}{x^3 + y^3 + z^3 - 3xyz} = \dfrac{3}{x+y+z}$

$$[\because \quad x^3 + y^3 + z^3 - 3xyz = (x+y+z)(x^2 + y^2 + z^2 - xy - yz - zx)]$$

Now, $\left(\dfrac{\partial}{\partial x} + \dfrac{\partial}{\partial y} + \dfrac{\partial}{\partial z}\right)^2 u = \left(\dfrac{\partial}{\partial x} + \dfrac{\partial}{\partial y} + \dfrac{\partial}{\partial z}\right)\left(\dfrac{\partial}{\partial x} + \dfrac{\partial}{\partial y} + \dfrac{\partial}{\partial z}\right) u$

$$= \left(\frac{\partial}{\partial x} + \frac{\partial}{\partial y} + \frac{\partial}{\partial z}\right)\left(\frac{\partial u}{\partial x} + \frac{\partial u}{\partial y} + \frac{\partial u}{\partial z}\right) = \left(\frac{\partial}{\partial x} + \frac{\partial}{\partial y} + \frac{\partial}{\partial z}\right)\left(\frac{3}{x+y+z}\right)$$

$$= -\frac{3}{(x+y+z)^2} - \frac{3}{(x+y+z)^2} - \frac{3}{(x+y+z)^2} = -\frac{9}{(x+y+z)^2} \qquad \ldots(1)$$

(*ii*) $\left(\dfrac{\partial}{\partial x} + \dfrac{\partial}{\partial y} + \dfrac{\partial}{\partial z}\right)^2 u = \left(\dfrac{\partial}{\partial x} + \dfrac{\partial}{\partial y} + \dfrac{\partial}{\partial z}\right)\left(\dfrac{\partial}{\partial x} + \dfrac{\partial}{\partial y} + \dfrac{\partial}{\partial z}\right) u$

$$= \left(\frac{\partial}{\partial x} + \frac{\partial}{\partial y} + \frac{\partial}{\partial z}\right)\left(\frac{\partial u}{\partial x} + \frac{\partial u}{\partial y} + \frac{\partial u}{\partial z}\right)$$

$$= \frac{\partial}{\partial x}\left(\frac{\partial u}{\partial x}+\frac{\partial u}{\partial y}+\frac{\partial u}{\partial z}\right)+\frac{\partial}{\partial y}\left(\frac{\partial u}{\partial x}+\frac{\partial u}{\partial y}+\frac{\partial u}{\partial z}\right)+\frac{\partial}{\partial z}\left(\frac{\partial u}{\partial x}+\frac{\partial u}{\partial y}+\frac{\partial u}{\partial z}\right)$$

$$= \frac{\partial^2 u}{\partial x^2}+\frac{\partial^2 u}{\partial x\partial y}+\frac{\partial^2 u}{\partial x\partial z}+\frac{\partial^2 u}{\partial y\partial x}+\frac{\partial^2 u}{\partial y^2}+\frac{\partial^2 u}{\partial y\partial z}+\frac{\partial^2 u}{\partial z\partial x}+\frac{\partial^2 u}{\partial z\partial y}+\frac{\partial^2 u}{\partial z^2}$$

$$= \frac{\partial^2 u}{\partial x^2}+\frac{\partial^2 u}{\partial y^2}+\frac{\partial^2 u}{\partial z^2}+2\frac{\partial^2 u}{\partial y\partial z}+2\frac{\partial^2 u}{\partial z\partial x}+2\frac{\partial^2 u}{\partial x\partial y}$$

$$\left[\because \frac{\partial^2 u}{\partial y\partial x}=\frac{\partial^2 u}{\partial x\partial y}, \frac{\partial^2 u}{\partial z\partial y}=\frac{\partial^2 u}{\partial y\partial z}, \frac{\partial^2 u}{\partial x\partial z}=\frac{\partial^2 u}{\partial z\partial x}\right]$$

$$\therefore \quad \frac{\partial^2 u}{\partial x^2}+\frac{\partial^2 u}{\partial y^2}+\frac{\partial^2 u}{\partial z^2}+2\frac{\partial^2 u}{\partial y\partial z}+2\frac{\partial^2 u}{\partial z\partial x}+2\frac{\partial^2 u}{\partial x\partial y}=\left(\frac{\partial}{\partial x}+\frac{\partial}{\partial y}+\frac{\partial}{\partial z}\right)^2 u=\frac{-9}{(x+y+z)^2}.$$

[From (1)]

Example 7. *If* $x^x y^y z^z = c$, *show that at* $x = y = z$, $\frac{\partial^2 z}{\partial x\partial y} = -(x \log ex)^{-1}$.

Sol. $x^x y^y z^z = c$ defines z as a function of x and y.

Taking log, $x \log x + y \log y + z \log z = \log c$

Differentiating partially w.r.t. y, we have

$$y \cdot \frac{1}{y} + 1 \cdot \log y + z \cdot \frac{1}{z}\frac{\partial z}{\partial y} + \log z \cdot \frac{\partial z}{\partial y} = 0$$

or
$$1 + \log y + (1 + \log z)\frac{\partial z}{\partial y} = 0 \quad \text{...(1)}$$

or
$$\frac{\partial z}{\partial y} = -\frac{1+\log y}{1+\log z}$$

Similarly,
$$\frac{\partial z}{\partial x} = -\frac{1+\log x}{1+\log z} \quad \text{...(2)}$$

Differentiating (1) partially w.r.t. x, we have

$$\left(\frac{1}{z}\frac{\partial z}{\partial x}\right)\frac{\partial z}{\partial y} + (1+\log z)\frac{\partial^2 z}{\partial x\partial y} = 0 \quad \text{or} \quad \frac{\partial^2 z}{\partial x\partial y} = -\frac{1}{z(1+\log z)}\frac{\partial z}{\partial x}\cdot\frac{\partial z}{\partial y} \quad \text{...(3)}$$

When $x = y = z$

From (2), $\frac{\partial z}{\partial y} = -1, \frac{\partial z}{\partial x} = -1$

From (3), $\frac{\partial^2 z}{\partial x\partial y} = -\frac{1}{x(1+\log x)}(-1)(-1) = -\frac{1}{x(\log e + \log x)} = -\frac{1}{x(\log ex)} = -(x \log ex)^{-1}$.

EXERCISE 1

1. Find the first order partial derivatives of the following functions :

   (i) $u = y^x$ (ii) $u = \log (x^2 + y^2)$

   (iii) $u = x^2 \sin \dfrac{y}{x}$ (iv) $u = \dfrac{x}{y} \tan^{-1}\left(\dfrac{y}{x}\right)$.

2. If $u = x^2 + y^2 + z^2$, prove that $xu_x + yu_y + zu_z = 2u$.

3. If $z = \log (x^2 + xy + y^2)$, prove that $x \dfrac{\partial z}{\partial x} + y\dfrac{\partial z}{\partial y} = 2$.

4. If $u = x^2y + y^2z + z^2x$, prove that $u_x + u_y + u_z = (x + y + z)^2$.

5. If $u = \dfrac{y}{z} + \dfrac{z}{x} + \dfrac{x}{y}$, show that $x \dfrac{\partial u}{\partial x} + y\dfrac{\partial u}{\partial y} + z\dfrac{\partial u}{\partial z} = 0$.

6. If $f(x, y) = x^3y - xy^3$, find $\left[\dfrac{1}{\dfrac{\partial f}{\partial x}} + \dfrac{1}{\dfrac{\partial f}{\partial y}}\right]_{\substack{x=1\\y=2}}$.

7. If $u = \log (\tan x + \tan y)$, prove that $\sin 2x \dfrac{\partial u}{\partial x} + \sin 2y\dfrac{\partial u}{\partial y} = 2$.

8. If $f(x, y, z) = \begin{vmatrix} x^2 & y^2 & z^2 \\ x & y & z \\ 1 & 1 & 1 \end{vmatrix}$, prove that $f_x + f_y + f_z = 0$.

9. Verify that $\dfrac{\partial^2 u}{\partial x \partial y} = \dfrac{\partial^2 u}{\partial y \partial x}$ for the following functions :

   (i) $u = ax^2 + 2hxy + by^2$ (ii) $u = \tan^{-1}\left(\dfrac{x}{y}\right)$ (iii) $u = \log\left(\dfrac{x^2 + y^2}{xy}\right)$

   (iv) $u = e^{ax} \sin by$ (v) $u = \log (x \sin y + y \sin x)$.

10. If $z = \log (e^x + e^y)$, show that $rt - s^2 = 0$; where $r = \dfrac{\partial^2 z}{\partial x^2}$, $t = \dfrac{\partial^2 z}{\partial y^2}$, $s = \dfrac{\partial^2 z}{\partial x \partial y}$.

11. If $u = \tan^{-1} \dfrac{xy}{\sqrt{1+x^2+y^2}}$, show that $\dfrac{\partial^2 u}{\partial x \partial y} = (1 + x^2 + y^2)^{-3/2}$.

12. If $u = e^{xyz}$, prove that $\dfrac{\partial^3 u}{\partial x \partial y \partial z} = (1 + 3xyz + x^2y^2z^2)e^{xyz}$.

13. If $u = \log (x^2 + y^2) + \tan^{-1} \dfrac{y}{x}$, prove that $\dfrac{\partial^2 u}{\partial x^2} + \dfrac{\partial^2 u}{\partial y^2} = 0$.

14. If $u = \tan^{-1}\left(\dfrac{2xy}{x^2 - y^2}\right)$, prove that $\dfrac{\partial^2 u}{\partial x^2} + \dfrac{\partial^2 u}{\partial y^2} = 0$.

15. Verify that $f_{xy} = f_{yx}$ when f is equal to

    (i) $\sin^{-1}\left(\dfrac{y}{x}\right)$ (ii) $\log x \tan^{-1} (x^2 + y^2)$.

16. Find the value of n so that the equation $V = r^n (3 \cos^2 \theta - 1)$ satisfies the relation

$$\frac{\partial}{\partial r}\left(r^2 \frac{\partial V}{\partial r}\right) + \frac{1}{\sin\theta}\frac{\partial}{\partial \theta}\left(\sin \theta \frac{\partial V}{\partial \theta}\right) = 0.$$

17. If $z = \tan(y + ax) - (y - ax)^{3/2}$, show that $\dfrac{\partial^2 z}{\partial x^2} = a^2 \dfrac{\partial^2 z}{\partial y^2}$.

18. If $V = (x^2 + y^2 + z^2)^{-1/2}$, prove that $\dfrac{\partial^2 V}{\partial x^2} + \dfrac{\partial^2 V}{\partial y^2} + \dfrac{\partial^2 V}{\partial z^2} = 0$.

19. If $V = r^m$, where $r^2 = x^2 + y^2 + z^2$, show that $V_{xx} + V_{yy} + V_{zz} = m(m + 1)\, r^{m-2}$.

20. If $u = \log(x^2 + y^2 + z^2)$, prove that $x \dfrac{\partial^2 u}{\partial y \partial z} = y \dfrac{\partial^2 u}{\partial z \partial x} = z \dfrac{\partial^2 u}{\partial x \partial y}$.

21. If $u = \log \sqrt{x^2 + y^2 + z^2}$, prove that $(x^2 + y^2 + z^2)\left(\dfrac{\partial^2 u}{\partial x^2} + \dfrac{\partial^2 u}{\partial y^2} + \dfrac{\partial^2 u}{\partial z^2}\right) = 1$.

22. If $u = x^2 \tan^{-1} \dfrac{y}{x} - y^2 \tan^{-1} \dfrac{x}{y}$, find the value of $\dfrac{\partial^2 u}{\partial x \partial y}$.

23. If $x^2 + y^2 + z^2 = \dfrac{1}{u^2}$, prove that $\dfrac{\partial^2 u}{\partial x^2} + \dfrac{\partial^2 u}{\partial y^2} + \dfrac{\partial^2 u}{\partial z^2} = 0$.

24. If $u = \sqrt{x^2 + y^2 + z^2}$, show that

(*i*) $\left(\dfrac{\partial u}{\partial x}\right)^2 + \left(\dfrac{\partial u}{\partial y}\right)^2 + \left(\dfrac{\partial u}{\partial z}\right)^2 = 1$ (*ii*) $\dfrac{\partial^2 u}{\partial x^2} + \dfrac{\partial^2 u}{\partial y^2} + \dfrac{\partial^2 u}{\partial z^2} = \dfrac{2}{u}$.

25. If $u = e^{x-at} \cos(x - at)$, show that $\dfrac{\partial^2 u}{\partial t^2} = a^2 \dfrac{\partial^2 u}{\partial x^2}$.

26. If $v = \dfrac{1}{\sqrt{t}}\, e^{\frac{-x^2}{4a^2 t}}$, prove that $\dfrac{\partial v}{\partial t} = a^2 \dfrac{\partial^2 v}{\partial x^2}$.

27. If $u = (1 - 2xy + y^2)^{-1/2}$, prove that $x \dfrac{\partial u}{\partial x} - y \dfrac{\partial u}{\partial y} = y^2 u^3$.

28. If $u = e^x(x \cos y - y \sin y)$, show that $\dfrac{\partial^2 u}{\partial x^2} + \dfrac{\partial^2 u}{\partial y^2} = 0$.

29. If $x = r \cos \theta$, $y = r \sin \theta$, prove that

(*i*) $\dfrac{\partial^2 r}{\partial x^2} \cdot \dfrac{\partial^2 r}{\partial y^2} = \left(\dfrac{\partial^2 r}{\partial x \partial y}\right)^2$ (*ii*) $\dfrac{\partial^2 r}{\partial x^2} + \dfrac{\partial^2 r}{\partial y^2} = \dfrac{1}{r}\left[\left(\dfrac{\partial r}{\partial x}\right)^2 + \left(\dfrac{\partial r}{\partial y}\right)^2\right]$.

30. If $x^2 = au + bv$, $y^2 = au - bv$, prove that $\left(\dfrac{\partial u}{\partial x}\right)_y \cdot \left(\dfrac{\partial x}{\partial u}\right)_v = \dfrac{1}{2} = \left(\dfrac{\partial v}{\partial y}\right)_x \cdot \left(\dfrac{\partial y}{\partial v}\right)_u$.

31. Show that $\dfrac{\partial^2 z}{\partial x^2} - 2 \dfrac{\partial^2 z}{\partial x \partial y} + \dfrac{\partial^2 z}{\partial y^2} = 0$, where $z = xf(x + y) + yg(x + y)$.

32. If $u = f(ax^2 + 2hxy + by^2)$, $v = \phi(ax^2 + 2hxy + by^2)$, prove that $\dfrac{\partial}{\partial y}\left(u \dfrac{\partial v}{\partial x}\right) = \dfrac{\partial}{\partial x}\left(u \dfrac{\partial v}{\partial y}\right)$.

[**Hint.** Given $u = f(z)$, $v = \phi(z)$, where $z = ax^2 + 2hxy + by^2$.

We have to prove that $\dfrac{\partial u}{\partial y} \cdot \dfrac{\partial v}{\partial x} + u \dfrac{\partial^2 v}{\partial y \partial x} = \dfrac{\partial u}{\partial x} \cdot \dfrac{\partial v}{\partial y} + u \dfrac{\partial^2 v}{\partial x \partial y}$ or $\dfrac{\partial u}{\partial y} \cdot \dfrac{\partial v}{\partial x} = \dfrac{\partial u}{\partial x} \cdot \dfrac{\partial v}{\partial y}$]

$\left(\because \dfrac{\partial^2 v}{\partial y \partial x} = \dfrac{\partial^2 v}{\partial x \partial y}\right)$

33. If $u = \log (x^3 + y^3 - x^2y - xy^2)$, show that $\dfrac{\partial^2 u}{\partial x^2} + 2\dfrac{\partial^2 u}{\partial x \partial y} + \dfrac{\partial^2 u}{\partial y^2} = -4(x+y)^{-2}$.

[**Hint.** $u = \log \{x^2(x-y) - y^2(x-y)\} = \log (x-y)(x^2-y^2) = \log (x-y)^2(x+y) = 2\log (x-y) + \log (x+y)$].

34. If $V = f(r)$ and $r^2 = x^2 + y^2 + z^2$, prove that $V_{xx} + V_{yy} + V_{zz} = f''(r) + \dfrac{2}{r} f'(r)$.

35. If $z = f(x + ay) + \phi(x - ay)$, prove that $\dfrac{\partial^2 z}{\partial y^2} = a^2 \dfrac{\partial^2 z}{\partial x^2}$.

36. Find p and q, if $x = \sqrt{a}\,(\sin u + \cos v)$, $y = \sqrt{a}\,(\cos u - \sin v)$, $z = 1 + \sin (u - v)$ where p and q mean $\dfrac{\partial z}{\partial x}$ and $\dfrac{\partial z}{\partial y}$ respectively.

$\left[\textbf{Hint. } x^2 + y^2 = 2az, \ \therefore\ z = \dfrac{x^2+y^2}{2a}\right]$

Answers

1. (*i*) $y^x \log y,\ xy^{x-1}$ (*ii*) $\dfrac{2x}{x^2+y^2}, \dfrac{2y}{x^2+y^2}$ (*iii*) $2x \sin \dfrac{y}{x} - y \cos \dfrac{y}{x},\ x \cos \dfrac{y}{x}$

(*iv*) $\dfrac{-x}{x^2+y^2} + \dfrac{1}{y}\tan^{-1}\dfrac{y}{x},\ \dfrac{x^2}{y(x^2+y^2)} - \dfrac{x}{y^2}\tan^{-1}\dfrac{y}{x}$ **6.** $-\dfrac{13}{22}$

16. $n = 2, -3$ **22.** $\dfrac{x^2-y^2}{x^2+y^2}$ **36.** $p = \dfrac{x}{a},\ q = \dfrac{y}{a}$.

Partial Derivative at Origin:

$$f_x = \left(\frac{\partial f}{\partial x}\right)_{(0,0)} = \operatorname*{Lt}_{\Delta x \to 0} \frac{f(\Delta x, 0) - f(0,0)}{\Delta x}$$

$$f_y = \left(\frac{\partial f}{\partial y}\right)_{(0,0)} = \operatorname*{Lt}_{\Delta y \to 0} \frac{f(0, \Delta y) - f(0,0)}{\Delta y}$$

Example 1. *Show that* $f(x, y) = \begin{cases} \dfrac{xy}{x^2+2y^2}, & (x, y) \neq (0, 0) \\ 0, & (x, y) = (0, 0) \end{cases}$ *is not continuous at (0, 0) but its partial derivatives* f_x *and* f_y *exist at (0, 0).*

Sol.
$$\operatorname*{Lt}_{\substack{y \to mx \\ x \to 0}} f(x,y) = \operatorname*{Lt}_{x\to 0}\left\{\operatorname*{Lt}_{y \to mx} \frac{xy}{x^2+2y^2}\right\}$$

$$= \operatorname*{Lt}_{x \to 0} \frac{mx^2}{x^2 + 2m^2x^4} = \operatorname*{Lt}_{x\to 0} \frac{mx^2}{x^2(1+2m^2x^2)}$$

$$= \frac{m}{1+0} = m = \text{not unique value.}$$

It takes different values for different m.

$\therefore$ limit does not exist.

Hence $f(x, y)$ is not continuous at (0, 0).

$$\left(\frac{\partial f}{\partial x}\right)_{(0,0)} = \underset{\Delta x \to 0}{\text{Lt}} \frac{f(\Delta x, 0) - f(0,0)}{\Delta x} = \underset{\Delta x \to 0}{\text{Lt}} \frac{0-0}{\Delta x}$$

$$= \underset{\Delta x \to 0}{\text{Lt}} (0) = 0$$

$$\left(\frac{\partial f}{\partial y}\right)_{(0,0)} = \underset{\Delta y \to 0}{\text{Lt}} \frac{f(0, \Delta y) - f(0,0)}{\Delta y} = \underset{\Delta y \to 0}{\text{Lt}} \frac{0-0}{\Delta y}$$

$$= \underset{\Delta y \to 0}{\text{Lt}} (0) = 0$$

$\therefore$ f_x and f_y are exist at (0, 0).

Example 2. *Show that the function* $f(x, y) = \begin{cases} (x+y)\sin\left(\dfrac{1}{x+y}\right), & x+y \neq 0 \\ 0, & x+y=0 \end{cases}$, *is continuous at (0, 0) but its partial derivatives f_x and f_y do not exist at (0, 0).* (OU-May 2008)

Sol. We have

$$|f(x, y) - f(0, 0)| = \left|(x+y)\sin\left(\frac{1}{x+y}\right)\right| \leq |x+y|$$

$$\leq |x| + |y| \leq 2\sqrt{x^2+y^2} < \in$$

Take $\delta < \dfrac{\in}{2}$, then

$|f(x, y) - 0| < \in$ whenever $0 < \sqrt{x^2+y^2} < \delta$

$$\therefore \quad \underset{(x,y)\to(0,0)}{\text{Lt}} f(x,y) = 0 = f(0, 0)$$

$\Rightarrow$ The given function $f(x, y)$ is continuous at (0, 0). Now we have f_x at (0, 0)

$$\left(\frac{\partial f}{\partial x}\right)_{(0,0)} = \underset{\Delta x \to 0}{\text{Lt}} \frac{f(\Delta x, 0) - f(0,0)}{\Delta x}$$

$$= \underset{\Delta x \to 0}{\text{Lt}} \frac{\Delta x \sin \dfrac{1}{\Delta x}}{\Delta x} = \underset{\Delta x \to 0}{\text{Lt}} \sin\left(\frac{1}{\Delta x}\right)$$

which is not exist.

$\therefore$ f_x does not exist at (0, 0).

Similarly, f_y at $(0, 0) = \left(\dfrac{\partial f}{\partial y}\right)_{(0,0)} = \underset{\Delta y \to 0}{\text{Lt}} \dfrac{f(0, \Delta y) - f(0,0)}{\Delta y}$

$$= \underset{\Delta y \to 0}{\text{Lt}} \frac{\Delta y \sin\left(\dfrac{1}{\Delta y}\right)}{\Delta y} = \underset{\Delta y \to 0}{\text{Lt}} \sin\left(\frac{1}{\Delta y}\right)$$

Also does not exist.

$\therefore$ f_y does not exist at (0, 0).

Example 3. *Show that the function* $f(x, y) = \begin{cases} \dfrac{x^2 + y^2}{|x| + |y|}, & (x, y) \neq (0, 0) \\ 0, & (x, y) = (0, 0) \end{cases}$ *is continuous at (0, 0) but its partial derivatives f_x and f_y do not exist at (0, 0).*

Sol. Consider

$$|f(x, y) - f(0, 0)| = \left| \frac{x^2 + y^2}{|x| + |y|} \right| \leq \frac{(|x| + |y|)^2}{|x| + |y|} = |x| + |y|$$

$$\leq 2\sqrt{x^2 + y^2} < \epsilon$$

Let $\delta < \dfrac{\epsilon}{2}$, then we have

$|f(x, y) - 0| < \epsilon$ whenever $0 < \sqrt{x^2 + y^2} < \delta$

$\therefore$ $\underset{(x,y)\to(0,0)}{\text{Lt}} f(x, y) = 0 = f(0, 0) \Rightarrow f(x, y)$ is continuous at (0, 0)

Now, we have

$$(f_x)_{(0,0)} = \underset{\Delta x \to 0}{\text{Lt}} \frac{f(\Delta x, 0) - f(0, 0)}{\Delta x}$$

$$= \underset{\Delta x \to 0}{\text{Lt}} \frac{\Delta x}{|\Delta x|} = \begin{cases} 1, & \text{if } \Delta x > 0 \\ -1, & \text{if } \Delta x < 0 \end{cases}$$

which is not unique.

$\therefore$ f_x does not exist at (0, 0).

Similarly, $$(f_y)(0, 0) = \underset{\Delta y \to 0}{\text{Lt}} \frac{f(0, \Delta y) - f(0, 0)}{\Delta y}$$

$$= \underset{\Delta y \to 0}{\text{Lt}} \frac{\Delta y}{|\Delta y|} = \begin{cases} 1, & \text{if } \Delta y > 0 \\ -1, & \text{if } \Delta y < 0 \end{cases}$$

which is also not unique.

Hence limit does not exist and f_y does not exist at (0, 0).

4.6. COMPOSITE FUNCTIONS

(*i*) If $u = f(x, y)$, where $x = \phi(t)$, $y = \psi(t)$

then u is called a composite function of (the **single variable**) t and we can find $\dfrac{du}{dt}$.

(*ii*) If $z = f(x, y)$, where $x = \phi(u, v)$, $y = \psi(u, v)$

then z is called a composite function of (**two variables**) u and v so that we can find $\dfrac{\partial z}{\partial u}$ and $\dfrac{\partial z}{\partial v}$.

4.7. TOTAL DERIVATIVE

If u is composite function of t, defined by the relations $u = f(x, y)$; $x = \phi(t)$, $y = \psi(t)$, *then*

$$\frac{du}{dt} = \frac{\partial u}{\partial x} \cdot \frac{dx}{dt} + \frac{\partial u}{\partial y} \cdot \frac{dy}{dt}.$$

Note 1. If $u = f(x, y, z)$ and x, y, z are function of t, then y is a composite function of t and

$$\frac{du}{dt} = \frac{\partial u}{\partial x} \cdot \frac{dx}{dt} + \frac{\partial u}{\partial y} \cdot \frac{dy}{dt} + \frac{\partial u}{\partial z} \cdot \frac{dz}{dt}.$$

Note 2. If $z = f(x, y)$ and x, y are functions of u and v, then

$$\frac{\partial z}{\partial u} = \frac{\partial z}{\partial x} \cdot \frac{\partial x}{\partial u} + \frac{\partial z}{\partial y} \cdot \frac{\partial y}{\partial u} ; \quad \frac{\partial z}{\partial v} = \frac{\partial z}{\partial x} \cdot \frac{\partial x}{\partial v} + \frac{\partial z}{\partial y} \cdot \frac{\partial y}{\partial v}.$$

Note 3. If $u = f(x, y)$, where $y = \phi(x)$, then since $x = \psi(x)$, u is a composite function of x.

$$\therefore \quad \frac{du}{dx} = \frac{\partial u}{\partial x} \cdot \frac{dx}{dx} + \frac{\partial u}{\partial y} \cdot \frac{dy}{dx} \quad \Rightarrow \quad \frac{du}{dx} = \frac{\partial u}{\partial x} + \frac{\partial u}{\partial y} \cdot \frac{dy}{dx}.$$

Note 4. Total derivative of u is also defined as

$$du = \frac{\partial u}{\partial x} dx + \frac{\partial u}{\partial y} dy$$

ILLUSTRATIVE EXAMPLES

Example 1. *If* $z = y + f(u)$, $u = \dfrac{x}{y}$, *show that* $u\dfrac{\partial z}{\partial x} + \dfrac{\partial z}{\partial y} = 1.$ (OU-Dec. 2013)

Sol. Given $z = y + f(u)$ and $u = \dfrac{x}{y}$

$$\frac{\partial z}{\partial x} = 0 + f'(u)\frac{\partial u}{\partial x} = f'(u)\left(\frac{1}{y}\right)$$

$$\frac{\partial z}{\partial y} = 1 + f'(u)\frac{\partial u}{\partial y} = 1 + f'(u)\left(-\frac{x}{y^2}\right)$$

$$\text{LHS} = u\frac{\partial z}{\partial x} + \frac{\partial z}{\partial y}$$

$$= uf'(u) \cdot \frac{1}{y} + 1 + f'(u)\left(-\frac{x}{y^2}\right) \qquad \left(\because \; u = \frac{x}{y}\right)$$

$$= \frac{x}{y^2} f'(u) + 1 - \frac{x}{y^2} f'(u) = 1.$$

Example 2. *If* $z = e^{ax+by} f(ax - by)$, *then prove that* $b\dfrac{\partial z}{\partial x} + a\dfrac{\partial z}{\partial y} = 2abz.$

(OU-June 2013, 2010, Apr. 2007)

Sol. Given $z = e^{ax+by} f(ax - by)$

$$\frac{\partial z}{\partial x} = e^{ax+by} \cdot f'(ax - by) \cdot a + f(ax - by) \cdot e^{ax+by} \cdot a$$

$$\Rightarrow \quad b\frac{\partial z}{\partial x} = abe^{ax+by} \cdot f'(ax - by) + abz \quad \ldots(1)$$

$$\frac{\partial z}{\partial y} = e^{ax+by} \cdot f'(ax - by)(-b) + f(ax - by)e^{ax+by} \cdot b$$

$$\Rightarrow \quad a\frac{\partial z}{\partial y} = -abe^{ax+by} f'(ax - by) + abz \quad \ldots(2)$$

$$(1) + (2) \Rightarrow b\frac{\partial z}{\partial x} + a\frac{\partial z}{\partial y} = 2abz.$$

Example 3. *Obtain the total differential of* $z = \tan^{-1}\left(\frac{x}{y}\right)$, $(x, y) \neq (0, 0)$.

(OU-June 2013, May 2008)

Sol. Let $z(x, y) = \tan^{-1}\left(\frac{x}{y}\right)$

$$\frac{\partial z}{\partial x} = \frac{1}{1+\left(\frac{x}{y}\right)^2}\left(\frac{1}{y}\right) = \frac{y}{x^2 + y^2}$$

$$\frac{\partial z}{\partial y} = \frac{1}{1+\left(\frac{x}{y}\right)^2}\left(-\frac{x}{y^2}\right) = -\frac{x}{x^2 + y^2}$$

By total derivative, we have

$$dz = \frac{\partial z}{\partial x} dx + \frac{\partial z}{\partial y} dy = \frac{1}{x^2 + y^2}(y\,dx - x\,dy).$$

Example 4. *For the function* $f(x, y) = \begin{cases} \dfrac{x^2 y(x - y)}{x^2 + y^2}, & (x, y) \neq (0, 0) \\ 0, & (x, y) = (0, 0) \end{cases}$ *show that* $\dfrac{\partial^2 f}{\partial x \partial y} \neq \dfrac{\partial^2 f}{\partial y \partial x}$ *at (0, 0).*

(OU-Dec. 2013)

Sol.

$$\left(\frac{\partial f}{\partial x}\right)_{(0,0)} = \underset{\Delta x \to 0}{\text{Lt}} \frac{f(\Delta x, 0) - f(0, 0)}{\Delta x} = 0,$$

$$\left(\frac{\partial f}{\partial y}\right)_{(0,0)} = \underset{\Delta y \to 0}{\text{Lt}} \frac{f(0, \Delta y) - f(0, 0)}{\Delta y} = 0$$

$$\left(\frac{\partial f}{\partial x}\right)_{(0,y)} = \underset{\Delta x \to 0}{\text{Lt}} \frac{f(\Delta x, y) - f(0, y)}{\Delta x} = \underset{\Delta x \to 0}{\text{Lt}} \frac{(\Delta x)^2 y(\Delta x - y)}{[(\Delta x)^2 + y^2]\Delta x} = 0$$

$$\left(\frac{\partial f}{\partial y}\right)_{(x,0)} = \underset{\Delta y \to 0}{\text{Lt}} \frac{f(x, \Delta y) - f(x, 0)}{\Delta y} = \underset{\Delta y \to 0}{\text{Lt}} \frac{x^2 \Delta y(x - \Delta y)}{[x^2 + (\Delta y^2)]\Delta y} = \frac{x^3 - 0}{x^2} = x$$

$$\left(\frac{\partial^2 f}{\partial x \partial y}\right)_{(0,0)} = \underset{\Delta x \to 0}{\text{Lt}} \frac{\left(\frac{\partial f}{\partial y}\right)_{(\Delta x,0)} - \left(\frac{\partial f}{\partial y}\right)_{(0,0)}}{\Delta x} = \underset{\Delta x \to 0}{\text{Lt}} \frac{\Delta x - 0}{\Delta x} = 1$$

$$\left(\frac{\partial^2 f}{\partial y \partial x}\right)_{(0,0)} = \underset{\Delta y \to 0}{\text{Lt}} \frac{\left(\frac{\partial f}{\partial x}\right)_{(0,\Delta y)} - \left(\frac{\partial f}{\partial x}\right)_{(0,0)}}{\Delta y} = \underset{\Delta y \to 0}{\text{Lt}} \frac{0 - 0}{\Delta y} = 0$$

Hence, $\frac{\partial^2 f}{\partial x \partial y} \neq \frac{\partial^2 f}{\partial y \partial x}$ at (0, 0).

Example 5. *If $z = f(x, y)$, $x = e^{2u} + e^{-2v}$, $y = e^{-2u} + e^{2v}$, then show that*

$$\frac{\partial f}{\partial u} - \frac{\partial f}{\partial v} = 2\left[x\frac{\partial f}{\partial x} - y\frac{\partial f}{\partial y}\right].$$ (OU-May 2008, 2009, June 2010)

Sol. Using the chain rule, we get

$$\frac{\partial f}{\partial u} = \frac{\partial f}{\partial x}\frac{\partial x}{\partial u} + \frac{\partial f}{\partial y}\frac{\partial y}{\partial u} = 2e^{2u}\frac{\partial f}{\partial x} - 2e^{-2u}\frac{\partial f}{\partial y} \quad \text{...(1)}$$

$$\frac{\partial f}{\partial v} = \frac{\partial f}{\partial x}\frac{\partial x}{\partial v} + \frac{\partial f}{\partial y}\frac{\partial y}{\partial v} = -2e^{-2v}\frac{\partial f}{\partial x} + 2e^{2v}\frac{\partial f}{\partial y} \quad \text{...(2)}$$

(1) – (2)

$$\Rightarrow \quad \frac{\partial f}{\partial u} - \frac{\partial f}{\partial v} = 2(e^{2u} + e^{-2v})\frac{\partial f}{\partial x} - 2(e^{-2u} + e^{2v})\frac{\partial f}{\partial y} = 2x\frac{\partial f}{\partial x} - 2y\frac{\partial f}{\partial y}$$

Example 6. *Find the first order partial derivatives of $f(x, y) = \sin(2x + 3y)$ from the first principles.* (OU-Dec. 2008)

Sol. Given $f = \sin(2x + 3y)$ from first principle, we have

$$\frac{\partial f}{\partial x} = \underset{\Delta x \to 0}{\text{Lt}} \frac{f(x + \Delta x, y) - f(x, y)}{\Delta x} = \underset{\Delta x \to 0}{\text{Lt}} \frac{\sin(2(x + \Delta x) + 3y) - \sin(2x + 3y)}{\Delta x}$$

$$\boxed{\sin C - \sin D = 2\cos\left(\frac{C + D}{2}\right)\sin\left(\frac{C - D}{2}\right)}$$

$$= \underset{\Delta x \to 0}{\text{Lt}} \frac{2\cos(2x + 3y + \Delta x)\sin \Delta x}{\Delta x} = 2\cos(2x + 3y)$$

$$\left(\because \underset{\Delta x \to 0}{\text{Lt}} \frac{\sin \Delta x}{\Delta x} = 1\right)$$

$$\frac{\partial f}{\partial y} = \underset{\Delta y \to 0}{\text{Lt}} \frac{f(x, y + \Delta y) - f(x, y)}{\Delta y}$$

$$= \underset{\Delta y \to 0}{\text{Lt}} \frac{\sin(2x + 3(y + \Delta y)) - \sin(2x + 3y)}{\Delta y}$$

$$= \underset{\Delta y \to 0}{\text{Lt}} \frac{2\cos\left(2x + 3y + \frac{3\Delta y}{2}\right)\sin\left(\frac{3\Delta y}{2}\right)}{\Delta y} = 3\cos(2x + 3y).$$

Example 7. *Find the total differential of* $u = \left(xz + \dfrac{x}{z}\right)^y$, $z \neq 0$. (OU-May 2009)

Sol. Let $$f(x, y, z) = \left(xz + \frac{x}{z}\right)^y$$

$$\frac{\partial f}{\partial x} = y\left(xz + \frac{x}{z}\right)^{y-1}\left(z + \frac{1}{z}\right)$$

$$\frac{\partial f}{\partial y} = \left(xz + \frac{x}{z}\right)^y \log\left(xz + \frac{x}{z}\right)$$

$$\frac{\partial f}{\partial z} = y\left(xz + \frac{x}{z}\right)^{y-1}\left(x - \frac{x}{z^2}\right)$$

∴ we get the total differential as

$$du = \frac{\partial f}{\partial x}dx + \frac{\partial f}{\partial y}dy + \frac{\partial f}{\partial z}dz$$

$$= \left(xz + \frac{x}{z}\right)^{y-1}\left[y\left(z + \frac{1}{z}\right)dx + xy\left(1 - \frac{1}{z^2}\right)dz\right] + \left[\left(xz + \frac{x}{z}\right)^y \log\left(xz + \frac{x}{z}\right)\right] dy.$$

Example 8. *If* $v = (x^2 + y^2 + z^2)^{-\frac{1}{2}}$, *find* $\dfrac{\partial^2 v}{\partial x^2} + \dfrac{\partial^2 v}{\partial y^2} + \dfrac{\partial^2 v}{\partial z^2}$. (OU-June 2011, Jan. 2012)

Sol. Given $$v = (x^2 + y^2 + z^2)^{-\frac{1}{2}}$$

$$\frac{\partial v}{\partial x} = -\frac{1}{2}(x^2 + y^2 + z^2)^{-\frac{3}{2}}(2x) = -x\,(x^2 + y^2 + z^2)^{-3/2}$$

$$\frac{\partial^2 v}{\partial x^2} = -\left\{x \cdot \frac{-3}{2}\left(x^2 + y^2 + z^2\right)^{-\frac{5}{2}}(2x) + 1 \cdot \left(x^2 + y^2 + z^2\right)^{-\frac{3}{2}}\right\}$$

$$\Rightarrow \quad \frac{\partial^2 v}{\partial x^2} = 3x^2(x^2 + y^2 + z^2)^{-\frac{5}{2}} - (x^2 + y^2 + z^2)^{-\frac{3}{2}} \quad ...(1)$$

Similarly, $$\frac{\partial^2 v}{\partial y^2} = 3y^2(x^2 + y^2 + z^2)^{-\frac{5}{2}} - (x^2 + y^2 + z^2)^{-\frac{3}{2}} \quad ...(2)$$

$$\frac{\partial^2 v}{\partial z^2} = 3z^2(x^2 + y^2 + z^2)^{-\frac{5}{2}} - (x^2 + y^2 + z^2)^{-\frac{3}{2}} \quad ...(3)$$

$$(1) + (2) + (3) \Rightarrow \frac{\partial^2 v}{\partial x^2} + \frac{\partial^2 v}{\partial y^2} + \frac{\partial^2 v}{\partial z^2} = 3\,(x^2 + y^2 + z^2)(x^2 + y^2 + z^2)^{-\frac{5}{2}} - 3\,(x^2 + y^2 + z^2)^{-\frac{3}{2}}$$

$$= 3\,(x^2 + y^2 + z^2)^{-\frac{3}{2}} - 3\,(x^2 + y^2 + z^2)^{-\frac{3}{2}} = 0.$$

Example 9. *If* $Z = \log(u^2 + v)$, $u = e^{x^2 + y^2}$, $v = x^2 + y$, *find* $\dfrac{\partial z}{\partial x}, \dfrac{\partial z}{\partial y}$. (OU-June 2011)

Sol. By chain rule, we get

$$\frac{\partial z}{\partial x} = \frac{\partial z}{\partial u}\cdot\frac{\partial u}{\partial x} + \frac{\partial z}{\partial v}\cdot\frac{\partial v}{\partial x}$$

$$= \frac{1}{u^2+v}(2u)\,.\,e^{x^2+y^2}(2x) + \frac{1}{u^2+v}(1)(2x)$$

$$= \frac{2x}{u^2+v}(1+2ue^{x^2+y^2})$$

$$\frac{\partial z}{\partial y} = \frac{\partial z}{\partial u}\cdot\frac{\partial u}{\partial y} + \frac{\partial z}{\partial v}\cdot\frac{\partial v}{\partial y}$$

$$= \frac{1}{u^2+v}(2u)\cdot e^{x^2+y^2}\cdot 2y + \frac{1}{u^2+v}(1)\cdot(1)$$

$$= \frac{1}{u^2+v}(1+4uye^{x^2+y^2}).$$

Example 10. *If* $u = x^2 \tan^{-1}\frac{y}{x} - y^2 \tan^{-1}\frac{x}{y}$, *then show that* $\frac{\partial^2 u}{\partial x \partial y} = \frac{x^2-y^2}{x^2+y^2}$.

(OU-Dec. 2007, May 2009)

Sol. Given $u = x^2 \tan^{-1}\frac{y}{x} - y^2 \tan^{-1}\frac{x}{y}$

$$\frac{\partial u}{\partial y} = x^2\,.\,\frac{1}{1+\left(\frac{y}{x}\right)^2}\left(\frac{1}{x}\right) - \left\{y^2\,.\,\frac{1}{1+\left(\frac{x}{y}\right)^2}\left(\frac{-x}{y^2}\right) + 2y\,.\,\tan^{-1}\frac{x}{y}\right\}$$

$$\frac{\partial u}{\partial y} = \frac{x^3}{x^2+y^2} + \frac{xy^2}{x^2+y^2} - 2y\tan^{-1}\frac{x}{y} = \frac{x(x^2+y^2)}{(x^2+y^2)} - 2y\tan^{-1}\frac{x}{y}$$

$$\frac{\partial u}{\partial y} = x - 2y\tan^{-1}\frac{x}{y}$$

$$\frac{\partial^2 u}{\partial x \partial y} = \frac{\partial}{\partial x}\left(\frac{\partial u}{\partial y}\right) = \frac{\partial}{\partial x}\left(x - 2y\tan^{-1}\frac{x}{y}\right)$$

$$= 1 - 2y\,.\,\frac{1}{1+\left(\frac{x}{y}\right)^2}\left(\frac{1}{y}\right) = 1 - \frac{2y^2}{x^2+y^2} = \frac{x^2+y^2-2y^2}{x^2+y^2} = \frac{x^2-y^2}{x^2+y^2}$$

Hence the result.

Example 11. *If* $x = u + v + w$, $y = uv + vw + wu$, $z = uvw$ *and f is a function of x, y, z, then show that* $u\frac{\partial f}{\partial u} + v\frac{\partial f}{\partial v} + w\frac{\partial f}{\partial w} = x\frac{\partial f}{\partial x} + 2y\frac{\partial f}{\partial y} + 3z\frac{\partial f}{\partial z}$. (OU-Dec. 2007)

Sol. By chain rule, we get

$$\frac{\partial f}{\partial u} = \frac{\partial f}{\partial x}\frac{\partial x}{\partial u} + \frac{\partial f}{\partial y}\frac{\partial y}{\partial u} + \frac{\partial f}{\partial z}\frac{\partial z}{\partial u} = \frac{\partial f}{\partial x}\cdot 1 + \frac{\partial f}{\partial y}(v+w) + \frac{\partial f}{\partial z}(vw)$$

$$\Rightarrow \quad u\frac{\partial f}{\partial u} = u\frac{\partial f}{\partial x} + (uv+uw)\frac{\partial f}{\partial y} + (uvw)\frac{\partial f}{\partial z} \quad ...(1)$$

Similarly, $$v\frac{\partial f}{\partial v} = v\frac{\partial f}{\partial x} + (uv + vw)\frac{\partial f}{\partial y} + (uvw)\frac{\partial f}{\partial z} \quad ...(2)$$

$$w\frac{\partial f}{\partial w} = w\frac{\partial f}{\partial x} + (uw + vw)\frac{\partial f}{\partial y} + (uvw)\frac{\partial f}{\partial z} \quad ...(3)$$

(1) + (2) + (3)

$$\Rightarrow \quad u\frac{\partial f}{\partial u} + v\frac{\partial f}{\partial v} + w\frac{\partial f}{\partial w} = (u + v + w)\frac{\partial f}{\partial x} + 2(uv + vw + uw)\frac{\partial f}{\partial y} + 3uvw\frac{\partial f}{\partial z}$$

$$= x\frac{\partial f}{\partial x} + 2y\frac{\partial f}{\partial y} + 3z\frac{\partial f}{\partial z}.$$

Example 12. *If* $z = x\,\phi\left(\frac{y}{x}\right) + \psi\left(\frac{y}{x}\right)$, *then prove that* $x^2\frac{\partial^2 z}{\partial x^2} + 2xy\frac{\partial^2 z}{\partial x\partial y} + y^2\frac{\partial^2 z}{\partial y^2} = 0.$

(OU-Dec. 2007)

Sol. Given $z = x\,\phi\left(\frac{y}{x}\right) + \psi\left(\frac{y}{x}\right)$

$$\frac{\partial z}{\partial x} = x\cdot\phi'\left(\frac{y}{x}\right)\left(\frac{-y}{x^2}\right) + \phi\left(\frac{y}{x}\right)\cdot 1 + \psi'\left(\frac{y}{x}\right)\left(-1\frac{y}{x^2}\right)$$

$$\frac{\partial z}{\partial x} = -\frac{y}{x}\phi'\left(\frac{y}{x}\right) + \phi\left(\frac{y}{x}\right) - \frac{y}{x^2}\psi'\left(\frac{y}{x}\right)$$

$$\frac{\partial^2 z}{\partial x^2} = -y\left[\frac{1}{x}\phi''\left(\frac{y}{x}\right)\left(\frac{-y}{x^2}\right) + \phi'\left(\frac{y}{x}\right)\left(-\frac{1}{x^2}\right)\right] + \phi'\left(\frac{y}{x}\right)\left(\frac{-y}{x^2}\right)$$
$$-y\left[\frac{1}{x^2}\psi''\left(\frac{y}{x}\right)\left(\frac{-y}{x^2}\right) + \psi'\left(\frac{y}{x}\right)\left(-\frac{2}{x^3}\right)\right]$$

$$= \frac{y^2}{x^3}\phi''\left(\frac{y}{x}\right) + \frac{y}{x^2}\phi'\left(\frac{y}{x}\right) - \frac{y}{x^2}\phi'\left(\frac{y}{x}\right) + \frac{y^2}{x^4}\psi''\left(\frac{y}{x}\right) + \frac{2y}{x^3}\psi'\left(\frac{y}{x}\right)$$

$$\frac{\partial z}{\partial y} = x\phi'\left(\frac{y}{x}\right)\left(\frac{1}{x}\right) + \psi'\left(\frac{y}{x}\right)\left(\frac{1}{x}\right) = \phi'\left(\frac{y}{x}\right) + \frac{1}{x}\psi'\left(\frac{y}{x}\right)$$

$$\frac{\partial^2 z}{\partial x\partial y} = \phi''\left(\frac{y}{x}\right)\left(\frac{-y}{x^2}\right) + \left[\frac{1}{x}\psi''\left(\frac{y}{x}\right)\left(\frac{-y}{x^2}\right) + \psi'\left(\frac{y}{x}\right)\left(\frac{-1}{x^2}\right)\right]$$

$$= \frac{-y}{x^2}\phi''\left(\frac{y}{x}\right) - \frac{y}{x^3}\psi''\left(\frac{y}{x}\right) - \frac{1}{x^2}\psi'\left(\frac{y}{x}\right)$$

$$\frac{\partial^2 z}{\partial y^2} = \phi''\left(\frac{y}{x}\right).\frac{1}{x} + \frac{1}{x}.\psi''\left(\frac{y}{x}\right).\frac{1}{x}.$$

Therefore, LHS $= \frac{y^2}{x}\phi''\left(\frac{y}{x}\right) + \frac{y^2}{x^2}\psi''\left(\frac{y}{x}\right) + \frac{2y}{x}\psi'\left(\frac{y}{x}\right) - \frac{2y^2}{x}\phi''\left(\frac{y}{x}\right) - \frac{2y^2}{x^2}\psi''\left(\frac{y}{x}\right)$

$$- \frac{2y}{x}\psi'\left(\frac{y}{x}\right) + \frac{y^2}{x}\phi''\left(\frac{y}{x}\right) + \frac{y^2}{x^2}\psi''\left(\frac{y}{x}\right) = 0$$

Example 13. *If* $u = f(x + ay) + \phi(x - ay)$ *prove that* $\frac{\partial^2 u}{\partial y^2} = a^2\frac{\partial^2 u}{\partial x^2}$.

Sol. Given $u = f(x + ay) + \phi\,(x - ay)$

$$\frac{\partial u}{\partial x} = f'(x + ay)\cdot 1 + \phi'(x - ay) \cdot 1$$

$$\frac{\partial^2 u}{\partial x^2} = f''(x + ay) + \phi''(x - ay)$$

$$a^2 \frac{\partial^2 u}{\partial x^2} = a^2 (f''(x + ay) + \phi''(x - ay)) \quad ...(1)$$

Similarly, $$\frac{\partial u}{\partial y} = f'(x + ay)\cdot a + \phi'(x - ay)(-a)$$

$$\frac{\partial^2 u}{\partial y^2} = f''(x + ay)\cdot a^2 + \phi''(x - ay)(-a)^2$$

$$= a^2 [f''(x + ay) + \phi''(x - ay)] = a^2 \frac{\partial^2 u}{\partial x^2} \quad [\because \text{ From (1)}]$$

Hence the result.

Example 14. *If $u = \sin^{-1}(x - y)$, $x = 3t$, $y = 4t^3$, show that* $\dfrac{du}{dt} = \dfrac{3}{\sqrt{1-t^2}}$.

Sol. The given equations define u as a composite function of t.

$$\therefore \quad \frac{du}{dt} = \frac{\partial u}{\partial x}\cdot\frac{dx}{dt} + \frac{\partial u}{\partial y}\cdot\frac{dy}{dt}$$

$$= \frac{1}{\sqrt{1-(x-y)^2}}\cdot 3 + \frac{1}{\sqrt{1-(x-y)^2}}(-1)\cdot 12t^2$$

$$= \frac{3(1-4t^2)}{\sqrt{1-(x-y)^2}} = \frac{3(1-4t^2)}{\sqrt{1-(3t-4t^3)^2}} = \frac{3(1-4t^2)}{\sqrt{1-9t^2+24t^4-16t^6}}$$

$$= \frac{3(1-4t^2)}{\sqrt{(1-t^2)(1-8t^2+16t^4)}} = \frac{3(1-4t^2)}{\sqrt{(1-t^2)(1-4t^2)^2}} = \frac{3}{\sqrt{1-t^2}}.$$

Example 15. *If z is a function of x and y, where $x = e^u + e^{-v}$ and $y = e^{-u} - e^v$, show that*

$$\frac{\partial z}{\partial u} - \frac{\partial z}{\partial v} = x\frac{\partial z}{\partial x} - y\frac{\partial z}{\partial y}.$$

Sol. Here z is a composite function of u and v.

$$\therefore \quad \frac{\partial z}{\partial u} = \frac{\partial z}{\partial x}\cdot\frac{\partial x}{\partial u} + \frac{\partial z}{\partial y}\cdot\frac{\partial y}{\partial u} = \frac{\partial z}{\partial x}\cdot e^u + \frac{\partial z}{\partial y}(-e^{-u})$$

and $$\frac{\partial z}{\partial v} = \frac{\partial z}{\partial x}\cdot\frac{\partial x}{\partial v} + \frac{\partial z}{\partial y}\cdot\frac{\partial y}{\partial v} = \frac{\partial z}{\partial x}(-e^{-v}) + \frac{\partial z}{\partial y}(-e^v)$$

Subtracting, $$\frac{\partial z}{\partial u} - \frac{\partial z}{\partial v} = (e^u + e^{-v})\frac{\partial z}{\partial x} - (e^{-u} - e^v)\frac{\partial z}{\partial y} = x\frac{\partial z}{\partial x} - y\frac{\partial z}{\partial y}.$$

Example 16. *If $u = f(y - z, z - x, x - y)$, prove that* $\dfrac{\partial u}{\partial x} + \dfrac{\partial u}{\partial y} + \dfrac{\partial u}{\partial z} = 0$. (OU-May 2015)

Sol. Here $u = f(X, Y, Z)$, where $X = y - z$, $Y = z - x$, $Z = x - y$

$\therefore$ u is a composite function of x, y and z.

$$\frac{\partial u}{\partial x} = \frac{\partial u}{\partial X}\cdot\frac{\partial X}{\partial x} + \frac{\partial u}{\partial Y}\cdot\frac{\partial Y}{\partial x} + \frac{\partial u}{\partial Z}\cdot\frac{\partial Z}{\partial x} = \frac{\partial u}{\partial X}(0) + \frac{\partial u}{\partial Y}(-1) + \frac{\partial u}{\partial Z}(1)$$

$$\frac{\partial u}{\partial y} = \frac{\partial u}{\partial X}\cdot\frac{\partial X}{\partial y} + \frac{\partial u}{\partial Y}\cdot\frac{\partial Y}{\partial y} + \frac{\partial u}{\partial Z}\cdot\frac{\partial Z}{\partial y} = \frac{\partial u}{\partial X}(1) + \frac{\partial u}{\partial Y}(0) + \frac{\partial u}{\partial Z}(-1)$$

$$\frac{\partial u}{\partial z} = \frac{\partial u}{\partial X}\cdot\frac{\partial X}{\partial z} + \frac{\partial u}{\partial Y}\cdot\frac{\partial Y}{\partial z} + \frac{\partial u}{\partial Z}\cdot\frac{\partial Z}{\partial z} = \frac{\partial u}{\partial X}(-1) + \frac{\partial u}{\partial Y}(1) + \frac{\partial u}{\partial Z}(0)$$

Adding, $\frac{\partial u}{\partial x} + \frac{\partial u}{\partial y} + \frac{\partial u}{\partial z} = 0.$

Example 17. *If $w = f(x, y)$, $x = r\cos\theta$, $y = r\sin\theta$, show that*

$$\left(\frac{\partial w}{\partial r}\right)^2 + \frac{1}{r^2}\left(\frac{\partial w}{\partial \theta}\right)^2 = \left(\frac{\partial f}{\partial x}\right)^2 + \left(\frac{\partial f}{\partial y}\right)^2.$$

Sol. The given equations define w as a composite function of r and θ.

$$\frac{\partial w}{\partial r} = \frac{\partial w}{\partial x}\cdot\frac{\partial x}{\partial r} + \frac{\partial w}{\partial y}\cdot\frac{\partial y}{\partial r} = \frac{\partial w}{\partial x}\cdot\cos\theta + \frac{\partial w}{\partial y}\cdot\sin\theta$$

or $$\frac{\partial w}{\partial r} = \frac{\partial f}{\partial x}\cos\theta + \frac{\partial f}{\partial y}\sin\theta \qquad \text{...(1)} \quad [\because \ w = f(x, y)]$$

Also, $$\frac{\partial w}{\partial \theta} = \frac{\partial w}{\partial x}\cdot\frac{\partial x}{\partial \theta} + \frac{\partial w}{\partial y}\cdot\frac{\partial y}{\partial \theta} = \frac{\partial w}{\partial x}(-r\sin\theta) + \frac{\partial w}{\partial y}(r\cos\theta)$$

or $$\frac{1}{r}\frac{\partial w}{\partial \theta} = -\frac{\partial f}{\partial x}\sin\theta + \frac{\partial f}{\partial y}\cos\theta \qquad \text{...(2)}$$

Squaring and adding (1) and (2), we get

$$\left(\frac{\partial w}{\partial r}\right)^2 + \frac{1}{r^2}\left(\frac{\partial w}{\partial \theta}\right)^2 = \left(\frac{\partial f}{\partial x}\right)^2 + \left(\frac{\partial f}{\partial y}\right)^2.$$

EXERCISE 2

1. (*a*) Find $\frac{du}{dt}$ when $u = x^2 + y^2$, $x = at^2$, $y = 2at$. Also verify by direct substitution.

   (*b*) If $u = x^2 + y^2 + z^2$ and $x = e^{2t}$, $y = e^{2t}\cos 3t$, $z = e^{2t}\sin 3t$; find $\frac{du}{dt}$ as a total derivative and verify the result by direct substitution.

2. If $u = \sin\frac{x}{y}$, $x = e^t$, $y = t^2$, find $\frac{du}{dt}$.

3. If $u = x^3 + y^3$, where $x = a\cos t$, $y = b\sin t$, find $\frac{du}{dt}$ and verify the result.

4. (*a*) If $z = u^2 + v^2$, $u = r\cos\theta$, $v = r\sin\theta$, find $\frac{\partial z}{\partial r}$ and $\frac{\partial z}{\partial \theta}$.

   (*b*) If $z = \log(u^2 + v)$, $u = e^{x^2+y^2}$, $v = x^2 + y$, find $\frac{\partial z}{\partial x}$ and $\frac{\partial z}{\partial y}$.

5. If $u = f(r, s)$, $r = x + y$, $s = x - y$, show that $\frac{\partial u}{\partial x} + \frac{\partial u}{\partial y} = 2\frac{\partial u}{\partial r}$.

6. If $z = e^{ax+by} f(ax - by)$, prove that $b\frac{\partial z}{\partial x} + a\frac{\partial z}{\partial y} = 2abz$.

7. If $x = u + v$, $y = uv$ and z is a function of x, y ; show that $u\dfrac{\partial z}{\partial u} + v\dfrac{\partial z}{\partial v} = x\dfrac{\partial z}{\partial x} + 2y\dfrac{\partial z}{\partial y}$.

8. If $u = f(r, s)$, $r = x + at$, $s = y + bt$, show that $\dfrac{\partial u}{\partial t} = a\dfrac{\partial u}{\partial x} + b\dfrac{\partial u}{\partial y}$.

9. If $u = x \log (xy)$, where $x^3 + y^3 + 3xy = 1$, find $\dfrac{du}{dx}$.

10. (*a*) If $u = f(r, s, t)$ and $r = \dfrac{x}{y}$, $s = \dfrac{y}{z}$, $t = \dfrac{z}{x}$, prove that $x\dfrac{\partial u}{\partial x} + y\dfrac{\partial u}{\partial y} + z\dfrac{\partial u}{\partial z} = 0$.

(*b*) If $x = u + v + w$, $y = vw + wu + uv$, $z = uvw$ and F is a function of x, y, z, show that

$$u\frac{\partial F}{\partial u} + v\frac{\partial F}{\partial v} + w\frac{\partial F}{\partial w} = x\frac{\partial F}{\partial x} + 2y\frac{\partial F}{\partial y} + 3z\frac{\partial F}{\partial z}.$$

(*c*) If $u = f(2x - 3y, 3y - 4z, 4z - 2x)$, prove that $\dfrac{1}{2}\dfrac{\partial u}{\partial x} + \dfrac{1}{3}\dfrac{\partial u}{\partial y} + \dfrac{1}{4}\dfrac{\partial u}{\partial z} = 0$.

(*d*) If $u = f(x^2 + 2yz, y^2 + 2zx)$, prove that $(y^2 - zx)\dfrac{\partial u}{\partial x} + (x^2 - yz)\dfrac{\partial u}{\partial y} + (z^2 - xy)\dfrac{\partial u}{\partial z} = 0$.

Answers

1. (*a*) $4a^2t(t^2 + 2)$ (*b*) $8e^{4t}$ **2.** $\dfrac{e^t(t-2)}{t^3}\cos\left(\dfrac{e^t}{t^2}\right)$

3. $-3a^3\cos^2 t \sin t + 3b^3 \sin^2 t \cos t$

4. (*a*) $2r, 0$ (*b*) $\dfrac{2x(2u^2+1)}{u^2+v}, \dfrac{4yu^2+1}{u^2+v}$

9. $1 + \log(xy) - \dfrac{x(x^2+y)}{y(y^2+x)}$

4.8. JACOBIANS

If u and v are functions of two independent variables x and y, then the determinant

$\begin{vmatrix} \dfrac{\partial u}{\partial x} & \dfrac{\partial u}{\partial y} \\ \dfrac{\partial v}{\partial x} & \dfrac{\partial v}{\partial y} \end{vmatrix}$ is called Jacobian of u, v with respect to x, y and is denoted by the symbol $J\left(\dfrac{u, v}{x, y}\right)$

or $\dfrac{\partial(u, v)}{\partial(x, y)}$.

Similarly, if u, v, w be functions of x, y, z, then the Jacobian of u, v, w with respect to x, y, z is

$$J\left(\frac{u, v, w}{x, y, z}\right) \text{ or } \frac{\partial(u, v, w)}{\partial(x, y, z)} = \begin{vmatrix} \dfrac{\partial u}{\partial x} & \dfrac{\partial u}{\partial y} & \dfrac{\partial u}{\partial z} \\ \dfrac{\partial v}{\partial x} & \dfrac{\partial v}{\partial y} & \dfrac{\partial v}{\partial z} \\ \dfrac{\partial w}{\partial x} & \dfrac{\partial w}{\partial y} & \dfrac{\partial w}{\partial z} \end{vmatrix}.$$

4.9. PROPERTIES OF JACOBIANS

I. *If u, v are functions of r, s where r, s are functions of x, y, then*

$$\frac{\partial(u, v)}{\partial(x, y)} = \frac{\partial(u, v)}{\partial(r, s)} \times \frac{\partial(r, s)}{\partial(x, y)}.$$

Proof. Since u, v are composite functions of x, y

$$\therefore \quad \left.\begin{aligned} \frac{\partial u}{\partial x} &= \frac{\partial u}{\partial r}\cdot\frac{\partial r}{\partial x} + \frac{\partial u}{\partial s}\cdot\frac{\partial s}{\partial x} = u_r r_x + u_s s_x \\ \frac{\partial u}{\partial y} &= \frac{\partial u}{\partial r}\cdot\frac{\partial r}{\partial y} + \frac{\partial u}{\partial s}\cdot\frac{\partial s}{\partial y} = u_r r_y + u_s s_y \\ \frac{\partial v}{\partial x} &= \frac{\partial v}{\partial r}\cdot\frac{\partial r}{\partial x} + \frac{\partial v}{\partial s}\cdot\frac{\partial s}{\partial x} = v_r r_x + v_s s_x \\ \frac{\partial v}{\partial y} &= \frac{\partial v}{\partial r}\cdot\frac{\partial r}{\partial y} + \frac{\partial v}{\partial s}\cdot\frac{\partial s}{\partial y} = v_r r_y + v_s s_y \end{aligned}\right] \quad \text{...(A)}$$

Now,
$$\frac{\partial(u, v)}{\partial(r, s)} \times \frac{\partial(r, s)}{\partial(x, y)} = \begin{vmatrix} u_r & u_s \\ v_r & v_s \end{vmatrix} . \begin{vmatrix} r_x & r_y \\ s_x & s_y \end{vmatrix}$$

Interchanging rows and columns in the second determinant

$$= \begin{vmatrix} u_r & u_s \\ v_r & v_s \end{vmatrix} . \begin{vmatrix} r_x & s_x \\ r_y & s_y \end{vmatrix} = \begin{vmatrix} u_r r_x + u_s s_x & u_r r_y + u_s s_y \\ v_r r_x + v_s s_x & v_r r_y + v_s s_y \end{vmatrix} = \begin{vmatrix} \frac{\partial u}{\partial x} & \frac{\partial u}{\partial y} \\ \frac{\partial v}{\partial x} & \frac{\partial v}{\partial y} \end{vmatrix} \quad | \text{ Using (A)}$$

$$= \frac{\partial(u, v)}{\partial(x, y)}.$$

II. *If J_1 is the Jacobian of u, v with respect to x, y and J_2 is the Jacobian of x, y, with respect to u, v, then $J_1 J_2 = 1$ i.e.,* $\dfrac{\partial(u, v)}{\partial(x, y)} \cdot \dfrac{\partial(x, y)}{\partial(u, v)} = 1.$

Proof. Let $u = u(x, y)$ and $v = v(x, y)$, so that u and v are functions of x, y.
Differentiating partially w.r.t. u and v, we get

$$\left.\begin{aligned} 1 &= \frac{\partial u}{\partial x}\cdot\frac{\partial x}{\partial u} + \frac{\partial u}{\partial y}\cdot\frac{\partial y}{\partial u} = u_x x_u + u_y y_u \\ 0 &= \frac{\partial u}{\partial x}\cdot\frac{\partial x}{\partial v} + \frac{\partial u}{\partial y}\cdot\frac{\partial y}{\partial v} = u_x x_v + u_y y_v \\ 0 &= \frac{\partial v}{\partial x}\cdot\frac{\partial x}{\partial u} + \frac{\partial v}{\partial y}\cdot\frac{\partial y}{\partial u} = v_x x_u + v_y y_u \\ 1 &= \frac{\partial v}{\partial x}\cdot\frac{\partial x}{\partial v} + \frac{\partial v}{\partial y}\cdot\frac{\partial y}{\partial v} = v_x x_v + v_y y_v \end{aligned}\right] \quad \text{...(A)}$$

Now,
$$\frac{\partial(u, v)}{\partial(x, y)} \cdot \frac{\partial(x, y)}{\partial(u, v)} = \begin{vmatrix} u_x & u_y \\ v_x & v_y \end{vmatrix} \begin{vmatrix} x_u & x_v \\ y_u & y_v \end{vmatrix}$$

Interchanging rows and columns in the second determinant

$$= \begin{vmatrix} u_x & u_y \\ v_x & v_y \end{vmatrix} . \begin{vmatrix} x_u & y_u \\ x_v & y_v \end{vmatrix} = \begin{vmatrix} u_x x_u + u_y y_u & u_x x_v + u_y y_v \\ v_x x_u + v_y y_u & v_x x_v + v_y y_v \end{vmatrix}$$

$$= \begin{vmatrix} 1 & 0 \\ 0 & 1 \end{vmatrix} = 1. \quad | \text{ Using (A)}$$

ILLUSTRATIVE EXAMPLES

Example 1. *If* $r = \sqrt{x^2 + y^2}$, $\theta = \tan^{-1} \frac{y}{x}$, *evaluate* $\frac{\partial(r, \theta)}{\partial(x, y)}$.

Sol. $r = \sqrt{x^2 + y^2}$, $\theta = \tan^{-1} \frac{y}{x}$

$$\frac{\partial r}{\partial x} = \frac{x}{\sqrt{x^2 + y^2}}, \qquad \frac{\partial r}{\partial y} = \frac{y}{\sqrt{x^2 + y^2}}$$

$$\frac{\partial \theta}{\partial x} = \frac{1}{1 + \frac{y^2}{x^2}}\left(-\frac{y}{x^2}\right) = -\frac{y}{x^2 + y^2}, \qquad \frac{\partial \theta}{\partial y} = \frac{1}{1 + \frac{y^2}{x^2}}\left(\frac{1}{x}\right) = \frac{x}{x^2 + y^2}$$

$$\frac{\partial(r, \theta)}{\partial(x, y)} = \begin{vmatrix} \frac{\partial r}{\partial x} & \frac{\partial r}{\partial y} \\ \frac{\partial \theta}{\partial x} & \frac{\partial \theta}{\partial y} \end{vmatrix} = \begin{vmatrix} \frac{x}{\sqrt{x^2 + y^2}} & \frac{y}{\sqrt{x^2 + y^2}} \\ -\frac{y}{x^2 + y^2} & \frac{x}{x^2 + y^2} \end{vmatrix}$$

$$= \frac{x^2}{(x^2 + y^2)^{3/2}} + \frac{y^2}{(x^2 + y^2)^{3/2}} = \frac{x^2 + y^2}{(x^2 + y^2)^{3/2}} = \frac{1}{\sqrt{x^2 + y^2}}.$$

Example 2. *If* $u = x^2 - 2y$, $v = x + y$, *prove that* $\frac{\partial(u, v)}{\partial(x, y)} = 2x + 2$.

Sol. $u = x^2 - 2y \Rightarrow \frac{\partial u}{\partial x} = 2x, \frac{\partial u}{\partial y} = -2$

$v = x + y \Rightarrow \frac{\partial v}{\partial x} = 1, \frac{\partial v}{\partial y} = 1$

$$\frac{\partial(u, v)}{\partial(x, y)} = \begin{vmatrix} u_x & u_y \\ v_x & v_y \end{vmatrix} = \begin{vmatrix} 2x & -2 \\ 1 & 1 \end{vmatrix} = 2x + 2.$$

Example 3. *If* $u = x(1 - y)$, $v = xy$, *prove that* $JJ' = 1$.

Sol. $u = x - xy \Rightarrow u_x = 1 - y, u_y = -x$ $\qquad \left(\because u_x = \frac{\partial u}{\partial x}\right)$

$v = xy \Rightarrow v_x = y, \quad v_y = x$

$$J = \frac{\partial(u, v)}{\partial(x, y)} = \begin{vmatrix} u_x & u_y \\ v_x & v_y \end{vmatrix} = \begin{vmatrix} 1 - y & -x \\ y & x \end{vmatrix}$$

$$= x(1 - y) + xy = x$$

From above, $u + v = x$ and $y = \frac{v}{x} = \frac{v}{u + v}$

Now, $x = u + v \Rightarrow \frac{\partial x}{\partial u} = 1, \frac{\partial x}{\partial v} = 1$

$$y = \frac{v}{u+v} \Rightarrow \frac{\partial y}{\partial u} = \frac{(u+v)(0) - v(1)}{(u+v)^2} = \frac{-v}{(u+v)^2}$$

and
$$\frac{\partial y}{\partial v} = \frac{(u+v)(1) - v(1)}{(u+v)^2} = \frac{u}{(u+v)^2}$$

$$J' = \frac{\partial(x, y)}{\partial(u, v)} = \begin{vmatrix} \frac{\partial x}{\partial u} & \frac{\partial x}{\partial v} \\ \frac{\partial y}{\partial u} & \frac{\partial y}{\partial v} \end{vmatrix} = \begin{vmatrix} 1 & 1 \\ \frac{-v}{(u+v)^2} & \frac{u}{(u+v)^2} \end{vmatrix}$$

$$= \frac{1}{(u+v)^2} \begin{vmatrix} 1 & 1 \\ -v & u \end{vmatrix} = \frac{1}{(u+v)^{\not 2}} \cancel{(u+v)} = \frac{1}{(u+v)} = \frac{1}{x}$$

Now, $$JJ' = x \cdot \frac{1}{x} = 1.$$

Example 4. *If* $u = xyz$, $v = x^2 + y^2 + z^2$, $w = x + y + z$, *find* $\frac{\partial(u, v, w)}{\partial(x, y, z)}$. (OU-Dec. 2012)

Sol.
$$u = xyz \Rightarrow u_x = yz,\ u_y = xz,\ u_z = xy$$
$$v = x^2 + y^2 + z^2 \Rightarrow v_x = 2x,\ v_y = 2y,\ v_z = 2z$$
$$w = x + y + z \Rightarrow w_x = 1,\ w_y = 1,\ w_z = 1$$

$$\frac{\partial(u, v, w)}{\partial(x, y, z)} = \begin{vmatrix} u_x & u_y & u_z \\ v_x & v_y & v_z \\ w_x & w_y & w_z \end{vmatrix}$$

$$= \begin{vmatrix} yz & xz & xy \\ 2x & 2y & 2z \\ 1 & 1 & 1 \end{vmatrix} = 2\begin{vmatrix} yz & xz & xy \\ x & y & z \\ 1 & 1 & 1 \end{vmatrix}$$

$$= 2\,[yz(y - z) - xz(x - z) + xy\,(x - y)]$$
$$= 2\,[y^2z - yz^2 - x^2z + xz^2 + x^2y - xy^2]$$
$$= 2\,[(y^2 - x^2)z + z^2\,(x - y) + xy(x - y)]$$
$$= 2(x - y)\,[-(x + y)z + z^2 + xy]$$
$$= 2(x - y)\,[-xz - yz + z^2 + xy]$$
$$= 2(x - y)\,[x(y - z) - z\,(y - z)]$$
$$= 2(x - y)\,(y - z)\,(x - z).$$

Example 5. *If* $x = r \sin\theta \cos\phi$, $y = r \sin\theta \sin\phi$, $z = r\cos\theta$, *show that* $\frac{\partial(x, y, z)}{\partial(r, \theta, \phi)} = r^2 \sin\theta$.

Sol. $$\frac{\partial(x, y, z)}{\partial(r, \theta, \phi)} = \begin{vmatrix} \frac{\partial x}{\partial r} & \frac{\partial x}{\partial \theta} & \frac{\partial x}{\partial \phi} \\ \frac{\partial y}{\partial r} & \frac{\partial y}{\partial \theta} & \frac{\partial y}{\partial \phi} \\ \frac{\partial z}{\partial r} & \frac{\partial z}{\partial \theta} & \frac{\partial z}{\partial \phi} \end{vmatrix} = \begin{vmatrix} \sin\theta\cos\phi & r\cos\theta\cos\phi & -r\sin\theta\sin\phi \\ \sin\theta\sin\phi & r\cos\theta\cos\phi & r\sin\theta\sin\phi \\ \cos\theta & -r\sin\theta & 0 \end{vmatrix}$$

Taking out common factors (r from second column and $r \sin\theta$ from third column)

$$= r^2 \sin\theta \begin{vmatrix} \sin\theta\cos\phi & \cos\theta\cos\phi & -\sin\phi \\ \sin\theta\sin\phi & \cos\theta\sin\phi & \cos\phi \\ \cos\theta & -\sin\theta & 0 \end{vmatrix}$$

Expanding by third row

$$= r^2 \sin\theta \left\{ \cos\theta \begin{vmatrix} \cos\theta\cos\phi & -\sin\phi \\ \cos\theta\sin\phi & \cos\phi \end{vmatrix} + \sin\theta \begin{vmatrix} \sin\theta\cos\phi & -\sin\phi \\ \sin\theta\sin\phi & \cos\phi \end{vmatrix} \right\}$$

$$= r^2 \sin\theta\, [\cos\theta\,(\cos\theta\cos^2\phi + \cos\theta\sin^2\phi) + \sin\theta\,(\sin\theta\cos^2\phi + \sin\theta\sin^2\phi)]$$

$$= r^2 \sin\theta\,(\cos^2\theta + \sin^2\theta) = r^2\sin\theta.$$

EXERCISE 3

1. If $x = r\cos\theta$, $y = r\sin\theta$, verify that $\dfrac{\partial(x, y)}{\partial(r, \theta)} \cdot \dfrac{\partial(r, \theta)}{\partial(x, y)} = 1.$

   $\left[\textbf{Hint. } r = \sqrt{x^2 + y^2},\ \theta = \tan^{-1}\dfrac{y}{x}.\right]$

2. If $x = r\cos\theta$, $y = r\sin\theta$, $z = z$, evaluate $\dfrac{\partial(x, y, z)}{\partial(r, \theta, z)}$.

3. If $u = \dfrac{yz}{x}$, $v = \dfrac{zx}{y}$ and $w = \dfrac{xy}{z}$, show that $\dfrac{\partial(u, v, w)}{\partial(x, y, z)} = 4$.

4. If $u = \dfrac{y^2}{2x}$, $v = \dfrac{x^2 + y^2}{2x}$, find $\dfrac{\partial(u, v)}{\partial(x, y)}$.

5. If $x = a\cos\xi\cosh\eta$, $y = a\sinh\xi\sin\eta$, show that $\dfrac{\partial(x, y)}{\partial(\xi, \eta)} = \dfrac{a^2}{2}(\cosh 2\xi - \cos 2\eta)$.

   $\left[\textbf{Hint. } \cosh^2 x = \dfrac{\cosh 2x + 1}{2},\ \sinh^2 x = \dfrac{\cosh 2x - 1}{2}\right]$

6. If $F = xu + v - y$, $G = u^2 + vy + w$, $H = zu - v + vw$, compute $\dfrac{\partial(F, G, H)}{\partial(u, w, v)}$.

7. If $u = \dfrac{x + y}{1 - xy}$ and $v = \tan^{-1} x + \tan^{-1} y$, find $\dfrac{\partial(u, v)}{\partial(x, y)}$.

8. If $u = x^2 - 2y$, $v = x + y + z$, $w = x - 2y + 3z$, find $\dfrac{\partial(u, v, w)}{\partial(x, y, z)}$.

9. If $u = x + y + z$, $uv = y + z$, $uvw = z$, show that $\dfrac{\partial(x, y, z)}{\partial(u, v, w)} = u^2 v$.

10. If $u = xyz$, $v = xy + yz + zx$, $w = x + y + z$, show that $\dfrac{\partial(u, v, w)}{\partial(x, y, z)} = (x - y)(y - z)(z - x)$.

11. If $u = \dfrac{x}{y - z}$, $v = \dfrac{y}{z - x}$, $w = \dfrac{z}{x - y}$, show that $\dfrac{\partial(u, v, w)}{\partial(x, y, z)} = 0$.

Answers

2. r **4.** $-\dfrac{y}{2x}$ **6.** $xw - x - xyv + 2uv - z$

7. 0 **8.** $10x + 4$.

4.10. TAYLOR'S THEOREM FOR A FUNCTION OF TWO VARIABLES

We know that by Taylor's theorem for a function $f(x)$ of single variable x,

$$f(x+h) = f(x) + hf'(x) + \frac{h^2}{2!}f''(x) + \frac{h^3}{3!}f'''(x) + \ldots$$

Now let $f(x, y)$ be a function of two independent variables x and y. **If y is kept constant,** then by Taylor's theorem for a function of single variable x, we have

$$f(x+h, \mathbf{y+k}) = f(x, \mathbf{y+k}) + h\frac{\partial}{\partial x}f(x, \mathbf{y+k}) + \frac{h^2}{2!}\frac{\partial^2}{\partial x^2}f(x, \mathbf{y+k}) + \frac{h^3}{3!}\frac{\partial^3}{\partial x^3}f(x, \mathbf{y+k}) + \ldots.. \quad \ldots(1)$$

Now **keeping x constant** and applying Taylor's theorem for a function of a single variable y, we have

$$f(\mathbf{x}, y+k) = f(\mathbf{x}, y) + k\frac{\partial}{\partial y}f(\mathbf{x}, y) + \frac{k^2}{2!}\frac{\partial^2}{\partial y^2}f(\mathbf{x}, \mathbf{y}) + \frac{k^3}{3!}\frac{\partial^3}{\partial y^3}f(\mathbf{x}, y) + \ldots\ldots \quad \ldots(2)$$

Using (2), we can write (1) as

$$\begin{aligned}
f(x+h, y+k) &= \left[f(x,y) + k\frac{\partial}{\partial y}f(x,y) + \frac{k^2}{2!}\frac{\partial^2}{\partial y^2}f(x,y) + \frac{k^3}{3!}\frac{\partial^3}{\partial y^3}f(x,y) + \ldots\ldots\right] \\
&\quad + h\frac{\partial}{\partial x}\left[f(x,y) + k\frac{\partial}{\partial y}f(x,y) + \frac{k^2}{2!}\frac{\partial^2}{\partial y^2}f(x,y) + \ldots\ldots\right] \\
&\quad + \frac{h^2}{2!}\frac{\partial^2}{\partial x^2}\left[f(x,y) + k\frac{\partial}{\partial y}f(x,y) + \ldots\ldots\right] + \frac{h^3}{3!}\frac{\partial^3}{\partial x^3}[f(x,y) + \ldots.] + \ldots\ldots \\
&= \left[f(x,y) + k\frac{\partial f}{\partial y} + \frac{k^2}{2!}\frac{\partial^2 f}{\partial y^2} + \frac{k^3}{3!}\frac{\partial^3 f}{\partial y^3} + \ldots\ldots\right] \\
&\quad + \left[h\frac{\partial f}{\partial x} + hk\frac{\partial^2 f}{\partial x\partial y} + \frac{hk^2}{2!}\frac{\partial^3 f}{\partial x\partial y^2} + \ldots\ldots\right] \\
&\quad + \left[\frac{h^2}{2!}\frac{\partial^2 f}{\partial x^2} + \frac{h^2k}{2!}\frac{\partial^3 f}{\partial x^2\partial y} + \ldots\ldots\right] + \frac{h^3}{3!}\frac{\partial^3 f}{\partial x^3} + \ldots\ldots \\
&= f(x,y) + \left(h\frac{\partial f}{\partial x} + k\frac{\partial f}{\partial y}\right) + \left(\frac{h^2}{2!}\frac{\partial^2 f}{\partial x^2} + hk\frac{\partial^2 f}{\partial x\partial y} + \frac{k^2}{2!}\frac{\partial^2 f}{\partial y^2}\right) \\
&\quad + \left(\frac{h^3}{3!}\frac{\partial^3 f}{\partial x^3} + \frac{h^2k}{2!}\frac{\partial^3 f}{\partial x^2\partial y} + \frac{hk^2}{2!}\frac{\partial^3 f}{\partial x\partial y^2} + \frac{k^3}{3!}\frac{\partial^3 f}{\partial y^3}\right) + \ldots. \\
&= f(x,y) + \left(h\frac{\partial f}{\partial x} + k\frac{\partial f}{\partial y}\right) + \frac{1}{2!}\left(h^2\frac{\partial^2 f}{\partial x^2} + 2hk\frac{\partial^2 f}{\partial x\partial y} + k^2\frac{\partial^2 f}{\partial y^2}\right) \\
&\quad + \frac{1}{3!}\left(h^3\frac{\partial^3 f}{\partial x^3} + 3h^2k\frac{\partial^3 f}{\partial x^2\partial y} + 3hk^2\frac{\partial^3 f}{\partial x\partial y^2} + k^3\frac{\partial^3 f}{\partial y^3}\right) + \ldots.
\end{aligned}$$

$$= f(x, y) + \left(h\frac{\partial}{\partial x} + k\frac{\partial}{\partial y}\right)f + \frac{1}{2!}\left(h\frac{\partial}{\partial x} + k\frac{\partial}{\partial y}\right)^2 f + \frac{1}{3!}\left(h\frac{\partial}{\partial x} + k\frac{\partial}{\partial y}\right)^3 f + ...$$

Cor. 1. Putting $x = a$ and $y = b$, we have

$$f(a + h, b + k) = f(a, b) + [hf_x(a, b) + kf_y(a, b)] + \frac{1}{2!}[h^2 f_{xx}(a, b) + 2hkf_{xy}(a, b) + k^2 f_{yy}(a, b)]$$
$$+ \frac{1}{3!}[h^3 f_{xxx}(a, b) + 3h^2 k f_{xxy}(a, b) + 3hk^2 f_{xyy}(a, b) + k^3 f_{yyy}(a, b)] + ...$$

Cor. 2. In Cor. 1, putting $a + h = x$ and $b + k = y$ so that $h = x - a$ and $k = y - b$, we have

$$f(x, y) = f(a, b) + [(x - a) f_x(a, b) + (y - b) f_y(a, b)] + \frac{1}{2!}[(x - a)^2 f_{xx}(a, b) + 2(x - a)(y - b)$$
$$f_{xy}(a, b) + (y - b)^2 f_{yy}(a, b)] + ...$$

Cor. 3. Putting $a = 0$, $b = 0$ in Cor. 2, we have

$$f(x, y) = f(0, 0) + [xf_x(0, 0) + yf_y(0, 0)] + \frac{1}{2!}[x^2 f_{xx}(0, 0) + 2xyf_{xy}(0, 0) + y^2 f_{yy}(0, 0)] + ...$$

This is called Maclaurin's theorem for two variables.

Note. Cor. 3 is used to expand $f(x, y)$ in powers of x and y [or to expand $f(x, y)$ in the neighbourhood of origin (0, 0)].

Cor. 2 is used to expand $f(x, y)$ in the neighbourhood of (a, b).

ILLUSTRATIVE EXAMPLES

Example 1. *Expand $e^x \sin y$ in powers of x and y as far as terms of the third degree.*

Sol. Here

$f(x, y) = e^x \sin y$; $\quad f(0, 0) = 0$

$f_x(x, y) = e^x \sin y$, $\quad f_x(0, 0) = 0$

$f_y(x, y) = e^x \cos y$, $\quad f_y(0, 0) = 1$

$f_{xx}(x, y) = e^x \sin y$, $\quad f_{xx}(0, 0) = 0$

$f_{xy}(x, y) = e^x \cos y$, $\quad f_{xy}(0, 0) = 1$

$f_{yy}(x, y) = - e^x \sin y$, $\quad f_{yy}(0, 0) = 0$

$f_{xxx}(x, y) = e^x \sin y$, $\quad f_{xxx}(0, 0) = 0$

$f_{xxy}(x, y) = e^x \cos y$, $\quad f_{xxy}(0, 0) = 1$

$f_{xyy}(x, y) = - e^x \sin y$, $\quad f_{xyy}(0, 0) = 0$

$f_{yyy}(x, y) = - e^x \cos y$, $\quad f_{yyy}(0, 0) = - 1$

....

$\therefore e^x \sin y = f(x, y)$

$$= f(0, 0) + [xf_x(0, 0) + yf_y(0, 0)] + \frac{1}{2!}[x^2 f_{xx}(0, 0) + 2xyf_{xy}(0, 0) + y^2 f_{yy}(0, 0)]$$
$$+ \frac{1}{3!}[x^3 f_{xxx}(0, 0) + 3x^2 y f_{xxy}(0, 0) + 3xy^2 f_{xyy}(0, 0) + y^3 f_{yyy}(0, 0)] + ...$$
$$= 0 + [x \cdot 0 + y \,.\, 1] + \frac{1}{2!}[x^2 \cdot 0 + 2xy \cdot 1 + y^2 \cdot 0] + \frac{1}{3!}[x^3 \cdot 0 + 3x^2 y \cdot 1 + 3xy^2 \cdot 0$$
$$+ y^3(- 1)] + ...$$

$$= y + xy + \frac{1}{2} x^2y - \frac{1}{6} y^3 + ...$$

Example 2. *Expand* $\tan^{-1} \frac{y}{x}$ *in the neighbourhood of (1, 1).*

Sol. Here $f(x, y) = \tan^{-1} \frac{y}{x}$, $\qquad f(1, 1) = \tan^{-1} 1 = \frac{\pi}{4}$

$$f_x(x, y) = \frac{1}{1+\frac{y^2}{x^2}} \cdot \left(-\frac{y}{x^2}\right) = -\frac{y}{x^2+y^2}, \qquad f_x(1, 1) = -\frac{1}{2}$$

$$f_y(x, y) = \frac{1}{1+\frac{y^2}{x^2}} \cdot \frac{1}{x} = \frac{x}{x^2+y^2}, \qquad f_y(1, 1) = \frac{1}{2}$$

$$f_{xx}(x, y) = -y(-1)(x^2+y^2)^{-2} \cdot 2x = \frac{2xy}{(x^2+y^2)^2}, \qquad f_{xx}(1, 1) = \frac{1}{2}$$

$$f_{xy}(x, y) = \frac{(x^2+y^2)\cdot 1 - x \cdot 2x}{(x^2+y^2)^2} = \frac{y^2-x^2}{(x^2+y^2)^2}, \qquad f_{xy}(1, 1) = 0$$

$$f_{yy}(x, y) = x(-1)(x^2+y^2)^{-2} \cdot 2y = -\frac{2xy}{(x^2+y^2)^2}, \qquad f_{yy}(1, 1) = -\frac{1}{2}$$

$$\therefore \quad \tan^{-1} \frac{y}{x} = f(x, y) = f(1, 1) + [(x-1) f_x(1, 1) + (y-1) f_y(1, 1)]$$

$$+ \frac{1}{2!} [(x-1)^2 f_{xx}(1, 1) + 2(x-1)(y-1) f_{xy}(1, 1) + (y-1)^2 f_{yy}(1, 1)] + ...$$

$$= \frac{\pi}{4} + \left[(x-1)\cdot\left(-\frac{1}{2}\right) + (y-1)\frac{1}{2}\right]$$

$$+ \frac{1}{2!}\left[(x-1)^2 \cdot \frac{1}{2} + 2(x-1)(y-1)\cdot 0 + (y-1)^2 \cdot \left(-\frac{1}{2}\right)\right] + ...$$

$$= \frac{\pi}{4} - \frac{1}{2}(x-1) + \frac{1}{2}(y-1) + \frac{1}{4}(x-1)^2 - \frac{1}{4}(y-1)^2 + ...$$

Example 3. *Find the first six terms of the expansion of the function* $e^x \log(1 + y)$ *in a Taylor series in the neighbourhood of the point (0, 0).* (OU-May 2009, Nov. 2006)

Sol. Here $f(x, y) = e^x \log(1 + y)$, $\qquad f(0, 0) = 0$

$$f_x(x, y) = e^x \log(1 + y), \qquad f_x(0, 0) = 0$$

$$f_y(x, y) = \frac{e^x}{1+y}, \qquad f_y(0, 0) = 1$$

$$f_{xx}(x, y) = e^x \log(1 + y), \qquad f_{xx}(0, 0) = 0$$

$$f_{xy}(x, y) = \frac{e^x}{1+y}, \qquad f_{xy}(0, 0) = 1$$

$$f_{yy}(x, y) = -\frac{e^x}{(1+y)^2}, \qquad f_{yy}(0, 0) = -1$$

$$f_{xxx}(x, y) = e^x \log(1 + y), \qquad f_{xxx}(0, 0) = 0$$

$$f_{xxy}(x, y) = \frac{e^x}{1+y}, \qquad f_{xxy}(0, 0) = 1$$

$$f_{xyy}(x, y) = -\frac{e^x}{(1+y)^2}, \qquad f_{xyy}(0, 0) = -1$$

$$f_{yyy}(x, y) = \frac{2e^x}{(1+y)^3}, \qquad f_{yyy}(0, 0) = 2$$

....

$\therefore \quad e^x \log(1 + y) = f(x, y)$

$$= f(0, 0) + [xf_x(0, 0) + yf_y(0, 0)] + \frac{1}{2!}[x^2 f_{xx}(0, 0) + 2xyf_{xy}(0, 0) + y^2 f_{yy}(0, 0)]$$

$$+ \frac{1}{3!}[x^3 f_{xxx}(0, 0) + 3x^2y f_{xxy}(0, 0) + 3xy^2 f_{xyy}(0, 0) + y^3 f_{yyy}(0, 0)] + ...$$

$$= 0 + [x.0 + y.1] + \frac{1}{2}[x^2.0 + 2xy.1 + y^2(-1)] + \frac{1}{6}[x^3.0 + 3x^2y.1 + 3xy^2.(-1) + y^3.2] + ...$$

$$= y + xy - \frac{1}{2}y^2 + \frac{1}{2}x^2y - \frac{1}{2}xy^2 + \frac{1}{3}y^3 ......$$

Example 4. *Expand* $\frac{(x+h)(y+k)}{x+h+y+k}$ *in powers of h, k upto and inclusive of the second degree terms.*

Sol. Here $$f(x+h, y+k) = \frac{(x+h)(y+k)}{x+h+y+k}$$

Putting $h = k = 0$, we have $f(x, y) = \frac{xy}{x+y}$

$$f_x = \frac{(x+y)\cdot y - xy\cdot 1}{(x+y)^2} = \frac{y^2}{(x+y)^2}\ ; \qquad f_y = \frac{x^2}{(x+y)^2}, \text{ by symmetry}$$

$$f_{xx} = -\frac{2y^2}{(x+y)^3}, \; f_{yy} = -\frac{2x^2}{(x+y)^3}$$

$$f_{xy} = \frac{(x+y)^2\cdot 2x - x^2\cdot 2(x+y)}{(x+y)^4} = \frac{2x(x+y) - 2x^2}{(x+y)^3} = \frac{2xy}{(x+y)^3}$$

$$\therefore \quad \frac{(x+h)(y+k)}{x+h+y+k} = f(x+h, y+k) = f(x, y) + [hf_x + kf_y] + \frac{1}{2!}[h^2 f_{xx} + 2hk\, f_{xy} + k^2 f_{yy}] + ...$$

$$= \frac{xy}{x+y} + \left[h\cdot\frac{y^2}{(x+y)^2} + k\cdot\frac{x^2}{(x+y)^2}\right]$$

$$+ \frac{1}{2}\left[h^2\cdot\frac{-2y^2}{(x+y)^3} + 2hk\frac{2xy}{(x+y)^3} + k^2\cdot\frac{-2x^2}{(x+y)^3}\right] + ...$$

$$= \frac{xy}{x+y} + \frac{y^2}{(x+y)^2}\cdot h + \frac{x^2}{(x+y)^2}\cdot k - \frac{y^2}{(x+y)^3}\cdot h^2$$

$$+ \frac{2xy}{(x+y)^3}\cdot hk - \frac{x^2}{(x+y)^3}\cdot k^2 + ...$$

Example 5. *Expand $x^2y + 3y - 2$ in powers of $(x - 1)$ and $(y + 2)$ using Taylor's Theorem.*

Sol. Expansion of $f(x, y)$ in powers of $(x - a)$ and $(y - b)$ is given by

$$f(x, y) = f(a, b) + [(x - a)f_x(a, b) + (y - b)f_y(a, b)] + \frac{1}{2!}[(x - a)^2 f_{xx}(a, b)$$

$$+ 2(x - a)(y - b) f_{xy}(a, b) + (y - b)^2 f_{yy}(a, b)] + \frac{1}{3!}[(x - a)^3 f_{xxx}(a, b)$$

$$+ 3(x - a)^2(y - b)f_{xxy}(a, b) + 3(x - a)(y - b)^2 f_{xyy}(a, b) + (y - b)^3 f_{yyy}(a, b)]$$

$$+ \ldots \qquad \ldots(1)$$

Here $f(x, y) = x^2y + 3y - 2$, $a = 1, b = -2$

$f(1, -2) = 1^2 \times (-2) + 3(-2) - 2 = -10$

$f_x = 2xy$, $f_x(1, -2) = 2(1)(-2) = -4$; $f_y = x^2 + 3$, $f_y(1, -2) = 1^2 + 3 = 4$

$f_{xx} = 2y$, $f_{xx}(1, -2) = 2(-2) = -4$; $f_{xy} = 2x$, $f_{xy}(1, -2) = 2(1) = 2$

$f_{yy} = 0$, $f_{yy}(1, -2) = 0$; $f_{xxx} = 0$, $f_{xxx}(1, -2) = 0$

$f_{xxy} = 2$, $f_{xxy}(1, -2) = 2$; $f_{xyy} = 0$, $f_{xyy}(1, -2) = 0$

$f_{yyy} = 0$, $f_{yyy}(1, -2) = 0$

All higher order partial derivatives vanish.

$\therefore$ From (1), we have

$$x^2y + 3y - 1 = f(x, y)$$

$$= -10 + [(x - 1)(-4) + (y + 2)(4)] + \frac{1}{2}[(x - 1)^2(-4) + 2(x - 1)(y + 2)(2) + (y + 2)^2(0)]$$

$$+ \frac{1}{6}[(x - 1)^3(0) + 3(x - 1)^2(y + 2)(2) + 3(x - 1)(y + 2)^2(0) + (y + 2)^3(0)]$$

$$= -10 - 4(x - 1) + 4(y + 2) - 2(x - 1)^2 + 2(x - 1)(y + 2) + (x - 1)^2(y + 2).$$

Example 6. *Expand $f(x, y) = x^2 + 3y^2 - 9x - 9y + 26$ in Taylor series of maximum order about (2, 2).* (OU-June 2014)

Sol. Taylor series about (2, 2) is

$$f(x, y) = f(2, 2) + [(x - 2)f_x(2, 2) + (y - 2)f_y(2, 2)] + \frac{1}{2!}[(x - 2)^2 f_{xx}(2, 2)$$

$$+ 2(x - 2)(y - 2)f_{xy}(2, 2) + (y - 2)^2 f_{yy}(2, 2)] + \ldots \qquad \ldots(1)$$

Now, $f(x, y) = x^2 + 3y^2 - 9x - 9y + 26 \Rightarrow f(2, 2) = 6$

$f_x(x, y) = 2x - 9 \Rightarrow f_x(2, 2) = -5$

$f_y(x, y) = 6y - 9 \Rightarrow f_y(2, 2) = 3$

$f_{xx}(x, y) = 2 \Rightarrow f_{xx}(2, 2) = 2$

$f_{xy}(x, y) = 0 \Rightarrow f_{xy}(2, 2) = 0$

$f_{yy}(x, y) = 6 \Rightarrow f_{yy}(2, 2) = 6$

Substitute these values in (1), we get

$$f(x, y) = 6 + [(x - 2)(-5) + (y - 2)(3)] + \frac{1}{2!}[(x - 2)^2(2) + (y - 2)^2(6)]$$

EXERCISE 4

1. Show that $e^y \log(1 + x) = x + xy - \dfrac{x^2}{2}$ approximately. *(OU-Dec. 2007)*

   [**Hint.** Find the expansion at (0, 0)]
2. Expand $e^x \cos y$ in powers of x and y as far as the terms of third degree.
3. Expand $e^{ax} \sin by$ in powers of x and y as far as the terms of third degree.
4. Expand e^{xy} at (1, 1).
5. Expand $e^x \cos y$ at $\left(1, \dfrac{\pi}{4}\right)$.
6. Expand $(1 + x + y^2)^{1/2}$ at (1, 0).
7. Expand $\sin(x + h)(y + k)$ by Taylor's Theorem.
8. Obtain the expansion of $\tan^{-1}\left(\dfrac{y}{x}\right)$ about (1, 1) upto the third degree terms.

Answers

2. $1 + x + \dfrac{1}{2!}(x^2 - y^2) + \dfrac{1}{3!}(x^3 - 3xy^2) + \ldots..$ 3. $by + abxy + \dfrac{1}{3!}(3a^2 bx^2y - b^3y^3) + \ldots..$

4. $e\left\{1 + (x-1) + (y-1) + \dfrac{1}{2!}\left((x-1)^2 + 4(x-1)(y-1) + (y-1)^2\right) + \ldots...\right\}$

5. $\dfrac{e}{\sqrt{2}}\left\{1 + (x-1) - \left(y - \dfrac{\pi}{4}\right) + \dfrac{(x-1)^2}{2} - (x-1)\left(y - \dfrac{\pi}{4}\right) - \dfrac{1}{2}\left(y - \dfrac{\pi}{4}\right)^2 + \ldots.\right\}$

6. $\sqrt{2}\left\{1 + \dfrac{x-1}{4} - \dfrac{(x-1)^2}{32} + \dfrac{y^2}{4} + \ldots\right\}$

7. $\sin xy + (hy + kx)\cos xy + hk\cos xy - \dfrac{1}{2}(hy + kx)^2 \sin xy + \ldots...$

8. $\dfrac{\pi}{4} - \dfrac{1}{2}(x-1) + \dfrac{1}{2}(y-1) + \dfrac{1}{4}(x-1)^2 - \dfrac{1}{4}(y-1)^2 - \dfrac{1}{12}(x-1)^3$

   $- \dfrac{1}{4}(x-1)^2(y-1) + \dfrac{1}{4}(x-1)(y-1)^2 + \dfrac{1}{12}(y-1)^3 + \ldots$

4.11. MAXIMA AND MINIMA OF FUNCTIONS OF TWO VARIABLES

A function $f(x, y)$ is said to have a **maximum value at x = a, y = b if f(a, b) > f(a + h, b + k),** for small and independent values of h and k, positive or negative.

A function $f(x, y)$ is said to have a **minimum value at x = a, y = b** if $f(a, b) < f(a + h, b + k)$, for small and independent values of h and k, positive or negative.

Thus, $f(x, y)$ has a maximum or minimum value at a point (a, b) according as

$$\Delta f = f(a + h, b + k) - f(a, b) < \text{or} > 0.$$

Geometrically, the surface $z = f(x, y)$ has a maximum at the point (a, b) if $f(a, b)$ is the greatest in the small neighbourhood of the point (a, b). As such, the surface descends in all directions at this point. The point of maxima may well be compared with the highest point of a dome. The surface has a minimum at the point (a, b) if $f(a, b)$ is the smallest in the small neighbourhood of the point (a, b). As such, the surface ascends in all directions at this point. The point of minima may well be compared with the lowest point of a bowl. However, sometimes, at the point (a, b), the tangent plane is horizontal and the surface descends in certain directions and ascends in other directions. Such a point is called a *saddle point*.

A maximum or a minimum value of a function is called its *extreme value*.

Note.

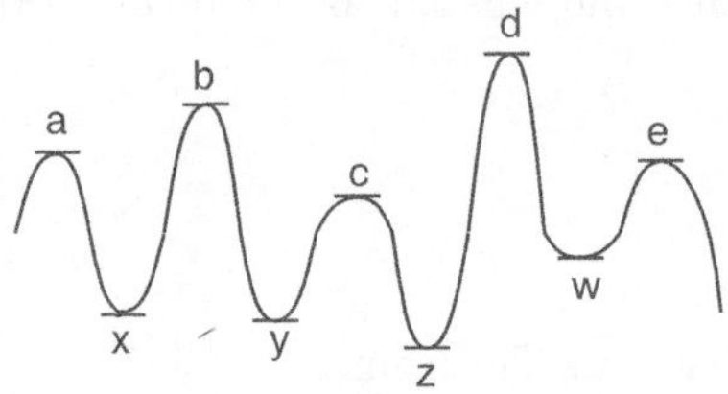

Here local maximum exists at a, b, c, d, e and global maximum occurs at d. Similarly local minimum exists at x, y, z, w, and global minimum at z.

RULE TO FIND THE EXTREME VALUES OF A FUNCTION $f(x)$:

(*i*) Solve $f'(x) = 0$, find x

(*ii*) At each x, find $f''(x)$

(*a*) If $f''(x) > 0$, then $f(x)$ has minimum value at x.

(*b*) If $f''(x) < 0$, then $f(x)$ has maximum value at x.

Example. *Find the local maximum and minimum values of the function* $f(x) = 3x^4 - 2x^3 - 6x^2 + 6x + 1$ *in* $[0, 2]$. (OU-May 2015)

Sol. Take $f'(x) = 0 \Rightarrow 12x^3 - 6x^2 - 12x + 6 = 0$ (Dividing by 6)

$$2x^3 - x^2 - 2x + 1 = 0$$

By synthetic division

$x = 1$	2	–1	–2	1
	0	2	1	–1
$x = -1$	2	1	–1	0
	0	–2	1	
	2	–1	0	

$$2x - 1 = 0 \Rightarrow x = \frac{1}{2}$$

We have, $f''(x) = 36x^2 - 12x - 12$

At $x = 1$, $f''(1) = 36 - 12 - 12 = 12 > 0$

$f(x)$ has local minimum value at $x = 1$

local minimum value of

$$f = 3 - 2 - 6 + 6 + 1 = 2 \quad (\because \text{ Use given } f(x))$$

At $x = -1$, $f''(-1) = 36 + 12 - 12 = 36 > 0$

$f(x)$ has local minimum at $x = -1$

local minima of $f = 3 + 2 - 6 - 6 + 1 = -6$

At $x = \frac{1}{2}$, $f''\left(\frac{1}{2}\right) = \frac{36}{4} - \frac{12}{2} - 12 = 9 - 6 - 12 = -9 < 0$

$f(x)$ has maximum value at $x = \dfrac{1}{2}$

local maxima of $\quad f = \dfrac{3}{16} - \dfrac{2}{8} - \dfrac{6}{4} + \dfrac{6}{2} + 1 = \dfrac{3 - 4 - 24 + 48 + 16}{16} = \dfrac{39}{16}.$

4.12. CONDITIONS FOR f(x, y) TO BE MAXIMUM OR MINIMUM

By Taylor's theorem, we have

$$\Delta f = f(a + h, b + k) - f(a, b)$$

$$= [hf_x(a, b) + kf_y(a, b)] + \frac{1}{2!}[h^2 f_{xx}(a, b) + 2hkf_{xy}(a, b) + k^2 f_{yy}(a, b)] + \ldots\ldots \quad \ldots(1)$$

For small values of h and k, the second and higher order terms are still smaller and may be neglected.

Thus, $\quad$ sign of Δf = sign of $[hf_x(a, b) + kf_y(a, b)]$

Taking $h = 0$, the sign of Δf changes with the sign of k. Similarly taking $k = 0$, the sign of Δf changes with the sign of h. Since Δf changes sign with h and k, $f(x, y)$ cannot have a maximum or a minimum value at (a, b) unless $f_x(a, b) = 0 = f_y(a, b)$.

Hence the necessary conditions for (x, y) to have a maximum or a minimum value at (a, b) are

$$f_x(a, b) = 0, f_y(a, b) = 0$$

If these conditions are satisfied, then for small values of h and k, we have, from (1),

$$\Delta f = \frac{1}{2!}[h^2 f_{xx}(a, b) + 2hkf_{xy}(a, b) + k^2 f_{yy}(a, b)] + \ldots\ldots$$

$$= \frac{1}{2!}[h^2 r + 2hks + k^2 t] + \ldots\text{where } r = f_{xx}(a, b),\ s = f_{xy}(a, b),\ t = f_{yy}(a, b)$$

or $$\Delta f = \frac{1}{2r}[h^2 r^2 + 2hkrs + k^2 rt] + \ldots\ldots = \frac{1}{2r}[(h^2 r^2 + 2hkrs + k^2 s^2) + k^2 rt - k^2 s^2)] + \ldots$$

$$= \frac{1}{2r}[(hr + ks)^2 + k^2(rt - s^2)] + \ldots\ldots \quad \ldots(2)$$

Now $(hr + ks)^2$ is always positive and $k^2(rt - s^2)$ will be positive if $rt - s^2 > 0$. In this case, Δf will have the same sign as that of r for all values of h and k.

Hence if $rt - s^2 > 0$, then $f(x, y)$ has a maximum or a minimum at (a, b) according as $r < 0$ or $r > 0$.

If $rt - s^2 < 0$, then Δf changes sign with h and k. Hence there is neither a maximum nor a minimum value at (a, b). The point (a, b) is a saddle point in this case.

If $rt - s^2 = 0$, no conclusion can be drawn about a maximum or a minimum value at (a, b) and hence further investigation is required.

Note. The point (a, b) is called a *stationary point* if $f_x(a, b) = 0$, $f_y(a, b) = 0$. The value $f(a, b)$ is called a *stationary value.* Thus every extreme value is a stationary value but the converse may not be true.

4.13. RULE TO FIND THE EXTREME VALUES OF A FUNCTION z = f(x, y)

(*i*) Find $\dfrac{\partial z}{\partial x}$ and $\dfrac{\partial z}{\partial y}$.

(*ii*) Solve $\frac{\partial z}{\partial x} = 0$ and $\frac{\partial z}{\partial y} = 0$ simultaneously.

Let (a, b); (c, d) ...... be the solutions of these equations.

(*iii*) For each solution in step (*ii*), find $r = \frac{\partial^2 f}{\partial x^2}$, $s = \frac{\partial^2 f}{\partial x \partial y}$, $t = \frac{\partial^2 f}{\partial y^2}$.

(*iv*) (*a*) If $rt - s^2 > 0$ and $r < 0$ for a particular solution (a, b) of step (*ii*), then z has a maximum value at (a, b).

(*b*) If $rt - s^2 > 0$ and $r > 0$ for a particular solution (a, b) of step (*ii*), then z has a minimum value at (a, b).

(*c*) If $rt - s^2 < 0$ for a particular solution (a, b) of step (*ii*), then z has no extreme value at (a, b).

(*d*) If $rt - s^2 = 0$, the case is doubtful and requires further investigation.

ILLUSTRATIVE EXAMPLES

Example 1. *Examine the function $x^3 + y^3 - 3axy$ for maxima and minima.*

(OU-May 2012, June 2010)

Sol. Here
$$f(x, y) = x^3 + y^3 - 3axy$$
$$f_x = 3x^2 - 3ay, \quad f_y = 3y^2 - 3ax,$$
$$r = f_{xx} = 6x, \quad s = f_{xy} = -3a, \quad t = f_{yy} = 6y$$

Now, $f_x = 0$ and $f_y = 0$

$$\Rightarrow \quad x^2 - ay = 0 \quad \text{...(1)}$$
$$y^2 - ax = 0 \quad \text{...(2)}$$

From (1), $y = \frac{x^2}{a}$

$\therefore$ From (2), $\frac{x^4}{a^2} - ax = 0$ or $x(x^3 - a^3) = 0$ or $x = 0, a$

when $x = 0$, $y = 0$; when $x = a$, $y = a$

$\therefore$ There are two stationary points $(0, 0)$ and (a, a).

Now, $rt - s^2 = 36xy - 9a^2$

At (0, 0), $rt - s^2 = -9a^2 < 0$

$\Rightarrow$ There is no extreme value at $(0, 0)$.

At (a, a), $rt - s^2 = 36a^2 - 9a^2 = 27a^2 > 0$

$\Rightarrow$ $f(x, y)$ has extreme value at (a, a)

Now, $r = 6a$

If $a > 0$, $r > 0$ so that $f(x, y)$ has a minimum value at (a, a).

Minimum value $= f(a, a) = a^3 + a^3 - 3a^3 = -a^3$

If $a < 0$, $r < 0$ so that $f(x, y)$ has a maximum value at (a, a).

Maximum value $= f(-a, -a) = -a^3 - a^3 + 3a^3 = a^3$.

Example 2. *Discuss the maxima and minima of $x^3y^2(1 - x - y)$.*

Sol. Here $f(x, y) = x^3y^2(1 - x - y) = x^3y^2 - x^4y^2 - x^3y^3$

$$f_x = 3x^2y^2 - 4x^3y^2 - 3x^2y^3,\ f_y = 2x^3y - 2x^4y - 3x^3y^2$$

$$r = f_{xx} = 6xy^2 - 12x^2y^2 - 6xy^3\ ;\ s = f_{xy} = 6x^2y - 8x^3y - 9x^2y^2\ ;\ t = f_{yy} = 2x^3 - 2x^4 - 6x^3y$$

Now, $\quad f_x = 0$ and $f_y = 0$

$\Rightarrow \quad x^2y^2(3 - 4x - 3y) = 0 \quad ...(1) \qquad x^3y(2 - 2x - 3y) = 0 \quad ...(2)$

Solving (1) and (2), the stationary points are (0, 0) and $\left(\frac{1}{2}, \frac{1}{3}\right)$.

Now, $\quad rt - s^2 = 6xy^2(1 - 2x - y)\,2x^3(1 - x - 3y) - [x^2y(6 - 8x - 9y)]^2$

$$= 12x^4y^2(1 - 2x - y)(1 - x - 3y) - x^4y^2(6 - 8x - 9y)^2$$

At (0, 0), $\quad rt - s^2 = 0$

$\therefore$ At (0, 0) further investigation is needed.

Consider $f(a + h, b + k) - f(a, b) = f(h, k) - f(0, 0) = h^3k^2(1 - h - k)$

$= h^3k^2$ (neglecting $h^4k^2 + h^3k^3$, which is very small as compared to h^3k^2)

which is > 0 if $h > 0$ and < 0 if $h < 0$

$\therefore$ $f(a + h, b + k) - f(a, b)$ does not keep same sign for all small values of h and k (positive or negative).

$\therefore$ There is no extreme value at (0, 0).

At $\left(\frac{1}{2}, \frac{1}{3}\right)$, $rt - s^2 = 12\cdot\frac{1}{16}\cdot\frac{1}{9}\left(-\frac{1}{3}\right)\left(-\frac{1}{2}\right) - \frac{1}{16}\cdot\frac{1}{9}(1) = \frac{1}{144}(2 - 1) = \frac{1}{14} > 0.$

Also, $\quad r = 6\cdot\frac{1}{2}\cdot\frac{1}{9} - 12\cdot\frac{1}{4}\cdot\frac{1}{9} - 6\cdot\frac{1}{2}\cdot\frac{1}{8} = \frac{1}{3} - \frac{1}{3} - \frac{3}{8} = -\frac{3}{8} < 0.$

$\therefore$ $f(x, y)$ has a maximum value at $\left(\frac{1}{2}, \frac{1}{3}\right)$

Maximum value $= f\left(\frac{1}{2}, \frac{1}{3}\right) = \frac{1}{8}\cdot\frac{1}{9}\left(1 - \frac{1}{2} - \frac{1}{3}\right) = \frac{1}{432}$.

Example 3. *Locate the stationary points of $x^4 + y^4 - 2x^2 + 4xy - 2y^2$ and determine their nature.*

Sol. Here $\quad f(x, y) = x^4 + y^4 - 2x^2 + 4xy - 2y^2$

$$f_x = 4x^3 - 4x + 4y,\quad f_y = 4y^3 + 4x - 4y$$

$$r = f_{xx} = 12x^2 - 4,\quad s = f_{xy} = 4,\ t = f_{yy} = 12y^2 - 4$$

Now, $\quad f_x = 0$ and $f_y = 0$

$\Rightarrow \quad x^3 - x + y = 0 \quad ...(1)$

$\quad y^3 + x - y = 0 \quad ...(2)$

Adding (1) and (2), $\quad x^3 + y^3 = 0 \Rightarrow y = -x.$

Putting $y = -x$ in (1), we get $x^3 - 2x = 0$ or $x(x^2 - 2) = 0 \quad \therefore \quad x = 0, \pm\sqrt{2}$.

Since $y = -x$, the stationary points are (0, 0), $(\sqrt{2}, -\sqrt{2})$ and $(-\sqrt{2}, \sqrt{2})$.

Now, $\quad rt - s^2 = (12x^2 - 4)(12y^2 - 4) - 16$

At (0, 0), $\quad rt - s^2 = 16 - 16 = 0.$

$\therefore$ At (0, 0), further investigation is needed.

Now, $\quad f(0, 0) = 0$

$$f(x, y) = x^4 + y^4 - 2(x^2 - 2xy + y^2) = x^4 + y^4 - 2(x - y)^2$$

When h and k are small, $f(h, k) = 2h^4 > 0$ for $h = k$

and $$f(h, k) = -2(h-k)^2 < 0 \text{ for } h \neq k$$

(neglecting $h^4 + k^4$, which is very small)

$\Rightarrow$ $f(0, 0) < f(h, k)$ for $h = k$

$f(0, 0) > f(h, k)$ for $h \neq k$

$\Rightarrow$ There is no extreme value at (0, 0).

At $(\sqrt{2}, -\sqrt{2})$, $rt - s^2 = (12 \times 2 - 4)^2 - 16 = 384 > 0$

$$r = 12 \times 2 - 4 = 20 > 0$$

$\therefore$ $f(x, y)$ has a minimum value at $(\sqrt{2}, -\sqrt{2})$.

Minimum value $= (\sqrt{2})^4 + (-\sqrt{2})^4 - 2(\sqrt{2})^2 + 4\sqrt{2}(-\sqrt{2}) - 2(-\sqrt{2})^2$

$= 4 + 4 - 4 - 8 - 4 = -8.$

Similarly, $f(x, y)$ has a minimum value -8 at $(-\sqrt{2}, \sqrt{2})$.

Example 4. *Examine for minimum and maximum values : sin x + sin y + sin (x + y).*

Sol. Here $f(x, y) = \sin x + \sin y + \sin (x + y)$

$$f_x = \cos x + \cos (x + y), f_y = \cos y + \cos (x + y)$$

$$r = f_{xx} = -\sin x - \sin (x + y), s = f_{xy} = -\sin (x + y),$$

$$t = f_{yy} = -\sin y - \sin (x + y)$$

Now, $f_x = 0$ and $f_y = 0$

$\Rightarrow$ $\cos x + \cos (x + y) = 0$ ...(1)

$\cos y + \cos (x + y) = 0$ ...(2)

Subtracting (2) from (1), $\cos x - \cos y = 0$ or $\cos x = \cos y$ $\therefore$ $x = y$

From (1), $\cos x + \cos 2x = 0$ or $\cos 2x = -\cos x = \cos (\pi - x)$ or $2x = \pi - x$ $\therefore$ $x = \frac{\pi}{3}$

$\therefore$ $x = y = \frac{\pi}{3}$ is a stationary point.

At $\left(\frac{\pi}{3}, \frac{\pi}{3}\right)$, $r = -\frac{\sqrt{3}}{2} - \frac{\sqrt{3}}{2} = -\sqrt{3} < 0$, $s = \frac{\sqrt{3}}{2}$, $t = -\frac{\sqrt{3}}{2} - \frac{\sqrt{3}}{2} = -\sqrt{3}$

or $$rt - s^2 = 3 - \frac{3}{4} = \frac{9}{4} > 0$$

Also, $r < 0$

$\therefore$ $f(x, y)$ has a maximum value at $\left(\frac{\pi}{3}, \frac{\pi}{3}\right)$.

Maximum value $= f\left(\frac{\pi}{3}, \frac{\pi}{3}\right) = \sin\frac{\pi}{3} + \sin\frac{\pi}{3} + \sin\frac{2\pi}{3} = \frac{\sqrt{3}}{2} + \frac{\sqrt{3}}{2} + \frac{\sqrt{3}}{2} = \frac{3\sqrt{3}}{2}$.

Example 5. *A rectangular box, open at the top, is to have a given capacity. Find the dimensions of the box requiring least material for its construction.*

Sol. Let x, y and z be the length, breadth and height respectively. Let V be the given capacity and S, the surface.

V is given $\Rightarrow$ V is constant

$$V = xyz \quad \text{or} \quad z = \frac{V}{xy} \quad ...(1)$$

$$S = xy + 2xz + 2yz = xy + \frac{2V}{y} + \frac{2V}{x} = f(x, y) \quad | \text{ Using (1)}$$

$$f_x = y - \frac{2V}{x^2}, f_y = x - \frac{2V}{y^2}$$

$$r = f_{xx} = \frac{4V}{x^3}, s = f_{xy} = 1, t = f_{yy} = \frac{4V}{y^3}$$

Now, $f_x = 0$ and $f_y = 0$

$$\Rightarrow \quad y - \frac{2V}{x^2} = 0 \quad ...(1)$$

$$x - \frac{2V}{y^2} = 0 \quad ...(2)$$

From (1), $y = \dfrac{2V}{x^2}$

$\therefore$ From (2), $x - 2V\cdot \dfrac{x^4}{4V^2} = 0$ or $x\left(1-\dfrac{x^3}{2V}\right) = 0$ or $x = (2V)^{1/3}$ $\quad | \because \; x \neq 0$

and $y = \dfrac{2V}{x^2} = \dfrac{2V}{(2V)^{3/2}} = (2V)^{1/3}$

$\therefore$ $x = y = (2V)^{1/3}$ is a stationary point.

At this point, $r = \dfrac{4V}{2V} = 2 > 0, s = 1, t = \dfrac{4V}{2V} = 2$

so that $rt - s^2 = 4 - 1 = 3 > 0$ and $r > 0$

$\Rightarrow$ S is minimum when $x = y = (2V)^{1/3}$

Also, $z = \dfrac{V}{xy} = \dfrac{V}{(2V)^{2/3}} = \dfrac{V^{1/3}}{2^{2/3}} = \dfrac{(2V)^{1/3}}{2} = \dfrac{y}{2}$

Hence S is minimum when $x = y = 2z = (2V)^{1/3}$.

Example 6. *Prove that if the perimeter of a triangle is constant, its area is maximum when the triangle is equilateral.*

Sol. Let a, b, c be the sides of a triangle whose perimeter $2s$ is constant.

Then $2s = a + b + c$ or $c = 2s - a - b$ $\quad ...(1)$

$$\Delta = \sqrt{s(s-a)(s-b)(s-c)} = \sqrt{s(s-a)(s-b)(a+b-s)} \quad | \text{ Using (1)}$$

Let $z = \Delta^2 = s(s-a)(s-b)(a+b-s) = f(a, b)$

$$f_a = s(s-b)\frac{\partial}{\partial a}[(s-a)(a+b-s)]$$

$$= s(s-b)[-(a+b-s)+(s-a)] = s(s-b)(2s-2a-b)$$

$$f_b = s(s-a)\frac{\partial}{\partial b}[(s-b)(a+b-s)]$$

$$= s(s-a)[-(a+b-s)+(s-b)] = s(s-a)(2s-a-2b)$$

$$r = f_{aa} = -2s(s-b), s = f_{ab} = s[-(2s-a-2b)-(s-a)] = s(2a+2b-3s)$$

$$t = f_{bb} = -2s(s-a)$$

Now, $f_a = 0$ and $f_b = 0$

$\Rightarrow$ $s(s-b)(2s-2a-b) = 0$ and $s(s-a)(2s-a-2b) = 0$

$$\Rightarrow \quad (s-b)(2s-2a-b) = 0 \quad ...(2)$$

$$(s-a)(2s-a-2b) = 0 \quad ...(3) \quad | \because \; s \neq 0$$

From (2), $s = b$ or $2s = 2a + b$

When $s = b$, from (3), $(b-a)(-a) = 0$ or $b = a$ $\quad | \because \; a \neq 0$

When $2s = 2a + b$, from (3), $\frac{b}{2}(a - b) = 0$ or $a = b$ $\quad | \because \; b \neq 0$

If we express z as a function of b and c, we similarly get $b = c$

$$\therefore \quad a = b = c = \frac{2s}{3}$$

$$r = -2s\left(\frac{s}{3}\right) = -\frac{2s^2}{3} < 0$$

$$\text{'}s\text{'} = s\left(\frac{4s}{3} + \frac{4s}{3} - 3s\right) = s\left(-\frac{s}{3}\right) = -\frac{s^2}{3}$$

$$t = -2s\left(\frac{s}{3}\right) = -\frac{2s^2}{3}$$

$$\text{'}rt - s^2\text{'} = \frac{4s^4}{9} - \frac{s^4}{9} = \frac{s^4}{3} > 0. \text{ Also } r < 0$$

$\therefore$ z and hence Δ is maximum when $a = b = c = \frac{2s}{3}$ *i.e.*, when the triangle is equilateral.

Example 7. *Show that the rectangular solid of maximum volume that can be inscribed in a given sphere is a cube.*

Sol. Let x, y, z be the length, breadth and height of the rectangular solid. If V is the volume of the solid, then

$$V = xyz \qquad \ldots(1)$$

Since each diagonal of the solid passes through the centre of the sphere

$\therefore$ each diagonal = diameter of sphere = d (say)

$$\Rightarrow \quad \sqrt{x^2 + y^2 + z^2} = d \quad \text{or} \quad x^2 + y^2 + z^2 = d^2 \quad \text{or} \quad z = \sqrt{d^2 - x^2 - y^2}$$

$\therefore$ From (1), $\quad V = xy\sqrt{d^2 - x^2 - y^2}$

or $\quad V^2 = x^2y^2(d^2 - x^2 - y^2) = d^2x^2y^2 - x^4y^2 - x^2y^4 = f(x, y)$

$$f_x = 2d^2xy^2 - 4x^3y^2 - 2xy^4 = 2xy^2(d^2 - 2x^2 - y^2)$$

$$f_y = 2d^2x^2y - 2x^4y - 4x^2y^3 = 2x^2y(d^2 - x^2 - 2y^2)$$

$$r = f_{xx} = 2d^2y^2 - 12x^2y^2 - 2y^4, \; s = f_{xy} = 4d^2xy - 8x^3y - 8xy^3$$

$$t = f_{yy} = 2d^2x^2 - 2x^4 - 12x^2y^2$$

Now, $\quad f_x = 0$ and $f_y = 0$

$$\Rightarrow \quad d^2 - 2x^2 - y^2 = 0 \qquad \ldots(2) \; [\because \; x \neq 0, y \neq 0]$$

$$d^2 - x^2 - 2y^2 = 0 \qquad \ldots(3)$$

Subtracting (3) from (2), $\quad -x^2 + y^2 = 0 \Rightarrow y = x$

$\therefore$ From (2), $\quad x = y = \frac{d}{\sqrt{3}}$

$$z = \sqrt{d^2 - \frac{d^2}{3} - \frac{d^2}{3}} = \frac{d}{\sqrt{3}}$$

At $x = y = \frac{d}{\sqrt{3}}$,

$$r = 2d^2 \cdot \frac{d^2}{3} - 12 \cdot \frac{d^2}{3} \cdot \frac{d^2}{3} - 2 \cdot \frac{d^4}{9} = -\frac{8d^4}{9} < 0$$

$$s = 4d^2 \cdot \frac{d^2}{3} - 8 \cdot \frac{d^3}{3\sqrt{3}} \cdot \frac{d}{\sqrt{3}} - 8 \cdot \frac{d}{\sqrt{3}} \cdot \frac{d^3}{3\sqrt{3}} = -\frac{4d^4}{9}$$

$$t = 2d^2 \cdot \frac{d^2}{3} - 2 \cdot \frac{d^4}{9} - 12 \cdot \frac{d^2}{3} \cdot \frac{d^2}{3} = -\frac{8d^4}{9}$$

$$rt - s^2 = \frac{64d^8}{81} - \frac{16d^8}{81} = \frac{16d^8}{27} > 0, \text{ Also } r < 0.$$

$\therefore$ V^2 and hence, V is maximum when $x = y = z$ *i.e.*, when the rectangular solid is a cube.

Example 8. *Find the volume of the largest rectangular parallelopiped that can be inscribed in the ellipsoid,* $\frac{x^2}{a^2} + \frac{y^2}{b^2} + \frac{z^2}{c^2} = 1.$

Sol. Let (x, y, z) be a vertex of the parallelopiped, then it lies on the ellipsoid

$$\frac{x^2}{a^2} + \frac{y^2}{b^2} + \frac{z^2}{c^2} = 1$$

Also its dimensions are $2x$, $2y$, $2z$ so that the volume V is given by $V = 2x \cdot 2y \cdot 2z = 8xyz$

$$\Rightarrow \quad V^2 = 64x^2y^2z^2 = 64x^2y^2c^2\left(1 - \frac{x^2}{a^2} - \frac{y^2}{b^2}\right) = 64c^2\left(x^2y^2 - \frac{x^4y^2}{a^2} - \frac{x^2y^4}{b^2}\right) = f(x, y)$$

$$f_x = 64c^2\left(2xy^2 - \frac{4x^3y^2}{a^2} - \frac{2xy^4}{b^2}\right); \qquad f_y = 64c^2\left(2x^2y - \frac{2x^4y}{a^2} - \frac{4x^2y^3}{b^2}\right)$$

$$r = f_{xx} = 64c^2\left(2y^2 - \frac{12x^2y^2}{a^2} - \frac{2y^4}{b^2}\right); \quad s = f_{xy} = 64c^2\left(4xy - \frac{8x^3y}{a^2} - \frac{8xy^3}{b^2}\right)$$

$$t = f_{yy} = 64c^2\left(2x^2 - \frac{2x^4}{a^2} - \frac{12x^2y^2}{b^2}\right)$$

Now, $\quad f_x = 0$ and $f_y = 0$

$$\Rightarrow \quad 128c^2xy^2\left(1 - \frac{2x^2}{a^2} - \frac{y^2}{b^2}\right) = 0 \text{ and } 128c^2x^2y\left(1 - \frac{x^2}{a^2} - \frac{2y^2}{b^2}\right) = 0$$

$$\Rightarrow \quad 1 - \frac{2x^2}{a^2} - \frac{y^2}{b^2} = 0 \qquad \text{...(1) } [\because x \neq 0, y \neq 0]$$

$$1 - \frac{x^2}{a^2} - \frac{2y^2}{b^2} = 0 \qquad \text{...(2)}$$

Subtracting (2) from (1), $-\frac{x^2}{a^2} + \frac{y^2}{b^2} = 0$ or $y = \frac{bx}{a}$

$\therefore$ From (1), $\quad 1 - \frac{2x^2}{a^2} - \frac{x^2}{a^2} = 0$ or $x^2 = \frac{a^2}{3}$ or $x = \frac{a}{\sqrt{3}}$

$$\therefore \quad y = \frac{b}{a} \cdot \frac{a}{\sqrt{3}} = \frac{b}{\sqrt{3}}$$

and $$z^2 = c^2\left(1-\frac{x^2}{a^2}-\frac{y^2}{b^2}\right) = c^2\left(1-\frac{1}{3}-\frac{1}{3}\right) = \frac{c^2}{3}$$

$\therefore$ $$z = \frac{c}{\sqrt{3}}$$

Thus $x = \frac{a}{\sqrt{3}}, y = \frac{b}{\sqrt{3}}$ is a stationary point.

At this point, $$r = 64c^2\left[\frac{2b^2}{3} - \frac{12}{a^2}\cdot\frac{a^2}{3}\cdot\frac{b^2}{3} - \frac{2}{b^2}\cdot\frac{b^4}{9}\right] = -\frac{512}{9}b^2c^2 < 0$$

$$s = 64c^2\left[4\cdot\frac{a}{\sqrt{3}}\cdot\frac{b}{\sqrt{3}} - \frac{8}{a^2}\cdot\frac{a^3}{3\sqrt{3}}\cdot\frac{b}{\sqrt{3}} - \frac{8}{b^2}\cdot\frac{a}{\sqrt{3}}\cdot\frac{b^3}{3\sqrt{3}}\right] = -\frac{256}{9}abc^2$$

$$t = 64c^2\left[2\cdot\frac{a^2}{3} - \frac{2}{a^2}\cdot\frac{a^4}{9} - \frac{12}{b^2}\cdot\frac{a^2}{3}\cdot\frac{b^2}{3}\right] = -\frac{512}{9}a^2c^2$$

$$rt - s^2 = \left(\frac{512}{9}\right)^2 a^2b^2c^4 - \left(\frac{256}{9}\right)^2 a^2b^2c^4 = \left(\frac{256}{9}\right)^2 a^2b^2c^4\,(4-1) > 0.$$

Also, $r < 0$

$\therefore$ V^2 and hence, V is maximum

When $x = \frac{a}{\sqrt{3}}, y = \frac{b}{\sqrt{3}}, z = \frac{c}{\sqrt{3}}$

Maximum volume $= 8\cdot\frac{a}{\sqrt{3}}\cdot\frac{b}{\sqrt{3}}\cdot\frac{c}{\sqrt{3}} = \frac{8abc}{3\sqrt{3}}$.

Example 9. *In a plane triangle ABC, find the maximum value of cos A cos B cos C.*

Sol. $\cos A \cos B \cos C = \cos A \cos B \cos[\pi - (A + B)]$ $\quad | \because \; A + B + C = \pi$

$$= -\cos A \cos B \cos(A + B) = f(A, B)$$

$$\frac{\partial f}{\partial A} = -\cos B[-\sin A \cos(A + B) - \cos A \sin(A + B)]$$

$$= \cos B \sin[A + (A + B)] = \cos B \sin(2A + B)$$

$$\frac{\partial f}{\partial B} = -\cos A[-\sin B \cos(A + B) - \cos B \sin(A + B)]$$

$$= \cos A \sin[B + (A + B)] = \cos A \sin(A + 2B)$$

$$r = 2\cos B \cos(2A + B)$$

$$s = -\sin A \sin(A + 2B) + \cos A \cos(A + 2B)$$

$$= \cos[A + (A + 2B)] = \cos(2A + 2B)$$

$$t = 2\cos A \cos(A + 2B)$$

Now, $$\frac{\partial f}{\partial A} = 0 \text{ and } \frac{\partial f}{\partial B} = 0$$

$\Rightarrow$ $$\cos B \sin(2A + B) = 0 \quad \text{...(1)}$$

$$\cos A \sin(A + 2B) = 0 \quad \text{...(2)}$$

If $\cos B = 0$, then $B = \frac{\pi}{2}$

From (2), $\cos A \sin (A + \pi) = 0$ or $\cos A(-\sin A) = 0$

$\Rightarrow$ either $\cos A = 0$ *i.e.*, $A = \frac{\pi}{4}$, which is not possible ($\because$ $A + B + C = \pi \Rightarrow C = 0$)

or $\sin A = 0$ *i.e.*, $A = 0$ or π, which is not possible.

$\therefore$ $\cos B \neq 0$. Similarly $\cos A \neq 0$

$\therefore$ From (1), $\sin (2A + B) = 0$ or $2A + B = \pi$

From (2), $\sin (A + 2B) = 0$ or $A + 2B = \pi$

Solving these equations, $A = B = \frac{\pi}{3}$

At $A = B = \frac{\pi}{3}$

$$r = 2 \cos \frac{\pi}{3} \cos \pi = -1, s = \cos \frac{4\pi}{3} = \cos\left(\pi + \frac{\pi}{3}\right) = -\cos \frac{\pi}{3} = -\frac{1}{2}$$

$$t = 2 \cos \frac{\pi}{3} \cos \pi = -1$$

$$rt - s^2 = 1 - \frac{1}{4} = \frac{3}{4} > 0$$

Also, $r = -1 < 0$

$\Rightarrow$ $f(A, B)$ is maximum at $A = B = \frac{\pi}{3}$

Maximum value $= f\left(\frac{\pi}{3}, \frac{\pi}{3}\right) = -\cos \frac{\pi}{3} \cos \frac{\pi}{3} \cos \frac{2\pi}{3} = -\left(\frac{1}{2}\right)\left(\frac{1}{2}\right)\left(-\frac{1}{2}\right) = \frac{1}{8}$.

Example 10. *Examine $f(x, y) = x^4 + 2x^2y - x^2 + 3y^2$ for maximum and minimum values.* (OU-Dec. 2013)

Sol. $\frac{\partial f}{\partial x} = 4x^3 + 4xy - 2x = 0 \Rightarrow 2x(2x^2 + 2y - 1) = 0$

$\Rightarrow$ $x = 0$ or $2x^2 + 2y - 1 = 0$ ...(1)

$\frac{\partial f}{\partial y} = 2x^2 + 6y = 0 \Rightarrow x^2 + 3y = 0$ ...(2)

When $x = 0$, from (2), we get $y = 0 \Rightarrow (0, 0)$

$(1) - 2 \times (2) \Rightarrow -4y - 1 = 0 \Rightarrow y = -\frac{1}{4}$

From (2), $x^2 - \frac{3}{4} = 0 \Rightarrow x^2 = \frac{3}{4} \Rightarrow x = \pm \frac{\sqrt{3}}{2}$

Hence stationary points are $(0, 0)$ $\left(\frac{\sqrt{3}}{2}, -\frac{1}{4}\right)$ and $\left(-\frac{\sqrt{3}}{2}, -\frac{1}{4}\right)$

$$r = \frac{\partial^2 f}{\partial x^2} = \frac{\partial}{\partial x}(4x^3 + 4xy - 2x) = 12x^2 + 4y - 2$$

$$s = \frac{\partial^2 f}{\partial x \partial y} = \frac{\partial}{\partial x}(2x^2 + 6y) = 4x$$

$$t = \frac{\partial^2 f}{\partial y^2} = \frac{\partial}{\partial y}(2x^2 + 6y) = 6$$

Case I: At (0, 0)

$$r = -2, s = 0, t = 6$$
$$rt - s^2 = -12 - 0 = -12 < 0$$

f has no minimum or maximum at (0, 0).

Case II: At $\left(\frac{\sqrt{3}}{2}, -\frac{1}{4}\right)$

$$r = 12\left(\frac{3}{4}\right) + 4\left(-\frac{1}{4}\right) - 2 = \frac{36-4-8}{4} = \frac{24}{4} = 6$$

$$s = 4\left(\frac{\sqrt{3}}{2}\right) = 2\sqrt{3}$$

$$t = 6$$
$$rt - s^2 = 36 - 12 = 24 > 0 \quad \text{and} \quad r > 0$$

Hence f has minimum at $\left(\frac{\sqrt{3}}{2}, -\frac{1}{4}\right)$

$$f_{\min} = \left(\frac{\sqrt{3}}{2}\right)^4 + 2\left(\frac{\sqrt{3}}{2}\right)^2\left(-\frac{1}{4}\right) - \left(\frac{\sqrt{3}}{2}\right)^2 + 3\left(-\frac{1}{4}\right)^2$$
$$= \frac{9}{16} - \frac{3}{8} - \frac{3}{4} + \frac{3}{16} = \frac{9-6-12+3}{16} = \frac{-6}{16} = \frac{-3}{8}$$

Case III: At $\left(-\frac{\sqrt{3}}{2}, -\frac{1}{4}\right)$

$$r = 12\left(\frac{-\sqrt{3}}{2}\right)^2 + 4\left(\frac{-1}{4}\right) - 2 = \frac{36}{4} - \frac{4}{4} - 2 = \frac{36-4-8}{4} = 6$$

$$s = 4x = 4\left(\frac{-\sqrt{3}}{2}\right) = -2\sqrt{3}$$

$$t = 6$$
$$rt - s^2 = 36 - 12 = 24 > 0 \quad \text{and } r > 0$$

Hence, $f_{\min} = \frac{-3}{8}$ (as case II).

Example 11. *Find the extremum values of $f(x, y) = x^3 + 3xy^2 - 3x^2 - 3y^2 + 4$.*

(OU-Apr. 2007)

Sol.
$$\frac{\partial f}{\partial x} = 3x^2 + 3y^2 - 6x = 0 \Rightarrow x^2 + y^2 - 2x = 0 \qquad ...(1)$$

$$\frac{\partial f}{\partial y} = 6xy - 6y = 0 \Rightarrow 6y(x-1) = 0 \Rightarrow y = 0 \text{ or } x = 1$$

At $y = 0$ from (1),

$$x^2 - 2x = 0 \Rightarrow x(x-2) = 0 \Rightarrow x = 0 \text{ or } 2$$

Hence, (0, 0), (2, 0)

At $x = 1$ from (1),

$$1 + y^2 - 2 = 0 \Rightarrow y^2 = 1 \Rightarrow y = \pm 1$$

Hence, (1, 1), (1, – 1)

$$r = \frac{\partial^2 f}{\partial x^2} = \frac{\partial}{\partial x}(3x^2 + 3y^2 - 6x) = 6x - 6$$

$$s = \frac{\partial^2 f}{\partial x \partial y} = \frac{\partial}{\partial x}(6xy - 6y) = 6y$$

$$t = \frac{\partial^2 f}{\partial y^2} = \frac{\partial}{\partial y}(6xy - 6y) = 6x - 6$$

Case I: At (0, 0)

$$r = -6, s = 0, t = -6$$

$$rt - s^2 = 36 > 0 \text{ and } r = -6 < 0$$

Hence, f has maximum at (0, 0).

$\therefore$ $f_{max} = 4$

Case II: At (2, 0)

$$r = 6, s = 0, t = -6$$

$$rt - s^2 = -36 < 0$$

$\therefore$ f has no min or max at (2, 0)

Case III: At (1, 1)

$$r = 0, s = 6, t = 0$$

$$rt - s^2 = 0 - 36 = -36 < 0$$

$\therefore$ f has no extreme value at (1, 1)

Case IV: At (1, –1)

$$r = 0, s = -6, t = 0$$

$$rt - s^2 = -36 < 0$$

$\therefore$ f has no extreme value at (1, –1).

EXERCISE 5

1. Examine for extreme values :

(i) $x^2 + y^2 + 6x + 12$ (ii) $x^3 + y^3 + 3xy$ (iii) $3x^2 - y^2 + x^3$

(iv) $x^2y^2 - 5x^2 - 8xy - 5y^2$ (v) $x^3 + 3xy^2 - 15x^2 - 15y^2 + 72x$

(vi) $x^2 - xy + y^2 + 3x - 2y + 1$ (vii) $x^3 + y^3 - 63(x + y) + 12xy$ (viii) $xy + \frac{a^3}{x} + \frac{a^3}{y}$

(ix) $\sin x \sin y \sin (x + y)$ (x) $\cos x + \cos y + \cos (x + y)$.

2. Find the stationary points of the function $z = x^3y^2 (12 - x - y)$ satisfying the condition $x > 0, y > 0$ and examine their nature.

Answers

1. (*i*) Min. value = 3 at (– 3, 0) (*ii*) Max. value = 1 at (– 1, – 1)

   (*iii*) Max. value = 4 at (– 2, 0) (*iv*) Max. value = 0 at (0, 0)

   (*v*) Max. value = 112 at (4, 0) ; Min. value = 108 at (6, 0)

   (*vi*) Min. value = $-\frac{4}{3}$ at $\left(-\frac{4}{3}, \frac{1}{3}\right)$

   (*vii*) Max. value = 784 at (– 7, – 7) ; Min. value = – 216 at (3, 3)

   (*viii*) Min. value = $3a^2$ at (a, a)

   (*ix*) Max. value = $\frac{3\sqrt{3}}{8}$ at $\left(\frac{\pi}{3}, \frac{\pi}{3}\right)$; Min. value = $-\frac{3\sqrt{3}}{8}$ at $\left(\frac{2\pi}{3}, \frac{2\pi}{3}\right)$

   (*x*) Min. value = $-\frac{3}{2}$ at $\left(\frac{2\pi}{3}, \frac{2\pi}{3}\right)$

2. Maximum at (6, 4)

4.14. DOUBLE INTEGRALS (This Topic is for Student Reference Only)

The definite integral $\int_a^b f(x)\,dx$ is defined as the limits of the sum $f(x_1)\delta x_1 + f(x_2)\delta x_2 + \ldots\ldots + f(x_n)\delta x_n$ when $n \to \infty$ and each of the lengths $\delta x_1, \delta x_2, \ldots\ldots, \delta x_n$ tends to zero. Here $\delta x_1, \delta x_2, \ldots\ldots, \delta x_n$ are n sub-intervals into which the range $b - a$ has been divided and $x_1, x_2, \ldots\ldots, x_n$ are values of x lying respectively in the first, second, ..., nth sub-interval.

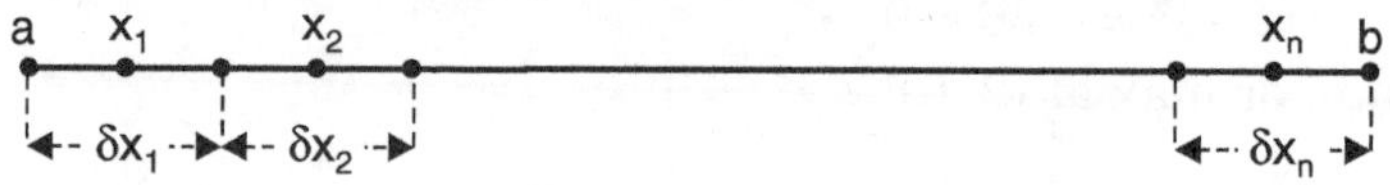

A double integral is its counterpart in two dimensions. Let a single-valued and bounded function $f(x, y)$ of two independent variables x, y be defined in a closed region R of the xy-plane. Divide the region R into sub-regions by drawing lines parallel to co-ordinate axes. Number the rectangles which lie entirely inside the region R, from 1 to n. Let (x_r, y_r) be any point inside the rth rectangle whose area is δA_r.

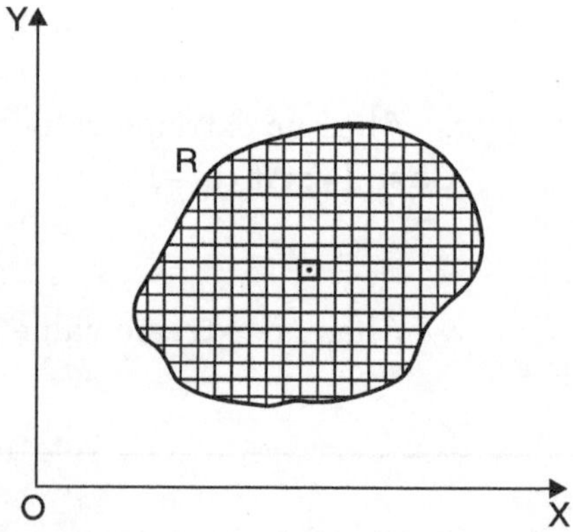

Consider the sum

$$f(x_1, y_1)\delta A_1 + f(x_2, y_2)\delta A_2 + \ldots\ldots + f(x_n, y_n)\delta A_n$$

$$= \sum_{r=1}^{n} f(x_r, y_r)\delta A_r \qquad \ldots(1)$$

Let the number of these sub-regions increase indefinitely, such that the largest linear dimension (*i.e.*, diagonal) of δA_r approaches zero. The limit of the sum (1), if it exists, irrespective of the mode of sub-division, is called the *double integral* of $f(x, y)$ over the region R and is denoted by $\iint_R f(x, y)\,dA$

In other words,

$$\lim_{\substack{n \to \infty \\ \delta A_r \to 0}} \sum_{r=1}^{n} f(x_r, y_r)\,\delta A_r = \iint_R f(x, y)\,dA$$

which is also expressed as

$$\iint_R f(x, y)\,dx\,dy \quad \text{or} \quad \iint_R f(x, y)\,dy\,dx$$

4.15. EVALUATION OF DOUBLE INTEGRALS (This Topic is for Student Reference Only)

The methods of evaluating the double integrals depend upon the nature of the curves bounding the region R. Let the region R be bounded by the curves $x = x_1$, $x = x_2$ and $y = y_1$, $y = y_2$.

(*i*) When x_1, x_2 are functions of y and y_1, y_2 are constants. Let AB and CD be the curves $x_1 = \phi_1(y)$ and $x_2 = \phi_2(y)$.

Take a horizontal strip PQ of width δy. Here the double integral is evaluated first w.r.t. x (treating y as a constant). The resulting expression which is a function of y is integrated w.r.t. y between the limits $y = y_1$ and $y = y_2$. Thus,

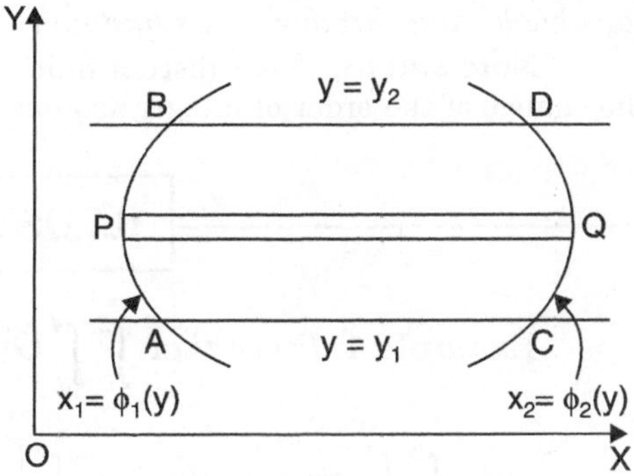

$$\iint_R f(x, y)\, dx\, dy = \boxed{\int_{y_1}^{y_2} \boxed{\int_{x_1 = \phi_1(y)}^{x_2 = \phi_2(y)} f(x, y)\, dx}\, dy}$$

the integration being carried from the inner to the outer rectangle. Geometrically, the integral in the inner rectangle indicates that the integration is performed along the horizontal strip PQ (keeping y constant) while the outer rectangle corresponds to the sliding of the strip PQ from AC to BD thus covering the entire region ABDC of integration.

(*ii*) When y_1, y_2 are functions of x and x_1, x_2 are constants. Let AB and CD be the curves $y_1 = \phi_1(x)$ and $y_2 = \phi_2(x)$. Take a vertical strip PQ of width δx. Here the double integral is evaluated first w.r.t. y (treating x as constant). The resulting expression which is a function of x is integrated w.r.t. x between the limits $x = x_1$ and $x = x_2$. Thus,

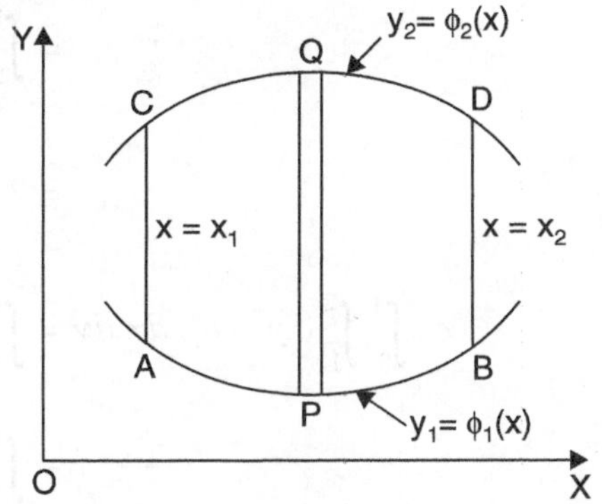

$$\iint_R f(x, y)\, dx\, dy = \boxed{\int_{x_1}^{x_2} \boxed{\int_{y_1 = \phi_1(x)}^{y_2 = \phi_2(x)} f(x, y)\, dy}\, dx}$$

the integration being carried from the inner to the outer rectangle. Geometrically, the integral in the inner rectangle indicates that the integration is performed along the vertical strip PQ (keeping x constant) while the outer rectangle corresponds to the sliding of the strip PQ from AC to BD thus covering the entire region ABDC of integration.

(*iii*) When x_1, x_2, y_1, y_2 are constants. Here the region of integration R is the rectangle ABDC. It is immaterial whether we integrate first along the horizontal strip PQ and then slide it from AB to CD ; or we integrate first along the vertical strip P′Q′ and then slide it from AC to BD. Thus the order of integration is immaterial, provided the limits of integration are changed accordingly.

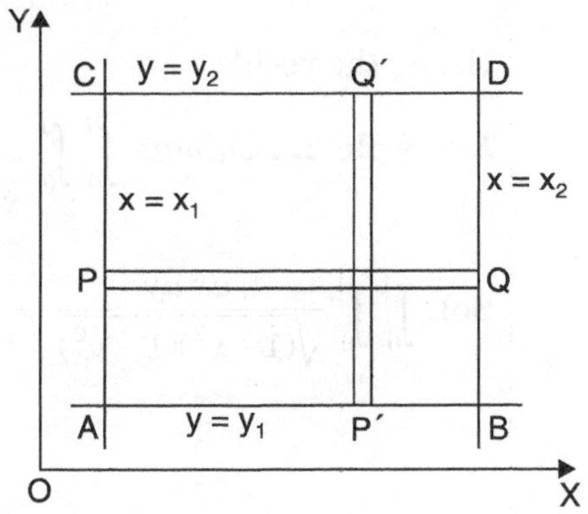

$$\iint_R f(x, y)\,dx\,dy = \boxed{\int_{y_1}^{y_2} \boxed{\int_{x_1}^{x_2} f(x, y)\,dx}\,dy} = \boxed{\int_{x_1}^{x_2} \boxed{\int_{y_1}^{y_2} f(x, y)\,dy}\,dx}$$

Note 1. From cases (*i*) and (*ii*) above, we observe that *integration is to be performed w.r.t. that variable having variable limits first and then w.r.t. the variable with constant limits.*

Note 2. If $f(x, y)$ has discontinuities within or on the boundary of the region of integration, then the change of the order of integration does not result into the same integrals.

ILLUSTRATIVE EXAMPLES

Example 1. *Prove that* $\int_1^2 \int_3^4 (xy + e^y)\,dy\,dx = \int_3^4 \int_1^2 (xy + e^y)\,dx\,dy.$

Sol. $$\int_1^2 \int_3^4 (xy + e^y)\,dy\,dx = \int_1^2 \left[\int_3^4 (xy + e^y)\,dy\right] dx$$

$$= \int_1^2 \left[\frac{xy^2}{2} + e^y\right]_3^4 dx = \int_1^2 \left(8x + e^4 - \frac{9}{2}x - e^3\right) dx$$

$$= \int_1^2 \left(\frac{7}{2}x + e^4 - e^3\right) dx = \left[\frac{7x^2}{4} + (e^4 - e^3)x\right]_1^2$$

$$= 7 + 2(e^4 - e^3) - \frac{7}{4} - (e^4 - e^3) = \frac{21}{4} + e^4 - e^3$$

$$\int_3^4 \int_1^2 (xy + e^y)\,dx\,dy = \int_3^4 \left[\int_1^2 (xy + e^y)\,dx\right] dy = \int_3^4 \left[\frac{yx^2}{2} + xe^y\right]_1^2 dy$$

$$= \int_3^4 \left(2y + 2e^y - \frac{y}{2} - e^y\right) dy = \int_3^4 \left(\frac{3y}{2} + e^y\right) dy$$

$$= \left[\frac{3y^2}{4} + e^y\right]_3^4 = 12 + e^4 - \frac{27}{4} - e^3 = \frac{21}{4} + e^4 - e^3$$

Hence the result.

Example 2. *Evaluate* $\int_0^1 \int_0^1 \frac{dx\,dy}{\sqrt{(1 - x^2)(1 - y^2)}}.$

Sol. $$\int_0^1 \int_0^1 \frac{dx\,dy}{\sqrt{(1 - x^2)(1 - y^2)}} = \int_0^1 \left[\int_0^1 \frac{dx}{\sqrt{(1 - x^2)(1 - y^2)}}\right] dy = \int_0^1 \frac{1}{\sqrt{1 - y^2}} \left[\sin^{-1} x\right]_0^1 dy$$

$$= \int_0^1 \frac{1}{\sqrt{1 - y^2}} \cdot \frac{\pi}{2}\,dy = \frac{\pi}{2}\left[\sin^{-1} y\right]_0^1 = \frac{\pi}{2} \cdot \frac{\pi}{2} = \frac{\pi^2}{4}.$$

Example 3. *Show that* $\int_0^1 dx \int_0^1 \frac{x-y}{(x+y)^3}\, dy \neq \int_0^1 dy \int_0^1 \frac{x-y}{(x+y)^3}\, dx.$

Sol. LHS $= \int_0^1 dx \int_0^1 \frac{2x-(x+y)}{(x+y)^3}\, dy = \int_0^1 dx \int_0^1 \left[\frac{2x}{(x+y)^3} - \frac{1}{(x+y)^2}\right] dy$

$$= \int_0^1 \left[2x \cdot \frac{(x+y)^{-2}}{-2} - \frac{(x+y)^{-1}}{-1}\right]_0^1 dx = \int_0^1 \left[\frac{-x}{(x+y)^2} + \frac{1}{x+y}\right]_0^1 dx$$

$$= \int_0^1 \left[\frac{-x}{(x+1)^2} + \frac{1}{x+1} + \frac{1}{x} - \frac{1}{x}\right] dx = \int_0^1 \frac{-x+x+1}{(x+1)^2}\, dx = \int_0^1 \frac{1}{(x+1)^2}\, dx$$

$$= \left[-\frac{1}{x+1}\right]_0^1 = -\frac{1}{2} + 1 = \frac{1}{2}$$

RHS $= \int_0^1 dy \int_0^1 \frac{(x+y)-2y}{(x+y)^3}\, dx = \int_0^1 dy \int_0^1 \left[\frac{1}{(x+y)^2} - \frac{2y}{(x+y)^3}\right] dx$

$$= \int_0^1 \left[\frac{(x+y)^{-1}}{-1} - 2y \,.\, \frac{(x+y)^{-2}}{-2}\right]_0^1 dy = \int_0^1 \left[-\frac{1}{x+y} + \frac{y}{(x+y)^2}\right]_0^1 dy$$

$$= \int_0^1 \left[-\frac{1}{1+y} + \frac{y}{(1+y)^2} + \frac{1}{y} - \frac{1}{y}\right] dy = \int_0^1 \frac{-1-y+y}{(1+y)^2}\, dy = -\int_0^1 \frac{1}{(1+y)^2}\, dy$$

$$= \left[\frac{1}{1+y}\right]_0^1 = \frac{1}{2} - 1 = -\frac{1}{2}$$

$\therefore$ The two integrals are not equal.

Example 4. *Evaluate* $\int_0^1 \int_0^{\sqrt{1+x^2}} \frac{dy\, dx}{1+x^2+y^2}.$

Sol. $\quad I = \int_0^1 \left[\int_0^{\sqrt{1+x^2}} \frac{1}{(1+x^2)+y^2}\, dy\right] dx$

$$= \int_0^1 \frac{1}{\sqrt{1+x^2}} \left[\tan^{-1} \frac{y}{\sqrt{1+x^2}}\right]_0^{\sqrt{1+x^2}} dx$$

$$= \int_0^1 \frac{1}{\sqrt{1+x^2}} [\tan^{-1} 1 - \tan^{-1} 0]\, dx = \frac{\pi}{4} \int_0^1 \frac{dx}{\sqrt{1+x^2}}$$

$$= \frac{\pi}{4} \left[\log(x + \sqrt{1+x^2})\right]_0^1 = \frac{\pi}{4} [\log(1+\sqrt{2}) - \log 1] = \frac{\pi}{4} \log(\sqrt{2}+1).$$

Example 5. *Evaluate* $\int_0^a \int_0^{\sqrt{a^2-y^2}} \sqrt{a^2-x^2-y^2}\, dx\, dy.$

Sol. $$I = \int_0^a \left[\int_0^{\sqrt{a^2-y^2}} \sqrt{(a^2-y^2)-x^2}\, dx \right] dy.$$

$$= \int_0^a \left[\frac{x\sqrt{a^2-y^2-x^2}}{2} + \frac{a^2-y^2}{2} \sin^{-1} \frac{x}{\sqrt{a^2-y^2}} \right]_0^{\sqrt{a^2-y^2}} dy$$

$$\left[\because \int \sqrt{a^2-x^2}\, dx = \frac{x\sqrt{a^2-x^2}}{2} + \frac{a^2}{2} \sin^{-1} \frac{x}{a} \right]$$

$$= \int_0^a \frac{a^2-y^2}{2} \sin^{-1} 1\, dy = \frac{\pi}{4} \int_0^a (a^2-y^2)\, dy$$

$$= \frac{\pi}{4} \left[a^2 y - \frac{y^3}{3} \right]_0^a = \frac{\pi}{4} \left[a^3 - \frac{a^3}{3} \right] = \frac{\pi a^3}{6}.$$

Example 6. *Evaluate* $\iint_R e^{2x+3y}\, dx\, dy$ *over the triangle bounded by* $x = 0$, $y = 0$ *and* $x + y = 1$.

Sol. The region R of integration is the triangle OAB. Here x varies from 0 to 1 and y varies from x-axis upto the line $x + y = 1$ *i.e.*, from 0 to $1 - x$.

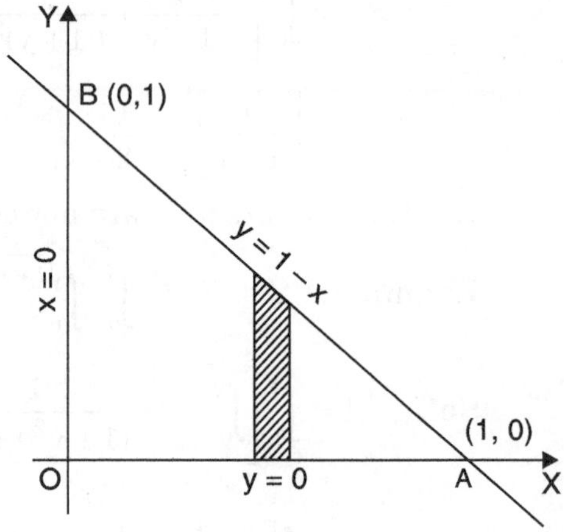

$\therefore$ The region R can be expressed as

$$0 \le x \le 1,\ 0 \le y \le 1 - x$$

$$\therefore \iint_R e^{2x+3y}\, dx\, dy$$

$$= \int_0^1 \int_0^{1-x} e^{2x+3y}\, dy\, dx$$

$$= \int_0^1 \left[\frac{1}{3} e^{2x+3y} \right]_0^{1-x} dx$$

$$= \frac{1}{3} \int_0^1 (e^{3-x} - e^{2x})\, dx$$

$$= \frac{1}{3} \left[-e^{3-x} - \frac{1}{2} e^{2x} \right]_0^1 = -\frac{1}{3} \left[\left(e^2 + \frac{1}{2} e^2 \right) - \left(e^3 + \frac{1}{2} \right) \right]$$

$$= -\frac{1}{3} \left[-e^2(e-1) + \frac{1}{2}(e^2 - 1) \right]$$

$$= \frac{1}{6}(e-1)[2e^2-(e+1)] = \frac{1}{6}(e-1)(2e^2-e-1)$$

$$= \frac{1}{6}(e-1)(e-1)(2e+1) = \frac{1}{6}(e-1)^2(2e+1).$$

Example 7. *Evaluate* $\iint_R y\,dxdy$, *where R is the region bounded by the parabolas* $y^2 = 4x$ *and* $x^2 = 4y$.

Sol. Solving $y^2 = 4x$ and $x^2 = 4y$, we have

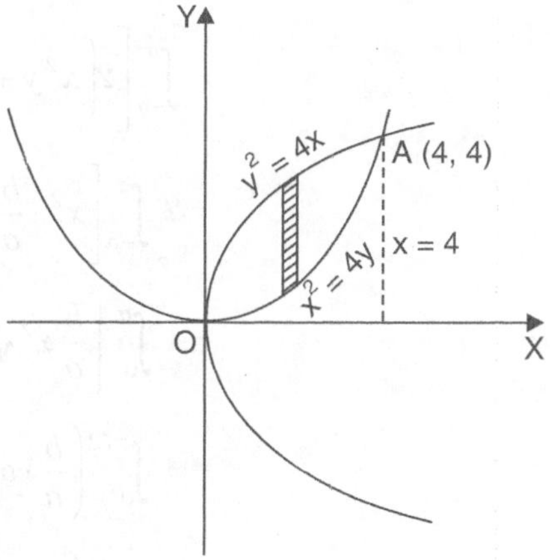

$$\left(\frac{x^2}{4}\right)^2 = 4x \quad \text{or} \quad x(x^3-64) = 0$$

$\therefore \quad x = 0, 4$

When $x = 4, y = 4$

$\therefore$ Co-ordinates of A are (4, 4)

The region R can be expressed as

$$0 \le x \le 4, \frac{x^2}{4} \le y \le 2\sqrt{x}$$

$$\therefore \quad \iint_R y\,dx\,dy = \int_0^4 \int_{x^2/4}^{2\sqrt{x}} y\,dy\,dx$$

$$= \int_0^4 \frac{1}{2}\left[y^2\right]_{x^2/4}^{2\sqrt{x}} dx = \frac{1}{2}\int_0^4 \left(4x - \frac{x^4}{16}\right) dx$$

$$= \frac{1}{2}\left[2x^2 - \frac{x^5}{80}\right]_0^4 = \frac{1}{2}\left[32 - \frac{1024}{80}\right] = \frac{48}{5}.$$

Example 8. *Evaluate* $\iint (x+y)^2\,dx\,dy$ *over the area bounded by the ellipse* $\frac{x^2}{a^2} + \frac{y^2}{b^2} = 1$.

Sol. For the ellipse

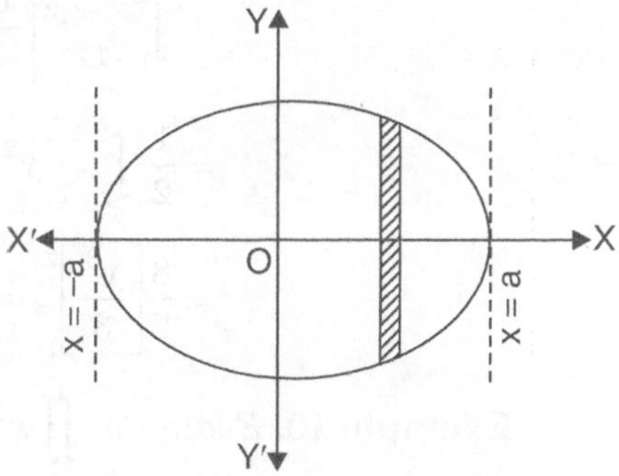

$$\frac{y}{b} = \pm\sqrt{1-\frac{x^2}{a^2}} \quad \text{or} \quad y = \pm\frac{b}{a}\sqrt{a^2-x^2}$$

$\therefore$ The region of integration R can be expressed as

$$-a \le x \le a, -\frac{b}{a}\sqrt{a^2-x^2} \le y \le \frac{b}{a}\sqrt{a^2-x^2}$$

$$\therefore \quad \iint_R (x+y)^2\,dxdy = \iint_R (x^2+y^2+2xy)\,dxdy$$

$$= \int_{-a}^{a} \int_{-\left(\frac{b}{a}\right)\sqrt{a^2-x^2}}^{\left(\frac{b}{a}\right)\sqrt{a^2-x^2}} (x^2+y^2+2xy)\,dydx$$

$$= \int_{-a}^{a} \int_{-\left(\frac{b}{a}\right)\sqrt{a^2-x^2}}^{\left(\frac{b}{a}\right)\sqrt{a^2-x^2}} (x^2+y^2)\, dydx + \int_{-a}^{a} \int_{-\left(\frac{b}{a}\right)\sqrt{a^2-x^2}}^{\left(\frac{b}{a}\right)\sqrt{a^2-x^2}} 2xydydx$$

$$= \int_{-a}^{a} \int_{0}^{\left(\frac{b}{a}\right)\sqrt{a^2-x^2}} 2(x^2+y^2)\, dy\, dx + 0$$

[Since $(x^2 + y^2)$ is an even function of y and $2xy$ is an odd function of y]

$$= \int_{-a}^{a} \left[2\left(x^2 y + \frac{y^3}{3} \right) \right]_0^{\left(\frac{b}{a}\right)\sqrt{a^2-x^2}} dx$$

$$= 2\int_{-a}^{a} \left[x^2 \cdot \frac{b}{a} \sqrt{a^2-x^2} + \frac{1}{3}\cdot\frac{b^3}{a^3}(a^2-x^2)^{3/2} \right] dx$$

$$= 4\int_{0}^{a} \left[\frac{b}{a} x^2 \sqrt{a^2-x^2} + \frac{b^3}{3a^3}(a^2-x^2)^{3/2} \right] dx$$

Put $x = a \sin \theta$
$\therefore \quad dx = a \cos \theta \, d\theta$

$$= 4\int_{0}^{\pi/2} \left(\frac{b}{a} \cdot a^2 \sin^2 \theta \cdot a \cos \theta + \frac{b^3}{3a^3} \cdot a^3 \cos^3 \theta \right) a \cos \theta \, d\theta$$

$$= 4\int_{0}^{\pi/2} \left(a^3 b \sin^2 \theta \cos^2 \theta + \frac{ab^3}{3} \cos^4 \theta \right) d\theta$$

$$= 4\left[a^3 b \cdot \frac{1\cdot 1}{4\cdot 2}\cdot\frac{\pi}{2} + \frac{ab^3}{3}\cdot\frac{3\cdot 1}{4\cdot 2}\cdot\frac{\pi}{2} \right] = \frac{\pi}{4}(a^3 b + ab^3) = \frac{\pi}{4} ab(a^2+b^2).$$

Example 9. *Evaluate* $\int_1^2 \int_1^3 xy^2 \, dx \, dy$. (OU-June 2013)

Sol. Consider $\int_{y=1}^{2} \boxed{\int_{x=1}^{3} xy^2 \, dx \; dy}$

$$= \int_{y=1}^{2} y^2 \left[\frac{x^2}{2} \right]_{x=1}^{3} dy$$

$$= \frac{1}{2} \int_{y=1}^{2} y^2 \, [9-1]\, dy$$

$$= \frac{8}{2}\left[\frac{y^3}{3} \right]_1^2 = \frac{4}{3}\; [8-1] = \frac{28}{3}.$$

Example 10. *Evaluate* $\iint x^2 y \, dx \, dy$ *over the first quadrant of* $x^2 + y^2 = 1$.

(OU-Jan. 2012)

Sol. y limits are from 0 to $\sqrt{1-x^2}$ x limits are from 0 to 1.

Hence, $\int_{x=0}^{1} \boxed{\int_{y=0}^{\sqrt{1-x^2}} x^2\, y\, dy} \, dx$

$$= \int_{x=0}^{1} x^2 \left[\frac{y^2}{2}\right]_{y=0}^{\sqrt{1-x^2}} dx$$

$$= \frac{1}{2}\int_{x=0}^{1} x^2\,[1 - x^2 - 0]\, dx$$

$$= \frac{1}{2}\int_0^1 (x^2 - x^4)\, dx$$

$$= \frac{1}{2}\left[\frac{x^3}{3} - \frac{x^5}{5}\right]_0^1 = \frac{1}{2}\left[\frac{1}{3} - \frac{1}{5}\right] = \frac{1}{2}\left[\frac{2}{15}\right] = \frac{1}{15}.$$

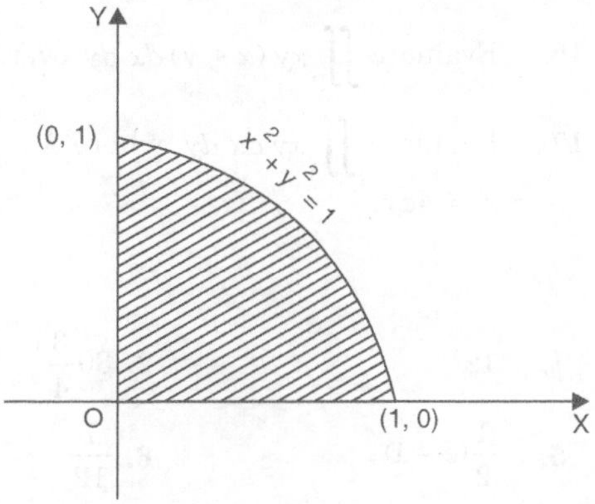

EXERCISE 6

Evaluate the following integrals (1–10) :

1. $\int_0^3\int_0^1 (x^2 + 3y^2)\, dy\, dx.$

2. $\int_0^3\int_1^2 xy\,(1 + x + y)\, dy\, dx.$

3. $\int_1^a\int_1^b \frac{dy\, dx}{xy}.$

4. $\int_1^2\int_0^x \frac{dy\, dx}{x^2 + y^2}.$

5. $\int_0^1 dx \int_0^x e^{y/x}\, dy.$

6. $\int_0^1\int_{x^2}^x (x^2 + 3y + 2)\, dy\, dx.$

7. $\int_0^1\int_0^{x^2} e^{y/x}\, dy\, dx.$

8. $\int_0^1\int_y^{y^2+1} x^2 y\, dx\, dy.$

9. $\int_0^{2a}\int_0^{\sqrt{2ax - x^2}} (x^2 + y^2)\, dy\, dx.$

10. $\int_0^1\int_x^{\sqrt{x}} (x^2 + y^2)\, dy\, dx.$

11. Evaluate $\iint (x^2 + y^2)\, dx\, dy$ over the region in the positive quadrant for which $x + y \le 1$.

12. Evaluate $\iint x^2y^2\, dx\, dy$ over the circle $x^2 + y^2 = 1$.

13. Evaluate $\iint xy\, dx\, dy$ over the positive quadrant of the circle $x^2 + y^2 = a^2$.

14. Compute the value of $\iint_R y\, dx\, dy$, where R is the region in the first quadrant bounded by the ellipse

$$\frac{x^2}{a^2} + \frac{y^2}{b^2} = 1.$$

15. Evaluate $\iint_S \sqrt{xy - y^2}\, dx\, dy$, where S is a triangle with vertices (0, 0), (10, 1) and (1, 1).

16. Evaluate $\iint_R xy\,(x+y)\,dx\,dy$ over the area between $y = x^2$ and $y = x$.

17. Evaluate $\iint_A xy\,dx\,dy$, where A is the domain bounded by x-axis, ordinate $x = 2a$ and the curve $x^2 = 4ay$.

Answers

1. 12	**2.** $30\frac{3}{4}$	**3.** $\log a \log b$	**4.** $\frac{\pi}{4}\log 2$
5. $\frac{1}{2}(e-1)$	**6.** $\frac{7}{12}$	**7.** $\frac{1}{2}$	**8.** $\frac{67}{120}$
9. $\frac{3\pi a^4}{4}$	**10.** $\frac{3}{35}$	**11.** $\frac{1}{6}$	**12.** $\frac{\pi}{24}$
13. $\frac{a^4}{8}$	**14.** $\frac{ab^2}{3}$	**15.** 6	**16.** $\frac{3}{56}$
17. $\frac{a^4}{3}$.			

4.16. EVALUATION OF DOUBLE INTEGRALS IN POLAR CO-ORDINATES (This Topic is for Student Reference Only)

To evaluate $\int_{\theta_1}^{\theta_2}\int_{r_1}^{r_2} f(r, \theta)\,dr\,d\theta$ over the region bounded by the straight lines $\theta = \theta_1$, $\theta = \theta_2$ and the curves $r = r_1$, $r = r_2$, we first integrate w.r.t. r between the limits $r = r_1$ and $r = r_2$ (treating θ as a constant). The resulting expression is then integrated w.r.t. θ between the limits $\theta = \theta_1$ and $\theta = \theta_2$.

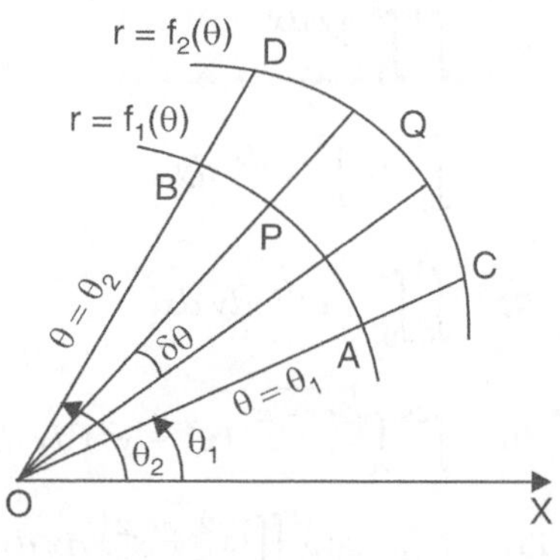

Geometrically, AB and CD are the curves $r = f_1(\theta)$ and $r = f_2(\theta)$ bounded by the lines $\theta = \theta_1$ and $\theta = \theta_2$ so that ACDB is the region of integration. PQ is a wedge of angular thickness $\delta\theta$. Then $\int_{r=r_1}^{r=r_2} f(r, \theta)\,dr$ indicates that the integration is performed along PQ (*i.e.*, r varies, θ is constant) and the integration w.r.t. θ means rotation of this strip PQ from AC to BD.

ILLUSTRATIVE EXAMPLES

Example 1. *Evaluate* $\int_0^{\pi/2}\left[\int_0^{a\cos\theta} r\sqrt{a^2 - r^2}\,dr\right]d\theta.$

Sol.
$$I = \int_0^{\pi/2}\left[\int_0^{a\cos\theta} -\frac{1}{2}(a^2 - r^2)^{1/2}(-2r)\,dr\right]d\theta$$

$$= \int_0^{\pi/2} \left[-\frac{1}{2} \cdot \frac{(a^2 - r^2)^{3/2}}{3/2} \right]_0^{a \cos \theta} d\theta$$

$$= -\frac{1}{3} \int_0^{\pi/2} (a^3 \sin^3 \theta - a^3)\, d\theta = -\frac{a^3}{3} \left[\frac{2}{3} - \frac{\pi}{2} \right] = \frac{a^3}{18} (3\pi - 4).$$

Example 2. *Evaluate* $\iint r^3\, dr d\theta$, *over the area bounded between the circles* $r = 2 \cos \theta$ *and* $r = 4 \cos \theta$.

Sol. The region of integration R is shown shaded. Here r varies from $2 \cos \theta$ to $4 \cos \theta$ while θ varies from $-\frac{\pi}{2}$ to $\frac{\pi}{2}$.

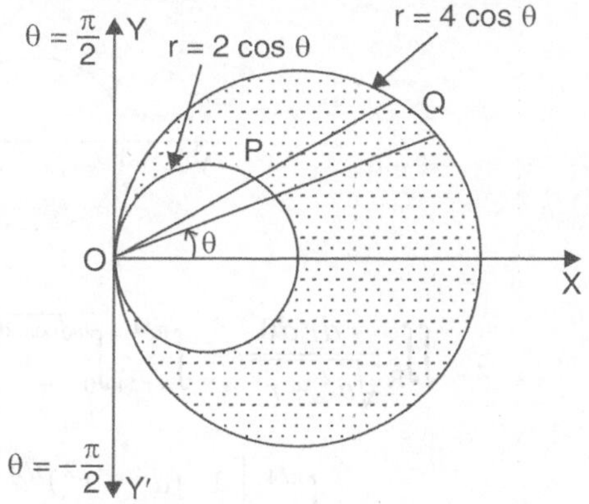

$$\therefore \quad \iint_R r^3\, dr\, d\theta = \int_{-\pi/2}^{\pi/2} \int_{2 \cos \theta}^{4 \cos \theta} r^3 dr\, d\theta$$

$$= \int_{-\pi/2}^{\pi/2} \left[\frac{r^4}{4} \right]_{2 \cos \theta}^{4 \cos \theta} d\theta$$

$$= \int_{-\pi/2}^{\pi/2} \frac{1}{4} (256 \cos^4 \theta - 16 \cos^4 \theta)\, d\theta$$

$$= 60 \int_{-\pi/2}^{\pi/2} \cos^4 \theta\, d\theta$$

$$= 120 \int_0^{\pi/2} \cos^4 \theta\, d\theta \qquad \text{(Since } \cos^4 \theta \text{ is an even function of } \theta)$$

$$= 120 \times \frac{3 \times 1}{4 \times 2} \cdot \frac{\pi}{2} = \frac{45}{2} \pi.$$

Example 3. *Evaluate* $\iint r \sin \theta\, dr d\theta$ *over the area of the cardioid* $r = a(1 + \cos \theta)$ *above the initial line.*

Sol. The region of integration R is covered by radial strips whose ends are $r = 0$ and $r = a\,(1 + \cos \theta)$, the strips starting from $\theta = 0$ and ending at $\theta = \pi$.

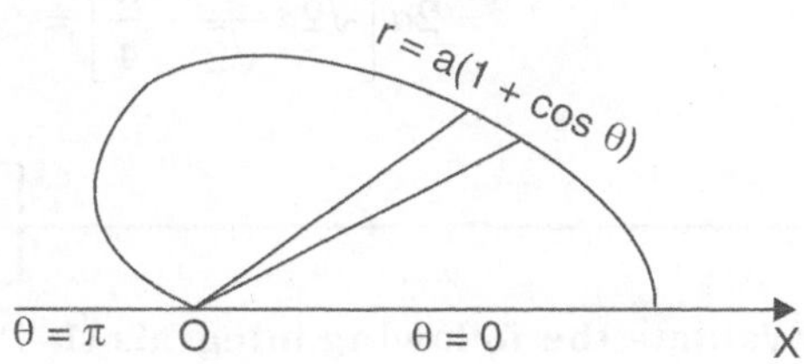

$$\therefore \quad \iint_R r \sin \theta\, dr\, d\theta = \int_0^{\pi} \int_0^{a(1 + \cos \theta)} r \sin \theta\, dr\, d\theta \cdot$$

$$= \int_0^{\pi} \sin \theta \left[\frac{r^2}{2} \right]_0^{a\,(1 + \cos \theta)} d\theta = \frac{1}{2} \int_0^{\pi} \sin \theta \,.\, a^2 (1 + \cos \theta)^2\, d\theta$$

$$= \frac{a^2}{2} \int_0^{\pi} 2 \sin \frac{\theta}{2} \cos \frac{\theta}{2} \cdot \left(2 \cos^2 \frac{\theta}{2} \right)^2 d\theta = 4a^2 \int_0^{\pi} \sin \frac{\theta}{2} \cos^5 \frac{\theta}{2}\, d\theta$$

$$= 4a^2 \int_0^{\pi/2} 2 \sin \phi \cos^5 \phi\, d\phi \qquad \left[\text{Putting } \frac{\theta}{2} = \phi \text{ and } d\theta = 2 d\phi \right]$$

$$= -8a^2 \int_0^{\pi/2} \cos^5 \phi \,(-\sin \phi)\, d\phi = -8a^2 \cdot \left[\frac{\cos^6 \phi}{6}\right]_0^{\pi/2} = -\frac{4a^2}{3}(0-1) = \frac{4a^2}{3}.$$

Example 4. *Evaluate* $\iint \frac{r\,dr\,d\theta}{\sqrt{a^2+r^2}}$ *over one loop of the lemniscate* $r^2 = a^2 \cos 2\theta$.

Sol. The region of integration R is covered by radial strips whose ends are $r = 0$ and $r = a\sqrt{\cos 2\theta}$, the strips starting from $\theta = -\frac{\pi}{4}$ and ending at $\theta = \frac{\pi}{4}$.

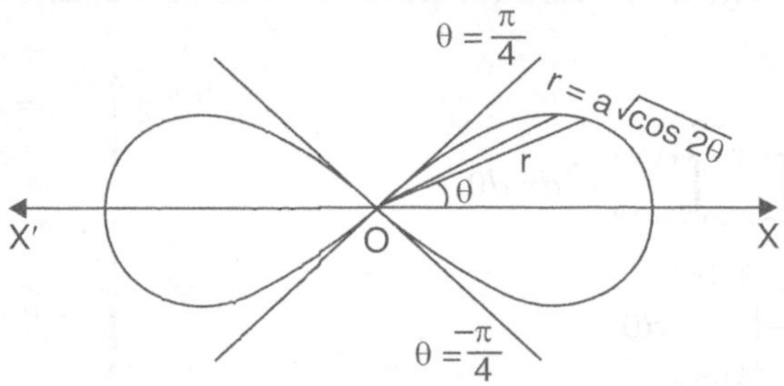

$$\therefore \iint_R \frac{r\,dr\,d\theta}{\sqrt{a^2+r^2}} = \int_{-\pi/4}^{\pi/4} \int_0^{a\sqrt{\cos 2\theta}} \frac{1}{2}(a^2+r^2)^{-1/2} \cdot 2r\,dr\,d\theta$$

$$= \int_{-\pi/4}^{\pi/4} \left[\frac{1}{2} \cdot \frac{(a^2+r^2)^{1/2}}{1/2}\right]_0^{a\sqrt{\cos 2\theta}} d\theta = \int_{-\pi/4}^{\pi/4} [(a^2 + a^2 \cos 2\theta)^{1/2} - a]\, d\theta$$

$$= a\int_{-\pi/4}^{\pi/4} [(1+\cos 2\theta)^{1/2} - 1]\, d\theta = a\int_{-\pi/4}^{\pi/4} [(2\cos^2\theta)^{1/2} - 1]\, d\theta$$

$$= a\int_{-\pi/4}^{\pi/4} (\sqrt{2}\cos\theta - 1)\, d\theta = 2a\int_0^{\pi/4} (\sqrt{2}\cos\theta - 1)\, d\theta = 2a\left[\sqrt{2}\sin\theta - \theta\right]_0^{\pi/4}$$

$$= 2a\left[\sqrt{2} \cdot \frac{1}{\sqrt{2}} - \frac{\pi}{4}\right] = 2a\left(1 - \frac{\pi}{4}\right).$$

EXERCISE 7

Evaluate the following integrals (1–4) :

1. $\int_0^{\pi} \int_0^{a \sin\theta} r\,dr\,d\theta.$
2. $\int_0^{\pi/2} \int_0^{a\cos\theta} r\sin\theta\,dr\,d\theta.$
3. $\int_0^{\pi/2} \int_{a(1+\cos\theta)}^{a} r\,dr\,d\theta.$
4. $\int_0^{\pi} \int_0^{a(1+\cos\theta)} r^2 \cos\theta\,dr\,d\theta.$
5. Show that $\iint_R r^2 \sin\theta\,dr\,d\theta = \frac{2a^3}{3}$, where R is the region bounded by the semi-circle $r = 2a \cos\theta$, above the initial line.
6. Evaluate $\iint r^3\,dr\,d\theta$ over the area included between the circles $r = 2\sin\theta$ and $r = 4\sin\theta$.

Answers

1. $\dfrac{\pi a^2}{4}$ 2. $\dfrac{a^2}{6}$ 3. $-a^2\left(1+\dfrac{\pi}{8}\right)$

4. $\dfrac{5}{8}\pi a^3$ 6. $\dfrac{45\pi}{2}$.

4.17. CHANGE OF ORDER OF INTEGRATION (This Topic is for Student Reference Only)

In a double integral, if the limits of integration are constant, then the order of integration is immaterial, provided the limits of integration are changed accordingly. Thus,

$$\int_c^d \int_a^b f(x, y)\, dx\, dy = \int_a^b \int_c^d f(x, y)\, dy\, dx\,.$$

But if the limits of integration are variable, a change in the order of integration necessitates change in the limits of integration. A rough sketch of the region of integration helps in fixing the new limits of integration.

ILLUSTRATIVE EXAMPLES

Example 1. *Change the order of integration in* $\int_0^a \int_y^a \dfrac{x\, dx\, dy}{x^2 + y^2}$ *and hence evaluate the same.*

Sol. From the limits of integration, it is clear that the region of integration is bounded by $x = y$, $x = a$, $y = 0$ and $y = a$. Thus the region of integration is the ΔOAB and is divided into horizontal strips. For changing the order of integration, we divide the region of integration into vertical strips. The new limits of integration become : y varies from 0 to x and x varies from 0 to a.

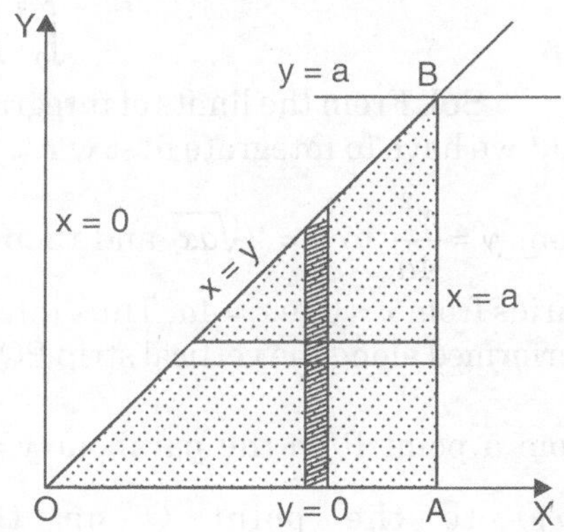

$$\int_0^a \int_y^a \frac{x\, dx\, dy}{x^2 + y^2} = \int_0^a \int_0^x \frac{x\, dy\, dx}{x^2 + y^2}$$

$$= \int_0^a x \cdot \left[\frac{1}{x} \tan^{-1} \frac{y}{x}\right]_0^x dx$$

$$= \int_0^a \frac{\pi}{4}\, dx = \frac{\pi}{4} \cdot \Big[x\Big]_0^a = \frac{\pi a}{4}.$$

Example 2. *Change the order of integration in* $I = \int_0^1 \int_{x^2}^{2-x} xy\, dy\, dx$ *and hence evaluate the same.*

Sol. From the limits of integration, it is clear that we have to integrate first with respect to y which varies from $y = x^2$ to $y = 2 - x$ and then with respect to x which varies from $x = 0$ to $x = 1$. The region of integration (shown shaded) is divided into vertical strips. For changing the order of integration, we divide the region of integration into horizontal strips.

Solving $y = x^2$ and $y = 2 - x$, the co-ordinates of A are (1, 1). Draw AM $\perp$ OY. The region of integration is divided into two parts, OAM and MAB.

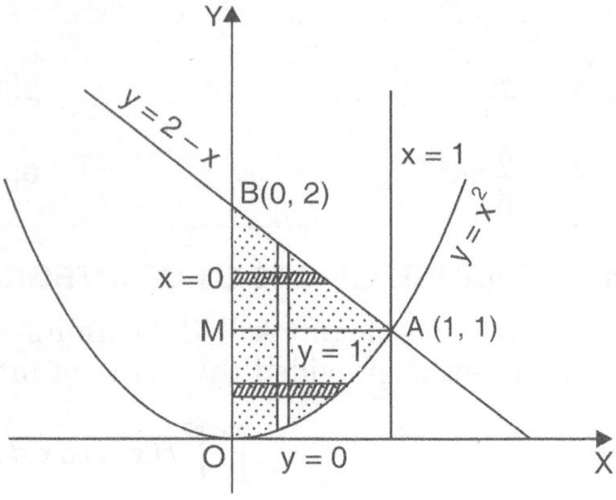

For the region OAM, x varies from 0 to $\sqrt{y}$ and y varies from 0 to 1. For the region MAB, x varies from 0 to $2 - y$ and y varies from 1 to 2.

$$\therefore \int_0^1 \int_{x^2}^{2-x} xy\, dx\, dy = \int_0^1 \int_0^{\sqrt{y}} xy\, dx\, dy + \int_1^2 \int_0^{2-y} xy\, dx\, dy$$

$$= \int_0^1 y \cdot \left[\frac{x^2}{2}\right]_0^{\sqrt{y}} dy + \int_1^2 y \cdot \left[\frac{x^2}{2}\right]_0^{2-y}$$

$$dy = \frac{1}{2}\int_0^1 y^2 dy + \frac{1}{2}\int_1^2 y(2-y)^2\, dy$$

$$= \frac{1}{2} \cdot \left[\frac{y^3}{3}\right]_0^1 + \frac{1}{2}\int_1^2 (4y - 4y^2 + y^3)\, dy = \frac{1}{6} + \frac{1}{2}\left[2y^2 - \frac{4y^3}{3} + \frac{y^4}{4}\right]_1^2$$

$$= \frac{1}{6} + \frac{1}{2}\left[\left(8 - \frac{32}{3} + 4\right) - \left(2 - \frac{4}{3} + \frac{1}{4}\right)\right] = \frac{3}{8}.$$

Example 3. *Change the order of integration in the following integral and evaluate :*

$$\int_0^{4a} \int_{x^2/4a}^{2\sqrt{ax}} dy\, dx.$$

Sol. From the limits of integration, it is clear that we have to integrate first w.r.t. y which varies from $y = \dfrac{x^2}{4a}$ to $y = 2\sqrt{ax}$ and then w.r.t. x which varies from $x = 0$ to $x = 4a$. Thus integration is first performed along the vertical strip PQ which extends from a point P on the parabola $y = \dfrac{x^2}{4a}$ (*i.e.*, $x^2 = 4ay$) to the point Q on the parabola $y = 2\sqrt{ax}$ (*i.e.*, $y^2 = 4ax$). Then the strip slides from O to A $(4a, 4a)$, the point of intersection of the two parabolas.

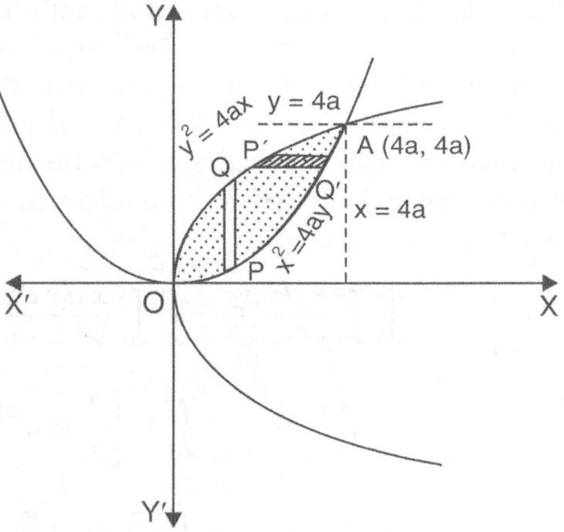

For changing the order of integration, we divide the region of integration OPAQO into horizontal strips P′Q′ which extend from P′ on the parabola $y^2 = 4ax$ *i.e.*, $x = \dfrac{y^2}{4a}$ to Q′ on the parabola $x^2 = 4ay$ *i.e.*, $x = 2\sqrt{ay}$. Then this strip slides from O to A$(4a, 4a)$, *i.e.*, varies from 0 to $4a$.

$$\therefore \int_0^{4a} \int_{x^2/4a}^{2\sqrt{ax}} dy\, dx = \int_0^{4a} \int_{y^2/4a}^{2\sqrt{ay}} dx\, dy$$

$$= \int_0^{4a} \left[x\right]_{y^2/4a}^{2\sqrt{ay}} dy = \int_0^{4a} \left(2\sqrt{ay} - \frac{y^2}{4a}\right) dy$$

$$= \left[2\sqrt{a}\cdot\frac{y^{3/2}}{3/2} - \frac{y^3}{12a}\right]_0^{4a} = \frac{4}{3}\sqrt{a}\cdot(4a)^{3/2} - \frac{64a^3}{12a}$$

$$= \frac{4}{3}\sqrt{a}\cdot 8a^{3/2} - \frac{16a^2}{3} = \frac{32a^2}{3} - \frac{16a^2}{3} = \frac{16a^2}{3}.$$

Example 4. *Change the order of integration in the integral* $\int_{-a}^{a}\int_{0}^{\sqrt{a^2-y^2}} f(x, y)\, dx\, dy.$

Sol. From the limits of integration, it is clear that we have to integrate first w.r.t. x which varies from $x = 0$ to $x = \sqrt{a^2 - y^2}$ and then w.r.t. y which varies from $y = -a$ to $y = a$. Thus integration is first performed along the horizontal strip PQ which extends from a point P on $x = 0$ (*i.e.*, y-axis) to the point Q on the circle $x = \sqrt{a^2 - y^2}$ (*i.e.*, $x^2 = a^2 - y^2$ or $x^2 + y^2 = a^2$). Then the strip slides from B′ to B.

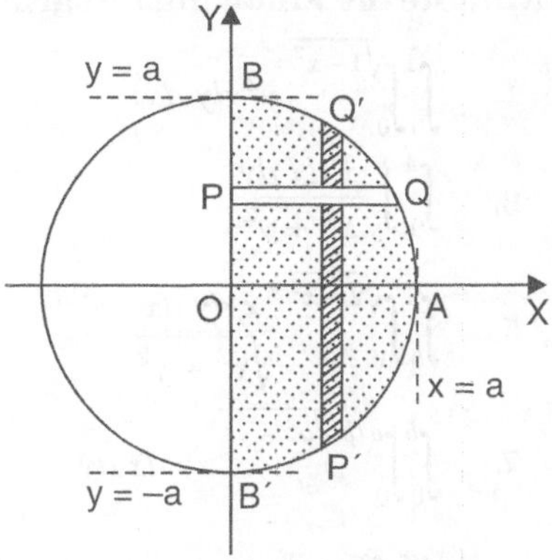

For changing the order of integration, we divide the region of integration B′AQBPB′ into vertical strips P′Q′ which extend from P′ on the circle $y = -\sqrt{a^2 - x^2}$ to Q′ on the circle $y = \sqrt{a^2 - x^2}$. $[x^2 + y^2 = a^2 \Rightarrow y = \pm\sqrt{a^2 - x^2}$; for points in the IV quadrant, $y = -\sqrt{a^2 - x^2}$ and for points in the first quadrant, $y = \sqrt{a^2 - x^2}]$. Then this strip slides from y-axis ($x = 0$) to A, where $x = a$.

$$\therefore \quad \int_{-a}^{a}\int_{0}^{\sqrt{a^2-x^2}} f(x,y)\,dx\,dy = \int_{0}^{a}\int_{-\sqrt{a^2-x^2}}^{\sqrt{a^2-x^2}} f(x,y)\,dy\,dx.$$

Example 5. *Evaluate* $\int_0^1\int_{2y}^{2} e^{x^2}\, dxdy$ *by changing the order of integration.*

(OU-Dec. 2013)

Sol. Given limits are

$$x = 2y \quad \text{to} \quad x = 2$$
$$y = 0 \quad \text{to} \quad y = 1$$

After changing the order of integration, we get new limits are

$$y = 0 \quad \text{to} \quad y = \frac{x}{2}$$
$$x = 0 \quad \text{to} \quad x = 2$$

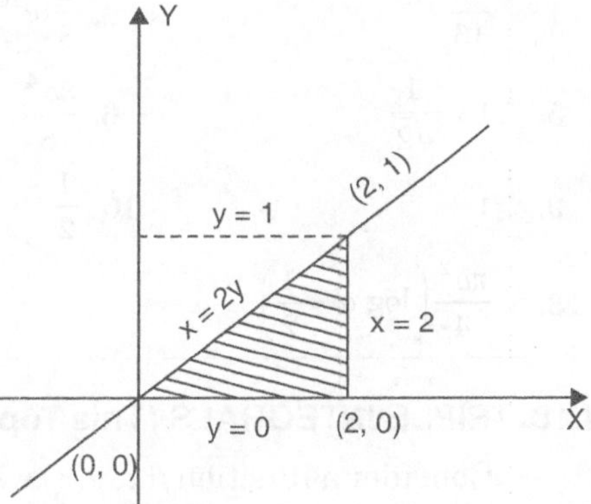

$$\therefore \quad \int_{x=0}^{2}\boxed{\int_{y=0}^{y=\frac{x}{2}} e^{x^2}\, dy}\, dx$$

$$= \int_{x=0}^{2} e^{x^2}\left[y\right]_{y=0}^{y=\frac{x}{2}} dx = \int_{x=0}^{2} e^{x^2}\cdot\frac{x}{2}\, dx$$

$$= \frac{1}{4} \int_{x=0}^{2} e^{x^2} (2x)\, dx \qquad \left(\int e^{f(x)} . f'(x)\, dx = e^{f(x)} \right)$$

$$= \frac{1}{4} \left[e^{x^2} \right]_0^2 = \frac{1}{4} \, [e^4 - e^0] = \frac{1}{4} \, [e^4 - 1].$$

EXERCISE 8

Evaluate the following integrals by changing the order of integration :

1. $\int_0^1 \int_0^{\sqrt{1-x^2}} y^2 \, dy \, dx.$

2. $\int_0^1 \int_{4y}^4 e^{x^2} dx \, dy.$

3. $\int_0^4 \int_y^4 \frac{x \, dx \, dy}{x^2 + y^2}.$

4. $\int_0^{a/\sqrt{2}} \int_y^{\sqrt{a^2 - y^2}} x \, dx \, dy.$

5. $\int_0^1 \int_x^{\sqrt{2 - x^2}} \frac{x \, dy \, dx}{\sqrt{x^2 + y^2}}.$

6. $\int_0^a \int_{x^2/a}^{2a - x} xy \, dy \, dx.$

7. $\int_0^b \int_0^{a/b \sqrt{b^2 - y^2}} xy \, dx \, dy.$

8. $\int_0^a \int_{\sqrt{ax}}^a \frac{y^2 \, dy \, dx}{\sqrt{y^4 - a^2 x^2}}.$

9. $\int_0^\infty \int_x^\infty \frac{e^{-y}}{y} \, dy \, dx.$

10. $\int_0^\infty \int_0^x x e^{-\frac{x^2}{y}} \, dy \, dx.$

11. $\int_0^3 \int_1^{\sqrt{4 - y}} (x + y) \, dx \, dy.$

12. $\int_0^a \int_{x/a}^{\sqrt{x/a}} (x^2 + y^2) \, dy \, dx.$

13. $\int_0^{\frac{a}{\sqrt{2}}} \int_y^{\sqrt{a^2 - y^2}} \log(x^2 + y^2) \, dx \, dy, \ (a > 0).$

Answers

1. $\frac{\pi}{16}$ 2. $\frac{1}{8}(e^{16} - 1)$ 3. π 4. $\frac{\sqrt{2}a^3}{6}$

5. $1 - \frac{1}{\sqrt{2}}$ 6. $\frac{3a^4}{8}$ 7. $\frac{1}{8} a^2 b^2$ 8. $\frac{\pi a^2}{6}$

9. 1 10. $\frac{1}{2}$ 11. $\frac{241}{60}$ 12. $\frac{a^3}{28} + \frac{a}{20}$

13. $\frac{\pi a^2}{4} \left(\log a - \frac{1}{2} \right).$

4.18. TRIPLE INTEGRALS (This Topic is for Student Reference Only)

Consider a function $f(x, y, z)$ which is continuous at every point of a finite region V of three dimensional space. Divide the region V into n sub-regions of respective volumes $\delta V_1, \delta V_2, ..., \delta V_n$.

Let (x_r, y_r, z_r) be an arbitrary point in the rth sub-region. Consider the sum $\sum_{r=1}^{n} f(x_r, y_r, z_r) \, \delta V_r$.

The limit of this sum as $n \to \infty$ and $\delta V_r \to 0$, if it exists, is called the *triple integral* of $f(x, y, z)$ over the region V and is denoted by $\iiint_V f(x, y, z)\, dV$.

For purposes of evaluation, it can be expressed as the repeated integral

$$\int_{z_1}^{z_2} \int_{y_1}^{y_2} \int_{x_1}^{x_2} f(x, y, z)\, dx\, dy\, dz, \qquad \ldots(1)$$

the order of integration depending upon the limits.

Let x_1, x_2 be function of y, z ; y_1, y_2 be function of z and z_1, z_2 be constants, *i.e.*,

Let $\quad x_1 = f_1(y, z), x_2 = f_2(y, z), y_1 = \phi_1(z), y_2 = \phi_2(z)$ and $z_1 = a, z_2 = b$.

Then the integral (1) is evaluated as follows :

$$\int_{z_1 = a}^{z_2 = b} \left[\int_{y_1 = \phi_1(z)}^{y_2 = \phi_2(z)} \left[\int_{x_1 = f_1(y, z)}^{x_2 = f_2(y, z)} f(x, y, z)\, dx \right] dy \right] dz$$

First $f(x, y, z)$ is integrated w.r.t. x (keeping y and z constant) between the limits x_1 and x_2. The resulting expression, which is a function of y and z is then integrated w.r.t. y (keeping z constant) between the limits y_1 and y_2. The resulting expression, which is a function of z only is then integrated w.r.t. z between the limits z_1 and z_2. The order of integration is from the innermost rectangle to the outermost rectangle.

Limits involving two variables are kept innermost, then the limits involving one variable and finally the constant limits.

If x_1, x_2 ; y_1, y_2 and z_1, z_2 are all constants, then the order of integration is immaterial, provided the limits are changed accordingly. Thus,

$$\int_{z_1}^{z_2} \int_{y_1}^{y_2} \int_{x_1}^{x_2} f(x, y, z)\, dx\, dy\, dz = \int_{x_1}^{x_2} \int_{z_1}^{z_2} \int_{y_1}^{y_2} f(x, y, z)\, dy\, dz\, dx$$

$$= \int_{x_1}^{x_2} \int_{y_1}^{y_2} \int_{z_1}^{z_2} f(x, y, z)\, dz\, dy\, dx, \text{ etc.}$$

ILLUSTRATIVE EXAMPLES

Example 1. *Evaluate* $\displaystyle \int_0^1 \int_0^{\sqrt{1-x^2}} \int_0^{\sqrt{1-x^2-y^2}} \frac{1}{\sqrt{1-x^2-y^2-z^2}}\, dz\, dy\, dx.$

Sol.

$$I = \int_0^1 \int_0^{\sqrt{1-x^2}} \int_0^{\sqrt{1-x^2-y^2}} \frac{1}{\sqrt{(1-x^2-y^2)-z^2}}\, dz\, dy\, dx$$

$$= \int_0^1 \int_0^{\sqrt{1-x^2}} \left[\sin^{-1} \frac{z}{\sqrt{1-x^2-y^2}} \right]_0^{\sqrt{1-x^2-y^2}} dy\, dx$$

$$= \int_0^1\int_0^{\sqrt{1-x^2}} (\sin^{-1} 1 - \sin^{-1} 0)\, dy\, dx$$

$$= \int_0^1\int_0^{\sqrt{1-x^2}} \frac{\pi}{2}\, dy\, dx = \int_0^1 \frac{\pi}{2}\Big[\, y \,\Big]_0^{\sqrt{1-x^2}} dx$$

$$= \frac{\pi}{2}\int_0^1 \sqrt{1-x^2}\, dx = \frac{\pi}{2}\left[\frac{x\sqrt{1-x^2}}{2} + \frac{1}{2}\sin^{-1} x\right]_0^1$$

$$= \frac{\pi}{4}[\sin^{-1} 1] = \frac{\pi}{4}\cdot\frac{\pi}{2} = \frac{\pi^2}{8}.$$

Example 2. *Evaluate* $\int_1^e \int_1^{\log y} \int_1^{e^x} \log z\, dz\, dx\, dy$. (OU-Apr. 2007)

Sol. $$I = \int_1^e \int_1^{\log y}\left[\int_1^{e^x} \log z\, dz\right] dx\, dy$$

$$= \int_1^e \int_1^{\log y}\left(\log z\,.\,z - \int \frac{1}{z}\,.\,z\, dz\right)_1^{e^x} dx\, dy$$

$$= \int_1^e \int_1^{\log y} \Big\{z\,(\log z - 1)\Big\}_1^{e^x} dx\, dy$$

$$= \int_1^e \int_1^{\log y} \{(x-1)\,e^x + 1)\}\, dx\, dy$$

Now, $\int_1^{\log y} [(x-1)\,e^x + 1]\, dx = \int_1^{\log y} (x-1)e^x\, dx + \Big[\, x \,\Big]_1^{\log y}$

$$= \Big[(x-1)\,e^x\Big]_1^{\log y} - \int_1^{\log y} 1\,.\,e^x\, dx + \log y - 1$$

$$= (\log y - 1)\,e^{\log y} - \Big[\, e^x \,\Big]_1^{\log y} + \log y - 1$$

$$= y(\log y - 1) - (e^{\log y} - e) + \log y - 1 \qquad [\because\ e^{\log y} = y]$$

$$= y(\log y - 1) - y + e + \log y - 1 = (y+1)\log y - 2y + e - 1$$

$\therefore$ $$I = \int_1^e [\log y\,.\,(y+1) - 2y + e - 1]\, dy$$

$$= \left[\log y\,.\left(\frac{y^2}{2} + y\right)\right]_1^e - \int_1^e \frac{1}{y}\left(\frac{y^2}{2} + y\right) dy - \Big[\, y^2 \,\Big]_1^e + (e-1)\Big[\, y \,\Big]_1^e$$

$$= \frac{e^2}{2} + e - \int_1^e \left(\frac{y}{2} + 1\right) dy - (e^2 - 1) + (e-1)^2$$

$$= \frac{e^2}{2} + e - \left[\frac{y^2}{4} + y\right]_1^e - 2e + 2 = \frac{e^2}{2} + e - \left[\left(\frac{e^2}{4} + e\right) - \left(\frac{1}{4} + 1\right)\right] - 2e + 2$$

$$= \frac{e^2}{4} - 2e + \frac{13}{4} = \frac{1}{4}(e^2 - 8e + 13).$$

EXERCISE 9

Evaluate the following integrals (1–11) :

1. $\int_0^1 \int_0^1 \int_0^1 e^{x+y+z}\, dx\, dy\, dz.$

2. $\int_0^a \int_0^a \int_0^a (yz + zx + xy)\, dx\, dy\, dz.$

3. $\int_{-1}^1 \int_0^z \int_{x-z}^{x+z} (x + y + z)\, dy\, dx\, dz.$

4. $\int_1^3 \int_{1/x}^1 \int_0^{\sqrt{xy}} xyz\, dz\, dy\, dx.$

5. $\int_0^1 \int_0^{1-x} \int_0^{1-x-y} xyz\, dz\, dy\, dx.$

6. $\int_0^1 \int_{y^2}^1 \int_0^{1-x} x\, dz\, dx\, dy.$

7. $\int_0^1 \int_0^{\sqrt{1-x^2}} \int_0^{\sqrt{1-x^2-y^2}} xyz\, dz\, dy\, dx.$

8. $\int_0^{\log 2} \int_0^x \int_0^{x+\log y} e^{x+y+z}\, dz\, dy\, dx.$

9. $\int_0^{\pi/2} d\theta \int_0^{a\sin\theta} dr \int_0^{(a^2-r^2)/a} r dz.$

10. $\int_0^{\pi/2} \int_0^{a\cos\theta} \int_0^{\sqrt{a^2-r^2}} r dz\, dr\, d\theta.$

11. $\int_0^a \int_0^x \int_0^{x-y} e^{x+y+z}\, dz\, dy\, dx.$

Answers

1. $(e-1)^3$ **2.** $\frac{3}{4}a^5$ **3.** 0 **4.** $\frac{13}{9} - \frac{1}{6}\log 3$

5. $\frac{1}{720}$ **6.** $\frac{4}{35}$ **7.** $\frac{1}{48}$ **8.** $\frac{8}{3}\log 2 - \frac{19}{9}$

9. $\frac{5\pi a^3}{64}$ **10.** $\frac{a^3}{3}\left(\frac{\pi}{2} - \frac{2}{3}\right)$ **11.** $\frac{1}{8}e^{4a} - \frac{3}{4}e^{2a} + e^a - \frac{3}{8}.$

4.19. CHANGE OF VARIABLES

Quite often, the evaluation of a double or triple integral is greatly simplified by a suitable change of variables.

Let the variables x, y in the double integral $\iint_R f(x, y)\, dx\, dy$ be changed to u, v by means of the relations $x = \phi(u, v)$, $y = \psi(u, v)$, then the double integral is transformed into

$$\iint_{R'} f\{\phi(u, v), \psi(u, v)\}\ |\,J\,|\ du\, dv, \quad \text{where } J = \frac{\partial(x, y)}{\partial(u, v)} = \begin{vmatrix} \frac{\partial x}{\partial u} & \frac{\partial x}{\partial v} \\ \frac{\partial y}{\partial u} & \frac{\partial y}{\partial v} \end{vmatrix}$$ is the Jacobian of

transformation from (x, y) to (u, v) co-ordinates and R′ is the region in the uv-plane which corresponds to the region R in the xy-plane.

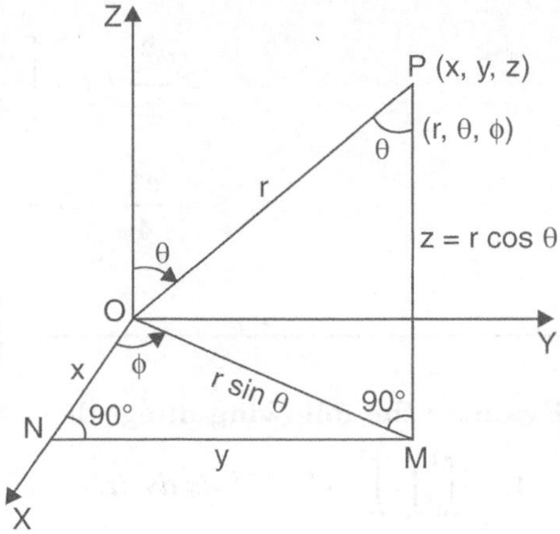

(i) To change cartesian co-ordinates (x, y) to polar co-ordinates (r, θ).

Here we have $x = r\cos\theta$, $y = r\sin\theta$ so that $x^2 + y^2 = r^2$

$$J = \begin{vmatrix} \dfrac{\partial x}{\partial r} & \dfrac{\partial x}{\partial \theta} \\ \dfrac{\partial y}{\partial r} & \dfrac{\partial y}{\partial \theta} \end{vmatrix}$$

$$= \begin{vmatrix} \cos\theta & -r\sin\theta \\ \sin\theta & r\cos\theta \end{vmatrix} = r(\cos^2\theta + \sin^2\theta) = r$$

$$\therefore \iint_R f(x, y)\, dx\, dy = \iint_{R'} f(r\cos\theta, r\sin\theta)\, r\, dr\, d\theta$$

i.e., replace x by $r\cos\theta$, y by $r\sin\theta$ and $dx\,dy$ by $r\,dr\,d\theta$.

(ii) To change cartesian co-ordinates (x, y, z) to spherical polar co-ordinates (r, θ, φ).

Here, we have

$$x = r\sin\theta\cos\phi$$
$$y = r\sin\theta\sin\phi$$
$$z = r\cos\theta$$

so that $x^2 + y^2 + z^2 = r^2$

$$J = \begin{vmatrix} \dfrac{\partial x}{\partial r} & \dfrac{\partial x}{\partial \theta} & \dfrac{\partial x}{\partial \phi} \\ \dfrac{\partial y}{\partial r} & \dfrac{\partial y}{\partial \theta} & \dfrac{\partial y}{\partial \phi} \\ \dfrac{\partial z}{\partial r} & \dfrac{\partial z}{\partial \theta} & \dfrac{\partial z}{\partial \phi} \end{vmatrix} = \begin{vmatrix} \sin\theta\cos\phi & r\cos\theta\cos\phi & -r\sin\theta\sin\phi \\ \sin\theta\sin\phi & r\cos\theta\sin\phi & r\sin\theta\cos\phi \\ \cos\theta & -r\sin\theta & 0 \end{vmatrix}$$

$$= r^2\sin\theta$$

$$\therefore \iiint_V f(x,y,z)\, dx\, dy\, dz = \iiint_{V'} f(r\sin\theta\cos\phi, r\sin\theta\sin\phi, r\cos\theta)\, r^2\sin\theta\, dr\, d\theta\, d\phi.$$

(iii) To change cartesian co-ordinates (x, y , z) to cylindrical co-ordinates (r, φ, z).

Here, we have

$$x = r\cos\phi$$
$$y = r\sin\phi$$
$$z = z$$

$$J = \begin{vmatrix} \dfrac{\partial x}{\partial r} & \dfrac{\partial x}{\partial \phi} & \dfrac{\partial x}{\partial z} \\ \dfrac{\partial y}{\partial r} & \dfrac{\partial y}{\partial \phi} & \dfrac{\partial y}{\partial z} \\ \dfrac{\partial z}{\partial r} & \dfrac{\partial z}{\partial \phi} & \dfrac{\partial z}{\partial z} \end{vmatrix} = \begin{vmatrix} \cos\phi & -r\sin\phi & 0 \\ \sin\phi & r\cos\phi & 0 \\ 0 & 0 & 1 \end{vmatrix}$$

$$= r(\cos^2\phi + \sin^2\phi) = r$$

$$\therefore \iiint_V f(x, y, z)\, dx\, dy\, dz$$

$$= \iiint_{V'} f(r\cos\phi, r\sin\phi, z)\, r\, dr\, d\phi\, dz.$$

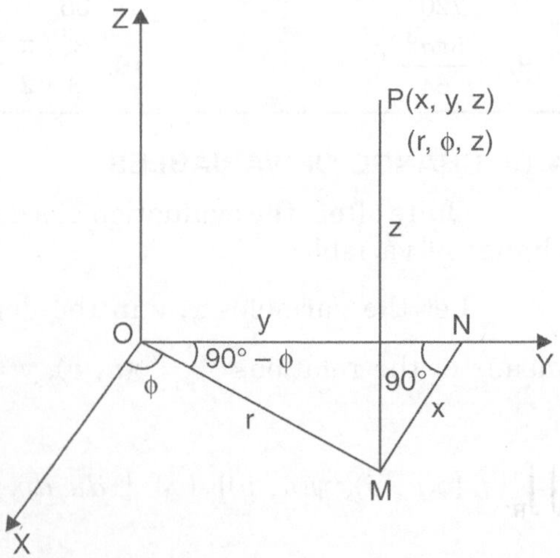

ILLUSTRATIVE EXAMPLES

Example 1. *Evaluate* $\iint \sqrt{\frac{1-x^2-y^2}{1+x^2+y^2}}\, dx\, dy$ *over the positive quadrant of the circle* $x^2 + y^2 = 1$.

Sol. Changing to polar co-ordinates by putting $x = r\cos\theta$, $y = r\sin\theta$; $x^2 + y^2 = 1$ transforms into $r = 1$. For the region of integration R, r varies from 0 to 1 and θ varies from 0 to $\frac{\pi}{2}$.

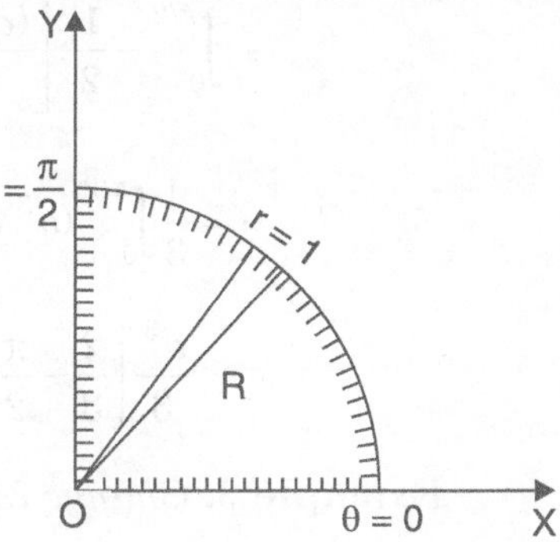

$$I = \iint_R \sqrt{\frac{1-x^2-y^2}{1+x^2+y^2}}\, dx\, dy$$

$$= \int_0^{\pi/2}\int_0^1 \sqrt{\frac{1-r^2}{1+r^2}}\, r\, dr\, d\theta$$

| $dx\, dy$ is replaced by $r\, dr\, d\theta$

$$= \int_0^{\pi/2}\int_0^1 \frac{r(1-r^2)}{\sqrt{1-r^4}}\, dr\, d\theta$$

Now, $\int_0^1 \frac{r(1-r^2)}{\sqrt{1-r^4}}\, dr = \int_0^1 \left(\frac{r}{\sqrt{1-r^4}} - \frac{r^3}{\sqrt{1-r^4}}\right) dr$

$$= \frac{1}{2}\int_0^1 \frac{2r}{\sqrt{1-r^4}}\, dr + \frac{1}{4}\int_0^1 -4r^3(1-r^4)^{-1/2}\, dr$$

$$= \frac{1}{2}\int_0^1 \frac{dt}{\sqrt{1-t^2}} + \frac{1}{4}\cdot\left[\frac{(1-r^4)^{1/2}}{1/2}\right]_0^1, \text{ where } t = r^2$$

$$= \frac{1}{2}\left[\sin^{-1} t\right]_0^1 + \frac{1}{2}(0-1) = \frac{1}{2}\left(\frac{\pi}{2}\right) - \frac{1}{2} = \frac{\pi}{4} - \frac{1}{2}$$

$$\therefore \quad I = \int_0^{\pi/2}\left(\frac{\pi}{4} - \frac{1}{2}\right) d\theta = \left(\frac{\pi}{4} - \frac{1}{2}\right)\left[\theta\right]_0^{\pi/2} = \left(\frac{\pi}{4} - \frac{1}{2}\right)\frac{\pi}{2} = \frac{\pi^2}{8} - \frac{\pi}{4}.$$

Example 2. *Evaluate* $\iint \sqrt{a^2 - x^2 - y^2}\, dx\, dy$ *over the semi-circle* $x^2 + y^2 = ax$ *in the positive quadrant.*

Sol. Changing to polar co-ordinates, $x^2 + y^2 = ax$ transforms into $r = a\cos\theta$. For the region of integration R, r varies from 0 to $a\cos\theta$ and θ varies from 0 to $\frac{\pi}{2}$.

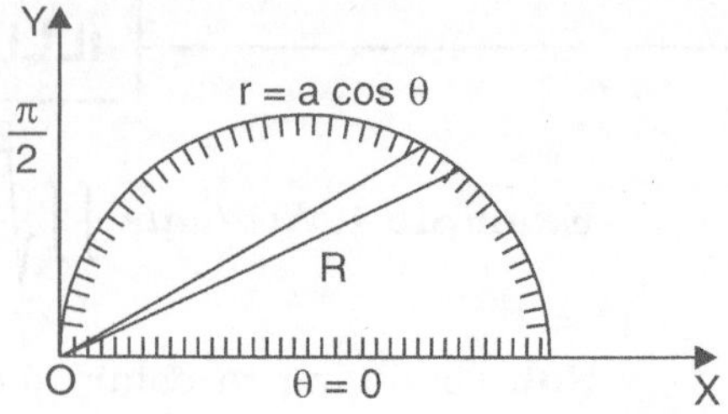

$$\therefore \iint_R \sqrt{a^2 - x^2 - y^2}\, dx\, dy$$

$$= \int_0^{\pi/2} \int_0^{a \cos\theta} \sqrt{a^2 - r^2}\,.\, r\, dr\, d\theta$$

$$= \int_0^{\pi/2} \int_0^{a \cos\theta} -\frac{1}{2}(a^2 - r^2)^{1/2}\,(-2r)\, dr d\theta$$

$$= \int_0^{\pi/2} -\frac{1}{2}.\left[\frac{(a^2 - r^2)^{3/2}}{3/2}\right]_0^{a\cos\theta} d\theta$$

$$= -\frac{1}{3}\int_0^{\pi/2} (a^3 \sin^3\theta - a^3)\, d\theta = -\frac{a^3}{3}\int_0^{\pi/2} (\sin^3\theta - 1)\, d\theta$$

$$= -\frac{a^3}{3}\left[\frac{2}{3} - \frac{\pi}{2}\right] = \frac{a^3}{3}\left(\frac{\pi}{2} - \frac{2}{3}\right).$$

Example 3. *Change into polar co-ordinates and evaluate* $\int_0^\infty \int_0^\infty e^{-(x^2 + y^2)}\, dy\, dx$.

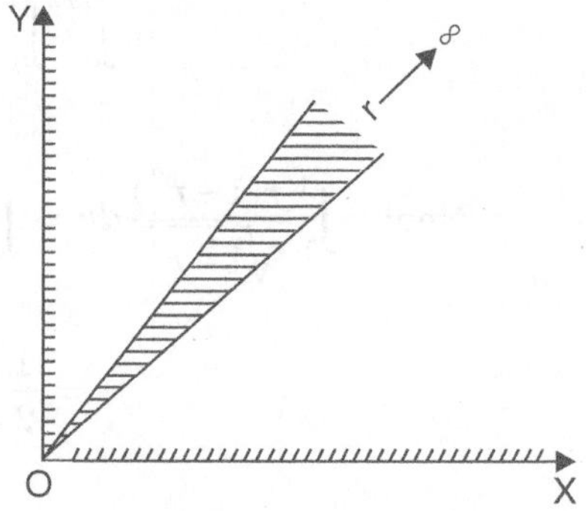

Sol. For the region of integration in cartesian co-ordinates, y varies from 0 to ∞ and x also varies from 0 to ∞. Thus the region of integration is the plane XOY. Changing to polar co-ordinates by putting $x = r\cos\theta$, $y = r\sin\theta$ so that $x^2 + y^2 = r^2$; for the region of integration r varies from 0 to ∞ and θ varies from 0 to $\frac{\pi}{2}$.

$$\therefore \int_0^\infty \int_0^\infty e^{-(x^2 + y^2)}\, dy\, dx = \int_0^{\pi/2} \int_0^\infty e^{-r^2} \cdot r\, dr\, d\theta$$

$$= \int_0^{\pi/2} \int_0^\infty \frac{1}{2} \cdot e^{-r^2} \cdot 2r\, dr\, d\theta$$

$$= \int_0^{\pi/2} \int_0^\infty \frac{1}{2} e^{-t} dt\, d\theta, \text{where } t = r^2$$

$$= \int_0^{\pi/2} \left[-\frac{1}{2}e^{-t}\right]_0^\infty d\theta = -\frac{1}{2}\int_0^{\pi/2} (0 - 1)\, d\theta = \frac{1}{2}\left[\theta\right]_0^{\pi/2} = \frac{\pi}{4}.$$

Example 4. *Evaluate* $\int_0^2 \int_0^{\sqrt{2x - x^2}} \frac{x\, dy\, dx}{\sqrt{x^2 + y^2}}$ *by changing to polar co-ordinates.*

Sol. In the given integral, y varies from 0 to $\sqrt{2x - x^2}$ and x varies from 0 to 2.

$$y = \sqrt{2x - x^2} \;\Rightarrow\; y^2 = 2x - x^2 \;\Rightarrow\; x^2 + y^2 = 2x.$$

In polar co-ordinates, we have $r^2 = 2r \cos\theta$ or $r = 2\cos\theta$.

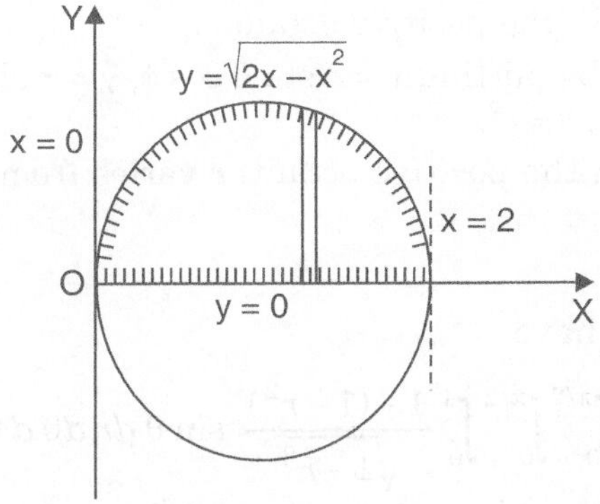

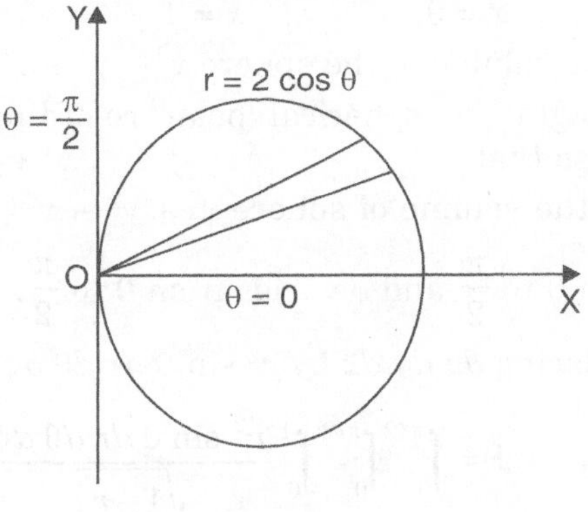

$\therefore$ For the region of integration, r varies from 0 to $2\cos\theta$ and θ varies from 0 to $\frac{\pi}{2}$.

In the given integral, replacing x by $r\cos\theta$, y by $r\sin\theta$, $dy\,dx$ by $r\,dr\,d\theta$, we have

$$I = \int_0^{\pi/2}\int_0^{2\cos\theta} \frac{r\cos\theta \cdot r\,dr\,d\theta}{r} = \int_0^{\pi/2}\int_0^{2\cos\theta} r\cos\theta\,dr\,d\theta$$

$$= \int_0^{\pi/2} \cos\theta \left[\frac{r^2}{2}\right]_0^{2\cos\theta} d\theta = \int_0^{\pi/2} 2\cos^3\theta\,d\theta = 2\cdot\frac{2}{3} = \frac{4}{3}.$$

Example 5. *Evaluate* $\iiint z(x^2 + y^2 + z^2)\,dx\,dy\,dz$ *through the volume of the cylinder* $x^2 + y^2 = a^2$ *intercepted by the planes* $z = 0$ *and* $z = h$.

Sol. Changing to cylindrical co-ordinates by changing x to $r\cos\phi$, y to $r\sin\phi$ and replacing $dx\,dy\,dz$ by $r\,dr\,d\phi\,dz$

$$I = \int_0^h\int_0^{2\pi}\int_0^a z(r^2 + z^2)\,r\,dr\,d\phi\,dz = \int_0^h\int_0^{2\pi}\int_0^a (zr^3 + z^3 r)\,dr\,d\phi\,dz$$

$$= \int_0^h\int_0^{2\pi}\left[z\cdot\frac{r^4}{4} + z^3\cdot\frac{r^2}{2}\right]_0^a d\phi\,dz = \int_0^h\int_0^{2\pi}\left(\frac{a^4}{4}z + \frac{a^2}{2}z^3\right)d\phi\,dz$$

$$= \int_0^h\left(\frac{a^4}{4}z + \frac{a^2}{2}z^3\right)\Big[\phi\Big]_0^{2\pi} dz = \int_0^h 2\pi\left(\frac{a^4}{4}z + \frac{a^2}{2}z^3\right)dz$$

$$= 2\pi\left[\frac{a^4z^2}{8} + \frac{a^2z^4}{8}\right]_0^h = \frac{\pi}{4}(a^4h^2 + a^2h^4) = \frac{\pi}{4}a^2h^2(a^2 + h^2).$$

Example 6. *Evaluate* $\int_0^1\int_0^{\sqrt{1-x^2}}\int_0^{\sqrt{1-x^2-y^2}} \frac{dz\,dy\,dx}{\sqrt{1-x^2-y^2-z^2}}$, *by changing to spherical polar co-ordinates.*

Sol. Here the region of integration is bounded by

$z = 0$, $\quad z = \sqrt{1 - x^2 - y^2}$ $\quad$ (*i.e.*, $x^2 + y^2 + z^2 = 1$)

$$y = 0, \qquad y = \sqrt{1 - x^2} \qquad (i.e., x^2 + y^2 = 1)$$
$$x = 0, \qquad x = 1$$

which is the volume of the sphere $x^2 + y^2 + z^2 = 1$ in the positive octant.

Changing to spherical polar co-ordinates by putting $x = r \sin\theta \cos\phi$, $y = r \sin\theta \sin\phi$, $z = r\cos\theta$ so that $x^2 + y^2 + z^2 = r^2$.

For the volume of sphere $x^2 + y^2 + z^2 = 1$ in the positive octant, r varies from 0 to 1, θ varies from 0 to $\frac{\pi}{2}$ and ϕ varies from 0 to $\frac{\pi}{2}$.

Replacing $dz\,dy\,dx$ by $r^2 \sin\theta\, dr\, d\theta\, d\phi$, we have

$$\text{I} = \int_0^{\pi/2}\int_0^{\pi/2}\int_0^1 \frac{r^2 \sin\theta\, dr\, d\theta\, d\phi}{\sqrt{1-r^2}} = \int_0^{\pi/2}\int_0^{\pi/2}\int_0^1 \frac{1-(1-r^2)}{\sqrt{1-r^2}} \sin\theta\, dr\, d\theta\, d\phi$$

$$= \int_0^{\pi/2}\int_0^{\pi/2}\int_0^1 \left(\frac{1}{\sqrt{1-r^2}} - \sqrt{1-r^2}\right)\sin\theta\, dr\, d\theta\, d\phi$$

$$= \int_0^{\pi/2}\int_0^{\pi/2} \sin\theta\left[\sin^{-1} r - \left(\frac{r\sqrt{1-r^2}}{2} + \frac{1}{2}\sin^{-1} r\right)\right]_0^1 d\theta\, d\phi$$

$$= \int_0^{\pi/2}\int_0^{\pi/2} \sin\theta\left[\frac{\pi}{2} - \frac{1}{2}\cdot\frac{\pi}{2}\right] d\theta\, d\phi = \int_0^{\pi/2} \frac{\pi}{4}\Big[-\cos\theta\Big]_0^{\pi/2} d\phi = \int_0^{\pi/2} \frac{\pi}{4}\, d\phi$$

$$= \frac{\pi}{4}\cdot\frac{\pi}{2} = \frac{\pi^2}{8}.$$

Note. For the whole volume of the sphere $x^2 + y^2 + z^2 = a^2$.

$$0 \le r \le a,\ 0 \le \theta \le \pi,\ 0 \le \phi \le 2\pi.$$

Example 7. *Evaluate* $\iint_R (x+y)^2\, dx\, dy$, *where R is the parallelogram in the xy-plane with vertices* (1, 0), (3, 1), (2, 2), (0, 1), *using the transformation* $u = x + y$ *and* $v = x - 2y$.

Sol. The vertices A(1, 0), B(3, 1), C(2, 2), D(0, 1) of the parallelogram ABCD in the xy-plane become A′(1, 1), B′(4, 1), C′(4, – 2), D′(1, – 2) in the uv-plane under the given transformation.

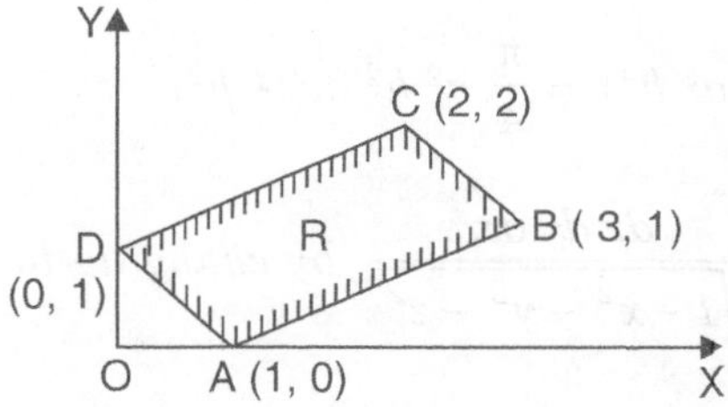

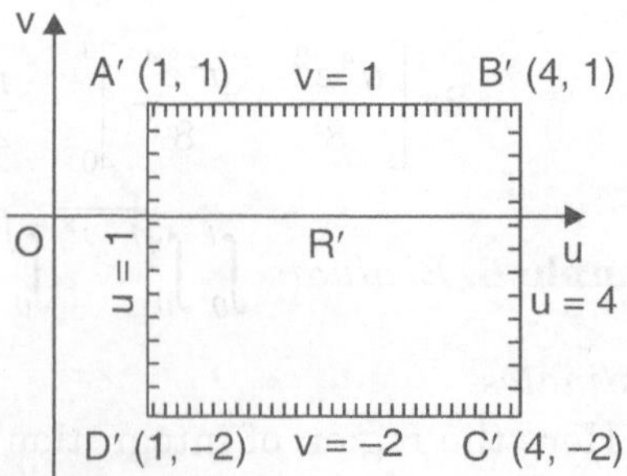

The region R in the xy-plane becomes the region R′ in the uv-plane which is a rectangle bounded by the line $u = 1$, $u = 4$ and $v = -2$, $v = 1$. Solving the given equations for x and y, we have $x = \frac{1}{3}(2u + v)$, $\quad y = \frac{1}{3}(u - v)$

$$\therefore \quad J = \frac{\partial(x, y)}{\partial(u, v)} = \begin{vmatrix} \frac{\partial x}{\partial u} & \frac{\partial v}{\partial x} \\ \frac{\partial y}{\partial u} & \frac{\partial y}{\partial v} \end{vmatrix} = \begin{vmatrix} \frac{2}{3} & \frac{1}{3} \\ \frac{1}{3} & -\frac{1}{3} \end{vmatrix} = -\frac{1}{3}$$

$$\therefore \quad \iint_R (x + y)^2 \, dx \, dy = \iint_{R'} u^2 \, |J| \, du \, dv = \int_{-2}^{1} \int_{1}^{4} u^2 \cdot \frac{1}{3} \, du \, dv$$

$$= \int_{-2}^{1} \frac{1}{3} \left[\frac{u^3}{3} \right]_1^4 dv = \int_{-2}^{1} 7 \, dv = 7 \left[v \right]_{-2}^{1} = 7 \times 3 = 21.$$

EXERCISE 10

1. Evaluate $\iint \sin \pi (x^2 + y^2) \, dx \, dy$ over the region bounded by the circle $x^2 + y^2 = 1$ by changing to polar co-ordinates.

2. Evaluate $\iint (a^2 - x^2 - y^2) \, dx \, dy$ over the semi-circle $x^2 + y^2 = ax$ in the positive quadrant by changing to polar co-ordinates.

3. Evaluate $\iint (x^2 + y^2)^{7/2} \, dx \, dy$ over the circle $x^2 + y^2 = 1$.

4. Evaluate $\iint xy \, (x^2 + y^2)^{3/2} \, dx \, dy$ over the positive quadrant of the circle $x^2 + y^2 = 1$.

5. Evaluate the following by changing into polar co-ordinates :

   (i) $\displaystyle\int_0^a \int_y^a \frac{x \, dx \, dy}{x^2 + y^2}$ (ii) $\displaystyle\int_{-a}^{a} \int_{-\sqrt{a^2 - x^2}}^{\sqrt{a^2 - x^2}} dy \, dx$ (iii) $\displaystyle\int_0^{4a} \int_{y^2/4a}^{y} \frac{x^2 - y^2}{x^2 + y^2} \, dx \, dy$

   (iv) $\displaystyle\int_0^a \int_0^{\sqrt{a^2 - y^2}} (x^2 + y^2) \, dx \, dy$ (v) $\displaystyle\int_0^a \int_0^{\sqrt{a^2 - y^2}} y^2 \sqrt{x^2 + y^2} \, dy \, dx$ (vi) $\displaystyle\int_0^a \int_y^a \frac{x^2 \, dx \, dy}{\sqrt{x^2 + y^2}}$.

6. Evaluate $\iint_D e^{-(x^2 + y^2)} \, dy \, dx$, where D is the region bounded by $x^2 + y^2 = a^2$.

7. Evaluate $\iint xy \, (x^2 + y^2)^{n/2} \, dx \, dy$ over the positive quadrant of $x^2 + y^2 = 4$, supposing $n + 3 > 0$.

8. Evaluate $\displaystyle\iint \sqrt{\frac{1 - \frac{x^2}{a^2} - \frac{y^2}{b^2}}{1 + \frac{x^2}{a^2} + \frac{y^2}{b^2}}} \, dx \, dy$ over the positive quadrant of the ellipse $\frac{x^2}{a^2} + \frac{y^2}{b^2} = 1$.

9. Transform the following to cartesian form and hence evaluate $\displaystyle\int_0^{\pi} \int_0^a r^3 \sin \theta \cos \theta \, dr \, d\theta$.

10. Evaluate $\iiint (x + y + z) \, dx \, dy \, dz$ over the tetrahedron bounded by the planes $x = 0$, $y = 0$, $z = 0$ and $x + y + z = 1$.

11. Evaluate $\iiint \frac{dx\,dy\,dz}{\sqrt{1-x^2-y^2-z^2}}$, the integral being extended to the positive octant of the sphere $x^2+y^2+z^2=1$.

12. Evaluate $\iiint \frac{dx\,dy\,dz}{(x+y+z+1)^3}$ over the tetrahedron bounded by the co-ordinate planes and the plane $x+y+z=1$.

13. Evaluate $\iiint z(x^2+y^2)\,dx\,dy\,dz$ over the volume of the cylinder $x^2+y^2=1$ intercepted by the planes $z=2$ and $z=3$.

14. Evaluate the following integrals through the volume of the sphere $x^2+y^2+z^2=1$, by changing into spherical polar co-ordinates :

(*i*) $\iiint z^2\,dx\,dy\,dz$ (*ii*) $\iiint (x^2+y^2+z^2)^m\,dx\,dy\,dz$. $(m>0)$

Answers

1. 2 **2.** $\frac{5\pi a^4}{64}$ **3.** $\frac{2\pi}{9}$ **4.** $\frac{1}{14}$

5. (*i*) $\frac{\pi a}{4}$ (*ii*) πa^2 (*iii*) $8\left(\frac{\pi}{2}-\frac{5}{3}\right)a^2$ (*iv*) $\frac{\pi a^4}{8}$

(*v*) $\frac{\pi a^5}{20}$ (*vi*) $\frac{a^3}{3}\log(\sqrt{2}+1)$ **6.** $\pi(1-e^{-a^2})$ **7.** $\frac{2^{n+3}}{n+4}$

8. $\frac{\pi ab}{8}(\pi-2)$ **9.** 0 **10.** $\frac{1}{8}$ **11.** $\frac{\pi^2}{8}$

12. $\frac{1}{2}\log 2-\frac{5}{16}$ **13.** $\frac{5\pi}{4}$ **14.** (*i*) $\frac{4\pi}{15}$ (*ii*) $\frac{4\pi}{2m+3}$.

UNIT 5
Vector Calculus

5.1. INTRODUCTION

In Mathematics and Physics, we generally deal with two types of quantities classified on the basis of two characteristics—magnitude and direction. A quantity which is completely specified by its magnitude only is called a *scalar*. To specify a scalar, we need a unit and a real number. For example, time, mass, length, temperature, work, electric charge, statistical data, are all scalar quantities. They obey all the laws of ordinary algebra.

A quantity which is completely specified by its magnitude as well as direction is called a *vector*. For example, weight, displacement, velocity, acceleration, force are all vector quantities.

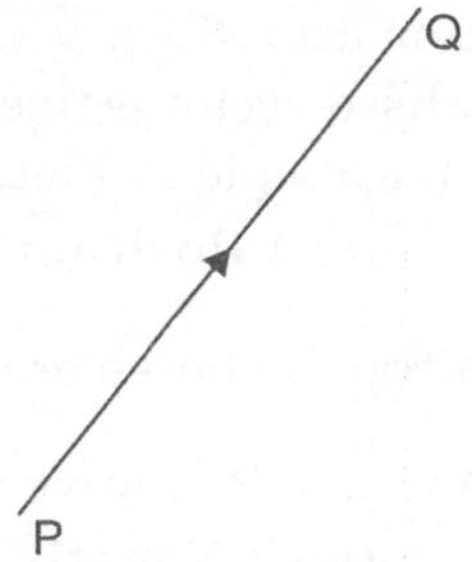

A vector is represented by a directed line segment. A vector whose magnitude is the length PQ and whose direction is from P to Q is denoted by $\overrightarrow{PQ}$. P and Q are called the initial and terminal points of the vector $\overrightarrow{PQ}$. The magnitude of vector $\overrightarrow{PQ}$ is denoted by $|\overrightarrow{PQ}|$ and $|\overrightarrow{PQ}| = PQ$.

To distinguish vectors from scalars, bold face type is used for vectors *e.g.*, **A, B, a, b, P, Q** etc. Other notations used to denote a vector **A** are $\vec{A}, \bar{A}, \tilde{A}, \hat{A}$ and its magnitude (or length) is denoted by | **A** | or A.

5.2. TYPES OF VECTORS

***(i)* Null Vector or Zero Vector.** A vector whose magnitude is zero is called a **null vector or zero vector** and is denoted by **0.** The initial and terminal points of a zero vector are coincident. Its direction is arbitrary.

A non-zero vector is called a **proper vector.**

***(ii)* Unit Vector.** A vector whose magnitude is unity is called a **unit vector.** If $\vec{a}$ is any vector and $\hat{a}$ is a unit vector in the direction of $\vec{a}$, then $\vec{a} = |\vec{a}|\,\hat{a} \Rightarrow \hat{a} = \dfrac{\vec{a}}{|\vec{a}|}$

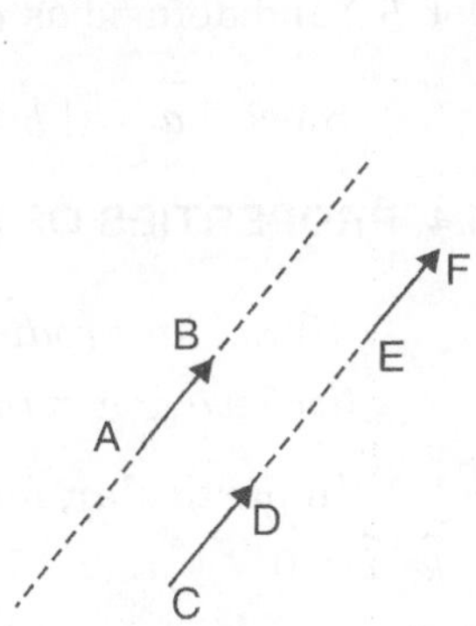

Unit vectors are distinguished from other vectors by putting a cap (^) over them.

***(iii)* Co-initial Vectors.** Two or more vectors having the same initial point are called **co-initial vectors.**

Thus, $\vec{AB}, \vec{AC}, \vec{AD}$, are co-initial vectors.

***(iv)* Equal Vectors.** Two vectors are said to be **equal** if they have the same magnitude and the same or parallel directions.

In the figure, $\vec{AB} = \vec{CD} = \vec{EF}$

***(v)* Parallel Vectors, Like and Unlike Vectors.** Two vectors are said to be **parallel** if they lie along the same or parallel lines, irrespective of their sense. However, if their sense is also the same, they are called *like parallel vectors* and if their sense is opposite, they are called *unlike parallel vectors.*

***(vi)* Free and Localised (or bound) Vectors.** Vectors having the same magnitude and direction and acting at different points in the same line or parallel lines are called *free vectors.* They are equivalent to each other. Thus a free vector is free to have any point as its initial point.

To specify some vectors completely, it is necessary to specify the point at which the vector acts. Such vectors are called localised or bound vectors. Thus the initial point of a localised vector is fixed and hence it is represented by a unique directed line segment. A force is an example of a localised vector.

***(vii)* Position Vector of a Point.** The position vector of a point P with respect to an arbitrarily chosen origin O is the vector $\vec{OP}$. The position vector of P specifies uniquely the position of P relative to O. If $\vec{OP} = \vec{r}$, then the point P is referred to as the point $P(\vec{r})$.

***(viii)* Negative of a Vector.** A vector whose magnitude is the same as that of a given vector $\vec{a}$ but has opposite direction is called the negative of $\vec{a}$ and is denoted by $-\vec{a}$. Clearly, if $\vec{a} = \vec{PQ}$, then $-\vec{a} = \vec{QP}$.

5.3. SCALAR PRODUCT (OR DOT PRODUCT) OF TWO VECTORS

Definition. The scalar product of two vectors $\vec{a}$ and $\vec{b}$ is denoted by $\vec{a} \cdot \vec{b}$ (read as $\vec{a}$ dot $\vec{b}$) and defined as $\vec{a} \cdot \vec{b} = |\vec{a}|\,|\vec{b}| \cos\theta = ab\cos\theta$ where θ is the angle between $\vec{a}$ and $\vec{b}$.

Since $|\vec{a}|, |\vec{b}|, \cos\theta$ are scalars, $\vec{a} \cdot \vec{b}$ is a scalar.

5.4. PROPERTIES OF SCALAR PRODUCT OF TWO VECTORS

(i) Scalar product of two vectors is commutative i.e., $\vec{a} \cdot \vec{b} = \vec{b} \cdot \vec{a}$ each $= ab\cos\theta$

(ii) Two non-zero vectors are perpendicular if and only if their scalar product is zero.

In particular, for the mutually perpendicular unit vectors $\hat{i}, \hat{j}, \hat{k}$, we have $\hat{i} \cdot \hat{j} = \hat{j} \cdot \hat{k} = \hat{k} \cdot \hat{i} = 0$.

(*iii*) *Scalar product of two vectors is positive, negative or zero according as the angle between the vectors is acute, obtuse or a right angle.*

(*iv*) If two vectors $\vec{a}$ and $\vec{b}$ are parallel, then $\vec{a} \cdot \vec{b} = ab$ or $-ab$ according as the vectors are like parallel ($\theta = 0$) or unlike parallel ($\theta = \pi$).

(*v*) $\vec{a} \cdot \vec{a} = aa \cos 0 = a^2$ and is written as $\vec{a}^2 = a^2$.

Thus, the square of any vector is the square of its magnitude.

In particular, for the mutually perpendicular unit vectors $\hat{i}, \hat{j}, \hat{k}$, we have $\hat{i} \cdot \hat{i} = \hat{j} \cdot \hat{j} = \hat{k} \cdot \hat{k} = 1^2 = 1$.

(*vi*) *Scalar product of two vectors is distributive i.e.,* $\vec{a} \cdot (\vec{b} + \vec{c}) = \vec{a} \cdot \vec{b} + \vec{a} \cdot \vec{c}$

(*vii*) If $\vec{a} = a_1\hat{i} + a_2\hat{j} + a_3\hat{k}$, $\vec{b} = b_1\hat{i} + b_2\hat{j} + b_3\hat{k}$,

Then $\vec{a} \cdot \vec{b} = a_1b_1 + a_2b_2 + a_3b_3$

$(\because \hat{i} \cdot \hat{i} = \hat{j} \cdot \hat{j} = \hat{k} \cdot \hat{k} = 1, \hat{i} \cdot \hat{j} = \hat{j} \cdot \hat{k} = \hat{k} \cdot \hat{i} = 0)$

Thus, the scalar product of two vectors is equal to the sum of the products of their corresponding components.

(*viii*) If $\vec{a} = a_1\hat{i} + a_2\hat{j} + a_3\hat{k}$, $\vec{b} = b_1\hat{i} + b_2\hat{j} + b_3\hat{k}$

then $\vec{a}$ and $\vec{b}$ are perpendicular if and only if $\vec{a} \cdot \vec{b} = 0$

i.e., if and only if $a_1b_1 + a_2b_2 + a_3b_3 = 0$

Also, $|\vec{a}| = \sqrt{a_1^2 + a_2^2 + a_3^2}$, $|\vec{b}| = \sqrt{b_1^2 + b_2^2 + b_3^2}$

Since, $\vec{a} \cdot \vec{b} = |\vec{a}|\,|\vec{b}| \cos\theta$

$$\therefore \quad \cos\theta = \frac{\vec{a} \cdot \vec{b}}{|\vec{a}||\vec{b}|} = \frac{a_1b_1 + a_2b_2 + a_3b_3}{\sqrt{a_1^2 + a_2^2 + a_3^2}\sqrt{b_1^2 + b_2^2 + b_3^2}}$$

5.5. VECTOR PRODUCT (OR CROSS PRODUCT) OF TWO VECTORS

The vector product of two vectors $\vec{a}$ and $\vec{b}$ is denoted by $\vec{a} \times \vec{b}$ (read as $\vec{a}$ cross $\vec{b}$) and defined as $\vec{a} \times \vec{b} = |\vec{a}|\,|\vec{b}| \sin\theta\, \hat{n}$

$= ab \sin\theta\, \hat{n}$

where θ is the angle between $\vec{a}$ and $\vec{b}$ and $\hat{n}$ is a unit vector normal to the plane of $\vec{a}$ and $\vec{b}$, such that the vectors $\vec{a}$, $\vec{b}$ and $\hat{n}$ form a right-handed system.

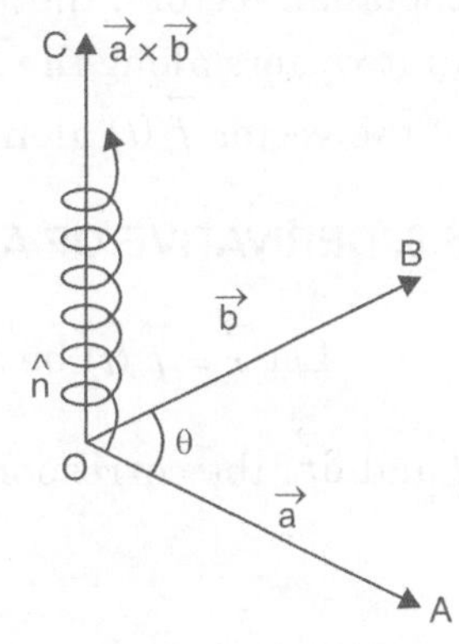

5.6. PROPERTIES OF VECTOR PRODUCT OF TWO VECTORS

(*i*) *Vector product is not commutative.*

$$\therefore \quad \vec{a} \times \vec{b} \neq \vec{b} \times \vec{a} \text{ rather } \vec{a} \times \vec{b} = -\,\vec{b} \times \vec{a}$$

(*ii*) *Vector product is distributive with respect to vector addition i.e.,*

$$\vec{a} \times (\vec{b} + \vec{c}) = \vec{a} \times \vec{b} + \vec{a} \times \vec{c}$$

and $$(\vec{b} + \vec{c}) \times \vec{a} = \vec{b} \times \vec{a} + \vec{c} \times \vec{a}.$$

(*iii*) *Vector product of two parallel vectors is zero.*

In particular, $$\vec{a} \times \vec{a} = \vec{0}$$

Also, $$\hat{i} \times \hat{i} = \hat{j} \times \hat{j} = \hat{k} \times \hat{k} = \hat{0}.$$

(*iv*) Since, $\hat{i}, \hat{j}, \hat{k}$ form a right-handed triad of mutually perpendicular vectors.

$$\Rightarrow \quad \hat{i} \times \hat{j} = \hat{k} \qquad \therefore\ \hat{j} \times \hat{i} = -\,\hat{k}$$

Similarly, $$\hat{j} \times \hat{k} = \hat{i} \qquad \therefore\ \hat{k} \times \hat{j} = -\,\hat{i}$$

$$\hat{k} \times \hat{i} = \hat{j} \qquad \therefore\ \hat{i} \times \hat{k} = -\,\hat{j}$$

(*v*) If $\vec{a} = a_1\hat{i} + a_2\hat{j} + a_3\hat{k}$ and $\vec{b} = b_1\hat{i} + b_2\hat{j} + b_3\hat{k}$

then $$\vec{a} \times \vec{b} = \begin{vmatrix} \hat{i} & \hat{j} & \hat{k} \\ a_1 & a_2 & a_3 \\ b_1 & b_2 & b_3 \end{vmatrix}.$$

5.7. VECTOR FUNCTIONS

If to each value of a scalar variable t, there corresponds a value of a vector $\vec{r}$, then $\vec{r}$ is called a vector function of the scalar variable t and we write $\vec{r} = \vec{r}\,(t)$ or $\vec{r} = \vec{f}\,(t)$.

For example, the position vector $\vec{r}$ of a particle moving along a curved path is a vector function of time t, a scalar.

Since every vector can be uniquely expressed as a linear combination of three fixed non-coplanar vectors, therefore, we may write $\vec{f}(t) = f_1(t)\,\hat{i} + f_2(t)\,\hat{j} + f_3\,(t)\,\hat{k}$ where $\hat{i}, \hat{j}, \hat{k}$ denote unit vectors along the axis of x, y, z respectively, $f_1(t)$, $f_2(t)$ and $f_3(t)$ are called the components of the vector $\vec{f}\,(t)$ along the coordinate axes.

5.8. DERIVATIVE OF A VECTOR FUNCTION WITH RESPECT TO A SCALAR

Let $\vec{r} = \vec{f}\,(t)$ be a vector function of the scalar variable t. Let δt be a small increment in t and $\delta\vec{r}$, the corresponding increment in $\vec{r}$.

Then $\vec{r} + \delta\vec{r} = \vec{f}(t + \delta t)$ so that $\delta\vec{r} = \vec{f}(t + \delta t) - \vec{f}(t)$

and $$\frac{\delta\vec{r}}{\delta t} = \frac{\vec{f}(t + \delta t) - \vec{f}(t)}{\delta t}$$

If $\underset{\delta t \to 0}{\text{Lt}} \frac{\delta\vec{r}}{\delta t} = \underset{\delta t \to 0}{\text{Lt}} \frac{\vec{f}(t + \delta t) - \vec{f}(t)}{\delta t}$ exists, then the value of this limit is denoted by $\frac{d\vec{r}}{dt}$ and is called the derivative of $\vec{r}$ with respect to t.

Since $\frac{d\vec{r}}{dt}$ is itself a vector function of t, its derivative is denoted by $\frac{d^2\vec{r}}{dt^2}$ and is called the second derivative of $\vec{r}$ with respect to t. Similarly, we can define higher order derivatives of $\vec{r}$.

5.9. GENERAL RULES FOR DIFFERENTIATION

If $\vec{a}, \vec{b}$ and $\vec{c}$ are vector functions of a scalar t and ϕ is a scalar function of t, then

(*i*) $\frac{d}{dt}(\vec{a} \pm \vec{b}) = \frac{d\vec{a}}{dt} \pm \frac{d\vec{b}}{dt}$

(*ii*) $\frac{d}{dt}(\vec{a} \cdot \vec{b}) = \vec{a} \cdot \frac{d\vec{b}}{dt} + \frac{d\vec{a}}{dt} \cdot \vec{b}$

(*iii*) $\frac{d}{dt}(\vec{a} \times \vec{b}) = \vec{a} \times \frac{d\vec{b}}{dt} + \frac{d\vec{a}}{dt} \times \vec{b}$

(*iv*) $\frac{d}{dt}(\phi\vec{a}) = \phi\frac{d\vec{a}}{dt} + \frac{d\phi}{dt}\vec{a}$

(*v*) $\frac{d}{dt}[\vec{a}\,\vec{b}\,\vec{c}] = \left[\frac{d\vec{a}}{dt}\,\vec{b}\,\vec{c}\right] + \left[\vec{a}\,\frac{d\vec{b}}{dt}\,\vec{c}\right] + \left[\vec{a}\,\vec{b}\,\frac{d\vec{c}}{dt}\right]$

(*vi*) $\frac{d}{dt}\{\vec{a} \times (\vec{b} \times \vec{c})\} = \frac{d\vec{a}}{dt} \times (\vec{b} \times \vec{c}) + \vec{a} \times \left(\frac{d\vec{b}}{dt} \times \vec{c}\right) + \vec{a} \times \left(\vec{b} \times \frac{d\vec{c}}{dt}\right)$.

Example 1. *Show that if $\vec{r} = \vec{a} \sin \omega t + \vec{b} \cos \omega t$, where $\vec{a}, \vec{b}, \omega$ are constants, then*

$$\frac{d^2\vec{r}}{dt^2} = -\omega^2\vec{r} \quad \text{and} \quad \vec{r} \times \frac{d\vec{r}}{dt} = -\omega\vec{a} \times \vec{b}.$$

Sol. We know that if $\vec{r} = \phi\vec{f}$, where ϕ is a scalar function of t, then $\frac{d\vec{r}}{dt} = \phi\frac{d\vec{f}}{dt} + \frac{d\phi}{dt}\vec{f}$.

But if $\vec{f}$ is a constant vector, then $\frac{d\vec{f}}{dt} = \vec{0}$

$$\therefore \quad \frac{d\vec{r}}{dt} = \frac{d\phi}{dt}\vec{f}$$

Now, $\vec{r} = \vec{a} \sin \omega t + \vec{b} \cos \omega t \Rightarrow \frac{d\vec{r}}{dt} = \vec{a}\,\omega \cos \omega t - \vec{b}\omega \sin \omega t$

$$\therefore \quad \frac{d^2\vec{r}}{dt^2} = -\vec{a}\,\omega^2 \sin \omega t - \vec{b}\,\omega^2 \cos \omega t = -\omega^2(\vec{a} \sin \omega t + \vec{b} \cos \omega t) = -\omega^2\vec{r}$$

Also, $\vec{r} \times \frac{d\vec{r}}{dt} = (\vec{a} \sin \omega t + \vec{b} \cos \omega t) \times (\vec{a}\, \omega \cos \omega t - \vec{b}\, \omega \sin \omega t)$

$$= \omega\,(-\vec{a} \times \vec{b} \sin^2 \omega t + \vec{b} \times \vec{a} \cos^2 \omega t) \qquad [\because\ \vec{a} \times \vec{a} = \vec{0} = \vec{b} \times \vec{b}\,]$$

$$= \omega\,(-\vec{a} \times \vec{b} \sin^2 \omega t - \vec{a} \times \vec{b} \cos^2 \omega t) = -\,\omega(\vec{a} \times \vec{b}).$$

Example 2. *If* $\vec{r} = a \cos t\hat{i} + a \sin t\hat{j} + at \tan \alpha \hat{k}$, *find* $\left|\frac{d\vec{r}}{dt} \times \frac{d^2\vec{r}}{dt^2}\right|$ *and* $\left[\frac{d\vec{r}}{dt}\ \frac{d^2\vec{r}}{dt^2}\ \frac{d^3\vec{r}}{dt^3}\right]$.

Sol.
$$\vec{r} = a \cos t\hat{i} + a \sin t\hat{j} + a\, t \tan \alpha \hat{k}$$

$$\Rightarrow \quad \frac{d\vec{r}}{dt} = -\,a \sin t\hat{i} + a \cos t\hat{j} + a \tan \alpha \hat{k}$$

$$\frac{d^2\vec{r}}{dt^2} = -\,a \cos t\hat{i} - a \sin t\hat{j}$$

$$\frac{d^3\vec{r}}{dt^3} = a \sin t\hat{i} - a \cos t\hat{j}$$

$$\therefore \quad \frac{d\vec{r}}{dt} \times \frac{d^2\vec{r}}{dt^2} = \begin{vmatrix} \hat{i} & \hat{j} & \hat{k} \\ -a \sin t & a \cos t & a \tan \alpha \\ -a \cos t & -a \sin t & 0 \end{vmatrix}$$

$$= a^2 \sin t \tan \alpha \hat{i} - a^2 \cos t \tan \alpha \hat{j} + a^2 \hat{k}$$

$$\therefore \quad \left|\frac{d\vec{r}}{dt} \times \frac{d^2\vec{r}}{dt^2}\right| = \sqrt{a^4 \sin^2 t \tan^2 \alpha + a^4 \cos^2 t \tan^2 \alpha + a^4}$$

$$= a^2 \sqrt{\tan^2 \alpha + 1} = a^2 \sec \alpha$$

Also, $$\left[\frac{d\vec{r}}{dt}\ \frac{d^2\vec{r}}{dt^2}\ \frac{d^3\vec{r}}{dt^3}\right] = \begin{vmatrix} -a \sin t & a \cos t & a \tan \alpha \\ -a \cos t & -a \sin t & 0 \\ a \sin t & -a \cos t & 0 \end{vmatrix} \qquad \text{(expanding by third column)}$$

$$= a \tan \alpha (a^2 \cos^2 t + a^2 \sin^2 t) = a^3 \tan \alpha.$$

5.10. SCALAR AND VECTOR FIELDS

A variable quantity whose value at any point in a region of space depends upon the position of the point, is called a *point function*. There are two types of point functions.

(*a*) *Scalar Point Function.* Let R be a region of space at each point of which a scalar $\phi = \phi(x, y, z)$ is given, then ϕ is called a *scalar function* and R is called a *scalar field.*

The temperature distribution in a medium, the distribution of atmospheric pressure in space are examples of scalar point functions.

(*b*) *Vector Point Function.* Let R be a region of space at each point of which a vector $\vec{v} = \vec{v}\,(x, y, z)$ is given, then $\vec{v}$ is called a *vector point function* and R is called a *vector field.* Each vector $\vec{v}$ of the field is regarded as a localised vector attached to the corresponding point (x, y, z).

The velocity of a moving fluid at any instant, the gravitational force are examples of vector point functions.

5.11. GRADIENT OF A SCALAR FIELD

Let $\phi(x, y, z)$ be a function defining a scalar field, then the vector $\hat{i}\frac{\partial \phi}{\partial x} + \hat{j}\frac{\partial \phi}{\partial y} + \hat{k}\frac{\partial \phi}{\partial z}$ is called the **gradient** of the scalar field ϕ and is denoted by grad ϕ.

Thus, $$\text{grad } \phi = \hat{i}\frac{\partial \phi}{\partial x} + \hat{j}\frac{\partial \phi}{\partial y} + \hat{k}\frac{\partial \phi}{\partial z}$$

The gradient of scalar field ϕ is obtained by operating on ϕ by the vector operator

$$\hat{i}\frac{\partial}{\partial x} + \hat{j}\frac{\partial}{\partial y} + \hat{k}\frac{\partial}{\partial z}$$

This operator is denoted by the symbol ∇, read as **del** (also called nabla).

Thus, grad $\phi = \nabla\phi$.

5.12. GEOMETRICAL INTERPRETATION OF GRADIENT

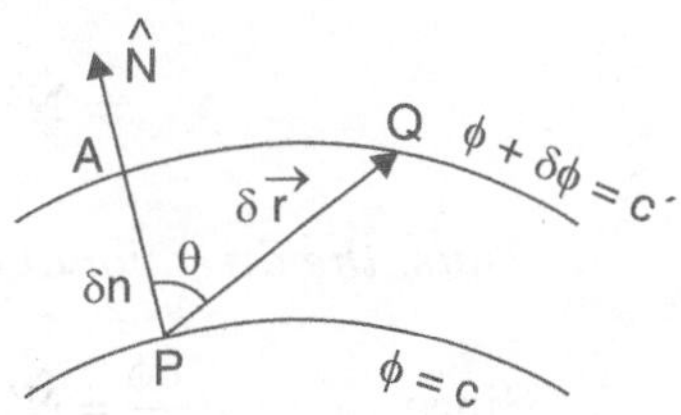

If a surface $\phi(x, y, z) = c$ is drawn through any point P such that at each point on the surface, the function has the same value as at P, then such a surface is called a *level surface* through P. For example, if $\phi(x, y, z)$ represents potential at the point (x, y, z), the **equipotential surface** $\phi(x, y, z) = c$ is a level surface.

Through any point passes one and only one level surface. Moreover, no two level surfaces can intersect.

Consider the level surface through P at which the function has value ϕ and another level surface through a neighbouring point Q where the value is $\phi + \delta\phi$.

Let $\vec{r}$ and $\vec{r} + \delta\vec{r}$ be the position vectors of P and Q respectively, then $\overrightarrow{PQ} = \delta\vec{r}$.

Now, $$\nabla\phi \cdot \delta\vec{r} = \left(\hat{i}\frac{\partial \phi}{\partial x} + \hat{j}\frac{\partial \phi}{\partial y} + \hat{k}\frac{\partial \phi}{\partial z}\right) \cdot (\hat{i}\,\delta x + \hat{j}\,\delta y + \hat{k}\,\delta z)$$

$$= \frac{\partial \phi}{\partial x}\delta x + \frac{\partial \phi}{\partial y}\delta y + \frac{\partial \phi}{\partial z}\delta z = \delta\phi \qquad \text{...(1)}$$

If Q lies on the same level surface as P, then $\delta\phi = 0$,

$\therefore$ (1) reduces to $\nabla\phi \cdot \delta\vec{r} = 0$.

Thus, $\nabla\phi$ is perpendicular to every $\delta\vec{r}$ lying in the surface.

Hence $\nabla\phi$ **is normal to the surface** $\phi(x, y, z) = c$.

Let $\nabla\phi = |\nabla\phi|\,\hat{N}$, where $\hat{N}$ is a unit vector normal to the surface. Let PA $= \delta n$ be the perpendicular distance between the two level surfaces through P and Q. Then the rate of change of ϕ in the direction of normal to the surface through P is

$$\frac{\partial \phi}{\partial n} = \underset{\delta n \to 0}{\text{Lt}} \frac{\delta \phi}{\delta n} = \underset{\delta n \to 0}{\text{Lt}} \frac{\nabla\phi \cdot \delta\vec{r}}{\delta n} \qquad \text{[by (1)]}$$

$$= \underset{\delta n \to 0}{\text{Lt}} \frac{|\nabla\phi|\hat{N}\cdot\vec{\delta r}}{\delta n} = |\nabla\phi| \qquad (\because \quad \hat{N}\cdot\vec{\delta r} = |\hat{N}||\vec{\delta r}|\cos\theta = |\vec{\delta r}|\cos\theta = \delta n)$$

$$\therefore \quad |\nabla\phi| = \frac{\partial\phi}{\partial n}.$$

Hence the gradient of a scalar field ϕ is a vector normal to the surface $\phi = c$ and has a magnitude equal to the rate of change of ϕ along this normal.

5.13. DIRECTIONAL DERIVATIVE

Let $PQ = \delta r$, then $\underset{\delta r \to 0}{\text{Lt}} \frac{\delta\phi}{\delta r} = \frac{\partial\phi}{\partial r}$ is called the directional derivative of ϕ at P in the direction PQ.

Let $\hat{N}'$ be a unit vector in the direction PQ, then $\delta r = \frac{\delta n}{\cos\theta} = \frac{\delta n}{\hat{N}\cdot\hat{N}'}$

$$\therefore \qquad \frac{\partial\phi}{\delta r} = \underset{\delta r \to 0}{\text{Lt}} \left[\hat{N}\cdot\hat{N}'\frac{\delta\phi}{\delta n}\right] = \hat{N}\cdot\hat{N}'\frac{\partial\phi}{\partial n}$$

$$= \hat{N}'\cdot\hat{N}\frac{\partial\phi}{\partial n} = \hat{N}'\cdot\hat{N}|\nabla\phi| = \hat{N}'\cdot\nabla\phi \qquad \left(\because \ |\nabla\phi| = \frac{\partial\phi}{\partial n} \text{ and } \hat{N}|\nabla\phi| = \nabla\phi\right)$$

Thus, the directional derivative $\frac{\partial\phi}{\partial r}$ is the resolved part of $\nabla\phi$ in the direction $\hat{N}'$.

Since, $\qquad \frac{\partial\phi}{\partial r} = \hat{N}'\cdot\nabla\phi = |\nabla\phi|\cos\theta \le |\nabla\phi|.$

$\therefore$ $\nabla\phi$ *gives the maximum rate of change of ϕ and the magnitude of this maximum is* $|\nabla\phi|$.

Note. Directional derivative of f in the direction of $\overline{PQ} = \nabla f\,.\,N$ where $N = \frac{\overline{PQ}}{|\overline{PQ}|}$ *i.e.*, Directional derivative $= \nabla f\,.\,\frac{\overline{PQ}}{|\overline{PQ}|}$.

5.14. PROPERTIES OF GRADIENT

(*a*) If ϕ is a constant scalar point function, then $\nabla\phi = \vec{0}$

(*b*) If ϕ_1 and ϕ_2 are two scalar point functions, then

(*i*) $\nabla(\phi_1 \pm \phi_2) = \nabla\phi_1 \pm \nabla\phi_2$

(*ii*) $\nabla(c_1\phi_1 + c_2\phi_2) = c_1\nabla\phi_1 + c_2\nabla\phi_2$, where c_1, c_2 are constants

(*iii*) $\nabla(\phi_1\phi_2) = \phi_1\nabla\phi_2 + \phi_2\nabla\phi_1$

(*iv*) $\nabla\left(\frac{\phi_1}{\phi_2}\right) = \frac{\phi_2\nabla\phi_1 - \phi_1\nabla\phi_2}{{\phi_2}^2}$, $\phi_2 \ne 0$.

All the above results can be easily proved. For example,

$$(iii) \quad \nabla(\phi_1\phi_2) = \left(\hat{i}\frac{\partial}{\partial x} + \hat{j}\frac{\partial}{\partial y} + \hat{k}\frac{\partial}{\partial z}\right)(\phi_1\phi_2) = \hat{i}\frac{\partial}{\partial x}(\phi_1\phi_2) + \hat{j}\frac{\partial}{\partial y}(\phi_1\phi_2) + \hat{k}\frac{\partial}{\partial z}(\phi_1\phi_2)$$

$$= \hat{i}\left(\phi_1\frac{\partial\phi_2}{\partial x} + \phi_2\frac{\partial\phi_1}{\partial x}\right) + \hat{j}\left(\phi_1\frac{\partial\phi_2}{\partial y} + \phi_2\frac{\partial\phi_1}{\partial y}\right) + \hat{k}\left(\phi_1\frac{\partial\phi_2}{\partial z} + \phi_2\frac{\partial\phi_1}{\partial z}\right)$$

$$= \phi_1 \left(\hat{i} \frac{\partial \phi_2}{\partial x} + \hat{j} \frac{\partial \phi_2}{\partial y} + \hat{k} \frac{\partial \phi_2}{\partial z} \right) + \phi_2 \left(\hat{i} \frac{\partial \phi_1}{\partial x} + \hat{j} \frac{\partial \phi_1}{\partial y} + \hat{k} \frac{\partial \phi_1}{\partial z} \right)$$

$$= \phi_1 \nabla \phi_2 + \phi_2 \nabla \phi_1.$$

(*iv*) $$\nabla \left(\frac{\phi_1}{\phi_2} \right) = \left(\hat{i} \frac{\partial}{\partial x} + \hat{j} \frac{\partial}{\partial y} + \hat{k} \frac{\partial}{\partial z} \right) \left(\frac{\phi_1}{\phi_2} \right) = \hat{i} \frac{\partial}{\partial x} \left(\frac{\phi_1}{\phi_2} \right) + \hat{j} \frac{\partial}{\partial y} \left(\frac{\phi_1}{\phi_2} \right) + \hat{k} \frac{\partial}{\partial z} \left(\frac{\phi_1}{\phi_2} \right)$$

$$= \hat{i} \frac{\phi_2 \frac{\partial \phi_1}{\partial x} - \phi_1 \frac{\partial \phi_2}{\partial x}}{{\phi_2}^2} + \hat{j} \frac{\phi_2 \frac{\partial \phi_1}{\partial y} - \phi_1 \frac{\partial \phi_2}{\partial y}}{{\phi_2}^2} + \hat{k} \frac{\phi_2 \frac{\partial \phi_1}{\partial z} - \phi_1 \frac{\partial \phi_2}{\partial z}}{{\phi_2}^2}$$

$$= \frac{1}{{\phi_2}^2} \left[\phi_2 \left(\hat{i} \frac{\partial \phi_1}{\partial x} + \hat{j} \frac{\partial \phi_1}{\partial y} + \hat{k} \frac{\partial \phi_1}{\partial z} \right) - \phi_1 \left(\hat{i} \frac{\partial \phi_2}{\partial x} + \hat{j} \frac{\partial \phi_2}{\partial y} + \hat{k} \frac{\partial \phi_2}{\partial z} \right) \right]$$

$$= \frac{\phi_2 \nabla \phi_1 - \phi_1 \nabla \phi_2}{{\phi_2}^2}.$$

ILLUSTRATIVE EXAMPLES

Example 1. *Find grad ϕ when ϕ is given by $\phi = 3x^2y - y^3z^2$ at the point (1, – 2, – 1).*

Sol. $$\text{grad } \phi = \nabla \phi = \left(\hat{i} \frac{\partial}{\partial x} + \hat{j} \frac{\partial}{\partial y} + \hat{k} \frac{\partial}{\partial z} \right) (3x^2y - y^3z^2)$$

$$= \hat{i} \frac{\partial}{\partial x} (3x^2y - y^3z^2) + \hat{j} \frac{\partial}{\partial y} (3x^2y - y^3z^2) + \hat{k} \frac{\partial}{\partial z} (3x^2y - y^3z^2)$$

$$= \hat{i}\ (6xy) + \hat{j} (3x^2 - 3y^2z^2) + \hat{k} (-2y^3z)$$

$$= -\mathbf{12}\hat{i} - \mathbf{9}\hat{j} - \mathbf{16}\hat{k} \text{ at the point } (1, -2, -1).$$

Example 2. *If $\vec{r} = x\hat{i} + y\hat{j} + z\hat{k}$, show that*

(*i*) *grad* $r = \frac{\vec{r}}{r}$ (*ii*) *grad* $\left(\frac{1}{r} \right) = -\frac{\vec{r}}{r^3}$

(*iii*) $\nabla r^n = nr^{n-2} \vec{r}$ (*iv*) $\nabla (\vec{a} \cdot \vec{r}) = \vec{a}$, *where* $\vec{a}$ *is a constant vector.*

(OU-Dec. 2009)

Sol. $r = |\vec{r}| = \sqrt{x^2 + y^2 + z^2}$, or $r^2 = x^2 + y^2 + z^2$

Differentiating partially w.r.t. x, we have $2r \frac{\partial r}{\partial x} = 2x$ or $\frac{\partial r}{\partial x} = \frac{x}{r}$

Similarly, $\frac{\partial r}{\partial y} = \frac{y}{r}$ and $\frac{\partial r}{\partial z} = \frac{z}{r}$

(*i*) $$\text{grad } r = \nabla r = \left(\hat{i}\frac{\partial}{\partial x} + \hat{j}\frac{\partial}{\partial y} + \hat{k}\frac{\partial}{\partial z}\right) r = \hat{i}\frac{\partial r}{\partial x} + \hat{j}\frac{\partial r}{\partial y} + \hat{k}\frac{\partial r}{\partial z}$$

$$= \hat{i}\left(\frac{x}{r}\right) + \hat{j}\left(\frac{y}{r}\right) + \hat{k}\left(\frac{z}{r}\right) = \frac{x\hat{i} + y\hat{j} + z\hat{k}}{r} = \frac{\vec{r}}{r}.$$

(*ii*) $$\text{grad}\left(\frac{1}{r}\right) = \nabla\left(\frac{1}{r}\right) = \left(\hat{i}\frac{\partial}{\partial x} + \hat{j}\frac{\partial}{\partial y} + \hat{k}\frac{\partial}{\partial z}\right)\left(\frac{1}{r}\right)$$

$$= \hat{i}\left(-\frac{1}{r^2}\frac{\partial r}{\partial x}\right) + \hat{j}\left(-\frac{1}{r^2}\frac{\partial r}{\partial y}\right) + \hat{k}\left(-\frac{1}{r^2}\frac{\partial r}{\partial z}\right)$$

$$= \hat{i}\left(-\frac{1}{r^2}\cdot\frac{x}{r}\right) + \hat{j}\left(-\frac{1}{r^2}\cdot\frac{y}{r}\right) + \hat{k}\left(-\frac{1}{r^2}\cdot\frac{z}{r}\right)$$

$$= -\frac{1}{r^3}(x\hat{i} + y\hat{j} + z\hat{k}) = -\frac{\vec{r}}{r^3}.$$

(*iii*) $$\nabla r^n = \left(\hat{i}\frac{\partial}{\partial x} + \hat{j}\frac{\partial}{\partial y} + \hat{k}\frac{\partial}{\partial z}\right) r^n = \hat{i}\left(nr^{n-1}\frac{\partial r}{\partial x}\right) + \hat{j}\left(nr^{n-1}\frac{\partial r}{\partial y}\right) + \hat{k}\left(nr^{n-1}\frac{\partial r}{\partial z}\right)$$

$$= \hat{i}\left(nr^{n-1}\cdot\frac{x}{r}\right) + \hat{j}\left(nr^{n-1}\cdot\frac{y}{r}\right) + \hat{k}\left(nr^{n-1}\cdot\frac{z}{r}\right) = nr^{n-2}(x\hat{i} + y\hat{j} + z\hat{k}) = nr^{n-2}\,\vec{r}.$$

(*iv*) Let $\vec{a} = a_1\hat{i} + a_2\hat{j} + a_3\hat{k}$, where a_1, a_2, a_3 are constants.

$$\vec{a}\cdot\vec{r} = a_1x + a_2y + a_3z$$

$$\therefore \quad \nabla(\vec{a}\cdot\vec{r}) = \left(\hat{i}\frac{\partial}{\partial x} + \hat{j}\frac{\partial}{\partial y} + \hat{k}\frac{\partial}{\partial z}\right)(a_1x + a_2y + a_3z)$$

$$= \hat{i}\frac{\partial}{\partial x}(a_1x + a_2y + a_3z) + \hat{j}\frac{\partial}{\partial y}(a_1x + a_2y + a_3z) + \hat{k}\frac{\partial}{\partial z}(a_1x + a_2y + a_3z)$$

$$= a_1\hat{i} + a_2\hat{j} + a_3\hat{k} = \vec{a}.$$

Example 3. *Find a unit vector normal to the surface $x^3 + y^3 + 3xyz = 3$ at the point $(1, 2, -1)$.*

Sol. Let $\phi = x^3 + y^3 + 3xyz - 3$, then $\dfrac{\partial\phi}{\partial x} = 3x^2 + 3yz$, $\dfrac{\partial\phi}{\partial y} = 3y^2 + 3xz$, $\dfrac{\partial\phi}{\partial z} = 3xy$

$$\therefore \quad \nabla\phi = \hat{i}\frac{\partial\phi}{\partial x} + \hat{j}\frac{\partial\phi}{\partial y} + \hat{k}\frac{\partial\phi}{\partial z} = (3x^2 + 3yz)\hat{i} + (3y^2 + 3xz)\hat{j} + (3xy)\hat{k}$$

At $(1, 2, -1)$, $\nabla\phi = -3\hat{i} + 9\hat{j} + 6\hat{k}$

which is a vector normal to the given surface at $(1, 2, -1)$.

Hence a unit vector normal to the given surface at $(1, 2, -1)$

$$\frac{\nabla\phi}{|\nabla\phi|} = \frac{-3\hat{i} + 9\hat{j} + 6\hat{k}}{\sqrt{[(-3)^2 + (9)^2 + (6)^2]}} = \frac{-3\hat{i} + 9\hat{j} + 6\hat{k}}{3\sqrt{14}} = \frac{1}{\sqrt{14}}(-\hat{i} + 3\hat{j} + 2\hat{k}).$$

Example 4. *Find a unit normal vector to the surface $xy^3z^2 = 4$ at the point $(-1, -1, 2)$.*

(OU-June 2013)

Sol. Let $f = xy^3z^2 - 4$

$$\text{Normal vector} = \nabla f = \sum \hat{i}\frac{\partial f}{\partial x} = y^3z^2\hat{i} + 3xy^2z^2\hat{j} + 2xy^3z\hat{k} \text{ at } (-1, -1, 2)$$

$$\nabla f = -4\hat{i} - 12\hat{j} + 4\hat{k}$$

$$|\nabla f| = \sqrt{16 + 144 + 16} = \sqrt{176} = 4\sqrt{11}$$

$$\therefore \quad \text{Unit normal vector} = \frac{\nabla f}{|\nabla f|}$$

$$= \frac{1}{4\sqrt{11}}(-4\hat{i} - 12\hat{j} + 4\hat{k}) = -\frac{1}{\sqrt{11}}(\hat{i} + 3\hat{j} - \hat{k}).$$

Example 5. *Find the angle between the surfaces $x^2 + y^2 + z^2 = 9$ and $z = x^2 + y^2 - 3$ at the point $(2, -1, 2)$.*

Sol. Angle between two surfaces at a point is the angle between the normals to the surfaces at that point.

Let $\phi_1 = x^2 + y^2 + z^2 - 9$ and $\phi_2 = x^2 + y^2 - z - 3$

Then $\text{grad } \phi_1 = 2x\hat{i} + 2y\hat{j} + 2z\hat{k}$ and $\text{grad } \phi_2 = 2x\hat{i} + 2y\hat{j} - \hat{k}$

Let $\vec{n_1} = \text{grad } \phi_1$ at the point $(2, -1, 2)$ and $\vec{n_2} = \text{grad } \phi_2$ at the point $(2, -1, 2)$. Then

$$\vec{n_1} = 4\hat{i} - 2\hat{j} + 4\hat{k} \quad \text{and} \quad \vec{n_2} = 4\hat{i} - 2\hat{j} - \hat{k}$$

The vectors $\vec{n_1}$ and $\vec{n_2}$ are along normals to the two surfaces at the point $(2, -1, 2)$. If θ is the angle between these vectors, then

$$\cos\theta = \frac{\vec{n_1} \cdot \vec{n_2}}{|\vec{n_1}||\vec{n_2}|} = \frac{4(4) - 2(-2) + 4(-1)}{\sqrt{16 + 4 + 16} \cdot \sqrt{16 + 4 + 1}} = \frac{16}{6\sqrt{21}}$$

$$\therefore \quad \theta = \cos^{-1}\left(\frac{8}{3\sqrt{21}}\right).$$

Example 6. *Find the directional derivative of $f(x, y) = x^2 + y^2$ at $(1, 1)$ in the direction of $2\hat{i} - 4\hat{j}$.* (OU-June 2010)

Sol. Given $f = x^2 + y^2$

$$\nabla f = \hat{i}\frac{\partial f}{\partial x} + \hat{j}\frac{\partial f}{\partial y} = 2x\hat{i} + 2y\hat{j} \text{ at } (1, 1)$$

$$\nabla f = 2\hat{i} + 2\hat{j}$$

Let $\vec{a} = 2\hat{i} - 4\hat{j}$

$$|\vec{a}| = \sqrt{4 + 16} = \sqrt{20} = 2\sqrt{5}$$

$$\therefore \quad \text{Directional derivative of } f \text{ in the direction of } \vec{a} = \nabla f \cdot \frac{\vec{a}}{|\vec{a}|}$$

$$= (2\hat{i} + 2\hat{j}) \cdot \frac{(2\hat{i} - 4\hat{j})}{2\sqrt{5}} = \frac{1}{2\sqrt{5}}(4 - 8) = -\frac{2}{\sqrt{5}}.$$

Example 7. *Find the directional derivative of $f(x, y) = x^2 + y^3$ at (1, 1) in the direction of $3\hat{i} + 4\hat{j}$.* (OU-May 2012)

Sol. Given $f = x^2 + y^3$

$$\therefore \quad \text{grad } f = \nabla f = \hat{i}\frac{\partial f}{\partial x} + \hat{j}\frac{\partial f}{\partial y} = 2x\hat{i} + 3y^2\hat{j} \text{ at } (1, 1)$$

$$\nabla f = 2\hat{i} + 3\hat{j}$$

Let
$$\vec{a} = 3\hat{i} + 4\hat{j}$$

$$|\vec{a}| = \sqrt{9+16} = 5$$

$\therefore$ Directional derivative of f in the direction of $\vec{a}$ is $\nabla f \cdot \dfrac{\vec{a}}{|\vec{a}|}$

$$= (2\hat{i} + 3\hat{j}) \cdot \frac{3\hat{i} + 4\hat{j}}{5} = \frac{1}{5}(6+12) = \frac{18}{5}.$$

Example 8. *Find the directional derivative of $xyz^2 + xz$ at (1, 1, 1) in a direction of the normal to the surface $3xy^2 + y = z$ at (0, 1, 1).*

Sol. Let $f = 3xy^2 + y - z$

$$\nabla f = 3y^2\hat{i} + (6xy+1)\hat{j} - \hat{k} \text{ at } (0, 1, 1)$$

$$\nabla f = 3\hat{i} + \hat{j} - \hat{k}$$

$$|\nabla f| = \sqrt{9+1+1} = \sqrt{11}$$

$$\therefore \quad \text{Unit normal vector } \hat{n} = \frac{\nabla f}{|\nabla f|} = \frac{1}{\sqrt{11}}(3\hat{i} + \hat{j} - \hat{k})$$

Let $\phi = xyz^2 + xz$

$$\therefore \quad \nabla\phi = (yz^2 + z)\hat{i} + xz^2\hat{j} + (2xyz + x)\hat{k} \text{ at } (1, 1, 1)$$

$$\nabla\phi = 2\hat{i} + \hat{j} + 3\hat{k}$$

$\therefore$ Directional derivative of the given function in the direction of $\hat{n}$ at (1, 1, 1) $= \nabla\phi \cdot \hat{n}$

$$= (2\hat{i} + \hat{j} + 3\hat{k}) \cdot \frac{1}{\sqrt{11}}(3\hat{i} + \hat{j} - \hat{k}) = \frac{1}{\sqrt{11}}(6+1-3) = \frac{4}{\sqrt{11}}.$$

Example 9. *Find the directional derivative of the function $f = x^2 - y^2 + 2z^2$ at the point P(1, 2, 3) in the direction of the line PQ where Q is the point (5, 0, 4).*

In what direction it will be maximum ? Find also the magnitude of this maximum.

Sol. We have $\nabla f = \hat{i}\dfrac{\partial f}{\partial x} + \hat{j}\dfrac{\partial f}{\partial y} + \hat{k}\dfrac{\partial f}{\partial z} = 2x\hat{i} - 2y\hat{j} + 4z\hat{k} = 2\hat{i} - 4\hat{j} + 12\hat{k}$ at P(1, 2, 3)

Also,
$$\overrightarrow{PQ} = \overrightarrow{OQ} - \overrightarrow{OP} = (5\hat{i} + 4\hat{k}) - (\hat{i} + 2\hat{j} + 3\hat{k}) = 4\hat{i} - 2\hat{j} + \hat{k}$$

If $\hat{n}$ is a unit vector in the direction $\overrightarrow{PQ}$, then $\hat{n} = \dfrac{4\hat{i} - 2\hat{j} + \hat{k}}{\sqrt{16+4+1}} = \dfrac{1}{\sqrt{21}}(4\hat{i} - 2\hat{j} + \hat{k})$

$\therefore$ Directional derivative of f in the direction $\overrightarrow{\mathbf{PQ}} = (\nabla f) \cdot \hat{n}$

$$= (2\hat{i} - 4\hat{j} + 12\hat{k}) \cdot \frac{1}{\sqrt{21}}(4\hat{i} - 2\hat{j} + \hat{k}) = \frac{1}{\sqrt{21}}\,[2(4) - 4(-2) + 12(1)]$$

$$= \frac{28}{\sqrt{21}} = \frac{4}{3}\sqrt{21}$$

The directional derivative of f is maximum in the direction of the normal to the given surface *i.e.*, in the direction of $\nabla f = 2\hat{i} - 4\hat{j} + 12\hat{k}$

The maximum value of this directional derivative $= |\nabla f|$

$$= \sqrt{(2)^2 + (-4)^2 + (12)^2} = \sqrt{164} = 2\sqrt{41}.$$

Example 10. *Find the directional derivative of $f = xy + yz + zx$ in the direction of a vector $\hat{i} + 2\hat{j} + 2\hat{k}$ at the point (1, 2, 0).*

Sol. Given $f = xy + yz + zx$

$$\text{grad } f = \nabla f = \hat{i}\frac{\partial f}{\partial x} + \hat{j}\frac{\partial f}{\partial y} + \hat{k}\frac{\partial f}{\partial z}$$

$$= (y + z)\hat{i} + (z + x)\hat{j} + (x + y)\hat{k}$$

Let $\vec{a} = \hat{i} + 2\hat{j} + 2\hat{k}$

$$|\vec{a}| = \sqrt{1^2 + 2^2 + 2^2} = 3$$

If $\hat{n}$ is the unit vector in the direction of the vector $\vec{a}$, then

$$\hat{n} = \frac{\vec{a}}{|\vec{a}|} = \frac{1}{3}(\hat{i} + 2\hat{j} + 2\hat{k})$$

$\therefore$ Directional derivative of f in the direction of $\vec{a} = \hat{n} \cdot \nabla f$

$$= \frac{1}{3}(\hat{i} + 2\hat{j} + 2\hat{k}) \cdot [(y + z)\hat{i} + (z + x)\hat{j} + (x + y)\hat{k}]$$

$$= \frac{1}{3}[(y + z) + 2(z + x) + 2(x + y)] \text{ at } (1, 2, 0)$$

$$= \frac{10}{3}.$$

Example 11. *Find the greatest value of the directional derivative of the function $f = x^2yz^3$ at (2, 1, – 1).*

Sol. Given $f = x^2yz^3$

We have $\text{grad } f = \nabla f = \hat{i}\frac{\partial f}{\partial x} + \hat{j}\frac{\partial f}{\partial y} + \hat{k}\frac{\partial f}{\partial z}$

$$= 2xyz^3\hat{i} + x^2z^3\hat{j} + 3x^2yz^2\hat{k} \text{ at } (2, 1, -1)$$

$$\nabla f = -4\hat{i} - 4\hat{j} + 12\hat{k}$$

$\therefore$ Greatest value of the directional derivative of $f = |\nabla f| = \sqrt{16+16+144} = 4\sqrt{11}$.

Example 12. *Find* ∇f*, if* $f(x, y, z) = \log_e (x^2 + y^2 + z^2)$. (OU-June 2014)

Sol.
$$\frac{\partial f}{\partial x} = \frac{1}{x^2+y^2+z^2}(2x) = \frac{2x}{x^2+y^2+z^2}$$

$$\frac{\partial f}{\partial y} = \frac{2y}{x^2+y^2+z^2}, \quad \frac{\partial f}{\partial z} = \frac{2z}{x^2+y^2+z^2}$$

$$\nabla f = i\frac{\partial f}{\partial x} + j\frac{\partial f}{\partial y} + k\frac{\partial f}{\partial z}$$

$$= i\left(\frac{2x}{x^2+y^2+z^2}\right) + j\left(\frac{2y}{x^2+y^2+z^2}\right) + k\left(\frac{2z}{x^2+y^2+z^2}\right)$$

$$= \frac{2}{x^2+y^2+z^2}(xi + yj + zk) = \frac{2\vec{r}}{r^2}.$$

Example 13. *If* $\bar{r} = xi + yj + zk$ *show that* $(\bar{u} . \nabla)\bar{r} = \bar{u}$. (OU-June 2011)

Sol. Let
$$\bar{u} = u_1 i + u_2 j + u_3 k$$

$$\bar{u} . \nabla = (u_1 i + u_2 j + u_3 k) . \left(i\frac{\partial}{\partial x} + j\frac{\partial}{\partial y} + k\frac{\partial}{\partial z}\right)$$

$$= u_1\frac{\partial}{\partial x} + u_2\frac{\partial}{\partial y} + u_3\frac{\partial}{\partial z}$$

$$(\bar{u} . \nabla)\bar{r} = u_1\frac{\partial \bar{r}}{\partial x} + u_2\frac{\partial \bar{r}}{\partial y} + u_3\frac{\partial \bar{r}}{\partial z} \qquad (\because \ \bar{r} = xi + yj + zk)$$

$$= u_1 i + u_2 j + u_3 k = \bar{u}.$$

Example 14. *Find directional derivative of* $f = 2x^2 + y^2 + z^2$ *at (1, 2, 3) in the direction of the line* $\frac{x}{3} = \frac{y}{4} = \frac{z}{5}$.

Sol.
$$\nabla f = \sum i\left(\frac{\partial f}{\partial x}\right) = i(4x) + j(2y) + k(2z) \text{ at } (1, 2, 3)$$

$$\nabla f = 4i + 4j + 6k$$

Line
$$\frac{x}{3} = \frac{y}{4} = \frac{z}{5} \Rightarrow \overline{PQ} = 3i + 4j + 5k$$

$$|\overline{PQ}| = \sqrt{9+16+25} = \sqrt{50}$$

$\therefore$ Directional derivative

$$= \nabla f . \frac{\overline{PQ}}{|\overline{PQ}|} = (4i + 4j + 6k) . \frac{1}{\sqrt{50}}(3i + 4j + 5k)$$

$$= \frac{1}{\sqrt{50}}(12 + 16 + 30) = \frac{58}{\sqrt{50}}.$$

EXERCISE 1

1. Find grad ϕ when ϕ is given by

   (*i*) $\phi = x^2 + yz$ (*ii*) $\phi = x^3 + y^3 + 3xyz$ (*iii*) $\phi = \log (x^2 + y^2 + z^2)$.

2. If $r = |\vec{r}|$ where $\vec{r} = x\hat{i} + y\hat{j} + z\hat{k}$, prove that

   (*i*) $\nabla f(r) = f'(r)\, \nabla r$ (*ii*) $\nabla \log r = \dfrac{\vec{r}}{r^2}$

   (*iii*) $\nabla (e^{r^2}) = 2e^{r^2}\, \vec{r}$ (*iv*) grad $|\vec{r}|^2 = 2\vec{r}$.

3. If $u = x + y + z$, $v = x^2 + y^2 + z^2$, $w = yz + zx + xy$, prove that (grad u). {(grad v) × (grad w)} = 0.

4. Find a unit vector normal to the surface

   (*i*) $xy^3z^2 = 4$ at the point $(-1, -1, 2)$ (*ii*) $x^2y + 2xz = 4$ at the point $(2, -2, 3)$.

5. Find the directional derivative of the function

   (*i*) $f(x, y, z) = xy^2 + yz^3$ at the point $(2, -1, 1)$ in the direction of the vector $\hat{i} + 2\hat{j} + 2\hat{k}$.

   (*ii*) $f(x, y, z) = 2xy + z^2$ at the point $(1, -1, 3)$ in the direction of the vector $\hat{i} + 2\hat{j} + 2\hat{k}$.

   (*iii*) $\phi = 4xz^3 - 3x^2yz^2$ at $(2, -1, 2)$ along z-axis.

   (*iv*) $\phi = xy^2 + yz^3$ at the point $(2, -1, 1)$ in the direction of the normal to the surface $x \log z - y^2 + 4 = 0$ at $(-1, 2, 1)$.

6. In what direction from $(3, 1, -2)$ is the directional derivative of $\phi = x^2y^2z^4$ maximum and what is its magnitude ?

7. What is the greatest rate of increase of $u = x^2 + yz^2$ at the point $(1, -1, 3)$?

8. The temperature at a point (x, y, z) in space is given by $T(x, y, z) = x^2 + y^2 - z$. A mosquito located at $(1, 1, 2)$ desires to fly in such a direction that it will get warm as soon as possible. In what direction should it fly ?

9. Calculate the angle between the normals to the surface $xy = z^2$ at the points $(4, 1, 2)$ and $(3, 3, -3)$.

10. If θ is the acute angle between the surfaces $xy^2z = 3x + z^2$ and $3x^2 - y^2 + 2z = 1$ at the point $(1, -2, 1)$, show that

$$\cos \theta = \frac{3}{7\sqrt{6}}.$$

11. Find the values of constants λ and μ so that the surfaces $\lambda x^2 - \mu yz = (\lambda + 2)x$ and $4x^2y + z^3 = 4$ may intersect orthogonally at the point $(1, -1, 2)$.

    [**Hint.** The point P(1, – 1, 2) lies on both the surfaces and $(\text{grad } \phi_1)_P . (\text{grad } \phi_2)_P = 0$.]

12. Find the angle between the tangent planes to the surfaces $x \log z = y^2 - 1$ and $x^2y = 2 - z$ at the point $(1, 1, 1)$.

Answers

1. (*i*) $2x\hat{i} + z\hat{j} + y\hat{k}$ (*ii*) $3(x^2 + yz)\hat{i} + 3(y^2 + xz)\hat{j} + 3xy\hat{k}$ (*iii*) $\dfrac{2(x\hat{i} + y\hat{j} + z\hat{k})}{x^2 + y^2 + z^2}$

4. (*i*) $-\dfrac{1}{\sqrt{11}}(\hat{i} + 3\hat{j} - \hat{k})$ (*ii*) $\dfrac{1}{3}(-\hat{i} + 2\hat{j} + 2\hat{k})$ 5. (*i*) $-\dfrac{11}{3}$ (*ii*) $\dfrac{14}{3}$

(*iii*) 144 (*iv*) $\frac{15}{\sqrt{17}}$ **6.** $96(\hat{i}+3\hat{j}-3\hat{k})$; $96\sqrt{19}$

7. 11 **8.** $\frac{1}{3}(2\hat{i}+2\hat{j}-\hat{k})$ **9.** $\cos^{-1}\left(\frac{-1}{\sqrt{22}}\right)$

11. $\lambda = 2.5, \mu = 1$ **12.** $\cos^{-1}\left(\frac{-1}{\sqrt{30}}\right)$.

5.15. DIVERGENCE OF A VECTOR POINT FUNCTION

The divergence of a differentiable vector point function $\vec{V}$ is denoted by div $\vec{V}$ and is defined as

$$\text{div } \vec{V} = \nabla \cdot \vec{V} = \left(\hat{i}\frac{\partial}{\partial x} + \hat{j}\frac{\partial}{\partial y} + \hat{k}\frac{\partial}{\partial z}\right) \cdot \vec{V} = \hat{i} \cdot \frac{\partial \vec{V}}{\partial x} + \hat{j} \cdot \frac{\partial \vec{V}}{\partial y} + \hat{k} \cdot \frac{\partial \vec{V}}{\partial z}.$$

Obviously, the divergence of a vector point function is a scalar point function.

If $\vec{V} = V_1\hat{i} + V_2\hat{j} + V_3\hat{k}$, then

$$\text{div } \vec{V} = \nabla \cdot \vec{V} = \left(\hat{i}\frac{\partial}{\partial x} + \hat{j}\frac{\partial}{\partial y} + \hat{k}\frac{\partial}{\partial z}\right) \cdot [V_1\hat{i} + V_2\hat{j} + V_3\hat{k}] = \frac{\partial V_1}{\partial x} + \frac{\partial V_2}{\partial y} + \frac{\partial V_3}{\partial z}.$$

Since $\hat{i}\cdot\hat{i} = \hat{j}\cdot\hat{j} = \hat{k}\cdot\hat{k} = 1$ and $\hat{i}\cdot\hat{j} = \hat{j}\cdot\hat{k} = \hat{k}\cdot\hat{i} = 0$.

5.16. CURL OF A VECTOR POINT FUNCTION

The curl (or rotation) of a differentiable vector point function $\vec{V}$ is denoted by curl $\vec{V}$ and is defined as

$$\text{curl } \vec{V} = \nabla \times \vec{V} = \left(\hat{i}\frac{\partial}{\partial x} + \hat{j}\frac{\partial}{\partial y} + \hat{k}\frac{\partial}{\partial z}\right) \times \vec{V} = \hat{i} \times \frac{\partial \vec{V}}{\partial x} + \hat{j} \times \frac{\partial \vec{V}}{\partial y} + \hat{k} \times \frac{\partial \vec{V}}{\partial z}.$$

Obviously, the curl of a vector point function is a vector point function.

If $\vec{V} = V_1\hat{i} + V_2\hat{j} + V_3\hat{k}$

then curl $\vec{V} = \nabla \times \vec{V} = \left(\hat{i}\frac{\partial}{\partial x} + \hat{j}\frac{\partial}{\partial y} + \hat{k}\frac{\partial}{\partial z}\right) \times [V_1\hat{i} + V_2\hat{j} + V_3\hat{k}]$

$$= \begin{vmatrix} \hat{i} & \hat{j} & \hat{k} \\ \frac{\partial}{\partial x} & \frac{\partial}{\partial y} & \frac{\partial}{\partial z} \\ V_1 & V_2 & V_3 \end{vmatrix} = \hat{i}\left(\frac{\partial V_3}{\partial y} - \frac{\partial V_2}{\partial z}\right) + \hat{j}\left(\frac{\partial V_1}{\partial z} - \frac{\partial V_3}{\partial x}\right) + \hat{k}\left(\frac{\partial V_2}{\partial x} - \frac{\partial V_1}{\partial y}\right).$$

ILLUSTRATIVE EXAMPLES

Example 1. *If* $\vec{r} = x\hat{i} + y\hat{j} + z\hat{k}$, *show that*

(*i*) *div* $\vec{r} = 3$ (*ii*) *curl* $\vec{r} = \vec{0}$.

Sol. (*i*) div $\vec{r} = \nabla \cdot \vec{r} = \frac{\partial}{\partial x}(x) + \frac{\partial}{\partial y}(y) + \frac{\partial}{\partial z}(z) = 1 + 1 + 1 = 3.$

(*ii*) curl $\vec{r} = \nabla \times \vec{r} = \begin{vmatrix} \hat{i} & \hat{j} & \hat{k} \\ \frac{\partial}{\partial x} & \frac{\partial}{\partial y} & \frac{\partial}{\partial z} \\ x & y & z \end{vmatrix}$

$$= \hat{i}\left[\frac{\partial}{\partial y}(z) - \frac{\partial}{\partial z}(y)\right] + \hat{j}\left[\frac{\partial}{\partial z}(x) - \frac{\partial}{\partial x}(z)\right] + \hat{k}\left[\frac{\partial}{\partial x}(y) - \frac{\partial}{\partial y}(x)\right]$$

$$= \hat{i}(0) + \hat{j}(0) + \hat{k}(0) = \vec{0}.$$

Example 2. *Find the divergence and curl of the vector* $\vec{V} = (xyz)\hat{i} + (3x^2y)\hat{j} + (xz^2 - y^2z)\hat{k}$ *at the point (2, – 1, 1).*

Sol. div $\vec{V} = \frac{\partial}{\partial x}(xyz) + \frac{\partial}{\partial y}(3x^2y) + \frac{\partial}{\partial z}(xz^2 - y^2z)$

$$= yz + 3x^2 + 2xz - y^2 = -1 + 12 + 4 - 1 = 14 \text{ at } (2, -1, 1)$$

curl $\vec{V} = \begin{vmatrix} \hat{i} & \hat{j} & \hat{k} \\ \frac{\partial}{\partial x} & \frac{\partial}{\partial y} & \frac{\partial}{\partial z} \\ xyz & 3x^2y & xz^2 - y^2z \end{vmatrix} = \hat{i}(-2yz - 0) + \hat{j}(xy - z^2) + \hat{k}(6xy - xz)$

$$= 2\hat{i} - 3\hat{j} - 14\hat{k} \text{ at } (2, -1, 1).$$

Example 3. *Find div* $\vec{F}$ *and curl* $\vec{F}$, *where* $\vec{F} = grad\ (x^3 + y^3 + z^3 - 3xyz)$.

(OU-June 2013)

Sol. Let $\phi = x^3 + y^3 + z^3 - 3xyz$, then

$$\vec{F} = \text{grad}\ \phi = \left(\hat{i}\frac{\partial}{\partial x} + \hat{j}\frac{\partial}{\partial y} + \hat{k}\frac{\partial}{\partial z}\right)(x^3 + y^3 + z^3 - 3xyz)$$

$$= (3x^2 - 3yz)\hat{i} + (3y^2 - 3zx)\hat{j} + (3z^2 - 3xy)\hat{k}$$

$\therefore$ div $\vec{F} = \frac{\partial}{\partial x}(3x^2 - 3yz) + \frac{\partial}{\partial y}(3y^2 - 3zx) + \frac{\partial}{\partial x}(3z^2 - 3xy)$

$$= 6x + 6y + 6z = 6(x + y + z)$$

and $$\text{curl } \vec{F} = \begin{vmatrix} \hat{i} & \hat{j} & \hat{k} \\ \dfrac{\partial}{\partial x} & \dfrac{\partial}{\partial y} & \dfrac{\partial}{\partial z} \\ 3x^2 - 3yz & 3y^2 - 3zx & 3z^2 - 3xy \end{vmatrix}$$

$$= \hat{i}(-3x + 3x) + \hat{j}(-3y + 3y) + \hat{k}(-3z + 3z) = \vec{0}.$$

5.17. PHYSICAL INTERPRETATION OF DIVERGENCE

Consider a fluid having density $\rho = \rho(x, y, z, t)$ and velocity $\vec{v} = \vec{v}(x, y, z, t)$ at a point (x, y, z) at time t. Let $\vec{V} = \rho \vec{v}$, then $\vec{V}$ is a vector having the same direction as $\vec{v}$ and magnitude $\rho \mid \vec{v} \mid$. It is known as *flux*. Its direction gives the direction of the fluid flow, and its magnitude gives the mass of the fluid crossing per unit time a unit area placed perpendicular to the direction of flow.

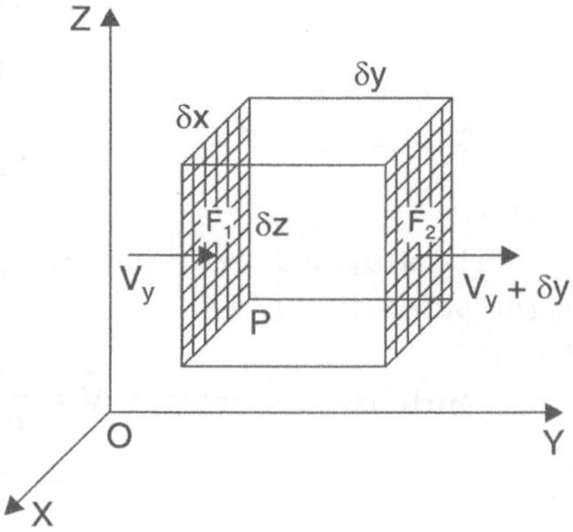

Consider the motion of the fluid having velocity $\vec{V} = V_x\hat{i} + V_y\hat{j} + V_z\hat{k}$ at a point P(x, y, z). Consider a small parallelopiped with edges δx, δy, δz parallel to the axes with one of its corners at P.

The mass of the fluid entering through the face F_1 per unit time is $V_y\, \delta x\, \delta z$ and that flowing out through the opposite face F_2 is $V_{y+\delta y}\, \delta x\, \delta z = \left(V_y + \dfrac{\partial V_y}{\partial y}\delta y\right)\delta x\, \delta z$ by using Taylor's series.

∴ The net decrease in the mass of fluid flowing across these two faces

$$= \left(V_y + \frac{\partial V_y}{\partial y}\delta y\right)\delta x\, \delta z - V_y\, \delta x\, \delta z = \frac{\partial V_y}{\partial y}\delta x\, \delta y\, \delta z$$

Similarly, considering the other two pairs of faces, we get the total decrease in the mass of fluid inside the parallelopiped per unit time $= \left(\dfrac{\partial V_x}{\partial x} + \dfrac{\partial V_y}{\partial y} + \dfrac{\partial V_z}{\partial z}\right)\delta x\, \delta y\, \delta z.$

Dividing this by the volume $\delta x\, \delta y\, \delta z$ of the parallelopiped, we have the rate of loss of fluid per unit time $= \dfrac{\partial V_x}{\partial x} + \dfrac{\partial V_y}{\partial y} + \dfrac{\partial V_z}{\partial z} = \text{div } \vec{V}$

Hence div $\vec{V}$ gives the rate of outflow per unit volume at a point of the fluid.

Solenoidal Vector. If **div** $\vec{V}$ **= 0** everywhere in some region R of space, then $\vec{V}$ is called **solenoidal vector point function.**

5.18. PHYSICAL INTERPRETATION OF CURL

Consider a rigid body rotating about a fixed axis through O with uniform angular velocity

$$\vec{\Omega} = \omega_1 \hat{i} + \omega_2 \hat{j} + \omega_3 \hat{k}$$

The velocity $\vec{V}$ of any point P(x, y, z) on the body is given by $\vec{V} = \vec{\Omega} \times \vec{r}$, where $\vec{r} = x\hat{i} + y\hat{j} + z\hat{k}$ is the position vector of P.

$$\therefore \quad \vec{V} = \vec{\Omega} \times \vec{r} = \begin{vmatrix} \hat{i} & \hat{j} & \hat{k} \\ \omega_1 & \omega_2 & \omega_3 \\ x & y & z \end{vmatrix}$$

$$= (\omega_2 z - \omega_3 y)\hat{i} + (\omega_3 x - \omega_1 z)\hat{j} + (\omega_1 y - \omega_2 x)\hat{k}$$

$$\text{curl } \vec{V} = \nabla \times \vec{V} = \begin{vmatrix} \hat{i} & \hat{j} & \hat{k} \\ \dfrac{\partial}{\partial x} & \dfrac{\partial}{\partial y} & \dfrac{\partial}{\partial z} \\ \omega_2 z - \omega_3 y & \omega_3 x - \omega_1 z & \omega_1 y - \omega_2 x \end{vmatrix}$$

$$= (\omega_1 + \omega_1)\hat{i} + (\omega_2 + \omega_2)\hat{j} + (\omega_3 + \omega_3)\hat{k} \qquad [\because \ \omega_1, \omega_2, \omega_3 \text{ are constants}]$$

$$= 2(\omega_1 \hat{i} + \omega_2 \hat{j} + \omega_3 \hat{k}) = 2\vec{\Omega}$$

$$\Rightarrow \quad \vec{\Omega} = \frac{1}{2} \text{curl } \vec{V}.$$

Thus, the angular velocity at any point is equal to half the curl of the linear velocity at that point of the body.

Irrotational Vector: If curl $\vec{V} = \vec{0}$ then $\vec{V}$ is said to be an **irrotational vector,** otherwise **rotational.**

5.19. PROPERTIES OF DIVERGENCE AND CURL

1. For a constant vector $\vec{a}$, div $\vec{a} = 0$, curl $\vec{a} = \vec{0}$

2. **div $(\vec{A} + \vec{B})$ = div $\vec{A}$ + div $\vec{B}$** or $\nabla \cdot (\vec{A} + \vec{B}) = \nabla \cdot \vec{A} + \nabla \cdot \vec{B}$

Proof. $\text{div}(\vec{A} + \vec{B}) = \nabla \cdot (\vec{A} + \vec{B}) = \left(\hat{i}\dfrac{\partial}{\partial x} + \hat{j}\dfrac{\partial}{\partial y} + \hat{k}\dfrac{\partial}{\partial z}\right) \cdot (\vec{A} + \vec{B})$

$$= \hat{i} \cdot \frac{\partial}{\partial x}(\vec{A} + \vec{B}) + \hat{j} \cdot \frac{\partial}{\partial y}(\vec{A} + \vec{B}) + \hat{k} \cdot \frac{\partial}{\partial z}(\vec{A} + \vec{B})$$

$$= \hat{i} \cdot \left(\frac{\partial \vec{A}}{\partial x} + \frac{\partial \vec{B}}{\partial x}\right) + \hat{j} \cdot \left(\frac{\partial \vec{A}}{\partial y} + \frac{\partial \vec{B}}{\partial y}\right) + \hat{k} \cdot \left(\frac{\partial \vec{A}}{\partial z} + \frac{\partial \vec{B}}{\partial z}\right)$$

$$= \left(\hat{i} \cdot \frac{\partial \vec{A}}{\partial x} + \hat{j} \cdot \frac{\partial \vec{A}}{\partial y} + \hat{k} \cdot \frac{\partial \vec{A}}{\partial z}\right) + \left(\hat{i} \cdot \frac{\partial \vec{B}}{\partial x} + \hat{j} \cdot \frac{\partial \vec{B}}{\partial y} + \hat{k} \cdot \frac{\partial \vec{B}}{\partial z}\right)$$

$$= \nabla \cdot \vec{A} + \nabla \cdot \vec{B} = \text{div } \vec{A} + \text{div } \vec{B}.$$

3. **Curl $(\vec{A}+\vec{B})$ = curl $\vec{A}$ + curl $\vec{B}$** or $\nabla \times (\vec{A}+\vec{B}) = \nabla \times \vec{A} + \nabla \times \vec{B}$

Proof. $\text{Curl}\,(\vec{A}+\vec{B}) = \nabla \times (\vec{A}+\vec{B}) = \left(\hat{i}\frac{\partial}{\partial x} + \hat{j}\frac{\partial}{\partial y} + \hat{k}\frac{\partial}{\partial z}\right) \times (\vec{A}+\vec{B})$

$$= \sum \hat{i} \times \frac{\partial}{\partial x}(\vec{A}+\vec{B}) = \sum \hat{i} \times \left(\frac{\partial \vec{A}}{\partial x} + \frac{\partial \vec{B}}{\partial x}\right)$$

$$= \sum \hat{i} \times \frac{\partial \vec{A}}{\partial x} + \sum \hat{i} \times \frac{\partial \vec{B}}{\partial x} = \nabla \times \vec{A} + \nabla \times \vec{B} = \text{curl}\,\vec{A} + \text{curl}\,\vec{B}.$$

4. If $\vec{A}$ is a vector function and ϕ is a scalar function, then

div $(\phi\vec{A}) = \phi$ **div** $\vec{A}$ + (**grad** ϕ) . $\vec{A}$ or $\nabla \cdot (\phi\vec{A}) = \phi(\nabla \cdot \vec{A}) + (\nabla\phi) \cdot \vec{A}$

(OU-Apr. 2007, Dec. 2008, May 2008)

Proof. $\text{div}\,(\phi\vec{A}) = \nabla . (\phi\vec{A}) = \left(\hat{i}\frac{\partial}{\partial x} + \hat{j}\frac{\partial}{\partial y} + \hat{k}\frac{\partial}{\partial z}\right) . (\phi\vec{A})$

$$= \hat{i} \cdot \frac{\partial}{\partial x}(\phi\vec{A}) + \hat{j} \cdot \frac{\partial}{\partial y}(\phi\vec{A}) + \hat{k} \cdot \frac{\partial}{\partial z}(\phi\vec{A})$$

$$= \sum \hat{i} \cdot \frac{\partial}{\partial x}(\phi\vec{A}) = \sum \hat{i} \cdot \left(\phi\frac{\partial \vec{A}}{\partial x} + \frac{\partial \phi}{\partial x}\vec{A}\right) = \phi\sum\left(\hat{i} \cdot \frac{\partial A}{\partial x}\right) + \sum\left(\hat{i}\frac{\partial \phi}{\partial x}\right) \cdot \vec{A}$$

$$= \phi(\nabla \cdot \vec{A}) + (\nabla\phi) \cdot \vec{A} = \phi\,\text{div}\,\vec{A} + (\text{grad}\,\phi) \cdot \vec{A}.$$

5. If A is a vector function and ϕ is a scalar function, then

curl $(\phi\vec{A})$ = (**grad** ϕ) $\times \vec{A}$ + ϕ **curl** $\vec{A}$

or $\nabla \times (\phi\vec{A}) = (\nabla\phi) \times \vec{A} + \phi(\nabla \times \vec{A})$ *(OU-Dec. 2008)*

Proof. $\text{Curl}\,(\phi\vec{A}) = \nabla \times (\phi\vec{A}) = \left(\hat{i}\frac{\partial}{\partial x} + \hat{j}\frac{\partial}{\partial y} + \hat{k}\frac{\partial}{\partial z}\right) \times (\phi\vec{A})$

$$= \sum \hat{i} \times \frac{\partial}{\partial x}(\phi\vec{A}) = \sum \hat{i} \times \left(\frac{\partial \phi}{\partial x}\vec{A} + \phi\frac{\partial \vec{A}}{\partial x}\right) = \sum \hat{i} \times \frac{\partial \phi}{\partial x}\vec{A} + \phi\sum \hat{i} \times \frac{\partial \vec{A}}{\partial x}$$

$$= \sum \frac{\partial \phi}{\partial x}\hat{i} \times \vec{A} + \phi\sum \hat{i} \times \frac{\partial \vec{A}}{\partial x} \qquad [\because \vec{a} \times (m\vec{b}) = (m\vec{a}) \times \vec{b} = m(\vec{a} \times \vec{b})]$$

$$= (\nabla\phi) \times \vec{A} + \phi(\nabla \times \vec{A}) = (\text{grad}\,\phi) \times \vec{A} + \phi\,\text{curl}\,\vec{A}.$$

6. $\nabla(\vec{A} \cdot \vec{B}) = (\vec{A} \cdot \nabla)\vec{B} + (\vec{B} \cdot \nabla)\vec{A} + \vec{A} \times (\nabla \times \vec{B}) + \vec{B} \times (\nabla \times \vec{A})$

Proof. $\nabla(\vec{A} \cdot \vec{B}) = \sum \hat{i}\frac{\partial}{\partial x}(\vec{A} \cdot \vec{B}) = \sum \hat{i}\left(\vec{A} \cdot \frac{\partial \vec{B}}{\partial x} + \frac{\partial \vec{A}}{\partial x} \cdot \vec{B}\right)$

$$= \sum \left(\vec{A} \cdot \frac{\partial \vec{B}}{\partial x} \right) \hat{i} + \sum \left(\vec{B} \cdot \frac{\partial \vec{A}}{\partial x} \right) \hat{i} \qquad ...(1)$$

Now, we know that $\vec{a} \times (\vec{b} \times \vec{c}) = (\vec{a} \cdot \vec{c})\vec{b} - (\vec{a} \cdot \vec{b})\vec{c}$

$$\therefore \qquad (\vec{a} \cdot \vec{b})\vec{c} = (\vec{a} \cdot \vec{c})\vec{b} - \vec{a} \times (\vec{b} \times \vec{c})$$

$$\therefore \qquad \left(\vec{A} \cdot \frac{\partial \vec{B}}{\partial x} \right) \hat{i} = (\vec{A} \cdot \hat{i}) \frac{\partial \vec{B}}{\partial x} - \vec{A} \times \left(\frac{\partial \vec{B}}{\partial x} \times \hat{i} \right) = (\vec{A} \cdot \hat{i}) \frac{\partial \vec{B}}{\partial x} + \vec{A} \times \left(\hat{i} \times \frac{\partial \vec{B}}{\partial x} \right)$$

$$\therefore \quad \sum \left(\vec{A} \cdot \frac{\partial \vec{B}}{\partial x} \right) \hat{i} = \left(\vec{A} \cdot \sum \hat{i} \frac{\partial}{\partial x} \right) \vec{B} + \vec{A} \times \sum \left(\hat{i} \times \frac{\partial \vec{B}}{\partial x} \right)$$

$$= (\vec{A} \cdot \nabla)\vec{B} + \vec{A} \times (\nabla \times \vec{B}) \qquad ...(2)$$

Interchanging $\vec{A}$ and $\vec{B} \cdot \sum \left(\vec{B} \cdot \frac{\partial \vec{A}}{\partial x} \right) \hat{i} = (\vec{B} \cdot \nabla)\vec{A} + \vec{B} \times (\nabla \times \vec{A})$ ...(3)

Substituting the values from (2) and (3) in (1), we get

$$\nabla(\vec{A} \cdot \vec{B}) = (\vec{A} \cdot \nabla)\vec{B} + (\vec{B} \cdot \nabla)\vec{A} + \vec{A} \times (\nabla \times \vec{B}) + \vec{B} \times (\nabla \times \vec{A}).$$

7. $\nabla \cdot (\vec{A} \times \vec{B}) = \vec{B} \cdot (\nabla \times \vec{A}) - \vec{A} \cdot (\nabla \times \vec{B})$

Proof. $\nabla \cdot (\vec{A} \times \vec{B}) = \sum \hat{i} \cdot \frac{\partial}{\partial x} (\vec{A} \times \vec{B}) = \sum \hat{i} \cdot \left(\frac{\partial \vec{A}}{\partial x} \times \vec{B} + \vec{A} \times \frac{\partial \vec{B}}{\partial x} \right)$

$$= \sum \hat{i} \cdot \left(\frac{\partial \vec{A}}{\partial x} \times \vec{B} \right) + \sum \hat{i} \cdot \left(\vec{A} \times \frac{\partial \vec{B}}{\partial x} \right) = \sum \hat{i} \cdot \left(\frac{\partial \vec{A}}{\partial x} \times \vec{B} \right) - \sum \hat{i} \cdot \left(\frac{\partial \vec{B}}{\partial x} \times \vec{A} \right)$$

$$= \sum \left(\hat{i} \times \frac{\partial \vec{A}}{\partial x} \right) \cdot \vec{B} - \sum \left(\hat{i} \times \frac{\partial \vec{B}}{\partial x} \right) \cdot \vec{A} \qquad \text{(Since } \vec{a} \cdot (\vec{b} \times \vec{c}) = (\vec{a} \times \vec{b}) \cdot \vec{c}\text{)}$$

$$= (\nabla \times \vec{A}) \cdot \vec{B} - (\nabla \times \vec{B}) \cdot \vec{A} = \vec{B} \cdot (\nabla \times \vec{A}) - \vec{A} \cdot (\nabla \times \vec{B}).$$

8. $\nabla \times (\vec{A} \times \vec{B}) = (\nabla \cdot \vec{B})\vec{A} - (\nabla \cdot \vec{A})\vec{B} + (\vec{B} \cdot \nabla)\vec{A} - (\vec{A} \cdot \nabla)\vec{B}$

Proof. $\nabla \times (\vec{A} \times \vec{B}) = \sum \hat{i} \times \frac{\partial}{\partial x} (\vec{A} \times \vec{B}) = \sum \hat{i} \times \left(\frac{\partial \vec{A}}{\partial x} \times \vec{B} + \vec{A} \times \frac{\partial \vec{B}}{\partial x} \right)$

$$= \sum \hat{i} \times \left(\frac{\partial \vec{A}}{\partial x} \times \vec{B} \right) + \sum \hat{i} \times \left(\vec{A} \times \frac{\partial \vec{B}}{\partial x} \right)$$

$$= \sum \left[(\hat{i} \cdot \vec{B}) \frac{\partial \vec{A}}{\partial x} - \left(\hat{i} \cdot \frac{\partial \vec{A}}{\partial x} \right) \vec{B} \right] + \sum \left[\left(\hat{i} \cdot \frac{\partial \vec{B}}{\partial x} \right) \vec{A} - (\hat{i} \cdot \vec{A}) \frac{\partial \vec{B}}{\partial x} \right]$$

(Since $\vec{a} \times (\vec{b} \times \vec{c}) = (\vec{a} \cdot \vec{c}) \vec{b} - (\vec{a} \cdot \vec{b}) \vec{c}$)

$$= \sum (\vec{B} \cdot \hat{i}) \frac{\partial \vec{A}}{\partial x} - \left(\sum \hat{i} \cdot \frac{\partial \vec{A}}{\partial x} \right) \vec{B} + \left(\sum \hat{i} \cdot \frac{\partial \vec{B}}{\partial x} \right) \vec{A} - \sum (\vec{A} \cdot \hat{i}) \frac{\partial \vec{B}}{\partial x}$$

$$= \left(\vec{B} \cdot \sum \hat{i} \frac{\partial}{\partial x} \right) \vec{A} - (\nabla \cdot \vec{A}) \vec{B} + (\nabla \cdot \vec{B}) \vec{A} - \left(\mathrm{A} \cdot \sum \hat{i} \frac{\partial}{\partial x} \right) \vec{B}$$

$$= (\vec{B} \cdot \nabla) \vec{A} - (\nabla \cdot \vec{A}) \vec{B} + (\nabla \cdot \vec{B}) \vec{A} - (\vec{A} \cdot \nabla) \vec{B}$$

$$= (\nabla \cdot \vec{B}) \vec{A} - (\nabla \cdot \vec{A}) \vec{B} + (\vec{B} \cdot \nabla) \vec{A} - (\vec{A} \cdot \nabla) \vec{B}.$$

5.20. REPEATED OPERATIONS BY ∇

Let ϕ (x, y, z) and $\vec{V}$ (x, y, z) be scalar and vector point functions respectively.

Since grad ϕ and curl $\vec{V}$ are also vector point functions, we can find their divergence as well as curl, whereas div $\vec{V}$ being a scalar point function, we can find its gradient only.

1. **div (grad ϕ) = $\nabla^2\phi$ where $\nabla^2 = \frac{\partial^2}{\partial \mathbf{x}^2} + \frac{\partial^2}{\partial \mathbf{y}^2} + \frac{\partial^2}{\partial \mathbf{z}^2}$**

Proof. div (grad ϕ) $= \nabla \cdot (\nabla\phi) = \left(\hat{i} \frac{\partial}{\partial x} + \hat{j} \frac{\partial}{\partial y} + \hat{k} \frac{\partial}{\partial z} \right) . \left(\hat{i} \frac{\partial \phi}{\partial x} + \hat{j} \frac{\partial \phi}{\partial y} + \hat{k} \frac{\partial \phi}{\partial z} \right)$

$$= \frac{\partial}{\partial x} \left(\frac{\partial \phi}{\partial x} \right) + \frac{\partial}{\partial y} \left(\frac{\partial \phi}{\partial y} \right) + \frac{\partial}{\partial z} \left(\frac{\partial \phi}{\partial z} \right) = \frac{\partial^2 \phi}{\partial x^2} + \frac{\partial^2 \phi}{\partial y^2} + \frac{\partial^2 \phi}{\partial z^2} = \left(\frac{\partial^2}{\partial x^2} + \frac{\partial^2}{\partial y^2} + \frac{\partial^2}{\partial z^2} \right) \phi$$

$$= \nabla^2 \phi, \quad \text{where} \quad \nabla^2 = \frac{\partial^2}{\partial x^2} + \frac{\partial^2}{\partial y^2} + \frac{\partial^2}{\partial z^2}$$

∇^2 is called the *Laplacian operator* and $\nabla^2\phi = 0$ is called *Laplace's equation.*

2. **Curl (grad ϕ) = $\nabla \times \nabla\phi = \vec{0}$**

Proof. Curl (grad ϕ) $= \nabla \times \nabla\phi = \nabla \times \left(\hat{i} \frac{\partial \phi}{\partial x} + \hat{j} \frac{\partial \phi}{\partial y} + \hat{k} \frac{\partial \phi}{\partial z} \right)$

$$= \begin{vmatrix} \hat{i} & \hat{j} & \hat{k} \\ \frac{\partial}{\partial x} & \frac{\partial}{\partial y} & \frac{\partial}{\partial z} \\ \frac{\partial \phi}{\partial x} & \frac{\partial \phi}{\partial y} & \frac{\partial \phi}{\partial z} \end{vmatrix} = \Sigma \hat{i} \left(\frac{\partial^2 \phi}{\partial y \partial z} - \frac{\partial^2 \phi}{\partial z \partial y} \right) = \vec{0}$$

3. **div (curl $\vec{V}$) = $\nabla \cdot (\nabla \times \vec{V}) = 0$.**

Proof. Let $\vec{V} = V_1\hat{i} + V_2\hat{j} + V_3\hat{k}$

then
$$\text{Curl } \vec{V} = \nabla \times \vec{V} = \begin{vmatrix} \hat{i} & \hat{j} & \hat{k} \\ \dfrac{\partial}{\partial x} & \dfrac{\partial}{\partial y} & \dfrac{\partial}{\partial z} \\ V_1 & V_2 & V_3 \end{vmatrix}$$

$$= \hat{i}\left(\frac{\partial V_3}{\partial y} - \frac{\partial V_2}{\partial z}\right) + \hat{j}\left(\frac{\partial V_1}{\partial z} - \frac{\partial V_3}{\partial x}\right) + \hat{k}\left(\frac{\partial V_2}{\partial x} - \frac{\partial V_1}{\partial y}\right)$$

$$\therefore \quad \text{div (curl } \vec{V}) = \nabla \cdot (\nabla \times \vec{V})$$

$$= \frac{\partial}{\partial x}\left(\frac{\partial V_3}{\partial y} - \frac{\partial V_2}{\partial z}\right) + \frac{\partial}{\partial y}\left(\frac{\partial V_1}{\partial z} - \frac{\partial V_3}{\partial x}\right) + \frac{\partial}{\partial z}\left(\frac{\partial V_2}{\partial x} - \frac{\partial V_1}{\partial y}\right)$$

$$= \left(\frac{\partial^2 V_3}{\partial x \partial y} - \frac{\partial^2 V_2}{\partial x \partial z}\right) + \left(\frac{\partial^2 V_1}{\partial y \partial z} - \frac{\partial^2 V_3}{\partial y \partial x}\right) + \left(\frac{\partial^2 V_2}{\partial z \partial x} - \frac{\partial^2 V_1}{\partial z \partial y}\right) = 0.$$

4. **Curl (curl $\vec{V}$) = grad div $\vec{V} - \nabla^2 \vec{V}$**

or $\quad \nabla \times (\nabla \times \vec{V}) = \nabla (\nabla \cdot \vec{V}) - \nabla^2 \vec{V}$. *(OU-Dec. 2013)*

Proof. Let $\vec{V} = V_1\hat{i} + V_2\hat{j} + V_3\hat{k}$

then
$$\text{curl } \vec{V} = \nabla \times \vec{V} = \begin{vmatrix} \hat{i} & \hat{j} & \hat{k} \\ \dfrac{\partial}{\partial x} & \dfrac{\partial}{\partial y} & \dfrac{\partial}{\partial z} \\ V_1 & V_2 & V_3 \end{vmatrix}$$

$$= \hat{i}\left(\frac{\partial V_3}{\partial y} - \frac{\partial V_2}{\partial z}\right) + \hat{j}\left(\frac{\partial V_1}{\partial z} - \frac{\partial V_3}{\partial x}\right) + \hat{k}\left(\frac{\partial V_2}{\partial x} - \frac{\partial V_1}{\partial y}\right)$$

$$\therefore \quad \text{Curl (curl } \vec{V}) = \nabla \times (\nabla \times \vec{V}) = \begin{vmatrix} \hat{i} & \hat{j} & \hat{k} \\ \dfrac{\partial}{\partial x} & \dfrac{\partial}{\partial y} & \dfrac{\partial}{\partial z} \\ \dfrac{\partial V_3}{\partial y} - \dfrac{\partial V_2}{\partial z} & \dfrac{\partial V_1}{\partial z} - \dfrac{\partial V_3}{\partial x} & \dfrac{\partial V_2}{\partial x} - \dfrac{\partial V_1}{\partial y} \end{vmatrix}$$

$$= \sum \hat{i}\left\{\frac{\partial}{\partial y}\left(\frac{\partial V_2}{\partial x} - \frac{\partial V_1}{\partial y}\right) - \frac{\partial}{\partial z}\left(\frac{\partial V_1}{\partial z} - \frac{\partial V_3}{\partial x}\right)\right\}$$

$$= \sum \hat{i}\left\{\frac{\partial}{\partial x}\left(\frac{\partial V_2}{\partial y} + \frac{\partial V_3}{\partial z}\right) - \left(\frac{\partial^2 V_1}{\partial y^2} + \frac{\partial^2 V_1}{\partial z^2}\right)\right\}$$

$$= \sum \hat{i}\left\{\frac{\partial}{\partial x}\left(\frac{\partial V_1}{\partial x} + \frac{\partial V_2}{\partial y} + \frac{\partial V_3}{\partial z}\right) - \left(\frac{\partial^2 V_1}{\partial x^2} + \frac{\partial^2 V_1}{\partial y^2} + \frac{\partial^2 V_1}{\partial z^2}\right)\right\}$$

$$= \sum \hat{i} \left\{ \frac{\partial}{\partial x} (\nabla \cdot \vec{V}) - (\nabla^2 V_1) \right\} = \Sigma \hat{i} \frac{\partial}{\partial x} (\nabla \cdot \vec{V}) - \nabla^2 \Sigma \hat{i} V_1$$

$$= \nabla(\nabla \cdot \vec{V}) - \nabla^2 \vec{V} = \text{grad}\ (\text{div}\ \vec{V}) - \nabla^2 \vec{V}.$$

Note 1. The above result can also be written as grad (div $\vec{V}$) = curl (curl $\vec{V}$) + $\nabla^2 \vec{V}$

or $\nabla(\nabla \cdot \vec{V}) = \nabla \times (\nabla \times \vec{V}) + \nabla^2 \vec{V}$.

Note 2. Treating ∇ as a vector, the results of repeated application of ∇ can be easily written down. Thus,

$\nabla \cdot \nabla \phi = \nabla^2 \phi$ $\quad (\because \vec{a} \cdot \vec{a} = a^2)$

$\nabla \times \nabla \phi = \vec{0}$ $\quad (\because \vec{a} \times \vec{a} = \vec{0})$

$\nabla \cdot (\nabla \times \vec{V}) = 0$ $\quad (\because \vec{a} \cdot (\vec{a} \times \vec{b}) = [\vec{a}\ \vec{a}\ \vec{b}] = 0)$

$\nabla \times (\nabla \times \vec{V}) = \nabla(\nabla \cdot \vec{V}) - \nabla^2 \vec{V}$ (by expanding as a vector triple product)

ILLUSTRATIVE EXAMPLES

Example 1. *If $\vec{f} = xy^2\hat{i} + 2x^2yz\hat{j} - 3yz^2k$, find div $\vec{f}$ at (1, – 1, 1).*

Sol. Given $\vec{f} = xy^2\hat{i} + 2x^2yz\hat{j} - 3yz^2\hat{k}$

Compare with $\vec{f} = f_1\hat{i} + f_2\hat{j} + f_3\hat{k}$, then

$$f_1 = xy^2,\ f_2 = 2x^2yz,\ f_3 = -3yz^2$$

$$\frac{\partial f_1}{\partial x} = y^2,\ \frac{\partial f_2}{\partial y} = 2x^2z,\ \frac{\partial f_3}{\partial z} = -6yz$$

Since, $\text{div}\ \vec{f} = \frac{\partial f_1}{\partial x} + \frac{\partial f_2}{\partial y} + \frac{\partial f_3}{\partial z} = y^2 + 2x^2z - 6yz$ at (1, – 1, 1)

$$\text{div}\ \vec{f} = \nabla \cdot \vec{f} = 1 + 2 + 6 = 9.$$

Example 2. *Show that $(x + 3y)\hat{i} + (y - 3z)\hat{j} + (x - 2z)\hat{k}$ is solenoidal.* (OU-May 2009)

Sol. Let $f = (x + 3y)\hat{i} + (y - 3z)\hat{j} + (x - 2z)\hat{k}$

$$= f_1\hat{i} + f_2\hat{j} + f_3\hat{k}$$

$$\text{div}\ f = \nabla \cdot f = \frac{\partial f_1}{\partial x} + \frac{\partial f_2}{\partial y} + \frac{\partial f_3}{\partial z}$$

$$= \frac{\partial}{\partial x}(x + 3y) + \frac{\partial}{\partial y}(y - 3z) + \frac{\partial}{\partial z}(x - 2z) = 1 + 1 - 2 = 0$$

$\therefore$ f is solenoidal.

Example 3. *If* $\vec{f} = (x + 3y)\hat{i} + (y - 2z)\hat{j} + (x + Pz)\hat{k}$ *is solenoidal, find P.*

Sol. Here, $f_1 = x + 3y$, $f_2 = y - 2z$, $f_3 = x + Pz$

$$\frac{\partial f_1}{\partial x} = 1, \ \frac{\partial f_2}{\partial y} = 1, \ \frac{\partial f_3}{\partial z} = P$$

Since $\vec{f}$ is solenoidal, div $\vec{f} = 0$

$$\Rightarrow \quad \frac{\partial f_1}{\partial x} + \frac{\partial f_2}{\partial y} + \frac{\partial f_3}{\partial z} = 0$$

$$\Rightarrow \quad 1 + 1 + P = 0 \Rightarrow P = -2.$$

Example 4. *Show that div* $(r^n\bar{r}) = (n + 3)r^n$. (OU-Apr. 2007)

Sol. Let $f = r^n\bar{r}$, where $\bar{r} = x\hat{i} + y\hat{j} + z\hat{k}$ and $r^2 = x^2 + y^2 + z^2$

Differentiating partially w.r.t. x, we get

$$2r\frac{\partial r}{\partial x} = 2x \Rightarrow \frac{\partial r}{\partial x} = \frac{x}{r}$$

$$2r\frac{\partial r}{\partial y} = 2y \Rightarrow \frac{\partial r}{\partial y} = \frac{y}{r}$$

$$2r\frac{\partial r}{\partial z} = 2z \Rightarrow \frac{\partial r}{\partial z} = \frac{z}{r}$$

$$\therefore \quad f = r^n(x\hat{i} + y\hat{j} + r\hat{k})$$

$$\text{div } f = \frac{\partial}{\partial x}(r^n x) + \frac{\partial}{\partial y}(r^n y) + \frac{\partial}{\partial z}(r^n z)$$

$$= r^n + nr^{n-1}\frac{\partial r}{\partial x}\cdot x + nr^{n-1}\frac{\partial r}{\partial y}\cdot y + r^n + nr^{n-1}\frac{\partial r}{\partial z}\cdot z + r^n$$

$$= nr^{n-1}\left[\frac{x^2}{r} + \frac{y^2}{r} + \frac{z^2}{r}\right] + 3r^n$$

$$= nr^{n-1}\left[\frac{r^2}{r}\right] + 3r^n = nr^n + 3r^n = (n + 3)r^n.$$

Example 5. *Prove that:*

(*i*) *div* $\left(\frac{\vec{r}}{r^3}\right) = 0$ (OU-May 2008) (*ii*) $\nabla^2(r^n) = n(n + 1)\,r^{n-2}$. (OU-May 2009)

where $\vec{r} = x\hat{i} + y\hat{j} + z\hat{k}$.

Sol. Here $r^2 = x^2 + y^2 + z^2$ so that $\frac{\partial r}{\partial x} = \frac{x}{r}, \frac{\partial r}{\partial y} = \frac{y}{r}, \frac{\partial r}{\partial z} = \frac{z}{r}$

$$\text{grad } r = \hat{i}\frac{\partial r}{\partial x} + \hat{j}\frac{\partial r}{\partial y} + \hat{k}\frac{\partial r}{\partial z} = \frac{\vec{r}}{r}$$

(*i*) Since div $(\phi\vec{A}) = \phi(\text{div } \vec{A}) + \vec{A} \cdot \text{grad } \phi$

$$\therefore \quad \text{div}\left(\frac{\vec{r}}{r^3}\right) = \text{div}\,(r^{-3}\,\vec{r}) = r^{-3}\,(\text{div } \vec{r}) + \vec{r} \cdot \text{grad } r^{-3} = 3r^{-3} + \vec{r} \cdot (-\,3r^{-4} \text{ grad } r)$$

$$[\because \quad \text{div } \vec{r} = 3]$$

$$= 3r^{-3} + \vec{r} \cdot \left(-\,3r^{-4}\,\frac{\vec{r}}{r}\right) = 3r^{-3} - 3r^{-5}\,(\vec{r} \cdot \vec{r}) = 3r^{-3} - 3r^{-5}\,(r^2) = 0.$$

(*ii*)
$$\nabla^2\,(r^n) = \nabla \cdot (\nabla r^n) = \nabla \cdot \left(n r^{n-1}\,\frac{\vec{r}}{r}\right) = n\nabla \cdot (r^{n-2}\,\vec{r}\,)$$

$$= n\,[(\nabla\, r^{n-2}) \cdot \vec{r} + r^{n-2}\,(\nabla \cdot \vec{r}\,)] \qquad [\because \quad \nabla \cdot (\phi\vec{A}) = (\nabla\,\phi) \cdot \vec{A} + \phi\,(\nabla \cdot \vec{A})]$$

$$= n\left[(n-2)\,r^{n-3}\,\frac{\vec{r}}{r} \cdot \vec{r} + r^{n-2}\,(3)\right] \qquad [\because \quad \nabla \cdot \vec{r} = 3]$$

$$= n\,[(n-2)\,r^{n-4}\,(r^2) + 3r^{n-2}] \qquad [\because \quad \vec{r} \cdot \vec{r} = r^2]$$

$$= n(n+1)r^{n-2}.$$

Aliter :

$$\nabla^2\,(r^n) = \left(\frac{\partial^2}{\partial x^2} + \frac{\partial^2}{\partial y^2} + \frac{\partial^2}{\partial z^2}\right) r^n$$

$$= \sum \frac{\partial^2}{\partial x^2}\,(r^n) = \sum \frac{\partial}{\partial x}\left(\frac{\partial r^n}{\partial x}\right)$$

$$= \sum \frac{\partial}{\partial x}\left(n r^{n-1}\,\frac{\partial r}{\partial x}\right) = \sum \frac{\partial}{\partial x}\left(n r^{n-1}\,\frac{x}{r}\right) = \sum n\,\frac{\partial}{\partial x}\,(r^{n-2}\,x)$$

$$= n\sum\left[(n-2)\,r^{n-3}\,\frac{\partial r}{\partial x} \cdot x + r^{n-2}\right] = n\sum\left[(n-2)\,r^{n-3}\,\frac{x}{r} \cdot x + r^{n-2}\right]$$

$$= n\sum\,[(n-2)\,r^{n-4}\,x^2 + r^{n-2}] = n[(n-2)\,r^{n-4}\,(x^2 + y^2 + z^2) + 3r^{n-2}]$$

$$= n\,[(n-2)\,r^{n-4}(r^2) + 3r^{n-2}] = n(n+1)r^{n-2}.$$

Example 6. *If* $\vec{a}$ *is a constant vector and* $\vec{r} = x\hat{i} + y\hat{j} + z\hat{k}$, *prove that curl* $(\vec{a} \times \vec{r}\,) = 2\vec{a}$.

(OU-June 2010, May 2012)

Sol. Let $\quad \vec{a} = a_1\hat{i} + a_2\hat{j} + a_3\hat{k}$, where the scalars a_1, a_2, a_3 are constants.

$$\vec{a} \times \vec{r} = \begin{vmatrix} \hat{i} & \hat{j} & \hat{k} \\ a_1 & a_2 & a_3 \\ x & y & z \end{vmatrix} = (a_2 z - a_3 y)\hat{i} + (a_3 x - a_1 z)\,\hat{j} + (a_1 y - a_2 x)\,\hat{k}$$

$$\text{curl }(\vec{a}\times\vec{r}) = \nabla\times(\vec{a}\times\vec{r}) = \begin{vmatrix} \hat{i} & \hat{j} & \hat{k} \\ \dfrac{\partial}{\partial x} & \dfrac{\partial}{\partial y} & \dfrac{\partial}{\partial z} \\ a_2z - a_3y & a_3x - a_1z & a_1y - a_2x \end{vmatrix}$$

$$= (a_1 + a_1)\hat{i} + (a_2 + a_2)\hat{j} + (a_3 + a_3)\hat{k} = 2(a_1\hat{i} + a_2\hat{j} + a_3\hat{k}) = 2\vec{a}.$$

Example 7. *Prove that the vector $f(r)\,\vec{r}$ is irrotational.*

Sol. The vector $f(r)\,\vec{r}$ will be irrotational if curl $[f(r)\,\vec{r}\,] = \vec{0}$

Since, $\quad \text{curl }(\phi\vec{A}) = (\text{grad }\phi)\times\vec{A} + \phi \text{ curl }\vec{A}$

$\therefore \quad \text{curl }[f(r)\,\vec{r}\,] = [\text{grad } f(r)]\times\vec{r} + f(r)\text{ curl }\vec{r}$

$$= [f'(r)\text{ grad } r]\times\vec{r} + f(r)\,\vec{0} \qquad [\because \text{ curl }\vec{r} = \vec{0}]$$

$$= \left[f'(r)\frac{\vec{r}}{r}\right]\times\vec{r} = \frac{f'(r)}{r}(\vec{r}\times\vec{r}) = \vec{0}, \text{ since } \vec{r}\times\vec{r} = \vec{0}.$$

$\therefore$ The vector $f(r)\,\vec{r}$ is irrotational.

Example 8. *A vector field is given by $\vec{A} = (x^2 + xy^2)\hat{i} + (y^2 + x^2y)\hat{j}$. Show that the field is irrotational and find the scalar potential.*

Sol. Field $\vec{A}$ is irrotational if curl $\vec{A} = \vec{0}$

Now, $$\text{curl }\vec{A} = \nabla\times\vec{A} = \begin{vmatrix} \hat{i} & \hat{j} & \hat{k} \\ \dfrac{\partial}{\partial x} & \dfrac{\partial}{\partial y} & \dfrac{\partial}{\partial z} \\ x^2 + xy^2 & y^2 + x^2y & 0 \end{vmatrix}$$

$$= \hat{i}(0 - 0) - \hat{j}(0 - 0) + \hat{k}(2xy - 2xy) = \vec{0}.$$

$\therefore$ Field $\vec{A}$ is irrotational.

If ϕ is the scalar potential, then $\vec{A} = \text{grad }\phi = \dfrac{\partial\phi}{\partial x}\hat{i} + \dfrac{\partial\phi}{\partial y}\hat{j} + \dfrac{\partial\phi}{\partial z}\hat{k}$

Since, $$\vec{A} = (x^2 + xy^2)\hat{i} + (y^2 + x^2y)\hat{j}$$

$\therefore$ $$\frac{\partial\phi}{\partial x} = x^2 + xy^2 \quad ...(1) \qquad \frac{\partial\phi}{\partial y} = y^2 + x^2y \quad ...(2)$$

$$\frac{\partial\phi}{\partial z} = 0. \quad ...(3)$$

Integrating (1) partially w.r.t. x treating y (and z) as constants $\phi = \dfrac{x^3}{3} + \dfrac{x^2y^2}{2} + \phi_1(y, z)$

Similarly from (2) and (3), we get $\phi = \frac{y^3}{3} + \frac{x^2y^2}{2} + \phi_2(z,x)$

$$\phi = \phi_3\,(x, y)$$

Comparing the three values of ϕ, we have

$\phi_1\,(y, z)$ = terms in ϕ independent of $x = \frac{y^3}{3}$

$\phi_2\,(z, x)$ = terms in ϕ independent of $y = \frac{x^3}{3}$

$\phi_3\,(x, y)$ = terms in ϕ independent of $z = \frac{x^3}{3} + \frac{y^3}{3} + \frac{x^2y^2}{2}$

Hence, $\phi = \frac{x^3}{3} + \frac{y^3}{3} + \frac{x^2y^2}{2}$.

Example 9. *Show that* $\vec{V} = 12x\hat{i} - 15y^2\hat{j} + \hat{k}$ *is irrotational and find a scalar function* $f(x, y, z)$ *such that* $\vec{V} = \text{grad } f$. (OU-Dec. 2013)

Sol. Given $\vec{V} = 12x\hat{i} - 15y^2\hat{j} + \hat{k}$

$$\text{Curl } \vec{V} = \nabla \times \vec{V} = \begin{vmatrix} \hat{i} & \hat{j} & \hat{k} \\ \frac{\partial}{\partial x} & \frac{\partial}{\partial y} & \frac{\partial}{\partial z} \\ 12x & -15y^2 & 1 \end{vmatrix}$$

$$= \hat{i}(0 + 0) - \hat{j}(0 - 0) + \hat{k}(0 - 0) = 0$$

$\therefore$ $\vec{V}$ is irrotational. Then $\vec{V} = \text{grad } f$.

$$\vec{V} = \text{grad } f = \hat{i}\frac{\partial f}{\partial x} + \hat{j}\frac{\partial f}{\partial y} + \hat{k}\frac{\partial f}{\partial z} = 12x\hat{i} - 15y^2\hat{j} + \hat{k}$$

Comparing, $\frac{\partial f}{\partial x} = 12x$ ...(1)

$\frac{\partial f}{\partial y} = -15y^2$ ...(2)

$\frac{\partial f}{\partial z} = 1$ ...(3)

We have, $df = \frac{\partial f}{\partial x}dx + \frac{\partial f}{\partial y}dy + \frac{\partial f}{\partial z}dz = 12\,x\,dx - 15y^2\,dy + dz$

Integrating, we get

$$f = 6x^2 - 5y^3 + z.$$

Example 10. *Evaluate* $\nabla \times \vec{F}$ *at (1, 2, 3) for* $\vec{F} = 3x^2\hat{i} + 5xy^2\hat{j} + xyz^3\hat{k}$. (OU-Jan. 2011)

Sol.
$$\nabla \times \vec{F} = \text{Curl } \vec{F} = \begin{vmatrix} \hat{i} & \hat{j} & \hat{k} \\ \frac{\partial}{\partial x} & \frac{\partial}{\partial y} & \frac{\partial}{\partial z} \\ 3x^2 & 5xy^2 & xyz^3 \end{vmatrix}$$
$$= \hat{i}(xz^3 - 0) - \hat{j}(yz^3 - 0) + \hat{k}(5y^2 - 0)$$
$$= xz^3\hat{i} - yz^3\hat{j} + 5y^2k \text{ at } (1, 2, 3) = 27\hat{i} - 54\hat{j} + 20\hat{k}.$$

Example 11. *If* $\vec{F} = (x + y - 1)\hat{i} + \hat{j} + (x + y)\hat{k}$, *find* $\vec{F} \cdot \text{curl } \vec{F}$. (OU-June 2011)

Sol.
$$\text{Curl } \vec{F} = \begin{vmatrix} \hat{i} & \hat{j} & \hat{k} \\ \frac{\partial}{\partial x} & \frac{\partial}{\partial y} & \frac{\partial}{\partial z} \\ x + y - 1 & 1 & x + y \end{vmatrix}$$
$$= \hat{i}(1 - 0) - \hat{j}(1 - 0) + \hat{k}(0 - 1) = \hat{i} - \hat{j} - \hat{k}$$
$$\vec{F} \cdot \textbf{curl } \vec{F} = (x + y - 1) - 1 - (x + y) = -2.$$

Example 12. *If* $u\vec{F} = \nabla v$, *where u, v are scalar fields and* $\vec{F}$ *is a vector field, show that* $\vec{F}$ *. curl* $\vec{F} = 0$.

Sol.
$$\text{curl } \vec{F} = \nabla \times \left(\frac{1}{u} \nabla v\right) \qquad \left[\because \ \vec{F} = \frac{1}{u}\nabla v\right]$$
$$= \nabla \frac{1}{u} \times \nabla v + \frac{1}{u} \nabla \times (\nabla v) \qquad [\because \ \nabla \times (\phi\vec{A}) = \Delta\phi \times \vec{A} + \phi\nabla \times \vec{A}]$$
$$= \nabla \frac{1}{u} \times \nabla v \qquad [\because \ \nabla \times \nabla v = \vec{0}]$$
$$\therefore \quad \vec{F} \cdot \text{curl } \vec{F} = \frac{1}{u} \nabla v \cdot \left(\nabla \frac{1}{u} \times \nabla v\right) = 0$$

being the scalar triple product in which two factors are equal.

Example 13. *Prove that:*

(i) $\nabla (\vec{a} \cdot \vec{u}) = (\vec{a} \cdot \nabla)\vec{u} + \vec{a} \times (\nabla \times \vec{u})$ (ii) $\nabla \times (\vec{a} \times \vec{u}) = (\nabla \cdot \vec{u})\vec{a} - (\vec{a} \cdot \nabla)\vec{u}$

where $\vec{a}$ *is a constant vector.*

Sol. (i)
$$\nabla(\vec{a} \cdot \vec{u}) = \sum i \frac{\partial}{\partial x} (\vec{a} \cdot \vec{u}) = \sum i \left(\vec{a} \cdot \frac{\partial \vec{u}}{\partial x}\right) \qquad \text{...(1)}$$

Now, $\vec{a} \times \left(\hat{i} \dfrac{\partial \vec{u}}{\partial x} \right) = \left(\vec{a} \cdot \dfrac{\partial \vec{u}}{\partial x} \right) \hat{i} - (\vec{a} \cdot \hat{i}) \dfrac{\partial \vec{u}}{\partial x}$

$$\Rightarrow \quad \left(\vec{a} \,.\, \frac{\partial \vec{u}}{\partial x} \right) \hat{i} = \vec{a} \times \left(\hat{i} \times \frac{\partial \vec{u}}{\partial x} \right) + (\vec{a} \,.\, \hat{i}) \frac{\partial \vec{u}}{\partial x}$$

$\therefore$ From (1), we have $\nabla(\vec{a} \cdot \vec{u}) = \sum \vec{a} \times \left(\hat{i} \times \dfrac{\partial \vec{u}}{\partial x} \right) + \sum (\vec{a} \cdot \hat{i}) \dfrac{\partial \vec{u}}{\partial x}$

$$= \vec{a} \times (\nabla \times \vec{u}) + (\vec{a} \cdot \nabla) \vec{u} = (\vec{a} \cdot \nabla) \vec{u} + \vec{a} \times (\nabla \times \vec{u}).$$

(ii) $\nabla \times (\vec{a} \times \vec{u}) = \sum \hat{i} \dfrac{\partial}{\partial x} \times (\vec{a} \times \vec{u}) = \sum \hat{i} \times \left(\vec{a} \times \dfrac{\partial \vec{u}}{\partial x} \right)$

$$= \sum \left(\hat{i} \cdot \frac{\partial \vec{u}}{\partial x} \right) \vec{a} - \sum (\hat{i} \cdot \vec{a}) \frac{\partial \vec{u}}{\partial x} = (\nabla \cdot \vec{u}) \vec{a} - (\vec{a} \cdot \nabla) \vec{u}.$$

Example 14. *Prove that:* $\nabla^2 f(r) = f''(r) + \dfrac{2}{r} f'(r)$.

Sol. $\nabla^2 f(r) = \nabla \cdot \{\nabla f(r)\} = \text{div}\,\{\text{grad}\, f(r)\} = \text{div}\,\{f'(r)\,\text{grad}\, r\} = \text{div}\left\{ f'(r) \dfrac{\vec{r}}{r} \right\}$

$$= \text{div}\left\{ \frac{1}{r} f'(r)\, \vec{r} \right\} = \frac{1}{r} f'(r)\, \text{div}\, \vec{r} + \vec{r} \cdot \textbf{grad} \left\{ \frac{1}{r} f'(r) \right\}$$

$$= \frac{3}{r} f'(r) + \vec{r} \cdot \left[\frac{d}{dr} \left(\frac{1}{r} f'(r) \right) \text{grad}\, r \right] = \frac{3}{r} f'(r) + \vec{r} \cdot \left[\left\{ -\frac{1}{r^3} f'(r) + \frac{1}{r} f''(r) \right\} \frac{\vec{r}}{r} \right]$$

$$= \frac{3}{r} f'(r) + \left[-\frac{1}{r^3} f'(r) + \frac{1}{r^2} f''(r) \right] (\vec{r} \cdot \vec{r}) = \frac{3}{r} f'(r) + \left[-\frac{1}{r^3} f'(r) + \frac{1}{r^2} f''(r) \right] r^2$$

$$= \frac{3}{r} f'(r) - \frac{1}{r} f'(r) + f''(r) = f''(r) + \frac{2}{r} f'(r).$$

Example 15. *Show that the vector field defined by the vector function,* $v = xyz\,(yzi + xzj + xyk)$ *is conservative.* (OU-Dec. 2012)

Sol. We shall prove that curl $\bar{v} = 0$

$$v = xy^2z^2\, i + x^2yz^2\, j + x^2\, y^2z\, k$$

$$\text{curl}\ \bar{v} = \begin{vmatrix} i & j & k \\ \dfrac{\partial}{\partial x} & \dfrac{\partial}{\partial y} & \dfrac{\partial}{\partial z} \\ xy^2z^2 & x^2yz^2 & x^2y^2z \end{vmatrix}$$

$$= i(2x^2\, yz - 2x^2yz) - j(2xy^2z - 2xy^2z) + k(2xyz^2 - 2xyz^2)$$

$$= 0$$

$\therefore$ v is conservative.

EXERCISE 2

1. Evaluate :

   (i) div $(3x^2\hat{i} + 5xy^2\hat{j} + xyz^3\hat{k})$ at the point (1, 2, 3).

   (ii) div $[(xy \sin z)\hat{i} + (y^2 \sin x)\hat{j} + (z^2 \sin xy)\hat{k}]$ at the point $\left(0, \frac{\pi}{2}, \frac{\pi}{2}\right)$.

2. Find the divergence and curl of the vectors

   (i) $\vec{V} = (xyz)\hat{i} + (3x^2y)\hat{j} + (xz^2 - y^2z)\hat{k}$ (ii) $\vec{R} = (x^2 + yz)\hat{i} + (y^2 + zx)\hat{j} + (z^2 + xy)\hat{k}$.

3. If $\vec{F} = (x + y + 1)\hat{i} + \hat{j} - (x + y)\hat{k}$, show that $\vec{F}$. Curl $\vec{F} = 0$.

4. If $\vec{V} = \dfrac{x\hat{i} + y\hat{j} + z\hat{k}}{\sqrt{x^2 + y^2 + z^2}}$, show that $\nabla . \vec{V} = \dfrac{2}{\sqrt{x^2 + y^2 + z^2}}$ and $\nabla \times \vec{V} = \vec{0}$.

5. If $\vec{A} = (3xz^2)\hat{i} - (yz)\hat{j} + (x + 2z)\hat{k}$, find curl (curl $\vec{A}$).

6. Show that the vector $\vec{V} = (x + 3y)\hat{i} + (y - 3z)\hat{j} + (x - 2z)\hat{k}$ is solenoidal.

7. If $u = x^2 + y^2 + z^2$ and $\vec{V} = x\hat{i} + y\hat{j} + z\hat{k}$, show that div $(u\vec{V}) = 5u$.

8. Show that the vector field $\vec{V} = (\sin y + z)\hat{i} + (x \cos y - z)\hat{j} + (x - y)\hat{k}$ is irrotational.

9. Show that the vector field $\vec{A}$, where $\vec{A} = (x^2 - y^2 + x)\hat{i} - (2xy + y)\hat{j}$ is irrotational, and find a scalar potential ϕ such that $\vec{A}$ = grad ϕ.

10. Show that the vector field defined by $\vec{F} = 2xyz^3\hat{i} + x^2z^3\hat{j} + 3x^2yz^2\hat{k}$ is irrotational. Find the scalar potential u such that $\vec{F}$ = grad u.

11. Find the constants a, b, c so that $\vec{F} = (x + 2y + az)\hat{i} + (bx - 3y - z)\hat{j} + (4x + cy + 2z)\hat{k}$ is irrotational.

    If $\vec{F}$ = grad ϕ, show that $\phi = \dfrac{x^2}{2} - \dfrac{3y^2}{2} + z^2 + 2xy + 4xz - yz$.

12. Show that $\vec{E} = \dfrac{\vec{r}}{r^2}$ is irrotational.

13. If $\vec{E}$ and $\vec{H}$ are irrotational, prove that $\vec{E} \times \vec{H}$ is solenoidal.

14. For a solenoidal vector $\vec{F}$, prove that curl curl curl curl $\vec{F} = \nabla^4\vec{F}$.

15. Find the directional derivative of $\nabla . (\nabla\phi)$ at the point (1, – 2, 1) in the direction of the normal to the surface

    $$xy^2z = 3x + z^2, \text{ where } \phi = 2x^3y^2z^4.$$

16. If $\vec{V}_1$ and $\vec{V}_2$ be the vectors joining the fixed points (x_1, y_1, z_1) and (x_2, y_2, z_2) respectively to a variable point (x, y, z) prove that

    (i) div $(\vec{V}_1 \times \vec{V}_2) = 0$ (ii) grad $(\vec{V}_1 . \vec{V}_2) = \vec{V}_1 + \vec{V}_2$

    (iii) curl $(\vec{V}_1 \times \vec{V}_2) = 2(\vec{V}_1 - \vec{V}_2)$.

17. If $\vec{a}$ is a constant vector and $\vec{r} = x\hat{i} + y\hat{j} + z\hat{k}$, prove that

(*i*) div $(\vec{a} \times \vec{r}) = 0$ (*ii*) curl $[(\vec{a} \cdot \vec{r})\vec{r}] = \vec{0}$

(*iii*) $\nabla(\vec{a} \cdot \vec{a}) = 2(\vec{a} \cdot \nabla)\vec{a} + 2\vec{a} \times (\nabla \times \vec{a})$.

18. Show that curl $\left(\hat{k} \times \text{grad}\, \frac{1}{r}\right) + \text{grad}\left(\hat{k} \cdot \text{grad}\, \frac{1}{r}\right) = \vec{0}$, where r is the distance of a point (x, y, z) from the origin and $\hat{k}$ is a unit vector in the direction of OZ.
[**Hint.** Use example 13 (*i*), (*ii*)]

19. If $\vec{r} = x\hat{i} + y\hat{j} + z\hat{k}$, prove that

(*i*) curl $(r^n \vec{r}) = \vec{0}$ (*ii*) $\frac{\vec{r}}{r^3}$ is solenoidal

(*iii*) $\nabla^2 \left(\frac{1}{r}\right) = 0$ (*iv*) $\nabla^2 (r^n \vec{r}) = n(n+3)r^{n-2}\vec{r}$

(*v*) $\nabla \cdot \left\{r\nabla\left(\frac{1}{r^3}\right)\right\} = \frac{3}{r^4}$ (*vi*) $\nabla^2 \left[\nabla \cdot \left(\frac{\vec{r}}{r^2}\right)\right] = 2r^{-4}$.

Answers

1. (*i*) 80 (*ii*) $\frac{\pi}{2}$

2. (*i*) $yz + 3x^2 + 2xz - y^2$; $-2yz\hat{i} + (xy - z^2)\hat{j} + x(6y - z)\hat{k}$ (*ii*) $2(x + y + z)$; $\vec{0}$

5. $-6x\hat{i} + (6z - 1)\hat{k}$ **9.** $\frac{1}{3}x^3 - xy^2 + \frac{1}{2}(x^2 - y^2)$ **10.** x^2yz^3

11. $a = 4, b = 2, c = -1$ **15.** $\frac{1724}{\sqrt{21}}$.

5.21. LINE INTEGRALS

Any integral which is to be evaluated along a curve is called a line integral.

Let $\vec{F}(P)$ be a continuous vector point function defined at every point of a curve C in space. Divide the curve C into n parts by the points

$$A = P_0, P_1, P_2, \ldots\ldots, P_n = B$$

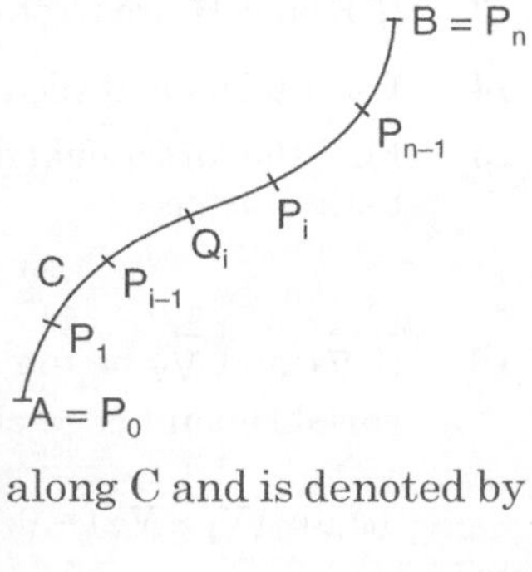

and let $\vec{R}_0, \vec{R}_1, \vec{R}_2, \ldots\ldots, \vec{R}_n$ be the position vectors of these points. Let Q_i be any point on the arc $P_{i-1} P_i$. Then the limit of the sum

$$\sum_{i=1}^{n} \vec{F}(Q_i) \cdot \delta\vec{R}_i, \text{ where } \delta\vec{R}_i = \vec{R}_i - \vec{R}_{i-1} \quad \ldots(1)$$

as $n \to \infty$ and every $|\delta\vec{R}_i| \to 0$, if it exists, is called a line integral of $\vec{F}$ along C and is denoted by

$$\int_C \vec{F} \cdot d\vec{R} \quad \text{or} \quad \int_C \vec{F} \cdot \frac{d\vec{R}}{dt}\, dt.$$

Clearly, it is a scalar. It is called the *tangential line integral* of $\vec{F}$ along C.

If the scalar products in (1) are replaced by vector products, then the corresponding line integral is defined as $\int_C \vec{F} \times d\vec{R}$ which is a vector.

If the vector function $\vec{F}$ is replaced by a scalar function ϕ, then the corresponding line integral is defined as $\int_C \phi \, d\vec{R}$, which is a vector.

If $\quad \vec{F}(x, y, z) = f_1\hat{i} + f_2\hat{j} + f_3\hat{k}$ and $\vec{R} = x\hat{i} + y\hat{j} + z\hat{k}$, then $d\vec{R} = \hat{i}\,dx + \hat{j}\,dy + \hat{k}\,dz$

$$\therefore \quad \int_C \vec{F} \cdot d\vec{R} = \int_C (f_1 dx + f_2 dy + f_3 dz)$$

If the parametric equations of the curve C are $x = x(t)$, $y = y(t)$, $z = z(t)$ and $t = t_1$ at A, $t = t_2$ at B, then

$$\int_C \vec{F} \cdot d\vec{R} = \int_{t_1}^{t_2} \left(f_1 \frac{dx}{dt} + f_2 \frac{dy}{dt} + f_3 \frac{dz}{dt} \right) dt$$

If C is a closed curve, then the integral sign $\int_C$ is replaced by $\oint_C$.

5.22. CIRCULATION

In fluid dynamics, if $\vec{V}$ represents the velocity of a fluid particle and C is a closed curve, then the integral $\oint_C \vec{V} \cdot d\vec{R}$ is called the *circulation* of $\vec{V}$ round the curve C.

If the circulation of $\vec{V}$ round every closed curve in a region D vanishes, then $\vec{V}$ is said to be *irrotational* in D.

5.23. WORK DONE BY A FORCE

Let $\vec{F}$ represent the force acting on a particle moving along an arc AB. The work done during a small displacement $\delta\vec{R}$ is $\vec{F} \cdot \delta\vec{R}$.

$\therefore$ The total work done by $\vec{F}$ during displacement from A to B is given by $\int_A^B \vec{F} \cdot d\vec{R}$

If the force $\vec{F}$ is conservative, then there exists a scalar function ϕ such that

$$\vec{F} = \nabla\phi = \hat{i}\frac{\partial \phi}{\partial x} + \hat{j}\frac{\partial \phi}{\partial y} + \hat{k}\frac{\partial \phi}{\partial z}$$

$\therefore$ The work done by $\vec{F}$ during displacement from A to B $= \int_A^B \vec{F} \cdot d\vec{R}$

$$= \int_A^B \left(\hat{i}\frac{\partial \phi}{\partial x} + \hat{j}\frac{\partial \phi}{\partial y} + \hat{k}\frac{\partial \phi}{\partial z} \right) \times (\hat{i}\,dx + \hat{j}\,dy + \hat{k}\,dz)$$

$$= \int_A^B \left(\frac{\partial \phi}{\partial x} dx + \frac{\partial \phi}{\partial y} dy + \frac{\partial \phi}{\partial z} dz \right) = \int_A^B d\phi = \Big[\phi \Big]_A^B = \phi_B - \phi_A$$

Thus, in a conservative field, the work done depends on the value of ϕ at the end points A and B, and not on the path joining A and B.

ILLUSTRATIVE EXAMPLES

Example 1. *If* $\vec{F} = 3xy\hat{i} - y^2\hat{j}$, *evaluate* $\int_C \vec{F} \cdot d\vec{r}$, *where C is the arc of the parabola* $y = 2x^2$ *from (0, 0) to (1, 2).*

Sol. Let $x = t$, then the parametric equations of the parabola $y = 2x^2$ are $x = t$, $y = 2t^2$.

At the point (0, 0), $x = 0$ and so $t = 0$

At the point (1, 2), $x = 1$ and so $t = 1$

If $\vec{r}$ is the position vector of any point (x, y) in C, then $\vec{r} = x\hat{i} + y\hat{j} = t\hat{i} + 2t^2\hat{j}$

Also in terms of t, $\vec{F} = (3t \times 2t^2)\hat{i} - (2t^2)^2\hat{j} = 6t^3\hat{i} - 4t^4\hat{j}$

$$\therefore \quad \int_C \vec{F} \cdot d\vec{r} = \int_C \left(\vec{F} \cdot \frac{d\vec{r}}{dt}\right) dt = \int_0^1 (6t^3\hat{i} - 4t^4\hat{j}) \cdot (\hat{i} + 4t\hat{j})\, dt$$

$$= \int_0^1 (6t^3 - 16t^5)\, dt = \left[\frac{6t^4}{4} - \frac{16t^6}{6}\right]_0^1 = \frac{3}{2} - \frac{8}{3} = -\frac{7}{6}.$$

Example 2. *A vector field is given by* $\vec{F} = (\sin y)\hat{i} + x(1 + \cos y)\hat{j}$. *Evaluate the line integral over the circular path given by* $x^2 + y^2 = a^2$, $z = 0$.

Sol. The parametric equations of the circular path are $x = a \cos t$, $y = a \sin t$, $z = 0$, where t varies from 0 to 2π.

Since the particle moves in the xy-plane ($z = 0$), we can take $\vec{r} = x\hat{i} + y\hat{j}$ so that $d\vec{r} = dx\hat{i} + dy\hat{j}$

$$\therefore \quad \oint_C \vec{F} \cdot d\vec{r} = \oint_C [\sin y\hat{i} + x(1 + \cos y)\hat{j}] \cdot (dx\hat{i} + dy\hat{j})$$

$$= \oint_C [\sin y\, dx + x(1 + \cos y)\, dy] = \oint_C [(\sin y\, dx + x \cos y\, dy) + x\, dy]$$

$$= \oint_C d(x \sin y) + \oint_C x\, dy = \int_0^{2\pi} d[a \cos t \sin(a \sin t)] + \int_0^{2\pi} a \cos t \cdot a \cos t\, dt$$

$$= \Big[a \cos t \sin(a \sin t)\Big]_0^{2\pi} + a^2 \int_0^{2\pi} \cos^2 t\, dt$$

$$= \frac{a^2}{2} \int_0^{2\pi} (1 + \cos 2t)\, dt = \frac{a^2}{2}\left[t + \frac{\sin 2t}{2}\right]_0^{2\pi} = \frac{a^2}{2}(2\pi) = \pi a^2.$$

Example 3. *If* $\vec{F} = 2y\hat{i} - z\hat{j} + x\hat{k}$, *evaluate* $\oint_C \vec{F} \times \vec{dr}$ *along the curve* $x = \cos t$, $y = \sin t$, $z = 2\cos t$ *from* $t = 0$ *to* $t = \frac{\pi}{2}$.

Sol. $$\vec{F} \times \vec{dr} = \begin{vmatrix} \hat{i} & \hat{j} & \hat{k} \\ 2y & -z & x \\ dx & dy & dz \end{vmatrix} = (-zdz - x\,dy)\hat{i} + (x\,dx - 2y\,dz)\hat{j} + (2y\,dy + z\,dx)\hat{k}$$

In terms of t,

$$\vec{F} \times \vec{dr} = [-2\cos t(-2\sin t)\,dt - \cos t(\cos t)\,dt]\hat{i}$$
$$+ [\cos t(-\sin t)\,dt - 2\sin t(-2\sin t)\,dt]\hat{j} + [2\sin t(\cos t)\,dt + 2\cos t(-\sin t)\,dt]\hat{k}$$
$$= [(4\cos t\sin t - \cos^2 t)\hat{i} + (4\sin^2 t - \cos t\sin t)\hat{j}]\,dt$$

$$\therefore \int_C \vec{F} \times \vec{dr} = \int_0^{\pi/2} [(4\cos t\sin t - \cos^2 t)\hat{i} + (4\sin^2 t - \cos t\sin t)\hat{j}]\,dt$$
$$= \left(4\cdot\frac{1}{2} - \frac{1}{2}\cdot\frac{\pi}{2}\right)\hat{i} + \left(4\cdot\frac{1}{2}\cdot\frac{\pi}{2} - \frac{1}{2}\right)\hat{j} = \left(2 - \frac{\pi}{4}\right)\hat{i} + \left(\pi - \frac{1}{2}\right)\hat{j}.$$

Example 4. *Find the work done by the force* $\vec{F} = x\hat{i} - z\hat{j} + 2y\hat{k}$ *in the displacement along the closed path C consisting of the segments* C_1, C_2 *and* C_3 *where*

on C_1,	$0 \le x \le 1$,	$y = x$,	$z = 0$
on C_2,	$0 \le z \le 1$,	$x = 1$,	$y = 1$
on C_3,	$1 \ge x \ge 0$,	$y = z = x$.	

Sol. Total work done

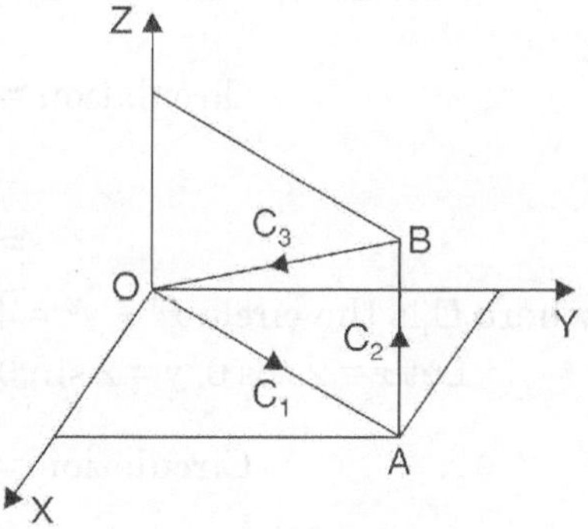

$$= \oint_C \vec{F} \cdot \vec{dr} = \oint_C (x\hat{i} - z\hat{j} + 2y\hat{k}).(dx\hat{i} + dy\hat{j} + dz\hat{k})$$
$$= \oint_C (xdx - zdy + 2ydz)$$
$$= \oint_{C_1} (xdx - zdy + 2ydz) + \oint_{C_2} (xdx - zdy + 2ydz) + \oint_{C_3} (xdx - zdy + 2ydz)$$
$$= W_1 + W_2 + W_3$$

where W_1, W_2, W_3 denote the work done in displacement along C_1, C_2 and C_3 respectively.

On C_1, $0 \le x \le 1$, $y = x$, $z = 0$, $dy = dx$, $dz = 0$

$$\therefore \quad W_1 = \int_{C_1} x\,dx = \int_0^1 x\,dx = \frac{1}{2}$$

On C_2, $0 \le z \le 1$, $x = 1$, $y = 1$, $dx = 0$, $dy = 0$

$$\therefore \quad W_2 = \int_{C_2} 2dz = 2\int_0^1 dz = 2$$

On C_3, $1 \geq x \geq 0$, $y = z = x$, $dy = dz = dx$

$$\therefore \quad W_3 = \int_{C_3} (x\,dx - x\,dx + 2\,x dx) = 2\int_1^0 x\,dx = -1.$$

$$\text{Total work done } = W_1 + W_2 + W_3 = \frac{1}{2} + 2 - 1 = \frac{3}{2}.$$

Example 5. *Find the work done by the force* $\vec{F} = (3x^2 - 6yz)\hat{i} + (2y + 3xz)\hat{j} + (1 - 4xyz^2)\hat{k}$ *in moving particle from the point (0, 0, 0) to the point (1, 1, 1) along the curve C:* $x = t$, $y = t^2$, $z = t^3$.

Sol. Given $f_1 = 3x^2 - 6yz$, $f_2 = 2y + 3xz$, $f_3 = 1 - 4xyz^2$

Also,

$$x = t \Rightarrow dx = dt$$
$$y = t^2 \Rightarrow dy = 2t\,dt$$
$$z = t^3 \Rightarrow dz = 3t^2\,dt$$

$$\text{Work done} = \int \vec{F} \cdot \vec{dr} = \int (f_1\,dx + f_2\,dy + f_3\,dz)$$

$$= \int [(3x^2 - 6yz)\,dx + (2y + 3xz)\,dy + (1 - 4xyz^2)\,dz]$$

$$= \int_{t=0}^{1} [(3t^2 - 6t^5)\,dt + (2t^2 + 3t^4)\,2t\,dt + (1 - 4t^9)\,3t^2\,dt]$$

$$= \int_0^1 (-12t^{11} + 4t^3 + 6t^2)\,dt = [-t^{12} + t^4 + 2t^3]_0^1 = 2.$$

Example 6. *Find the circulation of* $\vec{F} = (2x - y + 2z)\hat{i} + (x + y - z)\hat{j} + (3x - 2y - 5z)\hat{k}$ *along the circle* $x^2 + y^2 = 4$ *in the xy-plane.*

Sol. Let $f_1 = 2x - y + 2z$, $f_2 = x + y - z$, $f_3 = 3x - 2y - 5z$

$$\text{Circulation} = \int_C \vec{F} \cdot \vec{dr} = \int_C (f_1\,dx + f_2\,dy)$$

($\because$ In xy-plane $z = 0 \Rightarrow dz = 0$)

$$= \int_C (2x - y)\,dx + (x + y)\,dy$$

where C is the circle $x^2 + y^2 = 4$.

Let $x = 2\cos\theta$, $y = 2\sin\theta$ then $dx = -2\sin\theta\,d\theta$, $dy = 2\cos\theta\,d\theta$ and θ varies from 0 to 2π.

$$\therefore \quad \text{Circulation} = \int_0^{2\pi} (4\cos\theta - 2\sin\theta)(-2\sin\theta)\,d\theta + (2\cos\theta + 2\sin\theta)(2\cos\theta)\,d\theta$$

$$= \int_0^{2\pi} (-8\sin\theta\cos\theta + 4\sin^2\theta + 4\cos^2\theta + 4\sin\theta\cos\theta)\,d\theta$$

$$= \int_0^{2\pi} (4 - 4\sin\theta\cos\theta)\,d\theta = \int_0^{2\pi} (4 - 2\sin 2\theta)\,d\theta$$

$$= (4\theta + \cos 2\theta)_0^{2\pi} = 8\pi$$

Example 7. *If* $\vec{F} = (x^2 + y^2)\hat{i} - 2xy\hat{j}$ *evaluate* $\oint_C \vec{F}.\vec{dr}$, *where C is the rectangle in xy-plane bounded by* $y = 0$, $y = 1$, $x = 0$, $x = 2$.

Sol. In xy-plane $z = 0 \Rightarrow dz = 0$

$$\oint_C \vec{F}.\vec{dr} = \int_C (F_1\, dx + F_2\, dy) = \int_C (x^2 + y^2)\, dx - 2xy\, dy \qquad ...(1)$$

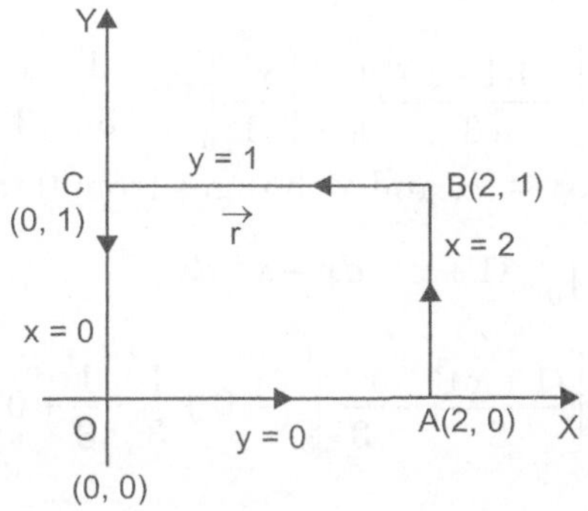

But $$\int_C \mathbf{F}\cdot dr = \int_{OA} \mathbf{F}\cdot dr + \int_{AB} \mathbf{F}\cdot dr + \int_{BC} \mathbf{F}\cdot dr + \int_{CO} \mathbf{F}\cdot dr \qquad ...(2)$$

(*i*) Along OA: $y = 0 \Rightarrow dy = 0$ and x varies from 0 to 2

$$\therefore \quad \int_{OA} \mathbf{F}\cdot dr = \int_0^2 x^2\, dx = \frac{8}{3}$$

(*ii*) Along AB: $x = 2 \Rightarrow dx = 0$ and y varies from 0 to 1

$$\therefore \quad \int_{AB} \mathbf{F}\cdot dr = \int_0^1 (-4y)\, dy = (-2y^2)_0^1 = -2$$

(*iii*) Along BC: $y = 1 \Rightarrow dy = 0$ and x varies from 2 to 0

$$\therefore \quad \int_{BC} \mathbf{F}\cdot dr = \int_2^0 (x^2 + 1)\, dx = \left[\frac{x^3}{3} + x\right]_2^0 = 0 - \left(\frac{8}{3} + 2\right) = -\frac{14}{3}$$

(*iv*) Along CO: $x = 0 \Rightarrow dx = 0$ and y varies from 1 to 0

$$\therefore \quad \int_{CO} \mathbf{F}\cdot dr = \int_1^0 0\, dy = 0$$

Hence from (1) and (2), we get

$$\oint_C \vec{F}.\vec{dr} = \frac{8}{3} - 2 - \frac{14}{3} + 0 = \frac{8 - 6 - 14}{2} = \frac{-12}{3} = -4.$$

Example 8. *Evaluate* $\int (y^2\, dx - x^2\, dy)$ *round the triangle whose vertices are (1, 0), (0, 1) and (– 1, 0) in the xy-plane.* (OU-May 2009)

Sol. In xy-plane $z = 0 \Rightarrow dz = 0$

$$\therefore \int \mathbf{F}\cdot dr = \int_C f_1\, dx + f_2\, dy$$

$$= \int_{AB} f\cdot dr + \int_{BC} f\cdot dr + \int_{CA} f\cdot dr \qquad ...(1)$$

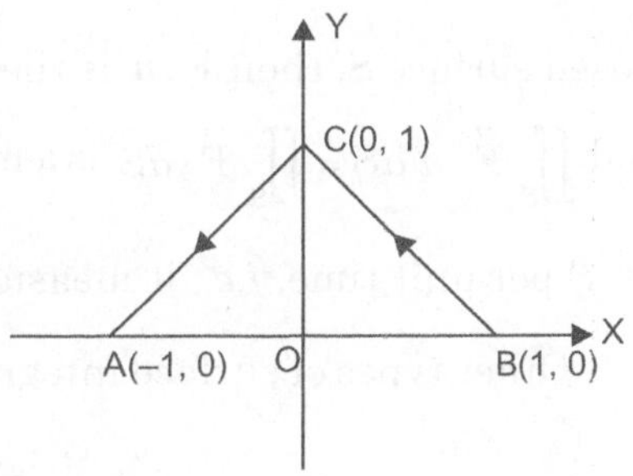

(*i*) Along AB: $y = 0 \Rightarrow dy = 0$ and x changes from -1 to 1

$$\int_{AB} (y^2\, dx - x^2\, dy) = \int_{-1}^{1} (0) = 0$$

(*ii*) Along BC: $x + y = 1 \Rightarrow y = 1 - x \Rightarrow dy = -dx$ and x changes from 1 to 0

$$\int_{BC} (y^2\,dx - x^2\,dy) = \int_{x=1}^{0} (1-x)^2\,dx - x^2\,(-dx)$$

$$= \left[\frac{-1(1-x)^3}{3}\right]_1^0 + \left[\frac{x^3}{3}\right]_1^0 = \frac{-1}{3} - \frac{1}{3} = \frac{-2}{3}$$

(*iii*) Along CA: $y = x + 1 \Rightarrow dy = dx$ and x changes from 0 to -1

$$\therefore \int_{CA} (y^2\,dx - x^2\,dy) = \int_0^{-1} (1+x)^2\,dx - x^2\,dx$$

$$= \left[\frac{(1+x)^3}{3} - \frac{x^3}{3}\right]_0^{-1} = 0 + \frac{1}{3} - \frac{1}{3} + 0 = 0$$

Hence from (1)

$$\int_C (y^2\,dx - x^2\,dy) = 0 - \frac{2}{3} + 0 = -\frac{2}{3}.$$

5.24. SURFACE INTEGRALS

Any integral which is to be evaluated over a surface is called a surface integral.

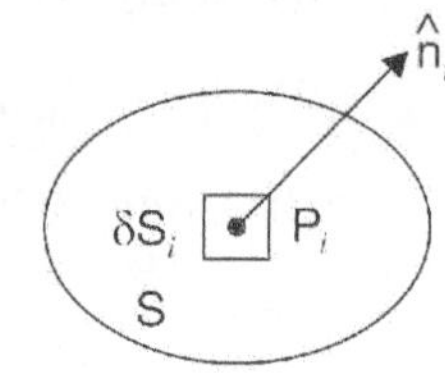

Let $\vec{F}(P)$ be a continuous vector point function and S a two sided surface. Divide S into a finite number of sub-surfaces $\delta S_1, \delta S_2, \ldots\ldots, \delta S_k$. Let P_i be any point in δS_i and $\hat{n}_i$ be the unit vector at P in the direction of outward drawn normal to the surface at P_i. Then the limit of the sum $\sum_{i=1}^{k} \vec{F}(P_i).\hat{n}_i\,\delta S_i$, as $k \to \infty$ and each $\delta S_i \to 0$ is called the *normal surface integral* of $\vec{F}(P)$ over S and is denoted by $\iint_S \vec{F}\cdot\hat{n}\,dS$

The surface element $\vec{\delta S}$ surrounding any point P can be regarded as a vector whose magnitude is area δS and the direction that of the outward drawn normal $\hat{n}$ *i.e.*, $\vec{\delta S} = \hat{n}\,\delta S$. The surface integral may alternatively be written as

$$\iint_S \vec{F}\cdot d\vec{S}.$$

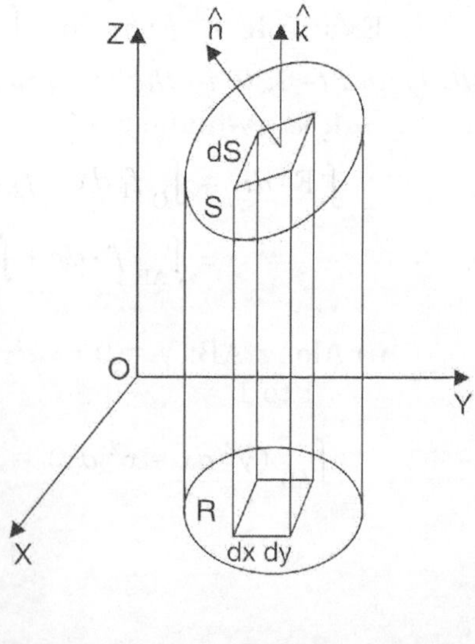

If $\vec{F}$ represents the velocity of a fluid at any point P on a closed surface S, then $\vec{F}.\hat{n}$ is the normal component of $\vec{F}$ at P and $\iint_S \vec{F}\cdot\hat{n}\,dS = \iint_S \vec{F}\cdot d\vec{S}$ is a measure of volume emerging from S per unit time, *i.e.*, it measures the *flux of $\vec{F}$ over S.*

Other types of surface integrals are $\iint_S \vec{F}\times d\vec{S}$, $\iint_S \phi\,d\vec{S}$.

Note. In order to evaluate surface integrals, it is convenient to express them as double integrals taken over the orthogonal projection of S on one of the coordinate planes. But this is possible only when any line perpendicular to the coordinate plane chosen meets the surface S is not more than one point.

Let R be the orthogonal projection of S on the xy-plane.

Let $\hat{n} = \cos\alpha\hat{i} + \cos\beta\hat{j} + \cos\gamma\hat{k}$ where $\cos\alpha$, $\cos\beta$, $\cos\gamma$ are the direction cosines of $\hat{n}$.

Now, $dxdy$ = projection of dS on the xy-plane = dS $\cos\gamma \Rightarrow d\text{S} = \dfrac{dx\,dy}{\cos\gamma}$

Also, $\quad |\hat{k}\,.\,\hat{n}| = \cos\gamma \quad \therefore \quad d\text{S} = \dfrac{dxdy}{|\hat{k}\,.\,\hat{n}|}$

Hence, $\iint_S \vec{F}\,.\,\hat{n}\,ds = \iint_R \vec{F}\,.\,\hat{n}\,\dfrac{dxdy}{|\bar{n}\,.\,\hat{k}|} = \iint \overline{F}\,.\,\bar{n}\,\dfrac{dydz}{|\bar{n}\,.\,\hat{i}|} = \iint \overline{F}\,.\,\bar{n}\,\dfrac{dxdz}{|\bar{n}\,.\,\hat{j}|}$.

5.25. VOLUME INTEGRALS

Any integral which is to be evaluated over a volume is called a volume integral.

If V is a volume bounded by a surface S, then the triple integrals

$$\iiint_V \phi\,dV \quad \text{and} \quad \iiint_V \vec{F}\,dV$$

are called volume integrals. The first of these is a scalar and the second is a vector.

If we sub-divide the volume V into small cuboids by drawing planes parallel to the coordinate planes, then $dV = dx\,dy\,dz$.

$$\therefore \quad \iiint_V \phi\,dV = \iiint_V \phi\,(x, y, z)\,dx\,dy\,dz$$

If $\vec{F} = F_1\hat{i} + F_2\hat{j} + F_3\hat{k}$, then

$$\iiint_V \vec{F}\,dV = \hat{i}\iiint_V F_1\,(x, y, z)\,dx\,dy\,dz + \hat{j}\iiint_V F_2\,(x, y, z)\,dx\,dy\,dz + \hat{k}\iiint_V F_3\,(x, y, z)\,dx\,dy\,dz.$$

ILLUSTRATIVE EXAMPLES

Example 1. *Evaluate* $\iint_S \vec{A}\,.\,\hat{n}\,dS$, *where* $\vec{A} = (x + y^2)\hat{i} - 2x\hat{j} + 2yz\hat{k}$ *and S is the surface of the plane* $2x + y + 2z = 6$ *in the first octant.*

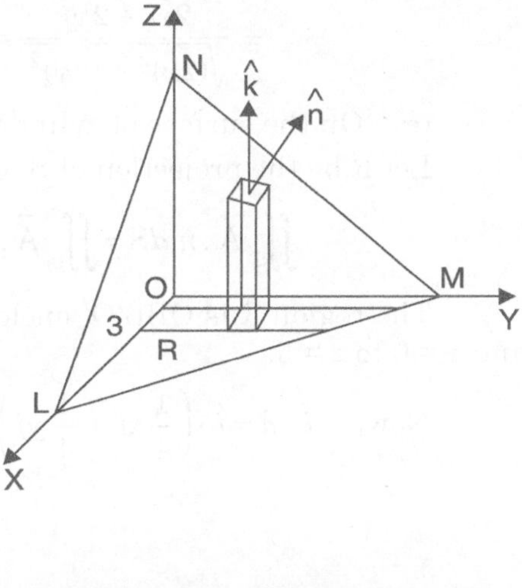

Sol. A vector normal to the surface S is given by

$$\nabla\,(2x + y + 2z) = 2\hat{i} + \hat{j} + 2\hat{k}$$

$\therefore \quad \hat{n}$ = a unit vector normal to surface S

$$= \frac{2\hat{i} + \hat{j} + 2\hat{k}}{\sqrt{(2)^2 + (1)^2 + (2)^2}} = \frac{2}{3}\hat{i} + \frac{1}{3}\hat{j} + \frac{2}{3}\hat{k}$$

$$\hat{k}\,.\,\hat{n} = \hat{k}\,.\left(\frac{2}{3}\hat{i} + \frac{1}{3}\hat{j} + \frac{2}{3}\hat{k}\right) = \frac{2}{3}$$

$$\therefore \quad \iint_S \vec{A}.\hat{n}\, dS = \iint_R \vec{A}.\hat{n} \frac{dxdy}{|\hat{k}.\hat{n}|},$$

where R is the projection of S, *i.e.* triangle LMN on the *xy*-plane. The region R, *i.e.*, triangle OLM is bounded by *x*-axis, *y*-axis and the line $2x + y = 6$, $z = 0$.

Now, $\vec{A}.\hat{n} = [(x + y^2)\hat{i} - 2x\hat{j} + 2yz\hat{k}].\left(\frac{2}{3}\hat{i} + \frac{1}{3}\hat{j} + \frac{2}{3}\hat{k}\right)$

$$= \frac{2}{3}(x + y^2) - \frac{2}{3}x + \frac{4}{3}yz = \frac{2}{3}y^2 + \frac{4}{3}yz$$

$$= \frac{2}{3}y^2 + \frac{4}{3}y\left(\frac{6 - 2x - y}{2}\right)$$

$$\left(\because \text{ On the plane } 2x + y + 2z = 6,\ z = \frac{6 - 2x - y}{2}\right)$$

$$= \frac{2}{3}y(y + 6 - 2x - y) = \frac{4}{3}y(3 - x)$$

Hence, $\iint_S \vec{A}.\hat{n}\, dS = \iint_R \vec{A}.\hat{n} \frac{dxdy}{|\hat{k}.\hat{n}|}$

$$= \iint_R \frac{4}{3}y(3 - x).\frac{3}{2}dxdy = \int_0^3 \int_0^{6-2x} 2y(3 - x)\, dydx$$

$$= \int_0^3 2(3 - x).\left[\frac{y^2}{2}\right]_2^{6-2x} dx = \int_0^3 (3 - x)(6 - 2x)^2\, dx$$

$$= 4\int_0^3 (3 - x)^3\, dx = 4.\left[\frac{(3 - x)^4}{4(-1)}\right]_0^3 = -(0 - 81) = 81.$$

Example 2. *Evaluate* $\iint_S \vec{A}.\hat{n}\, dS$, *where* $\vec{A} = z\hat{i} + x\hat{j} - 3y^2z\hat{k}$ *and S is the surface of the cylinder* $x^2 + y^2 = 16$ *included in the first octant between* $z = 0$ *and* $z = 5$.

Sol. A vector normal to the surface S is given by $\nabla(x^2 + y^2) = 2x\hat{i} + 2y\hat{j}$

$\therefore \quad \hat{n}$ = a unit vector normal to surface S

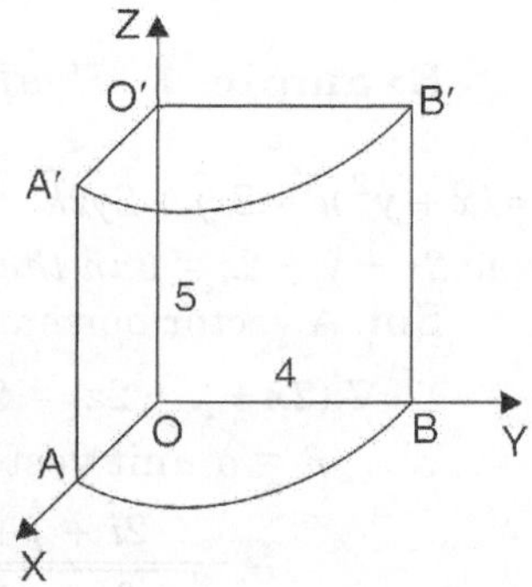

$$= \frac{2x\hat{i} + 2y\hat{j}}{\sqrt{(2x)^2 + (2y)^2}} = \frac{x\hat{i} + y\hat{j}}{\sqrt{x^2 + y^2}} = \frac{x\hat{i} + y\hat{j}}{4}$$

($\because$ On the surface of cylinder, $x^2 + y^2 = 16$)

Let R be the projection of S on *yz*-plane, then

$$\iint_S \vec{A}.\hat{n}\, dS = \iint_R \vec{A}.\hat{n} \frac{dydz}{|\hat{i}.\hat{n}|}$$

The region R is OBB′O′ enclosed by $y = 0$ to $y = 4$ and $z = 0$ to $z = 5$.

Now, $\hat{i}.\hat{n} = \hat{i}\cdot\left(\frac{1}{4}x\hat{i} + \frac{1}{4}y\hat{j}\right) = \frac{1}{4}x$

$$\vec{A}\cdot\hat{n} = (z\hat{i} + x\hat{j} - 3y^2 z\hat{k})\cdot\left(\frac{1}{4}x\hat{i} + \frac{1}{4}y\hat{j}\right)$$

$$= \frac{1}{4}zx + \frac{1}{4}xy = \frac{1}{4}x(y+z).$$

Hence, $$\iint_S \vec{A}.\hat{n}\,dS = \iint_R \vec{A}.\hat{n}\frac{dydz}{|\hat{i}.\hat{n}|} = \iint_R \frac{1}{4}x(y+z)\frac{dydz}{\frac{1}{4}x} = \int_0^5\int_0^4 (y+z)\,dy\,dz$$

$$= \int_0^5\left[\frac{y^2}{2} + zy\right]_0^4 dz = \int_0^5 (8+4z)\,dz = \left[8z + 2z^2\right]_0^5 = 40 + 50 = 90.$$

Example 3. *Evaluate* $\int_S \vec{F}\cdot n\,dS$, *where* $\vec{F} = 18z\hat{i} - 12\hat{j} + 3y\hat{k}$ *and S is the part of the surface of the plane* $2x + 3y + 6z = 12$ *located in the first octant.* (OU-Jan. 2012)

Sol. Let $\phi = 2x + 3y + 6z - 12$

$$\nabla\phi = \sum \hat{i}\frac{\partial\phi}{\partial x} = 2\hat{i} + 3\hat{j} + 6\hat{k}$$

$$|\nabla\phi| = \sqrt{4+9+36} = 7$$

$$\text{Unit normal} = \vec{n} = \frac{\nabla\phi}{|\nabla\phi|} = \frac{1}{7}(2\hat{i} + 3\hat{j} + 6\hat{k})$$

In xy-plane, we have

$$\int_S \vec{F}\cdot\vec{n}\,dS = \iint_R \vec{F}\cdot\vec{n}\frac{dx\,dy}{|\vec{n}\cdot\vec{k}|} \quad ...(1)$$

$$\vec{F}\cdot\vec{n} = (18z\hat{i} - 12\hat{j} + 3y\hat{k})\cdot\frac{1}{7}(2\hat{i} + 3\hat{j} + 6\hat{k})$$

$$= \frac{1}{7}(36z - 36 + 18y) = \frac{6}{7}(6z - 6 + 3y)$$

$$\vec{n}\cdot\vec{k} = \frac{1}{7}(2\hat{i} + 3\hat{j} + 6\hat{k})\cdot k = \frac{6}{7}$$

Given $2x + 3y + 6z = 12$

$$\Rightarrow \quad y = \frac{12-2x}{3} \quad (\because \text{ In } xy\text{-plane, } z = 0)$$

If $y = 0 \Rightarrow x = 6$

From (1),

$$\iint \vec{F}\cdot\vec{n}\frac{dx\,dy}{|\vec{n}\cdot\vec{k}|} = \iint_R \frac{6}{7}(6z - 6 + 3y)\frac{7}{6}dx\,dy$$

$$= \iint_R (12 - 2x - 3y - 6 + 3y)\,dx\,dy \quad (\because 6z = 12 - 2x - 3y)$$

$$= \int_{x=0}^{6}\int_{y=0}^{\frac{1}{3}(12-2x)} (6-2x)\,dy\,dx = \int_{x=0}^{6} (6y - 2xy)_{y=0}^{\frac{12-2x}{3}}\,dx$$

$$= \int_{x=0}^{6}\left[6\left(\frac{12-2x}{3}\right) - 2x\left(\frac{12-2x}{3}\right)\right]dx$$

$$= \int_0^6 \left(24 - 4x - \frac{24}{3}x + \frac{4}{3}x^2\right) dx = \left(24x - 2x^2 - 4x^2 + \frac{4}{9}x^3\right)_0^6$$

$$= 144 - 72 - 144 + \frac{4}{9} \times 216 = 96 - 72 = 24.$$

Example 4. *If* $\vec{F} = 4xz\hat{i} - y^2\hat{j} + yz\hat{k}$ *evaluate* $\int_S \vec{F} \cdot \vec{n}\, dS$, *where S is the surface of the cube bounded by* $x = 0, x = 1, y = 0, y = 1, z = 0, z = 1$.

Sol. We have,

$$\int_S \vec{F} \cdot \vec{n}\, dS = \int_{OABC} F \cdot n\, dS + \int_{PQRS} F \cdot n\, dS + \int_{OCSR} F \cdot n\, dS$$
$$+ \int_{ABPQ} F \cdot n\, dS + \int_{OAQR} F \cdot n\, dS + \int_{BCSP} F \cdot n\, dS \quad \ldots(1)$$

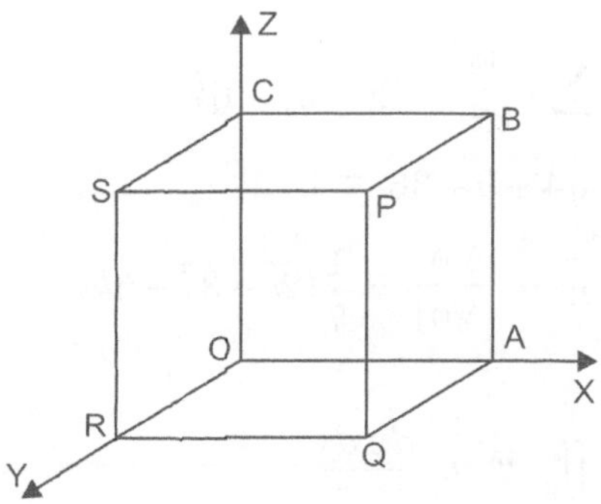

(*i*) For the face OABC, $\vec{n} = -\hat{j}$, $y = 0 \Rightarrow dy = 0$ and $dS = dx\, dz$

$$\int_{OABC} \vec{F} \cdot \vec{n}\, dS = \iint y^2\, dx\, dz = 0 \qquad (\because \; y = 0)$$

(*ii*) For the face PQRS, $\vec{n} = \hat{j}$, $y = 1$, $dS = dx\, dz$

$$\int_{PQRS} \vec{F} \cdot \vec{n}\, dS = \iint (4xz\hat{i} - y^2\hat{j} + yz\hat{k}) \cdot \hat{j}\, dx\, dz$$

$$= -\iint y^2\, dx\, dz = -\int_{x=0}^1 \int_{z=0}^1 dx\, dz \qquad (\because \; y = 1)$$

$$= -\int_{x=0}^1 (z)_{z=0}^1\, dx = -\int_{x=0}^1 1\, dx = -(x)_0^1 = -1$$

(*iii*) For the face OCSR, $\vec{n} = -\hat{i}$, $x = 0$, $dS = dy\, dz$

$$\int_{OCSR} \vec{F} \cdot \vec{n}\, dS = \iint (4xz\hat{i} - y^2\hat{j} + yz\hat{k}) \cdot \hat{i}\, dy\, dz$$

$$= -\iint (4xz\, dy\, dz) = 0 \qquad (\because \; x = 0)$$

(*iv*) For the face ABPQ, $\vec{n} = \hat{i}$, $x = 1$, $dS = dy\,dz$

$$\therefore \quad \int_{ABPQ} \vec{F} \cdot \vec{n}\, dS = \iint (4xz\hat{i} - y^2\hat{j} + yz\hat{k}) \cdot \hat{i}\, dy\, dz$$

$$= \int_{y=0}^{1} \int_{z=0}^{1} 4xz\, dy\, dz = \int_{y=0}^{1} 4\left(\frac{z^2}{2}\right)_{z=0}^{1} dy \qquad (\because \quad x = 1)$$

$$= 2\int_{y=0}^{1} dy = 2(y)_0^1 = 2$$

(*v*) For the face OAQR, $\vec{n} = -\hat{k}$, $z = 0$, $dS = dx\,dy$

$$\int_{OAQR} \vec{F} \cdot \vec{n}\, dS = 0 \qquad (\because \quad z = 0)$$

(*vi*) For the face BCSP, $\vec{n} = \hat{k}$, $z = 1$, $dS = dx\,dy$

$$\int_{BCSP} \vec{F} \cdot \vec{n}\, dS = \iint yz\, dx\, dy = \int_{x=0}^{1} \int_{y=0}^{1} y\, dx\, dy \qquad (\because \quad z = 1)$$

$$= \int_{x=0}^{1} \left(\frac{y^2}{2}\right)_0^1 dx = \frac{1}{2}\int_0^1 dx = \frac{1}{2}(x)_0^1 = \frac{1}{2}$$

From (1), we get

$$\int_S \vec{F} \cdot \vec{n}\, dS = 0 - 1 + 0 + 2 + 0 + \frac{1}{2} = \frac{3}{2}.$$

Example 5. *If* $\vec{F} = (2x^2 - 3z)\hat{i} - 2xy\hat{j} - 4x\hat{k}$, *then evaluate* $\iiint_V \nabla \cdot \vec{F}\, dV$, *where V is bounded by the planes* $x = 0$, $y = 0$, $z = 0$ *and* $2x + 2y + z = 4$.

Sol. $$\nabla \cdot \vec{F} = \frac{\partial}{\partial x}(2x^2 - 3z) + \frac{\partial}{\partial y}(-2xy) + \frac{\partial}{\partial z}(-4x) = 4x - 2x = 2x$$

$$\therefore \quad \iiint_V \nabla \cdot \vec{F}\, dV = \iiint_V 2x\, dx\, dy\, dz$$

$$= \int_0^2 \int_0^{2-x} \int_0^{4-2x-2y} 2x\, dz\, dy\, dx = \int_0^2 \int_0^{2-x} 2x \Big[z\Big]_0^{4-2x-2y} dy\, dx$$

$$= \int_0^2 \int_0^{2-x} 2x(4 - 2x - 2y)\, dy\, dx = \int_0^2 \int_0^{2-x} [4x(2 - x) - 4xy]\, dy\, dx$$

$$= \int_0^2 \Big[4x(2 - x)y - 2xy^2\Big]_0^{2-x} dx = \int_0^2 [4x(2 - x)^2 - 2x(2 - x)^2]\, dx$$

$$= \int_0^2 2x(2 - x)^2\, dx = 2\int_0^2 (4x - 4x^2 + x^3)\, dx$$

$$= 2\left[2x^2 - 4\frac{x^3}{3} + \frac{x^4}{4}\right]_0^2 = 2\left(8 - \frac{32}{3} + 4\right) = \frac{8}{3}.$$

Example 6. *If* $\vec{A} = 2xz\hat{i} - x\hat{j} + y^2\hat{k}$, *evaluate* $\iiint_V \vec{A}\,dV$, *where V is the region bounded by the surface x = 0, y = 0, x = 2, y = 6, z = x^2, z = 4.*

Sol. $\iiint_V \vec{A}\,dV = \iiint_V (2xz\hat{i} - x\hat{j} + y^2\hat{k})\; dx\,dy\,dz$

$$= \int_0^2 \int_0^6 \int_{x^2}^4 (2xz\hat{i} - x\hat{j} + y^2\hat{k})\; dz\,dy\,dx$$

$$= \hat{i}\int_0^2 \int_0^6 \int_{x^2}^4 2x\,z dz\,dy\,dx - \hat{j}\int_0^2 \int_0^6 \int_{x^2}^4 xdz\,dy\,dx + \hat{k}\int_0^2 \int_0^6 \int_{x^2}^4 y^2\,dz\,dy\,dx$$

$$= \hat{i}\int_0^2 \int_0^6 \Big[xz^2\Big]_{x^2}^4 dy\,dx - \hat{j}\int_0^2 \int_0^6 \Big[xz\Big]_{x^2}^4 dy\,dx + \hat{k}\int_0^2 \int_0^6 \Big[y^2 z\Big]_{x^2}^4 dy\,dx$$

$$= \hat{i}\int_0^2 \int_0^6 (16x - x^5)\,dy\,dx - \hat{j}\int_0^2 \int_0^6 (4x - x^3)\,dy\,dx + \hat{k}\int_0^2 \int_0^6 y^2\,(4 - x^2)\,dy\,dx$$

$$= \hat{i}\int_0^2 (16x - x^5)\Big[y\Big]_0^6 dx - \hat{j}\int_0^2 (4x - x^3)\Big[y\Big]_0^6 dx + \hat{k}\int_0^2 (4 - x^2)\left[\frac{y^3}{3}\right]_0^6 dx$$

$$= 6\hat{i}\int_0^2 (16x - x^5)\,dx - 6\hat{j}\int_0^2 (4x - x^3)\,dx + 72\hat{k}\int_0^2 (4 - x^2)\,dx$$

$$= 6\hat{i}\left[8x^2 - \frac{x^6}{6}\right]_0^2 - 6\hat{j}\left[2x^2 - \frac{x^4}{4}\right]_0^2 + 72\hat{k}\left[4x - \frac{x^3}{3}\right]_0^2 = 128\hat{i} - 24\hat{j} + 384\hat{k}.$$

EXERCISE 3

1. If $\vec{f}(t) = t\hat{i} + (t^2 - 2t)\hat{j} + (3t^2 + 3t^3)\hat{k}$, find $\int_0^1 \vec{f}(t)\,dt$.

2. If $\vec{r} = t\hat{i} - t^2\hat{j} + (t - 1)\hat{k}$ and $\vec{S} = 2t^2\hat{i} + 6t\hat{k}$, evaluate

   (*i*) $\int_0^2 \vec{r}\,.\,\vec{S}\,dt$ (*ii*) $\int_0^2 \vec{r} \times \vec{S}\,dt$.

3. Find the value of $\vec{r}$ satisfying the equation $\dfrac{d^2\vec{r}}{dt^2} = 6t\hat{i} - 24t^2\hat{j} + 4\sin t\hat{k}$, given that $\vec{r} = 2\hat{i} + \hat{j}$ and $\dfrac{d\vec{r}}{dt} = -\hat{i} - 3\hat{k}$ at $t = 0$.

4. The acceleration of a particle at any time t is given by $\vec{a} = 12\cos 2t\hat{i} - 8\sin 2t\hat{j} + 16t\hat{k}$. If the velocity $\vec{v}$ and displacement $\vec{r}$ are zero at $t = 0$, find $\vec{v}$ and $\vec{r}$ at any time t.

5. If $\phi = 2xyz^2$, $\vec{F} = xy\hat{i} - z\hat{j} + x^2\hat{k}$, and C is the curve $x = t^2$, $y = 2t$, $z = t^3$ from $t = 0$ to $t = 1$, evaluate the line integrals

   (*i*) $\int_C \phi\,d\vec{r}$ (*ii*) $\int_C \vec{F} \times d\vec{r}$.

6. Evaluate the line integrals $\int_C [(x^2 + xy)dx + (x^2 + y^2)dy]$, where C is the square formed by the lines $y = \pm 1$ and $x = \pm 1$.

7. If $\vec{F} = (3x^2 + 6y)\hat{i} - 14yz\hat{j} + 20xz^2\hat{k}$, evaluate the line integral $\int_C \vec{F} \cdot d\vec{r}$ from (0, 0, 0) to (1, 1, 1) along the following paths C :
(i) $x = t,\ y = t^2,\ z = t^3$ (ii) the straight line joining (0, 0, 0) to (1, 1, 1)
(iii) the straight line from (0, 0, 0) to (1, 0, 0) then to (1, 1, 0) and then to (1, 1, 1).

8. Show that $\int_C \vec{F} \cdot d\vec{r} = 3\pi$, given that $\vec{F} = z\hat{i} + x\hat{j} + y\hat{k}$ and C being the arc of the curve $\vec{r} = \cos t\hat{i} + \sin t\hat{j} + t\hat{k}$ from $t = 0$ to $t = 2\pi$.

9. Evaluate $\int_C [(2x^2y + y + z^2)\hat{i} + 2(1 + yz^3)\hat{j} + (2z + 3y^2z^2)\hat{k}] . d\vec{r}$ along the curve C : $y^2 + z^2 = a^2$, $x = 0$.

10. Evaluate $\int_C [(y + 3z)\, dx + (2z + x)\, dy + (3x + 2y)\, dz]$, where C is the arc of helix $x = a \cos \theta$, $y = a \sin \theta$, $z = \dfrac{2a\theta}{\pi}$ between the points $(a, 0, 0)$ and $(0, a, a)$.

11. Find the circulation of $\vec{F}$ round the curve C, where $\vec{F} = y\hat{i} + z\hat{j} + x\hat{k}$ and C is the circle $x^2 + y^2 = 1$, $z = 0$.

12. Find the circulation of $\vec{F}$ round the curve C, where $\vec{F} = e^x \sin y\hat{i} + e^x \cos y\hat{j}$ and C is the rectangle whose vertices are (0, 0), (1, 0), $\left(1, \dfrac{\pi}{2}\right)$ and $\left(0, \dfrac{\pi}{2}\right)$.

13. Find the work done when a force $\vec{F} = (x^2 - y^2 + x)\hat{i} - (2xy + y)\hat{j}$ moves a particle in the xy-plane from (0, 0) to (1, 1) along the parabola $y^2 = x$. Is the work done different when the path is the straight line $y = x$?

14. Find the work done in moving a particle once round the circle $x^2 + y^2 = 9$ in the xy-plane if the field of force is $\vec{F} = (2x - y - z)\hat{i} + (x + y - z^2)\hat{j} + (3x - 2y + 4z)\hat{k}$.

15. (i) Compute the work done by the force $\vec{F} = (2y + 3)\hat{i} + xz\hat{j} + (yz - x)\hat{k}$ when it moves a particle from the point (0, 0, 0) to the point (2, 1, 1) along the curve $x = 2t^2$, $y = t$, $z = t^3$.

(ii) Find the work done in moving a particle in the force field $\vec{F} = 3x^2\hat{i} + (2xz - y)\hat{j} + z\hat{k}$ along the curve defined by $x^2 = 4y$, $3x^3 = 8z$ from $x = 0$ to $x = 2$.

16. Evaluate $\iint_S \vec{F} \cdot \hat{n}\, dS$, where $\vec{F} = 18z\hat{i} - 12\hat{j} + 3y\hat{k}$ and S is the surface of the plane $2x + 3y + 6z = 12$ in the first octant.

17. Evaluate $\iint_S \vec{A} \cdot \hat{n}\, dS$, where $\vec{A} = 12x^2y\hat{i} - 3yz\hat{j} + 2z\hat{k}$ and S is the portion of the plane $x + y + z = 1$ included in the first octant.

18. Show that $\iint_S \vec{F} \cdot \hat{n}\, dS = \dfrac{3}{2}$, where $\vec{F} = 4xz\hat{i} - y^2\hat{j} + yz\hat{k}$ and S is the surface of the cube bounded by the planes

$$x = 0,\ x = 1,\ y = 0,\ y = 1,\ z = 0,\ z = 1.$$

19. Evaluate $\iint_S \vec{F} \cdot \hat{n}\, dS$, where $\vec{F} = yz\hat{i} + zx\hat{j} + xy\hat{k}$ and S is that part of the surface $x^2 + y^2 + z^2 = 1$ which lies in the first octant.

20. Evaluate $\iint_S \vec{F} \cdot \hat{n}\, dS$, where $\vec{F} = 4x\hat{i} - 2y^2\hat{j} + z^2\hat{k}$ and S is the surface bounding the region

$$x^2 + y^2 = 4,\ z = 0,\ z = 3.$$

21. Evaluate $\iiint_V \phi\, dV$, where $\phi = 45x^2y$ and V is the closed region bounded by the planes

$$4x + 2y + z = 8,\ x = 0,\ y = 0,\ z = 0.$$

22. If $\vec{F} = (2x^2 - 3z)\hat{i} - 2xy\hat{j} - 4x\hat{k}$, then evaluate $\iiint_V \nabla \times \vec{F} dV$, where V is the closed region bounded by the planes $x = 0,\ y = 0,\ z = 0$ and $2x + 2y + z = 4$.

Answers

1. $\frac{1}{2}\hat{i} - \frac{2}{3}\hat{j} + \frac{7}{4}\hat{k}$ 2. (*i*) 12 (*ii*) $-24\hat{i} - \frac{40}{3}\hat{j} + \frac{64}{5}\hat{k}$

3. $\vec{r} = (t^3 - t + 2)\hat{i} + (1 - 2t^4)\hat{j} + (t - 4\sin t)\hat{k}$

4. $\vec{v} = 6\sin 2t\hat{i} + 4(\cos 2t - 1)\hat{j} + 8t^2\hat{k},\ \vec{r} = 3(1 - \cos 2t)\hat{i} + 2(\sin 2t - 2t)\hat{j} + \frac{8}{3}t^3\hat{k}$

5. (*i*) $\frac{8}{11}\hat{i} + \frac{4}{5}\hat{j} + \hat{k}$ (*ii*) $-\frac{9}{11}\hat{i} - \frac{2}{3}\hat{j} - \frac{7}{5}\hat{k}$ 6. 0

7. (*i*) 5 (*ii*) $\frac{13}{3}$ (*iii*) $\frac{23}{3}$ 9. 0

10. $2a^2$ 11. $-\pi$ 12. 0

13. $-\frac{2}{3}$; No, it is same 14. 18π 15. (*i*) $8\frac{8}{35}$ (*ii*) 16 16. 24

17. $\frac{49}{120}$ 19. $\frac{3}{8}$ 20. 84π 21. 128

22. $\frac{8}{3}(\hat{j} - \hat{k})$.

5.26. DIVERGENCE THEOREM OF GAUSS (Relation between Surface and Volume Integrals)

If $\vec{F}$ is a vector point function having continuous first order partial derivatives in the region V bounded by a closed surface S, then $\iiint_V \nabla \cdot \vec{F}\, dV = \iint_S \vec{F} \cdot \hat{n}\, dS$, *where $\hat{n}$ is the outwards drawn unit normal vector to the surface S.*

[*i.e.*, the volume integral of the divergence of a vector point function $\vec{F}$ taken over the volume V enclosed by a surface S, is equal to the surface integral of the normal component of $\vec{F}$ taken over the closed surface S].

* **Proof.** Let $\vec{F} = F_1\hat{i} + F_2\hat{j} + F_3\hat{k}$, then

* Proof is for student reference only.

$$\nabla \cdot \vec{F} = \text{div } \vec{F} = \frac{\partial F_1}{\partial x} + \frac{\partial F_2}{\partial y} + \frac{\partial F_3}{\partial z}.$$

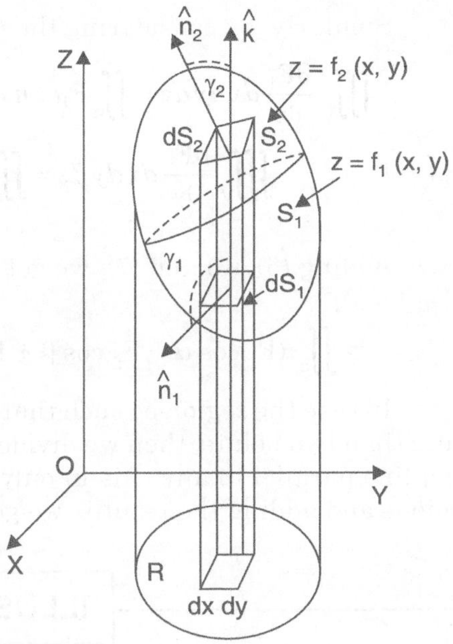

Let α, β, γ be the angles which the outwards drawn unit normal vector $\hat{n}$ makes with the positive directions of x, y, z-axes respectively, then $\cos\alpha, \cos\beta, \cos\gamma$ are the direction cosines of $\hat{n}$ and $\hat{n} = \cos\alpha\hat{i} + \cos\beta\hat{j} + \cos\gamma\hat{k}$.

$$\therefore \vec{F} \cdot \hat{n} = (F_1\hat{i} + F_2\hat{j} + F_3\hat{k}) \cdot (\cos\alpha\hat{i} + \cos\beta\hat{j} + \cos\gamma\hat{k})$$
$$= F_1\cos\alpha + F_2\cos\beta + F_3\cos\gamma$$

Therefore, the cartesian equivalent of divergence theorem is $\iiint_V \left(\frac{\partial F_1}{\partial x} + \frac{\partial F_2}{\partial y} + \frac{\partial F_3}{\partial z}\right) dx\, dy\, dz$

$$= \iint_S (F_1\cos\alpha + F_2\cos\beta + F_3\cos\gamma)\, dS \quad ...(1)$$

Suppose that S is such a closed surface that a line parallel to the co-ordinate axes meets it in two points only. Let S_1 and S_2 denote the lower and upper portions of S with equations $z = f_1(x, y)$ and $z = f_2(x, y)$ respectively.

Let R be the projection of S on the xy-plane, then

$$\iiint_V \frac{\partial F_3}{\partial z}\, dx\, dy\, dz = \iint_R \left[\int_{z=f_1(x,y)}^{z=f_2(x,y)} \frac{\partial F_3}{\partial z}\, dz\right] dx\, dy$$

$$= \iint_R \left[F_3(x, y, z)\right]_{f_1}^{f_2} dx\, dy = \iint_R [F_3(x, y, f_2) - F_3(x, y, f_1)]\, dx\, dy$$

$$= \iint_R F_3(x, y, f_2)\, dx\, dy - \iint_R F_3(x, y, f_1)\, dx\, dy \quad ...(2)$$

Now for the upper portion S_2 of S, the normal $\hat{n}_2$ to S_2 makes an acute angle γ_2 with $\hat{k}$.

$$\therefore \quad dx\, dy = \cos\gamma_2\, dS_2 = \hat{k} \cdot \hat{n}_2\, dS_2$$

For the lower portion S_1 of S, the normal $\hat{n}_1$ to S_1 makes an obtuse angle γ_1 with $\hat{k}$.

$$\therefore \quad dx\, dy = -\cos\gamma_1\, dS_1 = -\hat{k} \cdot \hat{n}_1\, dS_1$$

$$\therefore \quad \iint_R F_3(x, y, f_2)\, dx\, dy = \iint_{S_2} F_3\, \hat{k} \cdot \hat{n}_2\, dS_2 \quad ...(3)$$

and
$$\iint_R F_3(x, y, f_1)\, dx\, dy = -\iint_{S_1} F_3\, \hat{k} \cdot \hat{n}_1\, dS_1 \quad ...(4)$$

Using (3) and (4), (2) becomes $\iiint_V \frac{\partial F_3}{\partial z}\, dx\, dy\, dz = \iint_{S_2} F_3\, \hat{k} \cdot \hat{n}_2\, dS_2 + \iint_{S_1} F_3\, \hat{k} \cdot \hat{n}_1\, dS_1$

$$= \iint_S F_3\hat{k} \cdot \hat{n}\, dS = \iint_R F_3\cos\gamma\, dS \quad ...(5)$$

Similarly, by considering the projection of S on *yz* and *zx*-planes, we have

$$\iiint_V \frac{\partial F_1}{\partial x} dx\, dy\, dz = \iint_S F_1 \hat{i} \cdot \hat{n}\, dS = \iint_S F_1 \cos\alpha\, dS \quad ...(6)$$

and
$$\iiint_V \frac{\partial F_3}{\partial y} dx\, dy\, dz = \iint_S F_2 \hat{j} \cdot \hat{n}\, dS = \iint_S F_2 \cos\beta\, dS \quad ...(7)$$

Adding (5), (6) and (7), we get (1) *i.e.*, $\iiint_V \left(\frac{\partial F_1}{\partial x} + \frac{\partial F_2}{\partial y} + \frac{\partial F_3}{\partial z} \right) dx\, dy\, dz$

$$= \iint_S (F_1 \cos\alpha + F_2 \cos\beta + F_3 \cos\gamma)\, dS \quad \text{or} \quad \iiint_V \nabla . \vec{F}\, dV = \iint_R \vec{F} . \hat{n}\, dS$$

In case the region be such that the lines drawn parallel to the coordinate axes meet it in more than two points, then we divide the region into various sub-regions each of which is met by a line parallel to any axis in only two points. Applying the theorem to each of these sub-regions and adding the results, we get the volume integral over the whole region.

ILLUSTRATIVE EXAMPLES

Example 1. *For any closed surface S, prove that* $\iint_S \text{curl}\, \vec{F} . \hat{n}\, dS = 0$.

Sol. By the divergence theorem, we have $\iint_S \text{curl}\, \vec{F} . \hat{n}\, dS = \iiint_V (\text{div curl}\, \vec{F})\, dV = 0$

where V is the volume enclosed by S since div curl $\vec{F} = 0$.

Example 2. *Evaluate* $\iint_S \vec{r} . \hat{n}\, dS$, *where S is a closed surface.*

Sol. By the divergence theorem, we have $\iint_S \vec{r} . \hat{n}\, dS = \iiint_V \nabla . \vec{r}\, dV$,
where V is the volume enclosed by S

$$= \iiint_V 3dV = 3V, \text{ since } \nabla . \vec{r} = \text{div } \vec{r} = 3.$$

Example 3. *Use divergence theorem to show that* $\oint_S \nabla r^2\, d\vec{S} = 6\, V$, *where S is any closed surface enclosing a volume V.*

Sol. By the divergence theorem, we have $\oint_S \nabla r^2\, d\vec{S} = \int_V \text{div } (\nabla r^2)\, dV$

$$= \int_V \nabla . (\nabla r^2)\, dV = \int_V \nabla^2 r^2\, dV$$

$$= \int_V \left(\frac{\partial^2}{\partial x^2} + \frac{\partial^2}{\partial y^2} + \frac{\partial^2}{\partial z^2} \right) (x^2 + y^2 + z^2)\, dV = \int_V 6dV = 6V.$$

Example 4. *Verify divergence theorem for* $\vec{F} = (x^2 - yz)\,\hat{i} + (y^2 - zx)\,\hat{j} + (z^2 - xy)\,\hat{k}$ *taken over the rectangular parallelopiped* $0 \le x \le a,\ 0 \le y \le b,\ 0 \le z \le c$.

Sol. For verification of divergence theorem, we shall evaluate the volume and surface integrals separately and show that they are equal.

Now, $$\text{div } \vec{F} = \frac{\partial}{\partial x}(x^2 - yz) + \frac{\partial}{\partial y}(y^2 - zx) + \frac{\partial}{\partial z}(z^2 - xy)$$
$$= 2(x + y + z)$$

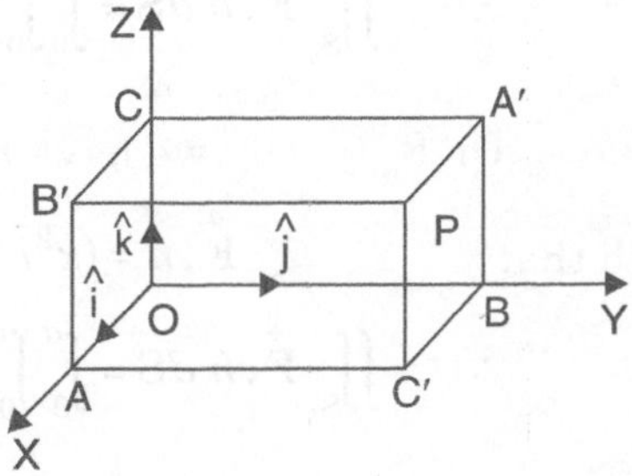

$$\therefore \iiint_V \text{div } \vec{F}\, dV$$
$$= \int_0^c\int_0^b\int_0^a 2(x+y+z)\, dx dy dz$$
$$= \int_0^c\int_0^b 2\left[\left(\frac{x^2}{2} + yx + zx\right)\right]_0^a dy\, dz$$
$$= \int_0^c\int_0^b 2\left(\frac{a^2}{2} + ya + za\right) dy\, dz = \int_0^c 2\left[\left(\frac{a^2}{2}y + \frac{ay^2}{2} + azy\right)\right]_0^b dz$$
$$= 2\int_0^c\left(\frac{a^2b}{2} + \frac{ab^2}{2} + abz\right) dz = 2\left[\frac{a^2b}{2}z + \frac{ab^2}{2}z + ab\frac{z^2}{2}\right]_0^c$$
$$= a^2bc + ab^2c + abc^2 = abc\,(a+b+c) \qquad \text{...(1)}$$

To evaluate the surface integral, divide the closed surface S of the rectangular parallelopiped into 6 parts.

S_1 : the face OAC′B, S_2 : the face CB′PA′, S_3 : the face OBA′C,
S_4 : the face AC′PB′, S_5 : the face OCB′A, S_6 : the face BA′PC′.

Also, $$\iint_S \vec{F}\cdot\hat{n}\, dS = \iint_{S_1}\vec{F}\cdot\hat{n}\, dS + \iint_{S_2}\vec{F}\cdot\hat{n}\, dS + \iint_{S_3}\vec{F}\cdot\hat{n}\, dS$$
$$+ \iint_{S_4}\vec{F}\cdot\hat{n}\, dS + \iint_{S_5}\vec{F}\cdot\hat{n}\, dS + \iint_{S_6}\vec{F}\cdot\hat{n}\, dS$$

On S_1 $(z = 0)$, we have $\hat{n} = -\hat{k}, \vec{F} = x^2\hat{i} + y^2\hat{j} - xy\hat{k}$

so that $$\vec{F}\cdot\hat{n} = (x^2\hat{i} + y^2\hat{j} - xy\hat{k})\cdot(-\hat{k}) = xy$$

$$\therefore \iint_S \vec{F}\cdot\hat{n}\, dS = \int_0^b\int_0^a xy\, dx\, dy = \int_0^b\left[y\frac{x^2}{2}\right]_0^a dy = \frac{a^2}{2}\int_0^b y\, dy = \frac{a^2b^2}{4}$$

On S_2 $(z = c)$, we have $\hat{n} = \hat{k}, \vec{F} = (x^2 - cy)\hat{i} + (y^2 - cx)\hat{j} + (c^2 - xy)\hat{k}$

so that $$\vec{F}\cdot\hat{n} = [(x^2 - cy)\hat{i} + (y^2 - cx)\hat{j} + (c^2 - xy)\hat{k}]\cdot\hat{k} = c^2 - xy$$

$$\iint_{S_2}\vec{F}\cdot\hat{n}\, dS = \int_0^b\int_0^a (c^2 - xy)\, dx\, dy = \int_0^b\left(c^2a - \frac{a^2}{2}y\right) dy = abc^2 - \frac{a^2b^2}{4}$$

On S_3 $(x = 0)$, we have $\hat{n} = -\hat{i}, \vec{F} = -yz\hat{i} + y^2\hat{j} + z^2\hat{k}$

so that $$\vec{F}\cdot\hat{n} = (-yz\hat{i} + y^2\hat{j} + z^2\hat{k})\cdot(-\hat{i}) = yz$$

$$\therefore \iint_{S_3}\vec{F}\cdot\hat{n}\, dS = \int_0^c\int_0^b yz\, dy\, dz = \int_0^c \frac{b^2}{2}z\, dz = \frac{b^2c^2}{4}$$

On S_4 $(x = a)$, we have $\hat{n} = \hat{i}, \vec{F} = (a^2 - yz)\hat{i} + (y^2 - az)\hat{j} + (z^2 - ay)\hat{k}$

so that $\vec{F} \cdot \hat{n} = [(a^2 - yz)\hat{i} + (y^2 - az)\hat{j} + (z^2 - ay)\hat{k}] \cdot \hat{i} = a^2 - yz$

$$\therefore \quad \iint_{S_4} \vec{F} \cdot \hat{n}\, dS = \int_0^c \int_0^b (a^2 - yz)\, dy\, dz = \int_0^c \left(a^2 b - \frac{b^2}{2} z\right) dz = a^2 bc - \frac{b^2 c^2}{4}$$

On S_5 $(y = 0)$, we have $\hat{n} = -\hat{j}, \vec{F} = x^2\hat{i} - zx\hat{j} + z^2\hat{k}$

so that $\vec{F} \cdot \hat{n} = (x^2\hat{i} - zx\hat{j} + z^2\hat{k}) \cdot (-\hat{j}) = zx$

$$\therefore \quad \iint_{S_5} \vec{F} \cdot \hat{n}\, dS = \int_0^a \int_0^c zx\, dz\, dx = \int_0^a \frac{c^2}{2} x\, dx = \frac{a^2 c^2}{4}$$

On S_6 $(y = b)$, we have $\hat{n} = \hat{j}, \vec{F} = (x^2 - bz)\hat{i} + (b^2 - zx)\hat{j} + (z^2 - bx)\hat{k}$

so that $\vec{F} \cdot \hat{n} = [(x^2 - bz)\hat{i} + (b^2 - zx)\hat{j} + (z^2 - bx)\hat{k}] \cdot \hat{j} = b^2 - zx$

$$\therefore \quad \iint_{S_6} \vec{F} \cdot \hat{n}\, dS = \int_0^a \int_0^c (b^2 - zx)\, dz\, dx = \int_0^a \left(b^2 c - \frac{c^2}{2} x\right) dx = ab^2 c - \frac{a^2 c^2}{4}$$

$$\therefore \quad \iint_S \vec{F} \cdot \hat{n}\, dS = \frac{a^2 b^2}{4} + abc^2 - \frac{a^2 b^2}{4} + \frac{b^2 c^2}{4} + a^2 bc - \frac{b^2 c^2}{4} + \frac{a^2 c^2}{4} + ab^2 c - \frac{a^2 c^2}{4}$$

$$= abc\,(a + b + c) \qquad \ldots(2)$$

The equality of (1) and (2) verifies divergence theorem.

Example 5. *Verify divergence theorem for $\vec{F} = 4x\hat{i} - 2y^2\hat{j} + z^2\hat{k}$ taken over the region bounded by the cylinder $x^2 + y^2 = 4$, $z = 0$, $z = 3$.* (OU-Dec. 2013)

Sol. Since div $\vec{F} = \frac{\partial}{\partial x}(4x) + \frac{\partial}{\partial y}(-2y^2) + \frac{\partial}{\partial z}(z^2) = 4 - 4y + 2z$

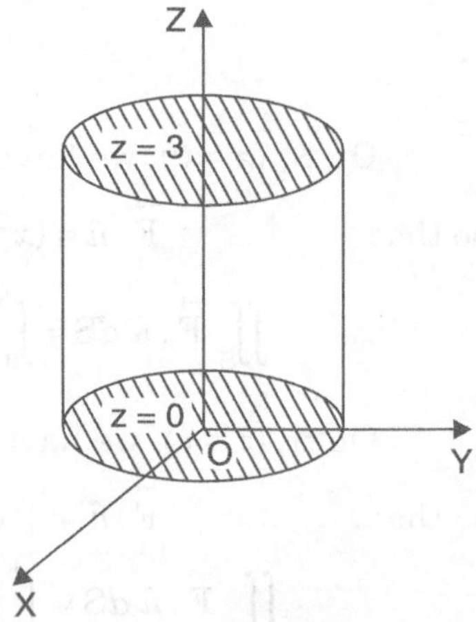

$$\therefore \quad \iiint_V \text{div}\, \vec{F}\, dV = \iiint_V (4 - 4y + 2z)\, dx\, dy\, dz$$

$$= \int_{-2}^{2} \int_{-\sqrt{4-x^2}}^{\sqrt{4-x^2}} \int_0^3 (4 - 4y + 2z)\, dz\, dy\, dx$$

$$= \int_{-2}^{2} \int_{-\sqrt{4-x^2}}^{\sqrt{4-x^2}} \left[4z - 4yz + z^2\right]_0^3 dy\, dx$$

$$= \int_{-2}^{2} \int_{-\sqrt{4-x^2}}^{\sqrt{4-x^2}} (12 - 12y + 9)\, dy\, dx$$

$$= \int_{-2}^{2} \int_{-\sqrt{4-x^2}}^{\sqrt{4-x^2}} 21\, dy\, dx,$$

(Since $12y$ is an odd function, $\int_{-a}^{a} 12y\, dy = 0$)

$$= \int_{-2}^{2} 42\sqrt{4 - x^2}\, dx = 84 \int_0^2 \sqrt{4 - x^2}\, dx = 84 \left[\frac{x\sqrt{4 - x^2}}{2} + \frac{4}{2} \sin^{-1} \frac{x}{2}\right]_0^2 = 84[2 \sin^{-1} 1]$$

$$= 84\left[2 \times \frac{\pi}{2}\right] = 84\pi \qquad \text{...(1)}$$

To evaluate the surface integral, divide the closed surface S of the cylinder into 3 parts.

S_1 : the circular base in the plane $z = 0$

S_2 : the circular top in the plane $z = 3$

S_3 : the curved surface of the cylinder, given by the equation $x^2 + y^2 = 4$.

Also, $$\iint_S \vec{F} \cdot \hat{n}\, dS = \iint_{S_1} \vec{F} \cdot \hat{n}\, dS + \iint_{S_2} \vec{F} \cdot \hat{n}\, dS + \iint_{S_3} \vec{F} \cdot \hat{n}\, dS$$

On S_1 ($z = 0$), we have $\hat{n} = -\hat{k}$, $\vec{F} = 4x\hat{i} - 2y^2\hat{j}$

so that $$\vec{F} \cdot \hat{n} = (4x\hat{i} - 2y^2\hat{j}) \cdot (-\hat{k}) = 0$$

$$\therefore \quad \iint_{S_1} \vec{F} \cdot \hat{n}\, dS = 0$$

On S_2 ($z = 3$), we have $\hat{n} = \hat{k}$, $\vec{F} = 4x\hat{i} - 2y^2\hat{j} + 9\hat{k}$

so that $$\vec{F} \cdot \hat{n} = (4x\hat{i} - 2y^2\hat{j} + 9\hat{k}) \cdot \hat{k} = 9$$

$$\therefore \quad \iint_{S_2} \vec{F} \cdot \hat{n}\, dS = \iint_{S_2} 9\,dx\,dy = 9 \iint_{S_2} dx\,dy$$

$$= 9 \times \text{area of surface } S_2 = 9\,(\pi \,.\, 2^2) = 36\pi$$

On S_3, $x^2 + y^2 = 4$

A vector normal to the surface S_3 is given by $\nabla(x^2 + y^2) = 2x\hat{i} + 2y\hat{j}$

$\therefore \quad \hat{n}$ = a unit vector normal to surface S_3

$$= \frac{2x\hat{i} + 2y\hat{j}}{\sqrt{4x^2 + 4y^2}} = \frac{2x\hat{i} + 2y\hat{j}}{\sqrt{4 \times 4}}, \text{ since } x^2 + y^2 = 4$$

$$= \frac{x\hat{i} + y\hat{j}}{2}$$

$$\vec{F} \cdot \hat{n} = (4x\hat{i} - 2y^2\hat{j} + z^2\hat{k}) \cdot \left(\frac{x\hat{i} + y\hat{j}}{2}\right) = 2x^2 - y^3$$

Also, on S_3, *i.e.*, $x^2 + y^2 = 4$, $x = 2\cos\theta$, $y = 2\sin\theta$ and $dS = 2d\theta\, dz$.

To cover the whole surface S_3, z varies from 0 to 3 and θ varies from 0 to 2π.

$$\therefore \quad \iint_{S_3} \vec{F} \cdot \hat{n}\, dS = \int_0^{2\pi} \int_0^3 [2(2\cos\theta)^2 - (2\sin\theta)^3]\, 2dz\, d\theta$$

$$= \int_0^{2\pi} 16\,(\cos^2\theta - \sin^3\theta) \times 3\, d\theta = 48 \int_0^{2\pi} (\cos^2\theta - \sin^3\theta)\, d\theta = 48\pi$$

$$\left(\text{Since } \int_0^{2\pi} \cos^2\theta\, d\theta = 2\int_0^{\pi} \cos^2\theta\, d\theta = 4\int_0^{\pi/2} \cos^2\theta\, d\theta = 4 \times \frac{1}{2} \times \frac{\pi}{2} = \pi, \int_0^{2\pi} \sin^3\theta\, d\theta = 0\right)$$

$$\therefore \quad \iint_S \vec{F} \cdot \hat{n}\, dS = 0 + 36\pi + 48\pi = 84\pi \qquad \text{...(2)}$$

The equality of (1) and (2) verifies divergence theorem.

Example 6. *Find* $\iint_S \vec{F}\,.\,\hat{n}\;dS$, *where* $\vec{F} = (2x+3z)\hat{i} - (xz+y)\hat{j} + (y^2+2z)\hat{k}$ *and S is the surface of the sphere having centre at (3, – 1, 2) and radius 3.*

Sol. Let V be the volume enclosed by the surface S. Then by Gauss divergence theorem, we have

$$\iint_S \vec{F}\,.\,\hat{n}\,dS = \iiint_V \text{div}\,\vec{F}\,dV$$

$$= \iiint_V \left[\frac{\partial}{\partial x}(2x+3z) + \frac{\partial}{\partial y}(-xz-y) + \frac{\partial}{\partial z}(y^2+2z)\right] dV$$

$$= \iiint_V (2-1+2)\,dV = 3\iiint_V dV = 3V$$

But V is the volume of a sphere of radius 3.

$$\therefore \qquad V = \frac{4}{3}\pi(3)^3 = 36\pi.$$

Hence $\iint_S \vec{F}\,.\,dS = 3 \times 36\pi = 108\pi.$

Example 7. *Evaluate* $\iint_S (y^2z^2\hat{i} + z^2x^2\hat{j} + z^2y^2\hat{k})\,.\,\hat{n}\;dS$, *where S is the part of the sphere* $x^2+y^2+z^2 = 1$ *above the xy-plane and bounded by this plane.*

Sol. Let V be the volume enclosed by the surface S. Then by divergence theorem, we have

$$\iint_S (y^2z^2\hat{i} + z^2x^2\hat{j} + z^2y^2\hat{k})\,.\,\hat{n}\,dS = \iiint_V \text{div}\,(y^2z^2\hat{i} + z^2x^2\hat{j} + z^2y^2\hat{k})\,dV$$

$$= \iiint_V \left[\frac{\partial}{\partial x}(y^2z^2) + \frac{\partial}{\partial y}(z^2x^2) + \frac{\partial}{\partial z}(z^2y^2)\right] dV$$

$$= \iiint_V 2zy^2\,dV = 2\iiint_V zy^2\,dV$$

Changing to spherical polar coordinates by putting

$$x = r\sin\theta\cos\phi,\; y = r\sin\theta\sin\phi,\; z = r\cos\theta$$

$$dV = r^2\sin\theta\,dr\,d\theta\,d\phi$$

To cover V, the limits of r will be 0 to 1, those of θ will be 0 to $\frac{\pi}{2}$ and those of ϕ will be 0 to 2π.

$$\therefore \quad 2\iiint_V zy^2\,dV = 2\int_0^{2\pi}\int_0^{\pi/2}\int_0^1 (r\cos\theta)(r^2\sin^2\theta\sin^2\phi)\,r^2\sin\theta\,dr\,d\theta\,d\phi$$

$$= 2\int_0^{2\pi}\int_0^{\pi/2}\int_0^1 r^5\sin^3\theta\cos\theta\sin^2\phi\,dr\,d\theta\,d\phi$$

$$= 2\int_0^{2\pi}\int_0^{\pi/2} \sin^3\theta\cos\theta\sin^2\phi\left[\frac{r^6}{6}\right]_0^1 d\theta\,d\phi$$

$$= \frac{2}{6}\int_0^{2\pi}\sin^2\phi\,.\,\frac{2}{4.2}\,d\phi = \frac{1}{12}\int_0^{2\pi}\sin^2\phi\,d\phi = \frac{\pi}{12}.$$

Example 8. *Evaluate* $\iint_S \vec{F}\,.\,\hat{n}\,dS$ *over the entire surface of the region above the xy-plane bounded by the cone* $z^2 = x^2+y^2$ *and the plane* $z = 4$, *if* $\vec{F} = 4xz\hat{i} + xyz^2\hat{j} + 3z\hat{k}$.

Sol. If V is the volume enclosed by S, then V is bounded by the surfaces $z = 0$, $z = 4$, $z^2 = x^2 + y^2$.

By divergence theorem, we have $\iint_S \vec{F} \cdot \hat{n}\, dS = \iiint_V \text{div}\, \vec{F}\, dV$

$$= \iiint_V \left[\frac{\partial}{\partial x}(4xz) + \frac{\partial}{\partial y}(xyz^2) + \frac{\partial}{\partial z}(3z)\right] dV = \iiint_V (4z + xz^2 + 3)\, dV$$

$$= \int_0^4 \int_{-z}^{z} \int_{-\sqrt{z^2-y^2}}^{\sqrt{z^2-y^2}} (4z + xz^2 + 3)\, dx\, dy\, dz$$

$$= 2\int_0^4 \int_{-z}^{z} \int_0^{\sqrt{z^2-y^2}} (4z + 3)\, dx\, dy\, dz, \text{ since } \int_{-\sqrt{z^2-y^2}}^{\sqrt{z^2-y^2}} x\, dx = 0$$

$$= 2\int_0^4 \int_{-z}^{z} (4z+3)\sqrt{z^2 - y^2}\, dy\, dz = 4\int_0^4 \int_0^z (4z+3)\sqrt{z^2-y^2}\, dy\, dz$$

$$= 4\int_0^4 (4z+3)\left[\frac{y\sqrt{z^2-y^2}}{2} + \frac{z^2}{2}\sin^{-1}\frac{y}{z}\right]_0^z dz$$

$$= 4\int_0^4 (4z+3)\left[\frac{z^2}{2}\sin^{-1} 1\right] dz = 4 \times \frac{\pi}{4}\int_0^4 (4z^3 + 3z^2)\, dz$$

$$= \pi\left[z^4 + z^3\right]_0^4 = \pi(256 + 64) = 320\pi.$$

Example 9. *Evaluate* $\iint_S (\nabla \times \vec{A}) \cdot \hat{n}\, dS$, *where S is the surface of the cone* $z = 2 - \sqrt{x^2 + y^2}$ *above the xy-plane and* $\vec{A} = (x - z)\hat{i} + (x^3 + yz)\hat{j} - 3xy^2\hat{k}$.

Sol. Here S is not a closed surface. The surface $z = 2 - \sqrt{x^2 + y^2}$ meets the xy-plane in a circle C given by $x^2 + y^2 = 4$, $z = 0$. Let S_1 be the plane region bounded by the circle C. If S′ is the surface consisting of the surfaces S and S_1, then *S′ is a closed surface.* By divergence theorem, we have

$$\iint_{S'} \text{curl}\, \vec{A} \cdot \hat{n}\, dS = 0 \qquad \text{[See Example 1]}$$

$$\Rightarrow \iint_S \text{curl}\, \vec{A} \cdot \hat{n}\, dS + \iint_{S_1} \text{curl}\, \vec{A} \cdot \hat{n}\, dS = 0 \qquad [\because \text{ S′ consists of S and } S_1]$$

$$\Rightarrow \iint_S \text{curl}\, \vec{A} \cdot \hat{n}\, dS - \iint_{S_1} \text{curl}\, \vec{A} \cdot \hat{k}\, dS = 0 \qquad [\because \text{ on } S_1,\ \hat{n} = -\hat{k}]$$

$$\Rightarrow \iint_S \text{curl}\, \vec{A} \cdot \hat{n}\, dS = \iint_{S_1} \text{curl}\, \vec{A} \cdot \hat{k}\, dS$$

Now, $$\text{curl}\, \vec{A} = \begin{vmatrix} \hat{i} & \hat{j} & \hat{k} \\ \dfrac{\partial}{\partial x} & \dfrac{\partial}{\partial y} & \dfrac{\partial}{\partial z} \\ x - z & x^3 + yz & -3xy^2 \end{vmatrix} = \hat{i}(-6xy - y) + \hat{j}(-1 + 3y^2) + \hat{k}(3x^2 - 0)$$

$\therefore \quad \text{curl}\ \vec{A}\,.\,\hat{k} = 3x^2$

$\therefore \quad \iint_S \text{curl}\ \vec{A}\,.\,\hat{n}\ dS = \iint_S 3x^2\ ds = \int_0^{2\pi}\int_0^2 3(r^2\cos^2\theta)\ r\ dr\ d\theta$, changing to polars

$$= 3\int_0^{2\pi}\int_0^2 r^3\cos^2\theta\ dr\ d\theta = 3\int_0^{2\pi}\left[\frac{r^4}{4}\right]_0^2 \cos^2\theta\ d\theta = 12\int_0^{2\pi}\cos^2\theta d\theta = 12\pi.$$

Example 10. *Use divergence theorem to evaluate* $\iint_S \vec{F}\cdot d\vec{S}$, *where* $\vec{F} = x^3\hat{i} + y^3\hat{j} + z^3\hat{k}$ *and S is the surface of the sphere* $x^2 + y^2 + z^2 = a^2$. (OU-Apr. 2007)

Sol. By Gauss divergence theorem

$$\iint_S \vec{F}\cdot dS = \iiint_V \text{div}\ \vec{F}\ dx\ dy\ dz \qquad \ldots(1)$$

$$\text{div}\ \vec{F} = \nabla\cdot\vec{F} = \frac{\partial(x^3)}{\partial x} + \frac{\partial(y^3)}{\partial y} + \frac{\partial(z^3)}{\partial z} = 3(x^2 + y^2 + z^2)$$

From (1), we have

$$\iint_S \vec{F}\cdot dS = \iiint 3(x^2 + y^2 + z^2)\ dx\ dy\ dz$$

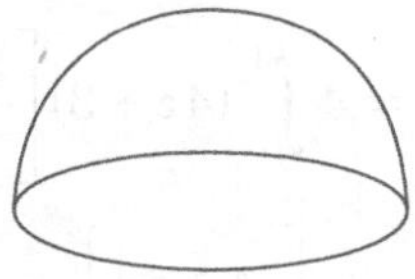

change into spherical polar coordinates $x = r\sin\theta\cos\phi$, $y = r\sin\theta\sin\phi$, $z = r\cos\theta$, where $\phi = 0$ to 2π, $\theta = 0$ to π and $r = 0$ to a

$$\therefore \quad \iint_S \vec{F}\cdot dS = 3\int_{r=0}^{a}\int_{\theta=0}^{\pi}\int_{\phi=0}^{2\pi} r^2(r^2\sin\theta)\ dr\ d\theta\ d\phi$$

$$= 3\int_{r=0}^{a}\int_{\theta=0}^{\pi} r^4\sin\theta\left[\int_{\phi=0}^{2\pi} d\phi\right] dr\ d\theta$$

$$= 3\int_{r=0}^{a}\int_{\theta=0}^{\pi} r^4\sin\theta\ (2\pi)\ dr\ d\theta$$

$$= 6\pi\int_{r=0}^{a} r^4\left[\int_{\theta=0}^{\pi}\sin\theta d\theta\right] dr$$

$$= 6\pi\int_0^a r^4\ (-\cos\pi + \cos 0)\ dr = 12\pi\int_0^a r^4\ dr$$

$$= 12\pi\left[\frac{r^5}{5}\right]_0^a = \frac{12\pi a^5}{5}.$$

Example 11. *By transforming into triple integral evaluate* $\iint_S (x^3\ dy\ dz + x^2 y\ dz\ dx + x^2 z\ dx\ dy)$, *where S is the closed surface consisting of the cylinder* $x^2 + y^2 = a^2$ *and the circular discs* $z = 0$, $z = b$.

Sol. By Gauss divergence theorem, we have

$$\iint_S (f_1\ dy\ dz + f_2\ dz\ dx + f_3\ dx\ dy) = \iiint_V \left(\frac{\partial f_1}{\partial x} + \frac{\partial f_2}{\partial y} + \frac{\partial f_3}{\partial z}\right) dx\ dy\ dz$$

$$= \iiint\left[\frac{\partial(x^3)}{\partial x} + \frac{\partial(x^2 y)}{\partial y} + \frac{\partial(x^2 z)}{\partial z}\right] dx\ dy\ dz$$

$$= 5\int_{x=-a}^{a}\int_{y=-\sqrt{a^2-x^2}}^{\sqrt{a^2-x^2}}\left[\int_{z=0}^{b} x^2\,dz\right]dy\,dx$$

$$= 20\int_{x=0}^{a}\int_{y=0}^{\sqrt{a^2-x^2}} [x^2 z]_{z=0}^{b}\,dy\,dx$$

$$= 20b\int_{x=0}^{a}\left[\int_{y=0}^{\sqrt{a^2-x^2}} x^2\,dy\right]dx$$

$$= 20b\int_{x=0}^{a} x^2[y]_{y=0}^{\sqrt{a^2-x^2}}\,dx$$

$$= 20b\int_{0}^{a} x^2\sqrt{a^2-x^2}\,dx$$

Put $x = a\sin\theta \Rightarrow dx = a\cos\theta\,d\theta$ and $\theta = 0$ to $\dfrac{\pi}{2}$

$$= 20a^4 b\int_{0}^{\pi/2}\sin^2\theta\cos^2\theta\,d\theta = 20^5 a^4 b\left[\frac{1\cdot 1}{4\cdot 2}\,\frac{\pi}{2}\right] = \frac{5}{4}\pi a^4 b.$$

$$\left(\because\ \int_0^{\pi/2}\sin^m\theta\cos^n\theta\,d\theta = \frac{(m-1)(m-3)\ldots(n-1)(n-3)\ldots}{(m+n)(m+n-2)\ldots}\cdot\frac{\pi}{2}\ \text{if } m, n \text{ are even}\right)$$

Example 12. *Verify divergence theorem for* $\vec{F} = x^2\hat{i} + y^2\hat{j} + z^2\hat{k}$ *over the surface S of the solid cut-off by the plane* $x + y + z = a$ *in the first octant.*

Sol. By Gauss divergence theorem

$$\int_S \vec{F}\cdot\vec{n}\,dS = \int_V \text{div}\,F\,dV$$

or

$$\iint_R \vec{F}\cdot\vec{n}\,\frac{dx\,dy}{|\vec{n}\cdot\vec{k}|} = \iiint\left(\frac{\partial \vec{F}_1}{\partial x} + \frac{\partial \vec{F}_2}{\partial y} + \frac{\partial \vec{F}_3}{\partial z}\right)dx\,dy\,dz$$

Let $\phi = x + y + z - a$, then $\nabla\phi = \sum i\dfrac{\partial\phi}{\partial x} = \hat{i} + \hat{j} + \hat{k}$ and $|\nabla\phi| = \sqrt{1^2 + 1^2 + 1^2} = \sqrt{3}$

$$\vec{n} = \frac{\nabla\phi}{|\nabla\phi|} = \frac{1}{\sqrt{3}}(\hat{i} + \hat{j} + \hat{k})$$

$$\vec{n}\cdot\vec{k} = \frac{1}{\sqrt{3}}$$

$$\text{LHS} = \iint \vec{F}\cdot\vec{n}\,\frac{dx\,dy}{|\vec{n}\cdot\vec{k}|} = \int_{x=0}^{a}\int_{y=0}^{a-x}\frac{\dfrac{x^2+y^2+z^2}{\sqrt{3}}}{1/\sqrt{3}}\,dx\,dy$$

$$= \int_{x=0}^{a}\int_{y=0}^{a-x}[x^2 + y^2 + (a - x - y)^2]\,dx\,dy$$

$$= \int_{x=0}^{a}\int_{y=0}^{a-x}[2x^2 + 2y^2 - 2ax + 2xy - 2ay + a^2]\,dx\,dy$$

$$= \int_{x=0}^{a}\left[2x^2 y + \frac{2y^3}{3} + xy^2 - 2axy - ay^2 + a^2 y\right]_{y=0}^{y=a-x} dx$$

$$= \int_0^a \left[-\frac{5}{3}x^3 + 3ax^2 - 2a^2x + \frac{2}{3}a^3 \right] dx \text{ , on simplification } = \frac{a^4}{4}$$

$$\text{RHS} = \iiint \left(\frac{\partial F_1}{\partial x} + \frac{\partial F_2}{\partial y} + \frac{\partial F_3}{\partial z} \right) dx\, dy\, dz$$

$$= 2 \int_{x=0}^{a} \int_{y=0}^{a-x} \left[\int_{z=0}^{a-x-y} (x + y + z)\, dz \right] dy\, dx$$

$$= 2 \int_{x=0}^{a} \int_{y=0}^{a-x} \left[z(x + y) + \frac{z^2}{2} \right]_{z=0}^{a-x-y} dy\, dx$$

$$= 2 \int_{x=0}^{a} \int_{y=0}^{a-x} (a - x - y) \left[x + y + \frac{a - x - y}{2} \right] dy\, dx$$

$$= \int_{x=0}^{a} \int_{y=0}^{a-x} (a - x - y)(a + x + y)\, dy\, dx$$

$$= \int_{x=0}^{a} \int_{y=0}^{a-x} [a^2 - (x - y)^2]\, dy\, dx$$

$$= \int_{x=0}^{a} \left[\int_{y=0}^{a-x} (a^2 - x^2 - y^2 - 2xy)\, dy \right] dx$$

$$= \int_0^a \left[a^2y - x^2y - \frac{y^3}{3} - xy^2 \right]_{y=0}^{a-x} dx$$

$$= \int_0^a (a - x)(2a^2 - x^2 - ax)\, dx = \frac{a^4}{4}$$

Since, LHS = RHS, divergence theorem is verified.

Example 13. *If S is any closed surface enclosing a volume V and* $\bar{F} = axi + byj + czk$, *prove that* $\int_S \bar{F} \,.\, \bar{N}\, dS = (a + b + c)\, V.$ (OU-June 2014)

Sol. Here $\bar{F} = axi + byj + czk$

$$\text{div } \bar{F} = \nabla \cdot \bar{F} = \frac{\partial f_1}{\partial x} + \frac{\partial f_2}{\partial y} + \frac{\partial f_3}{\partial z}$$

$$= \frac{\partial (ax)}{\partial x} + \frac{\partial (by)}{\partial y} + \frac{\partial (cz)}{\partial z} = a + b + c$$

By divergence theorem, we have

$$\int_S \bar{F} \,.\, \bar{N}\, dS = \int_V \text{div } \bar{F}\, dV$$

$$= \int_V (a + b + c)\, dV = (a + b + c)V$$

5.27. GREEN'S THEOREM IN THE PLANE

If M(x, y) and N(x, y) be continuous functions of x and y having continuous partial derivatives $\frac{\partial M}{\partial y}$ *and* $\frac{\partial N}{\partial x}$ *in a region R of the xy-plane bounded by a closed curve C, then*

$$\oint_C (Mdx + Ndy) = \iint_R \left(\frac{\partial N}{\partial x} - \frac{\partial M}{\partial y} \right) dxdy$$

where C is traversed in the counter clockwise direction. (Proof is for student reference only)

*Proof. Let us assume that the region R is such that any line parallel to either axes meets the boundary curve C in at most two points.

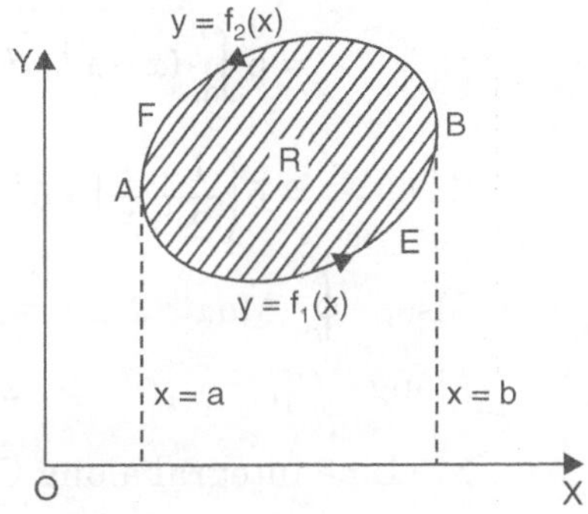

[The proof can be easily extended to other cases.]

Suppose the region R is bounded between the lines $x = a$, $x = b$ and two arcs AEB and BFA whose equations are $y = f_1(x)$ and $y = f_2(x)$ respectively such that $f_2(x) > f_1(x)$.

Now, $$\iint_R \frac{\partial M}{\partial y}\,dx\,dy = \int_a^b \left[\int_{f_1(x)}^{f_2(x)} \frac{\partial M}{\partial y}\,dy\right] dx$$

$$= \int_a^b \Big[M(x, y)\Big]_{f_1(x)}^{f_2(x)} dx = \int_a^b [M(x, f_2) - M(x, f_1)]\,dx$$

$$= \int_a^b M(x, f_2)\,dx - \int_a^b M(x, f_1)\,dx = -\int_b^a M(x, f_2)\,dx - \int_a^b M(x, f_1)\,dx$$

$$= -\left[\int_a^b M(x, f_1)\,dx + \int_b^a M(x, f_2)\,dx\right] = -\oint_C M dx$$

or $$\oint_C M dx = -\iint_R \frac{\partial M}{\partial y}\,dx\,dy \qquad ...(1)$$

Similarly, we can show that $$\oint_C N dy = \iint_R \frac{\partial N}{\partial x}\,dx\,dy \qquad ...(2)$$

Adding (1) and (2), we have $$\oint_C (M dx + N dy) = \iint_R \left(\frac{\partial N}{\partial x} - \frac{\partial M}{\partial y}\right) dx\,dy$$

This theorem is useful for changing a line integral around a closed curve C into a double integral over the region R enclosed by C.

ILLUSTRATIVE EXAMPLES

Example 1. *Verify Green's theorem in the plane for* $\oint_C (3x^2 - 8y^2)\,dx + (4y - 6xy)\,dy$, *where C is the boundary of the region defined by*

(a) $y = \sqrt{x}$, $y = x^2$ *(b)* $x = 0$, $y = 0$, $x + y = 1$.

Sol. *(a)* $y = \sqrt{x}$ *i.e.,* $y^2 = x$ and $y = x^2$ are two parabolas intersecting at O(0, 0) and A(1, 1).

Here $M = 3x^2 - 8y^2$, $N = 4y - 6xy$

$$\frac{\partial M}{\partial y} = -16y, \quad \frac{\partial N}{\partial x} = -6y$$

$$\therefore \quad \frac{\partial N}{\partial x} - \frac{\partial M}{\partial y} = 10y$$

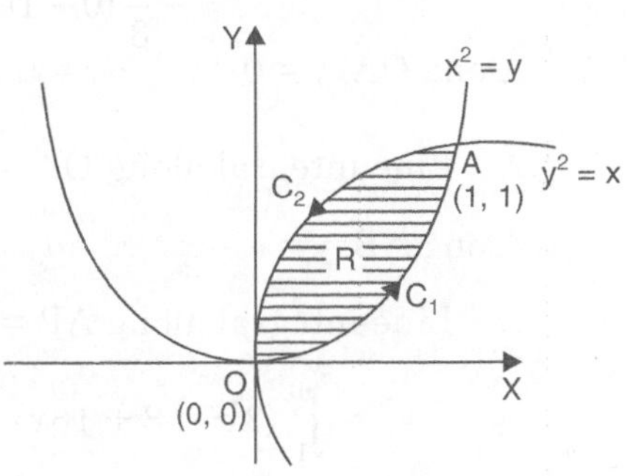

If R is the region bounded by C, then

$$\iint_R \left(\frac{\partial N}{\partial x} - \frac{\partial M}{\partial y}\right) dx\,dy$$

$$= \int_0^1 \int_{x^2}^{\sqrt{x}} 10y\,dy\,dx = \int_0^1 5\Big[y^2\Big]_{x^2}^{\sqrt{x}} dx$$

*Proof is for student reference only.

$$= 5\int_0^1 (x - x^4)\, dx = 5\left[\frac{x^2}{2} - \frac{x^5}{5}\right]_0^1$$

$$= 5\left(\frac{1}{2} - \frac{1}{5}\right) = 5\left(\frac{3}{10}\right) = \frac{3}{2} \qquad ...(1)$$

Also, $\oint_C (\text{M}dx + \text{N}dy) = \int_{C_1} (\text{M}dx + \text{N}dy) + \int_{C_2} (\text{M}dx + \text{N}dy)$

Along C_1, $x^2 = y$, $\therefore$ $2x\,dx = dy$ and the limits of x are from 0 to 1.

$\therefore$ Line integral along $C_1 = \int_{C_1} (\text{M}dx + \text{N}dy)$

$$= \int_0^1 (3x^2 - 8x^4)\,dx + (4x^2 - 6x\,.\,x^2)\,2x\,dx = \int_0^1 (3x^2 + 8x^3 - 20x^4)\,dx$$

$$= \left[x^3 + 2x^4 - 4x^5\right]_0^1 = -1$$

Along C_2, $y^2 = x$, $\therefore$ $2y\,dy = dx$ and the limits of y are from 1 to 0.

$\therefore$ Line integral along $C_2 = \int_{C_2} (\text{M}dx + \text{N}dy)$

$$= \int_1^0 (3y^4 - 8y^2)\,2y dy + (4y - 6y^2\,.\,y)\,dy$$

$$= \int_1^0 (4y - 22y^3 + 6y^5)\,dy = \left[2y^2 - \frac{11}{2}y^4 + y^6\right]_1^0 = \frac{5}{2}$$

$\therefore$ Line integral along $C = -1 + \frac{5}{2} = \frac{3}{2}$ *i.e.,* $\oint_C (\text{M}dx + \text{N}dy) = \frac{3}{2}$ ...(2)

The equality of (1) and (2) verifies Green's theorem in the plane.

(*b*) Here $\iint_R \left(\frac{\partial \text{N}}{\partial x} - \frac{\partial \text{M}}{\partial y}\right) dx\,dy = \int_0^1 \int_0^{1-x} 10y\,dy\,dx$

$$= \int_0^1 5\left[y^2\right]_0^{1-x} dx$$

$$= 5\int_0^1 (1-x)^2\,dx = 5\left[\frac{(1-x)^3}{-3}\right]_0^1$$

$$= -\frac{5}{3}(0-1) = \frac{5}{3} \qquad ...(1)$$

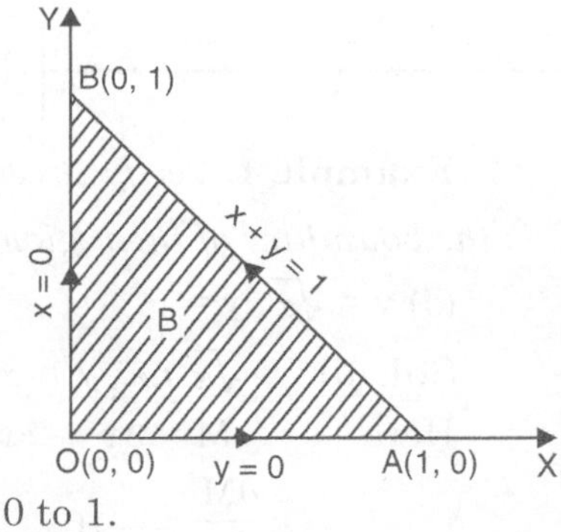

Along OA, $y = 0$ $\therefore$ $dy = 0$ and the limits of x are from 0 to 1.

$\therefore$ Line integral along OA $= \int_0^1 3x^2\,dx = \left[x^3\right]_0^1$

Along AB, $y = 1 - x$ $\therefore$ $dy = -dx$ and the limits of x are from 1 to 0.

$\therefore$ Line integral along AB $= \int_1^0 [3x^2 - 8(1-x)^2]\,dx + [4(1-x) - 6x(1-x)](-dx)$

$$= \int_1^0 (3x^2 - 8 + 16x - 8x^2 - 4 + 4x + 6x - 6x^2)\,dx = \int_1^0 (-12 + 26x - 11x^2)\,dx$$

$$= \left[-12x + 13x^2 - \frac{11}{3}x^3\right]_1^0 = -\left[-12 + 13 - \frac{11}{3}\right] = \frac{8}{3}$$

Along BO, $x = 0$ $\therefore$ $dx = 0$ and the limits of y are from 1 to 0.

$\therefore$ Line integral along BO $= \int_1^0 4y\,dy = \left[2y^2\right]_1^0 = -2$

$\therefore$ Line integral along C (*i.e.*, along OABO) $= 1 + \frac{8}{3} - 2 = \frac{5}{3}$

i.e., $$\oint_C (\text{M}dx + \text{N}dy) = \frac{5}{3} \qquad \text{...(2)}$$

The equality of (1) and (2) verifies Green's theorem in the plane.

Example 2. *Use Green's theorem in a plane to evaluate the integral* $\oint_C [(2x^2 - y^2)\,dx + (x^2 + y^2)\,dy]$, *where C is the boundary in the xy-plane of the area enclosed by the x-axis and the semi-circle $x^2 + y^2 = 1$ in the upper half xy-plane.* (OU-May 2012)

Sol. If R is the region bounded by the closed curve C, then by Green's theorem in the plane, we have

$$\oint_C (\text{M}dx + \text{N}dy) = \iint_R \left(\frac{\partial \text{N}}{\partial x} - \frac{\partial \text{M}}{\partial y}\right) dx\,dy$$

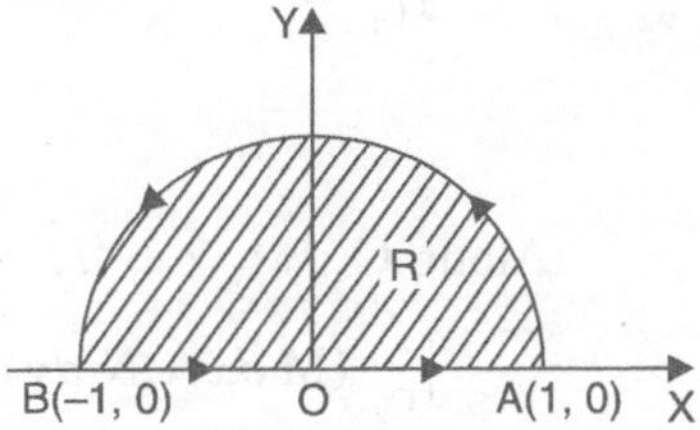

Here $\text{M} = 2x^2 - y^2,\ \text{N} = x^2 + y^2$

$$\frac{\partial \text{M}}{\partial y} = -2y,\ \frac{\partial \text{N}}{\partial x} = 2x$$

$$\therefore \quad \frac{\partial \text{N}}{\partial x} - \frac{\partial \text{M}}{\partial y} = 2(x + y)$$

The region R is bounded by

$$x = -1,\ x = 1,\ y = 0,\ y = \sqrt{1 - x^2}$$

$$\therefore \quad \oint_C [(2x^2 - y^2)\,dx + (x^2 + y^2)\,dy] = \iint_R 2(x + y)\,dx\,dy$$

$$= \int_{-1}^{1}\int_0^{\sqrt{1-x^2}} 2\,(x + y)\,dy\,dx = 2\int_{-1}^{1}\left[xy + \frac{y^2}{2}\right]_0^{\sqrt{1-x^2}} dx$$

$$= 2\int_{-1}^{1}\left[x\sqrt{1 - x^2} + \frac{1}{2}(1 - x^2)\right] dx$$

$$= \int_{-1}^{1}(1 - x^2)\,dx, \text{ since } \int_{-1}^{1} x\sqrt{1 - x^2}\,dx = 0$$

$$= 2\int_0^1 (1 - x^2)\,dx = 2\left[x - \frac{x^3}{3}\right]_0^1 = 2\left(1 - \frac{1}{3}\right) = \frac{4}{3}.$$

Example 3. *Verify Green's theorem for* $\int_C [(xy + y^2)\,dx + x^2\,dy]$, *where C is bounded by $y = x$ and $y = x^2$.* (OU-May 2009)

Sol. By Green's theorem

$$\int_C (\text{M}\,dx + \text{N}\,dy) = \iint_S \left(\frac{\partial \text{N}}{\partial x} - \frac{\partial \text{M}}{\partial y}\right) dx\,dy$$

Let $\text{M} = xy + y^2,\ \text{N} = x^2$

$$\text{LHS} = \int_C (M\,dx + N\,dy)$$

$$= \int_{C_1} (M\,dx + N\,dy) + \int_{C_2} (M\,dx + N\,dy) \quad \ldots(1)$$

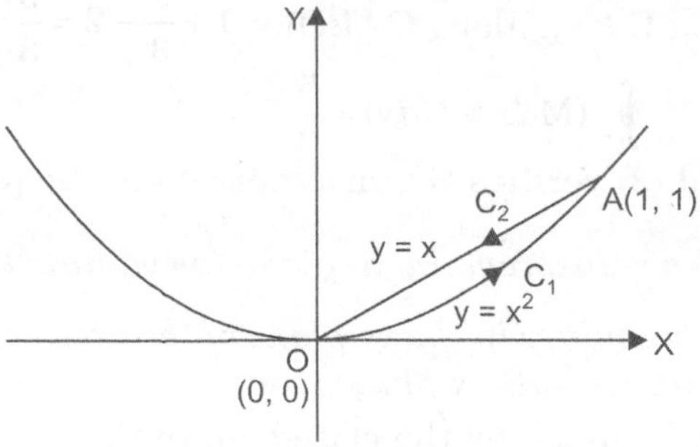

Along C_1: $y = x^2 \Rightarrow dy = 2x\,dx$ and x varies from 0 to 1

$$\int_{C_1} (M\,dx + N\,dy) = \int_0^1 [x(x^2) + x^4]\,dx + x^2 \cdot 2x\,dx = \int_0^1 (3x^3 + x^4)\,dx$$

$$= \left[3 \cdot \frac{x^4}{4} + \frac{x^5}{5}\right]_0^1 = \frac{3}{4} + \frac{1}{5} = \frac{19}{20}$$

Along C_2: $y = x \Rightarrow dy = dx$ and x varies from 1 to 0

$$\int_{C_2} (M\,dx + N\,dy) = \int_1^0 (x \cdot x + x^2)\,dx + x^2\,dx = 3\int_1^0 x^2\,dx = 3\left[\frac{x^3}{3}\right]_1^0 = -1$$

From (1), we have

$$\text{LHS} = \int_C (M\,dx + N\,dy) = \frac{19}{20} - 1 = \frac{-1}{20}$$

$$\text{RHS} = \iint_S \left(\frac{\partial N}{\partial x} - \frac{\partial M}{\partial y}\right) dx\,dy = \int_{x=0}^1 \left[\int_{y=x^2}^x (2x - x - 2y)\,dy\right] dx$$

$$= \int_0^1 [xy - y^2]_{y=x^2}^{y=x}\,dx = \int_0^1 [(x^2 - x^2) - (x^3 - x^4)]\,dx$$

$$= \int_0^1 (x^4 - x^3)\,dx = \left[\frac{x^5}{5} - \frac{x^4}{4}\right]_0^1 = \frac{1}{5} - \frac{1}{4} = \frac{-1}{20}$$

Since LHS = RHS, Green's theorem is verified.

Example 4. *Apply Green's theorem to evaluate* $\oint_C [(2xy - y^2)dx + (x^2 + y^2)\,dy]$, *where C is the closed curve bounded by* $y = x^2$ *and* $y^2 = x$. (OU-Apr. 2007)

Sol. Let $M = 2xy - y^2$, $N = x^2 + y^2$

$$\frac{\partial M}{\partial y} = 2x - 2y, \quad \frac{\partial N}{\partial x} = 2x$$

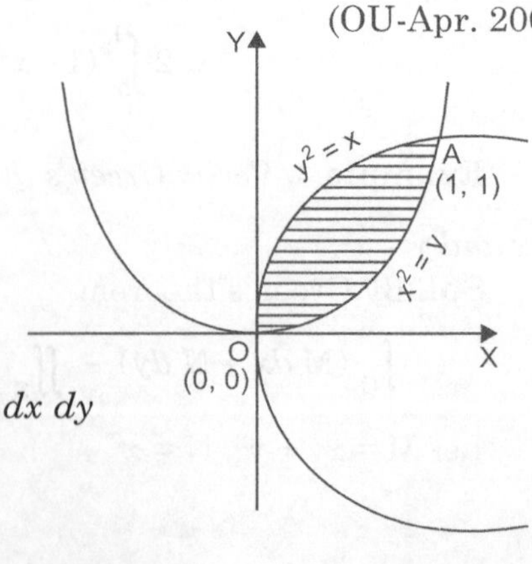

By Green's theorem, we have

$$\int_C (M\,dx + N\,dy) = \iint_S \left(\frac{\partial N}{\partial x} - \frac{\partial M}{\partial y}\right) dx\,dy$$

$$= \int_{x=0}^1 \int_{y=x^2}^{\sqrt{x}} [2x - (2x - 2y)]\,dx\,dy$$

$$= \int_0^1 \left[\int_{y=x^2}^{\sqrt{x}} 2y\,dy\right]dx = 2\int_0^1 \left[\frac{y^2}{2}\right]_{y=x^2}^{\sqrt{x}} dx$$

$$= \int_0^1 (x - x^4)\,dx = \left[\frac{x^2}{2} - \frac{x^5}{5}\right]_0^1 = \frac{1}{2} - \frac{1}{5} = \frac{3}{10}.$$

Example 5. *Find the area of the ellipse* $\frac{x^2}{a^2} + \frac{y^2}{b^2} = 1$ *using Green's theorem.*

(OU-Jan. 2012)

Sol. By Green's theorem, we have

$$\int_C (\mathrm{M}\,dx + \mathrm{N}\,dy) = \iint_S \left(\frac{\partial \mathrm{N}}{\partial x} - \frac{\partial \mathrm{M}}{\partial y}\right) dx\,dy$$

Then area $= \frac{1}{2}\oint_C (x\,dy - y\,dx)$

For the ellipse, we have

$x = a\cos\theta$, $y = b\sin\theta$ and $\theta = 0$ to 2π

$$\therefore \quad \text{Area A} = \frac{1}{2}\int (x\,dy - y\,dx)$$

$$= \frac{1}{2}\int_0^{2\pi} [a\cos\theta \cdot b\cos\theta - b\sin\theta(-a\sin\theta)]\,d\theta$$

$$= \frac{1}{2}ab\int_0^{2\pi} (\cos^2\theta + \sin^2\theta)\,d\theta = \frac{1}{2}ab\int_0^{2\pi} 1\,d\theta$$

$$= \frac{1}{2}ab\,[\theta]_0^{2\pi} = \frac{ab}{2}(2\pi) = \pi ab.$$

Example 6. *Find the area of the circle* $x^2 + y^2 = a^2$ *using Green's theorem.*

Sol. For the circle $x^2 + y^2 = a^2$, we have $x = a\cos\theta$, $y = a\sin\theta$ and $\theta = 0$ to 2π

By Green's theorem,

$$\text{Area,} \quad \mathrm{A} = \frac{1}{2}\int (x\,dy - y\,dx)$$

$$= \frac{1}{2}\int_0^{2\pi} [a\cos\theta \cdot a\cos\theta - a\sin\theta(-a\sin\theta)]\,d\theta$$

$$= \frac{1}{2}a^2\int_0^{2\pi} (\cos^2\theta + \sin^2\theta)\,d\theta = \frac{1}{2}a^2\,[\theta]_0^{2\pi}$$

$$= \frac{a^2}{2}[2\pi] = \pi a^2.$$

Example 7. *Evaluate by Green's theorem,* $\oint_C [(y - \sin x)\,dx + \cos x\,dy]$, *where C is the triangle enclosed by the lines* $y = 0$, $x = \frac{\pi}{2}$, $\pi y = 2x$.

Sol. Here $\mathrm{M} = y - \sin x$, $\mathrm{N} = \cos x$

By Green's theorem, we have

$$\oint_C [M\,dx + \mathrm{N}\,dy] = \iint \left(\frac{\partial \mathrm{N}}{\partial x} - \frac{\partial \mathrm{M}}{\partial y}\right) dx\,dy$$

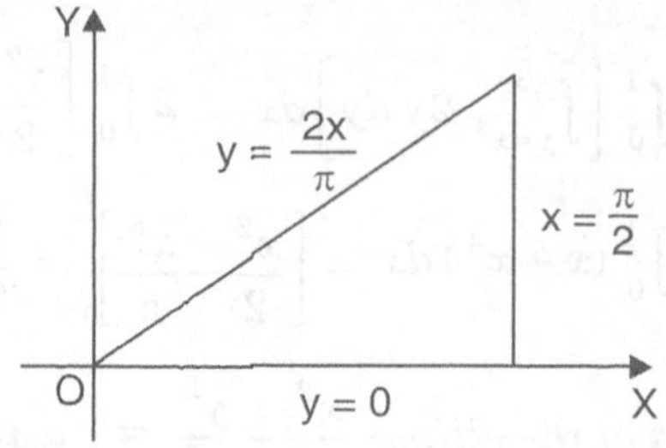

$$= -\int_{x=0}^{\pi/2}\left[\int_{y=0}^{2x/\pi}(1+\sin x)\,dy\right]dx = -\int_{x=0}^{\pi/2}(1+\sin x)\,[y]_{y=0}^{2x/\pi}\,dx$$

$$= -\int_{x=0}^{\pi/2}(1+\sin x)\left(\frac{2x}{\pi}\right)dx = -\frac{2}{\pi}\int_{0}^{\pi/2}(x + x\sin x)\,dx$$

$$= -\frac{2}{\pi}\left[\frac{x^2}{2} + x(-\cos x) - 1(-\sin x)\right]_0^{\pi/2} \quad \left(\because \int uv = uv_1 - u'v_2 + \ldots\right)$$

$$= -\frac{2}{\pi}\left[\frac{\pi^2}{8} - 0 + 1 - 0 - 0 - 0\right] = -\left(\frac{\pi}{4} + \frac{2}{\pi}\right).$$

5.28. STOKE'S THEOREM (Relation between Line and Surface Integrals)

If S be an open surface bounded by a closed curve C and $\vec{F} = F_1\hat{i} + F_2\hat{j} + F_3\hat{k}$ be any vector point function having continuous first order partial derivatives, then $\oint_C \vec{F}\cdot d\vec{r} = \iint_S \text{curl}\,\vec{F}\cdot\hat{n}\,dS$, where $\hat{n}$ is a unit normal vector at any point of S drawn in the sense in which a right handed screw would advance when rotated in the sense of description of C.

* **Proof.** Let $\hat{n}$ make angles α, β, γ with positive directions of x, y, z axes, then $\hat{n} = \cos\alpha\hat{i} + \cos\beta\hat{j} + \cos\gamma\hat{k}$.

Also, $\vec{r} = x\hat{i} + y\hat{j} + z\hat{k}$ so that $d\vec{r} = \hat{i}dx + \hat{j}dy + \hat{k}dz$

$$\text{curl}\,\vec{F} = \nabla\times\vec{F} = \begin{vmatrix} \hat{i} & \hat{j} & \hat{k} \\ \frac{\partial}{\partial x} & \frac{\partial}{\partial y} & \frac{\partial}{\partial z} \\ F_1 & F_2 & F_3 \end{vmatrix}$$

$$= \left(\frac{\partial F_3}{\partial y} - \frac{\partial F_2}{\partial z}\right)\hat{i} + \left(\frac{\partial F_1}{\partial z} - \frac{\partial F_3}{\partial x}\right)\hat{j} + \left(\frac{\partial F_2}{\partial x} - \frac{\partial F_1}{\partial y}\right)\hat{k}$$

$$\therefore \quad \text{curl}\,\vec{F}\cdot\hat{n} = \left(\frac{\partial F_3}{\partial y} - \frac{\partial F_2}{\partial z}\right)\cos\alpha + \left(\frac{\partial F_1}{\partial z} - \frac{\partial F_3}{\partial x}\right)\cos\beta + \left(\frac{\partial F_2}{\partial x} - \frac{\partial F_1}{\partial y}\right)\cos\gamma$$

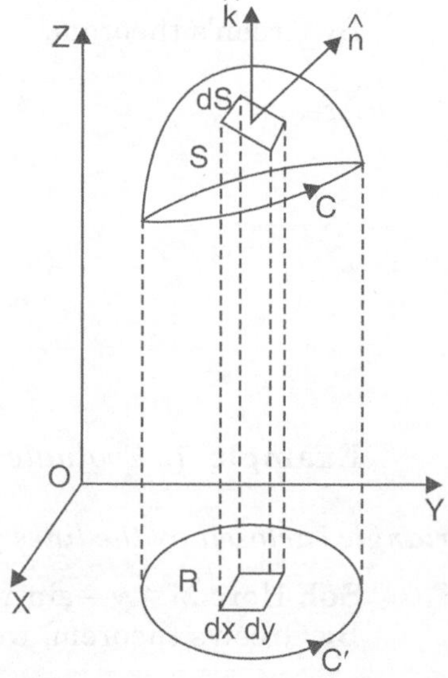

Also, $\vec{F}\cdot d\vec{r} = (F_1\hat{i} + F_2\hat{j} + F_3\hat{k})\times(\hat{i}dx + \hat{j}dy + \hat{k}dz)$

$$= F_1\,dx + F_2\,dy + F_3\,dz$$

* Proof is for student reference only.

$\therefore$ Stoke's theorem can be written as $\oint_C (F_1 dx + F_2 dy + F_3 dz)$

$$= \iint_S \left[\left(\frac{\partial F_3}{\partial y} - \frac{\partial F_2}{\partial z}\right)\cos\alpha + \left(\frac{\partial F_1}{\partial z} - \frac{\partial F_3}{\partial x}\right)\cos\beta + \left(\frac{\partial F_2}{\partial x} - \frac{\partial F_1}{\partial y}\right)\cos\gamma\right] dS \quad ...(1)$$

Let $z = \phi(x, y)$ be the equation of the surface S whose projection on the xy-plane is R. Then the projection of C on the xy-plane is the curve C′ which bounds the region R.

$$\therefore \quad \oint_C F_1(x, y, z)\, dx = \oint_{C'} F_1(x, y, \phi)\, dx = \oint_{C'} [F_1(x, y, \phi)\, dx + 0\, dy]$$

$$= -\iint_R \frac{\partial}{\partial y} F_1(x, y, \phi)\, dx\, dy \qquad | \text{ by Green's theorem in plane for the region R}$$

$$= -\iint_R \left[\frac{\partial F_1}{\partial y} + \frac{\partial F_1}{\partial z}\frac{\partial \phi}{\partial y}\right] dx\, dy \quad ...(2)$$

Now the direction ratios of the normal $\hat{n}$ to the surface $z = \phi(x, y)$ are $\frac{\partial \phi}{\partial x}, \frac{\partial \phi}{\partial y}, -1$

$$\Rightarrow \qquad \frac{\cos\alpha}{\frac{\partial \phi}{\partial x}} = \frac{\cos\beta}{\frac{\partial \phi}{\partial y}} = \frac{\cos\gamma}{-1} \quad \Rightarrow \quad \frac{\partial \phi}{\partial y} = -\frac{\cos\beta}{\cos\gamma}$$

Since $dx\, dy$ is the projection of dS on the xy-plane

$$\therefore \qquad dxdy = \cos\gamma\, dS$$

From (2), we have $\oint_C F_1(x, y, z)\, dx = -\iint_S \left[\frac{\partial F_1}{\partial y} - \frac{\partial F_1}{\partial z} \cdot \frac{\cos\beta}{\cos\gamma}\right] \cos\gamma\, dS$

$$= -\iint_S \left[\frac{\partial F_1}{\partial y}\cos\gamma - \frac{\partial F_1}{\partial z}\cos\beta\right] dS$$

$$= \iint_S \left[\frac{\partial F_1}{\partial z}\cos\beta - \frac{\partial F_1}{\partial y}\cos\gamma\right] dS \quad ...(3)$$

Similarly, we can prove that

$$\oint_C F_2(x, y, z)\, dy = \iint_S \left[\frac{\partial F_2}{\partial x}\cos\gamma - \frac{\partial F_2}{\partial z}\cos\alpha\right] dS \quad ...(4)$$

and $$\oint_C F_3(x, y, z)\, dz = \iint_S \left[\frac{\partial F_3}{\partial y}\cos\alpha - \frac{\partial F_3}{\partial x}\cos\beta\right] dS \quad ...(5)$$

Adding (3), (4) and (5), we get (1). Hence the theorem is proved. In words, Stoke's theorem states that "The line integral of the tangential component of a vector point function $\vec{F}$ taken around a closed curve C is equal to the surface integral of the normal component of $\vec{F}$ taken over any surface S having C as its boundary".

ILLUSTRATIVE EXAMPLES

Example 1. *Verify Stoke's theorem for* $\vec{F} = (x^2 + y^2)\hat{i} - 2xy\hat{j}$ *taken round the rectangle bounded by the lines* $x = \pm a$, $y = 0$, $y = b$.

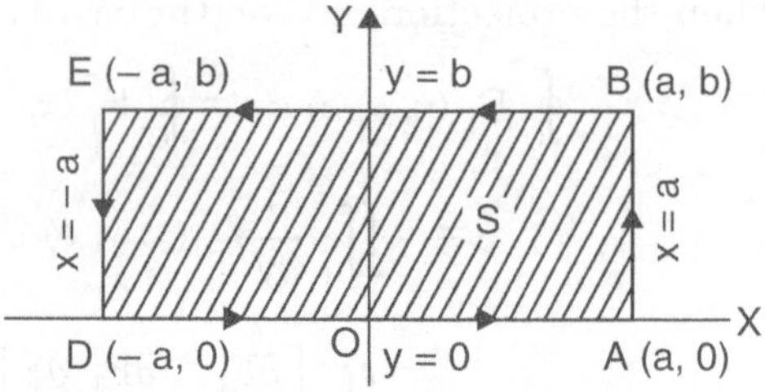

Sol. Let C denote the boundary of the rectangle ABED, then

$$\oint_C \vec{F}.d\vec{r} = \oint_C [(x^2 + y^2)\hat{i} - 2xy\hat{j}].(\hat{i}\,dx + \hat{j}\,dy)$$

$$= \oint_C [(x^2 + y^2)\,dx - 2xy\,dy]$$

The curve C consists of four lines AB, BE, ED and DA.

Along AB, $x = a$, $dx = 0$ and y varies from 0 to b.

$$\therefore \int_{AB} [(x^2 + y^2)\,dx - 2xy\,dy] = \int_0^b -2ay\,dy = -a\left[y^2\right]_0^b = -ab^2 \quad ...(1)$$

Along BE, $y = b$, $dy = 0$ and x varies from a to $-a$.

$$\therefore \int_{BE} [(x^2 + y^2)\,dx - 2xy\,dy] = \int_a^{-a} (x^2 + b^2)\,dx = \left[\frac{x^3}{3} + b^2x\right]_a^{-a} = -\frac{2a^2}{3} - 2ab^2 \quad ...(2)$$

Along ED, $x = -a$, $dx = 0$ and y varies from b to 0.

$$\therefore \int_{ED} [(x^2 + y^2)\,dx - 2xy\,dy] = \int_a^0 2ay\,dy = a\left[y^2\right]_b^0 = -ab^2 \quad ...(3)$$

Along DA, $y = 0$, $dy = 0$ and x varies from $-a$ to a.

$$\therefore \int_{DA} [(x^2 + y^2)\,dx - 2dx\,dy] = \int_{-a}^a x^2\,dx = \frac{2a^3}{3} \quad ...(4)$$

Adding (1), (2), (3) and (4), we get

$$\oint_C \vec{F}.d\vec{r} = -ab^2 - \frac{2a^3}{3} - 2ab^2 - ab^2 + \frac{2a^3}{3} = -4ab^2 \quad ...(5)$$

Now,
$$\text{curl}\,\vec{F} = \begin{vmatrix} \hat{i} & \hat{j} & \hat{k} \\ \frac{\partial}{\partial x} & \frac{\partial}{\partial y} & \frac{\partial}{\partial z} \\ x^2 + y^2 & -2xy & 0 \end{vmatrix} = (-2y - 2y)\hat{k} = -4y\hat{k}$$

For the surface S, $\hat{n} = \hat{k}$

$$\text{curl}\,\vec{F}.\hat{n} = -4y\hat{k}.\hat{k} = -4y$$

$$\therefore \iint_S \text{curl}\,\vec{F}.\hat{n}\,dS = \int_0^b \int_{-a}^a -4y\,dx\,dy = \int_0^b -4y\left[x\right]_{-a}^a dy$$

$$= -8a\int_0^b y\,dy = -8a\left[\frac{y^2}{2}\right]_0^b = -4ab^2 \quad ...(6)$$

The equality of (5) and (6) verifies Stoke's Theorem.

Example 2. *Verify Stoke's Theorem for the vector field $\vec{F} = (2x - y)\hat{i} - yz^2\hat{j} - y^2z\hat{k}$ over the upper half surface of $x^2 + y^2 + z^2 = 1$, bounded by its projection on the xy-plane.*

Sol. Let S be the upper half surface of the sphere $x^2 + y^2 + z^2 = 1$. The boundary C of S is a circle in the xy-plane of radius unity and centre O. The equations of C are $x^2 + y^2 = 1$, $z = 0$ whose parametric form is $x = \cos t$, $y = \sin t$, $z = 0$, $0 \le t < 2\pi$.

$$\oint_C \vec{F} \cdot d\vec{r} = \oint_C [(2x - y)\hat{i} - yz^2\hat{j} - y^2z\hat{k}] \cdot (\hat{i}\, dx + \hat{j}\, dy + \hat{k}\, dz)$$

$$= \oint_C [(2x - y)\, dx - yz^2\, dy - y^2z\, dz]$$

$$= \oint_C (2x - y)\, dx, \text{ since on C, } z = 0,\ dz = 0$$

$$= \int_0^{2\pi} (2\cos t - \sin t) \frac{dx}{dt}\, dt$$

$$= \int_0^{2\pi} (2\cos t - \sin t)(-\sin t)\, dt$$

$$= \int_0^{2\pi} (-\sin 2t + \sin^2 t)\, dt$$

$$= \int_0^{2\pi} \left(-\sin 2t + \frac{1 - \cos 2t}{2} \right) dt$$

$$= \left[\frac{\cos 2t}{2} + \frac{t}{2} - \frac{\sin 2t}{4} \right]_0^{2\pi} = \frac{1}{2} + \pi - \frac{1}{2} = \pi$$

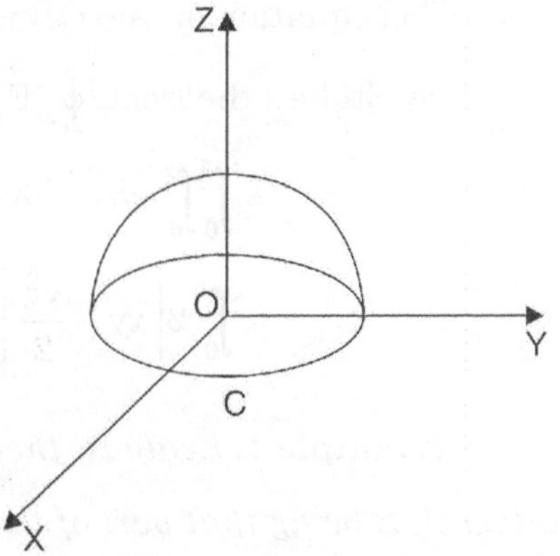

$$\text{Also, curl } \vec{F} = \begin{vmatrix} \hat{i} & \hat{j} & \hat{k} \\ \dfrac{\partial}{\partial x} & \dfrac{\partial}{\partial y} & \dfrac{\partial}{\partial z} \\ 2x - y & -yz^2 & -y^2z \end{vmatrix} = (-2yz + 2yz)\hat{i} + (0 - 0)\hat{j} + (0 + 1)\hat{k} = \hat{k}$$

$$\text{curl } \vec{F} \cdot \hat{n} = \hat{k} \cdot \hat{n} = \hat{n} \cdot \hat{k}$$

$$\iint_S \text{curl } \vec{F} \cdot \hat{n}\, dS = \iint_S \hat{n} \cdot \hat{k}\, dS = \iint_R \hat{n} \cdot \hat{k} \frac{dx\, dy}{|\hat{n} \cdot \hat{k}|}$$

where R is the projection of S on xy-plane.

$$= \iint_R dx\, dy = \int_{-1}^{1} \int_{-\sqrt{1-x^2}}^{\sqrt{1-x^2}} dy\, dx = \int_{-1}^{1} 2\sqrt{1 - x^2}\, dx = 4\int_0^1 \sqrt{1 - x^2}\, dx$$

$$= 4\left[\frac{x\sqrt{1 - x^2}}{2} + \frac{1}{2}\sin^{-1} x \right]_0^1 = 4\left[\frac{1}{2} \cdot \frac{\pi}{2} \right] = \pi$$

Since $\oint_C \vec{F} \cdot d\vec{r} = \iint_S \text{curl } \vec{F} \cdot \hat{n}\, dS$, Stoke's theorem is verified.

Example 3. *Evaluate $\oint_C \vec{F} \cdot d\vec{r}$ by Stoke's Theorem, where $\vec{F} = y^2\hat{i} + x^2\hat{j} - (x + z)\hat{k}$ and C is the boundary of the triangle with vertices at (0, 0, 0), (1, 0, 0) and (1, 1, 0).*

Sol. Since z-coordinates of each vertex of the triangle is zero, therefore, the triangle lies in the xy-plane and $\hat{n} = \hat{k}$

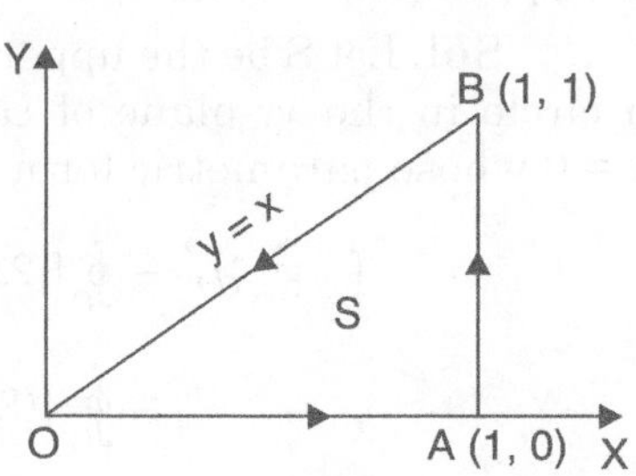

$$\text{curl } \vec{F} = \begin{vmatrix} \hat{i} & \hat{j} & \hat{k} \\ \dfrac{\partial}{\partial x} & \dfrac{\partial}{\partial y} & \dfrac{\partial}{\partial z} \\ y^2 & x^2 & -(x+z) \end{vmatrix} = \hat{j} + 2(x-y)\hat{k}$$

$\therefore$ $\text{curl } \vec{F} . \hat{n} = [\hat{j} + 2(x-y)\hat{k}] . \hat{k} = 2(x-y)$

The equation of line OB is $y = x$.

By Stoke's theorem, $\oint_C \vec{F} . d\vec{r} = \iint_S \text{curl } \vec{F} . \hat{n} \, dS$

$$= \int_0^1 \int_0^x 2(x-y) \, dy dx$$

$$= \int_0^1 2\left[xy - \frac{y^2}{2}\right]_0^x dx = 2\int_0^1 \left(x^2 - \frac{x^2}{2}\right) dx = \int_0^1 x^2 \, dx = \frac{1}{3}.$$

Example 4. *Evaluate the surface integral* $\iint_S \text{curl } \vec{F} . \hat{n} \, dS$ *by transforming it into a line integral, S being that part of the surface of the paraboloid* $z = 1 - x^2 - y^2$, *for which* $z \ge 0$ *and* $\vec{F} = y\hat{i} + z\hat{j} + x\hat{k}$.

Sol. The boundary C of the surface S is the circle $x^2 + y^2 = 1$, $z = 0$ whose parametric equations are

$$x = \cos t, \; y = \sin t, \; z = 0, \; 0 \le t < 2\pi.$$

By Stoke's theorem, we have $\iint_S \text{curl } \vec{F} . \hat{n} \, dS = \oint_C \vec{F} . d\vec{r}$

$$= \oint_C (y\hat{i} + z\hat{j} + x\hat{k}) . (\hat{i}\, dx + \hat{j}\, dy + \hat{k}\, dz)$$

$$= \oint_C (y\, dx + z\, dy + x\, dz) = \oint_C y\, dx, \text{ since on C, } z = 0, dz = 0$$

$$= \oint_C y \frac{dx}{dt} dt = \int_0^{2\pi} \sin t(-\sin t)\, dt = -\int_0^{2\pi} \sin^2 t\, dt$$

$$= -2\int_0^{\pi} \sin^2 t\, dt = -4\int_0^{\pi/2} \sin^2 t\, dt = -4 \times \frac{1}{2} \times \frac{\pi}{2} = -\pi.$$

Example 5. *Verify Stokes theorem for* $\vec{F} = -y^3\hat{i} + x^3\hat{j}$, *where S is the circular disc* $x^2 + y^2 \le 1$, $z = 0$.

Sol. By Stokes theorem, we have

$$\int_C \vec{F} \cdot dr = \int_S \text{Curl } \vec{F} \cdot \vec{n} \, dS$$

Given $x^2 + y^2 = 1$, $z = 0$

Let $x = \cos\theta$, $y = \sin\theta \Rightarrow dx = -\sin\theta\, d\theta$, $dy = \cos\theta\, d\theta$ and $\theta = 0$ to 2π

$$\text{LHS} = \oint_C \vec{F} . dr = \int_C F_1\, dx + F_2\, dy \qquad (\because \; z = 0 \Rightarrow dz = 0)$$

$$= \int_C (-y^3\, dx + x^3\, dy) = \int_0^{2\pi} [-\sin^3\theta\,(-\sin\theta) + \cos^3\theta\cos\theta]\, d\theta$$

$$= \int_0^{2\pi} (\cos^4\theta + \sin^4\theta)\, d\theta = 4\int_0^{\pi/2} (\cos^4\theta + \sin^4\theta)\, d\theta$$

$$= 4\left[\frac{3\cdot 1}{4\cdot 2}\frac{\pi}{2} + \frac{3\cdot 1}{4\cdot 2}\frac{\pi}{2}\right]$$

$$\left(\because \int_0^{\pi/2} \cos^n\theta\, d\theta = \int_0^{\pi/2} \sin^n\theta\, d\theta = \frac{(n-1)(n-3)\ldots}{n(n-2)\ldots}\,\frac{\pi}{2} \text{ if } n \text{ is even}\right)$$

$$= 4\times 2\times \frac{3\pi}{16} = \frac{3\pi}{2}$$

$$\text{Now, Curl } \vec{F} = \nabla\times\vec{F} = \begin{vmatrix} \hat{i} & \hat{j} & \hat{k} \\ \frac{\partial}{\partial x} & \frac{\partial}{\partial y} & \frac{\partial}{\partial z} \\ -y^3 & x^3 & 0 \end{vmatrix} = \hat{k}(3x^2 + 3y^2)$$

$$\text{RHS} = \int_S \text{Curl } \vec{F}\cdot\vec{n}\, dS = \iint (\nabla\times\vec{F})\cdot\vec{n}\,\frac{dx\, dy}{|\vec{n}\cdot\hat{k}|}$$

Since $z = 0 \Rightarrow n = k$ and $|\vec{n}\cdot\hat{k}| = |\hat{k}\cdot\hat{k}| = 1 = 3\iint (x^2 + y^2)\, dx\, dy$

Let $x = r\cos\phi$, $y = r\sin\phi$. Then $dx\, dy = r\, dr\, d\phi$ and $r = 0$ to 1, $\phi = 0$ to 2π

$$= 3\int_{\phi=0}^{2\pi}\int_{r=0}^{1} r^2\cdot r\, dr\, d\phi = 3\int_{\phi=0}^{2\pi}\left[\frac{r^4}{4}\right]_0^1 d\phi$$

$$= \frac{3}{4}\int_0^{2\pi} d\phi = \frac{3}{4}[\phi]_0^{2\pi} = \frac{3\pi}{2}$$

Since LHS = RHS, Stokes theorem is verified.

Example 6. *Apply Stokes theorem, to evaluate* $\oint_C (y\, dx + z dy + z\, xd)$, *where C is the curve of intersection of the sphere* $x^2 + y^2 + z^2 = a^2$ *and* $x + z = a$.

Sol. Let $\vec{F}\cdot dr = y\, dx + z\, dy + x\, dz$

$$\Rightarrow \qquad \vec{F} = y\hat{i} + z\hat{j} + x\hat{k}$$

$$\text{Curl F} = \begin{vmatrix} \hat{i} & \hat{j} & \hat{k} \\ \frac{\partial}{\partial x} & \frac{\partial}{\partial y} & \frac{\partial}{\partial z} \\ y & z & x \end{vmatrix} = -(\hat{i} + \hat{j} + \hat{k})$$

Let $\phi = x + z - a$

$$\Rightarrow \qquad \nabla\phi = \sum i\frac{\partial\phi}{\partial x} = \hat{i} + \hat{k}$$

$$|\nabla\phi| = \sqrt{1+1} = \sqrt{2}$$

$$\therefore \qquad \vec{n} = \frac{\nabla\phi}{|\nabla\phi|} = \frac{1}{\sqrt{2}}(\hat{i} + \hat{k})$$

By Stoke's theorem, we have

$$\oint_C \vec{F} \cdot d\vec{r} = \int_S \text{Curl } \vec{F} \cdot \vec{n}\, dS = -\int (\hat{i} + \hat{j} + \hat{k}) \cdot \frac{1}{\sqrt{2}}(\hat{i} + \hat{k})\, dS$$

$$= -\int_S \left(\frac{1}{\sqrt{2}} + \frac{1}{\sqrt{2}}\right) dS = -\sqrt{2}\int_S dS = -\sqrt{2}\,S$$

$$= -\sqrt{2} \times 4\pi a^2 = -4\sqrt{2}\pi a^2. \qquad (\because \text{ on a sphere } S = 4\pi r^2)$$

Example 7. *Verify Stokes theorem for* $\vec{F} = (y - z + 2)\hat{i} + (yz + 4)\hat{j} - xz\hat{k}$, *where S is the surface of the cube* $x = 0, y = 0, z = 0, x = 2, y = 2, z = 2$ *above the xy-plane.*

Sol. By Stokes theorem

$$\int_C \vec{F} \cdot d\vec{r} = \int_S \text{Curl } \vec{F} \cdot \vec{n}\, dS$$

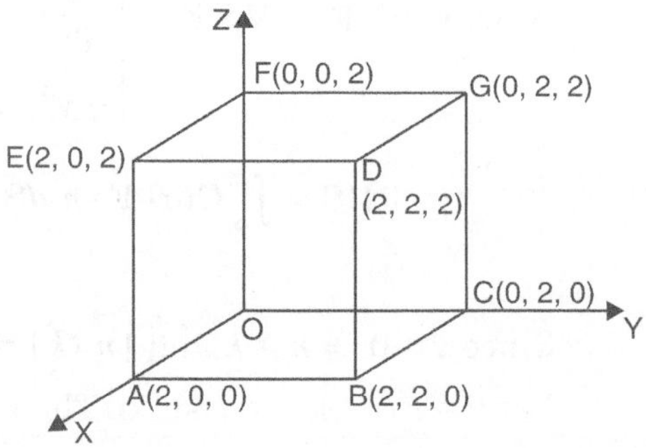

$$\text{Curl } \vec{F} = \nabla \times \vec{F} = \begin{vmatrix} \hat{i} & \hat{j} & \hat{k} \\ \dfrac{\partial}{\partial x} & \dfrac{\partial}{\partial y} & \dfrac{\partial}{\partial z} \\ y - z + 2 & yz + 4 & -xz \end{vmatrix}$$

$$= \hat{i}(0 - y) - \hat{j}(-z + 1) + \hat{k}(0 - 1)$$

$$= -y\hat{i} - (1 - z)\hat{j} - \hat{k}$$

To obtain surface integral, integrate over the five surfaces ABDE, OCGE, BCGD, OAEF, DEFG over the surface ABDE, $x = 2$, $\hat{n} = \hat{i}$, $ds = dy\, dz$.

$$\iiint_{ABDE} (\nabla \times \vec{F}) \cdot \hat{n}\, ds = \iint [-y\hat{i} + (z - 1)\hat{j} - \hat{k}] \cdot \hat{i}\, dx\, dy$$

$$= \int_0^2 \int_0^2 -y\, dy\, dz = -\int_0^2 \left[\frac{y^2}{2}\right]_0^2 dz = -2[z]_0^2 = -4$$

On OCGF, $x = 0$, $\hat{n} = -i$, $ds = dy\, dz$

$$\iiint_{OCGF} (\nabla \times \vec{F}) \cdot \hat{n}\, ds = \int_0^2 \int_0^2 [-y\hat{i} + (z - 1)\hat{j} - \hat{k}] \cdot (-\hat{i})\, dy\, dz = 4$$

On BCGD, $y = 2$, $\hat{n} = \hat{j}$, $ds = dx\, dz$

$$\iiint_{BCGD} (\nabla \times \vec{F}) \cdot \hat{n}\, ds = \int_0^2 \int_0^2 [-y\hat{i} + (z - 1)\hat{j} - \hat{k}] \cdot \hat{j}\, dx\, dz = 0$$

On OAEF, $y = 0$, $\hat{n} = -\hat{j}$, $ds = dx\, dz$

$$\iiint_{OAEF} (\nabla \times \vec{F}) \cdot \hat{n}\, ds = \int_0^2 \int_0^2 [-y\hat{i} + (z - 1)\hat{j} - \hat{k}] \cdot (-\hat{j})\, dx\, dz = 0$$

On DEFG, $z = 2$, $\hat{n} = \hat{k}$, $ds = dx\,dy$

$$\iiint_{\text{DEFG}} (\nabla \times \vec{F}) \cdot \hat{n}\, ds = \int_0^2 \int_0^2 [-y\hat{i} + (z-1)\hat{j} - \hat{k}] \cdot \hat{k}\, dx\, dy = -4$$

$$\text{RHS} = -4 + 4 + 0 + 0 - 4 = -4$$

Now, $\int_C \vec{F} \cdot d\vec{r} = \int (y - z + 2)\, dx + (yz + 4)\, dy - (xz)\, dz$

Equation of xy-plane is $z = 0 \Rightarrow dz = 0$

$$\therefore \quad \int_C \vec{F} \cdot d\vec{r} = \int_C [(y + 2)\, dx + 4\, dy]$$

$$\int_C \vec{F} \cdot d\vec{r} = \int_{OA} \vec{F} \cdot d\vec{r} + \int_{BC} \vec{F} \cdot d\vec{r} + \int_{AB} \vec{F} \cdot d\vec{r} + \int_{CO} \vec{F} \cdot d\vec{r}$$

Along $\vec{OA}$, $y = 0$, $z = 0 \Rightarrow dy = 0$, $dz = 0$ and $x = 0$ to 2

$$\int_{OA} \vec{F} \cdot d\vec{r} = \int_0^2 2\, dx = 2[x]_0^2 = 4$$

Along $\vec{BC}$, $y = 2$, $z = 0 \Rightarrow dy = 0$, $dz = 0$ and $x = 2$ to 0

$$\int_{BC} \vec{F} \cdot d\vec{r} = \int_2^0 4\, dx = 4[x]_2^0 = -8$$

Along $\vec{AB}$, $x = 2$, $z = 0 \Rightarrow dx = 0$, $dz = 0$ and $y = 0$ to 2

$$\int_{AB} \vec{F} \cdot d\vec{r} = \int_0^2 4\, dy = 4[y]_0^2 = 8$$

Along $\vec{CO}$, $x = 0$, $z = 0 \Rightarrow dx = 0$, $dz = 0$ and $y = 2$ to 0

$$\int_{CO} \vec{F} \cdot d\vec{r} = \int_2^0 4\, dy = 4[y]_2^0 = -8$$

Hence, LHS $\int_C \vec{F} \cdot d\vec{r} = 4 - 8 + 8 - 8 = -4$.

Since LHS = RHS, Stoke's theorem is verified.

Example 8. *Prove by Stokes theorem, Curl grad* $\phi = \vec{0}$.

Sol. By Stokes theorem

$$\int_S (\text{Curl grad } \phi) \cdot \vec{n}\, dS = \int_S (\nabla \times \nabla\phi) \cdot \vec{n}\, dS = \int_C \nabla\phi \cdot dr$$

$$= \int_C \left(\hat{i} \frac{\partial \phi}{\partial x} + \hat{j} \frac{\partial \phi}{\partial y} + \hat{k} \frac{\partial \phi}{\partial z} \right) \cdot (\hat{i}\, dx + \hat{j}\, dy + \hat{k}\, dz)$$

$$= \int_C \left(\frac{\partial \phi}{\partial x} dx + \frac{\partial \phi}{\partial y} dy + \frac{\partial \phi}{\partial z} dz \right) = \int_C d\phi$$

$= [\phi]_P = P - P = 0$, where P is any point on C

$\Rightarrow \quad$ Curl grad $\phi = 0$.

EXERCISE 4

1. Use divergence theorem to show that $\oint_C r^n \vec{r} \cdot d\vec{S} = (n+3)\int_V r^n \, dV \quad (n \neq -3)$ where S is any closed surface enclosing a volume V.

2. Verify divergence theorem for $\vec{F} = x^2\hat{i} + z\hat{j} + yz\hat{k}$ taken over the cube bounded by $x = 0, x = 1$; $y = 0, y = 1$; $z = 0, z = 1$.

3. Verify divergence theorem for $\vec{F} = 4xz\hat{i} - y^2\hat{j} + yz\hat{k}$ taken over the cube bounded by $x = 0, x = 1$; $y = 0, y = 1$; $z = 0, z = 1$.

4. Verify divergence theorem for $\vec{F} = (x^3 - yz)\hat{i} - 2x^2y\hat{j} + 2\hat{k}$ taken over the cube bounded by the planes $x = 0, x = a, y = 0, y = a, z = 0, z = a$.

5. Verify divergence theorem for $\vec{F} = y\hat{i} + x\hat{j} + z^2\hat{k}$ over the cylindrical region bounded by $x^2 + y^2 = a^2$, $z = 0$ and $z = h$.

6. Verify divergence theorem for $\vec{F} = 2x^2y\hat{i} - y^2\hat{j} + 4xz^2\hat{k}$ taken over the region in the first octant bounded by $y^2 + z^2 = 9$ and $x = 2$.

7. If S is any closed surface enclosing a volume V and $\vec{F} = x\hat{i} + 2y\hat{j} + 3z\hat{k}$, prove that $\iint_S \vec{F} \cdot \hat{n} \, dS = 6V$.

8. If $\vec{F} = ax\hat{i} + by\hat{j} + cz\hat{k}$, a, b, c are constants, show that $\iint_S \vec{F} \cdot d\vec{S} = \frac{4}{3}\pi(a + b + c)$, where S is the surface of a unit sphere.

9. Use divergence theorem to evaluate $\iint_S \vec{F} \cdot d\vec{S}$, where $\vec{F} = x^3\hat{i} + y^3\hat{j} + z^3\hat{k}$ and S is the surface of the sphere $x^2 + y^2 + z^2 = a^2$.

10. Use divergence theorem to evaluate $\iint_S \vec{F} \cdot \hat{n} \, dS$, where $\vec{F} = x^3\hat{i} + x^2y\hat{j} + x^2z\hat{k}$ and S is the surface bounding the region $x^2 + y^2 = a^2$, $z = 0$, $z = b$.

11. Evaluate $\iint_S (x\hat{i} + y\hat{j} + z^2\hat{k}) \cdot \hat{n} \, dS$, where S is the closed surface bounded by the cone $x^2 + y^2 = z^2$ and the plane $z = 1$.

12. If $\vec{F} = (x^2 + y - 4)\hat{i} + 3xy\hat{j} + (2xz + z^2)\hat{k}$, evaluate $\iint_S (\nabla \times \vec{F}) \cdot \hat{n} \, dS$, where S is the surface of the sphere $x^2 + y^2 + z^2 = 16$ above the xy-plane.

13. Verify Green's theorem in the plane for $\oint_C (2xy - x^2)\, dx + (x^2 + y^2)\, dy$, where C is the boundary of the region enclosed by $y = x^2$ and $y^2 = x$.

14. Verify Green's theorem in the xy-plane for $\oint_C (xy^2 - 2xy)\, dx + (x^2y + 3)\, dy$ around the boundary C of the region enclosed by $y^2 = 8x$ and $x = 2$.

15. Evaluate by Green's theorem $\oint_C e^{-x}(\sin y\, dx + \cos y\, dy)$, where C is the rectangle with vertices $(0, 0), (\pi, 0), \left(\pi, \frac{\pi}{2}\right)$ and $\left(0, \frac{\pi}{2}\right)$.

16. Apply Green's theorem to prove that the area enclosed by a plane curve is $\frac{1}{2}\int_C (xdy - ydx)$. Hence find the area of an ellipse whose semi major and minor axes are of lengths a and b.

17. Verify Stoke's theorem for the vector field $\vec{F} = (x^2 - y^2)\hat{i} + 2xy\hat{j}$ integrated round the rectangle in the plane $z = 0$ and bounded by the lines $x = 0, y = 0, x = a, y = b$.

18. Verify Stoke's theorem for the function $\vec{F} = x^2\hat{i} + xy\hat{j}$ integrated round the square whose sides are $x = 0, y = 0, x = a$ and $y = a$ in the plane $z = 0$.

19. Verify Stoke's theorem for $\vec{F} = (x^2 - y^2)\hat{i} + 2xy\hat{j}$ over the box bounded by the planes $x = 0, x = a$; $y = 0$; $y = b, z = 0, z = c$; if the face $z = 0$ is cut.

20. Verify Stoke's theorem for $\vec{F} = y\hat{i} + z\hat{j} + x\hat{k}$, where S is the upper half surface of the sphere $x^2 + y^2 + z^2 = 1$ and C is its boundary.

21. Evaluate by Stoke's theorem $\oint_C (yz\, dx + zx\, dy + xy\, dz)$, where C is the curve $x^2 + y^2 = 1, z = y^2$.

22. Use Stoke's theorem to evaluate $\iint_S (\nabla \times \vec{F}) . \hat{n}\, dS$, where $\vec{F} = y\hat{i} + (x - 2xz)\hat{j} - xy\hat{k}$ and S is the surface of the sphere $x^2 + y^2 + z^2 = a^2$ above the xy-plane.

23. Evaluate by Stoke's theorem $\oint_C (\sin z\, dx - \cos x\, dy + \sin y\, dz)$, where C is the boundary of the rectangle $0 \le x \le \pi, 0 \le y \le 1, z = 3$.

24. Evaluate $\oint_C \vec{F}.\vec{dr}$, where $\vec{F} = y\hat{i} + xz^3\hat{j} - zy^3\hat{k}$, C is the circle $x^2 + y^2 = 4, z = 1.5$.

25. Apply Stoke's theorem to evaluate $\oint_C [(x + y)dx + (2x - z)dy + (y + z)dz]$, where C is the boundary of the triangle with vertices (2, 0, 0), (0, 3, 0) and (0, 0, 6).

26. Apply Stoke's theorem to evaluate $\oint_C (ydx + zdy + xdz)$ where C is the curve of intersection of $x^2 + y^2 + z^2 = a^2$ and $x + z = a$.

Answers

9. $\frac{12}{5}\pi a^5$	**10.** $\frac{5}{4}\pi a^4 b$	**11.** 3π	**12.** -16π
15. $2(e^{-\pi} - 1)$	**16.** πab	**21.** 0	**22.** 0
23. 2	**24.** $\frac{19\pi}{2}$	**25.** 21	**26.** $-\frac{\pi a^2}{\sqrt{2}}$.

EXAMINATION PAPERS

FACULTY OF ENGINEERING

B.E. I-Year (Main) Examination, April 2016

Subject: Mathematics–I

(Code No. 5002)

Time: 3 hours | *Max. Marks: 75*

Note: Answer All questions from Part–A. Answer any Five questions from Part–B.

PART-A (25 Marks)

1. Test the convergence of $\sum_{n=1}^{\infty} \frac{1}{n^2+1}$. (3)

2. State Leibnitz's test for convergence of an alternating series. (2)

3. Find the radius of curvature of (0, 0) on the curve $y^2 = 4x$. (3)

4. Find the asymptotes of the curve $x^2y^2 - x^2y - xy^2 + xy + y + 1 = 0$ parallel to the x-axis. (2)

5. If $u = f(2x - 3y, 3y - 4z, 4z - 2x)$, prove that $\frac{1}{2}\frac{\partial u}{\partial x} + \frac{1}{3}\frac{\partial u}{\partial y} + \frac{1}{4}\frac{\partial u}{\partial z} = 0$. (3)

6. Determine $\lim_{(x,y)\to(0,0)} \frac{xy^2}{x^2+y^4}$ if it exists. (2)

7. Find the constant a, b and c such that

$\vec{V} = (3x + ay + z)\hat{i} + (2x - y + bz)\hat{j} + (x + cy + z)\hat{k}$ is irrotational. (3)

8. Find the directional derivative of $f = x^2 - y^2 + 2z^2$ at the point P(1, 2, 3) in the direction of the line PQ where Q is the point (5, 0, 4). (2)

9. Determine whether the vectors (2, 2, 0), (3, 0, 2) and (2, – 2, 2) and linearly independent. (3)

10. Find the rank of the matrix $\begin{bmatrix} 1 & 2 & 3 & 2 \\ -1 & 1 & 3 & -5 \\ 2 & 3 & 4 & 5 \end{bmatrix}$. (2)

PART–B (50 Marks)

11. (*a*) Discuss the convergence of the series. (5)

$$\frac{x}{1.2}+\frac{x^2}{3.4}+\frac{x^3}{5.6}+\frac{x^4}{7.8}+\ldots\ldots\infty.$$

(*b*) Test for convergence of series $\sum_{n=1}^{\infty}\frac{1}{\left(1+\frac{1}{n}\right)^{n^2}}$. (5)

12. (*a*) State Cauchy's mean value theorem and verify the theorem for $f(x) = \ln xe^x$ and $g(x) = 1/x$ in the interval $[1, e]$. (5)

(*b*) Find the radius of curvature of $x^3 + y^3 = 3axy$ at the point $\left(\frac{3a}{2},\frac{3a}{2}\right)$. (5)

13. (*a*) If $f(x, y) = \begin{cases}\frac{x^2+y^2}{x-y}, & (x,y)\neq(0,0)\\ 0, & (x,y)=(0,0)\end{cases}$, find $\frac{\partial f}{\partial x}$ and $\frac{\partial f}{\partial y}$ at (0, 0). (5)

(*b*) Expand $f(x, y) = \sin xy$ is powers of x and y, upto 3rd degree terms. (5)

14. (*a*) Prove that $\nabla \cdot (\nabla \times \vec{F}) = 0$, where $\vec{F}(x, y, z)$ is a vector field. (5)

(*b*) For any closed surface S, prove that

$$\int_S \left\lfloor x(y-z)\hat{i}+y(z-x)\hat{j}+z(x-y)\hat{k}\right\rfloor.\hat{n}\,ds = 0$$ (5)

15. (*a*) Using Cayley-Hamilton theorem, find the inverse of the matrix (5)

$$\begin{pmatrix}1 & 1 & 2\\ 0 & -2 & 0\\ 0 & 0 & 3\end{pmatrix}.$$

(*b*) Write the matrix of the quadratic form $x^2 + 5y^2 + z^2 + 2xy + 2yz + 6zx$ and find its nature. (5)

16. (*a*) If $x = r\cos\theta$, $y = r\sin\theta$, $z = z$, find $\frac{\partial(x,y,z)}{\partial(r,\theta,z)}$. (5)

(*b*) Using the line internal, find the work done by the force

$\vec{F} = (2y+3)\hat{i} + xz\hat{j} + (yz-x)\hat{k}$ when it moves a particle from the point (0, 0, 0) to the point (2, 1, 1) along the curve $x = 2t^2$, $y = t$, $z = t^3$. (5)

17. (*a*) Sketch the curve $a^2y^2 = x^2(a^2 - x^2)$, $a > 0$. (5)

(*b*) If $z = f(x, y)$ and $x = e^u + e^{-v}$, $y = e^{-u} - e^v$, prove that $\frac{\partial z}{\partial u}-\frac{\partial z}{\partial v} = x\frac{\partial z}{\partial x}-y\frac{\partial z}{\partial y}$. (5)

FACULTY OF ENGINEERING

B.E. I-Year (New) (Suppl.) Examination, December 2015

Subject: Mathematics–I

(Code No. 5002/N)

Time: 3 hours *Max. Marks: 75*

Note: Answer All questions from Part–A. Answer any Five questions from Part–B.

PART-A (25 Marks)

1. Step p-series test. (2)

2. Show that the series $\sum \frac{\cos n\pi}{n^2}$ is absolutely convergent. (3)

3. Find the value of C of Cauchy mean value theorem for $f(x) = x^2$ and $g(x) = x^3$ in [1, 4]. (2)

4. Obtain all the asymptotes of the curve $\frac{4}{x^2} + \frac{9}{y^2} = 1$. (3)

5. Determine $\lim_{(x, y) \to (1, -1)} x^2 - y^2$. (2)

6. If $f(x, y) = \frac{x+y}{x-y}$, find $\frac{\partial f}{\partial x}$ at (2, 3). (3)

7. Evaluate $\nabla\left(\frac{1}{r}\right)$, where $r = |\bar{r}|$, $\bar{r} = xi + yj + zk$. (2)

8. Evaluate $\int_C \bar{F} . d\bar{r}$, where $\bar{F} = x^2 i + y^3 j$ and C is the arc of the parabola $y = x^2$ in the xy plane from (0, 0) to (1, 1). (3)

9. Define linear transformation. (2)

10. Obtain the symmetric matrix A for the quadratic form $Q = x^2 + 2y^2 + 3z^2 + 4xy + 8yz + 6xz$. (3)

PART-B (50 Marks)

11. (*a*) Discuss the convergence of the series $\sum \left[\sqrt{n^2+1} - \sqrt{n^2-1}\right]$. (5)

(*b*) State Leibnitz test and use it to test the convergence of the series $\sum (-1)^n \frac{n}{2n+1}$. (5)

12. (*a*) State and prove Rolle's theorem. (5)

(*b*) Find the Taylor series expansion of $f(x) = x^3 + 3x^2 + 2x + 1$ about $x = -1$. (5)

13. (*a*) If $w = x^2 + y^2 + z^2$, $x = \cos t$, $y = \ln(t + 1)$, $z = e^t$, find $\frac{dw}{dt}$ at $t = 0$. (5)

(*b*) Evaluate $\int_0^\infty \int_0^\infty e^{-(x^2 + y^2)} dx\, dy$ by changing to polar coordinates. (5)

14. (*a*) Find the values of a and b such that the surface $5x^2 - 2yz - 9z = 0$ intersects the surface $ax^2 + by^3 = 4$ orthogonally at $(1, -2, 1)$. (5)

(*b*) Use Stoke's theorem to evaluate $\int_C (x + y)\, dx + (2x - z)\, dy + (y + z)\, dz$, where C is the boundary of the triangle with vertices (2, 0, 0), (0, 3, 0) and (0, 0, 6). (5)

15. (*a*) Show that the vectors (1, 3, 5), (2, – 1, 4), (– 2, 8, 2) are linearly dependent. (5)

(*b*) Verify Cayley-Hamilton theorem for the matrix $A = \begin{pmatrix} 1 & 4 \\ 3 & 2 \end{pmatrix}$. (5)

16. (*a*) Find the radius of curvature of the curve $x^2 = 8y$ at (4, 2). (5)

(*b*) Find the extreme values of xyz when $x + y + z = 3$. (5)

17. (*a*) Find the directional derivative of $f(x, y, z) = x^2 + y^2 + 2z^2$ at (1, 1, 2) in the direction of grad f. (5)

(*b*) Determine the values of k for which the system of equations $x - ky + z = 0$, $kx + 3y - kz = 0$, $3x + y - z = 0$ has (*i*) only zero solution (*ii*) non-zero solution. (5)

FACULTY OF ENGINEERING AND INFORMATICS

B.E. I-Year (Old) Examination, December 2015

Subject: Mathematics–I

(Code No. 5010/O)

Time: 3 hours *Max. Marks: 75*

Note: Answer All questions from Part–A. Answer any Five questions from Part–B.

PART-A (25 Marks)

1. Discuss the applicability of Rolle's theorem for $f(x) = \tan x$ in $[0, \pi]$. (3)

2. Find the envelope of the family of lines $y = cx + c^2$. (2)

3. Determine $\lim_{(x, y) \to (0, 0)} \frac{x^2 - y^2}{x^2 + y^2}$, if it exists. (3)

4. Find the total derivative of $f(x, y) = xy^3 + x^3y$. (2)

5. Evaluate $\int_0^\infty \int_0^\infty \overline{e}^{(x^2 + y^2)} \, dx \, dy$. (3)

6. State Green's theorem. (2)

7. Find the value of λ so that the system of equations $2x + y = 2z = 0, x + y + 3z = 0, 4x + 3y + \lambda z = 0$ has a non-zero solution. (3)

8. Are the vectors (1, 2), (2, 3), (4, 3) linearly dependent? (2)

9. Show that the geometric series $1 + r + r^2 + r^3 + ...$ is convergent if $| r | < 1$. (3)

10. Test the convergence of the series $\frac{1}{2} - \frac{2}{3} + \frac{3}{4} - \frac{4}{5} + ...$. (2)

Part–B (50 Marks)

11. (*a*) Find the Taylor series expansion of $f(x) = x^2 \ln x$ about $x = 1$. (5)

(*b*) Find the radius of curvature of the curve $x = t - \sin t, y = 1 - \cos t$ at $t = \pi$. (5)

12. (*a*) If $f(x, y) = \frac{x - y}{x + y}$, find f_{xy} and f_{yx} at (1, 1). (5)

(*b*) Find the extreme values of xy on the line segment $x + 2y = 2, x \geq 0, y \geq 0$. (5)

13. Verify Gauss's divergence theorem for $\overline{F} = 2x^2yi - y^2j + 4xz^2k$ over the cube bounded by $x = 0, x = 1, y = 0, y = 1, z = 0$ and $z = 1$. (10)

14. (*a*) If $A = \begin{pmatrix} 2 & 1 & 1 \\ -1 & 2 & -1 \\ 1 & -1 & 2 \end{pmatrix}$, find A^{-1} by Cayley-Hamilton theorem. (5)

(*b*) Reduce the quadratic form $Q = x^2 + 4y^2 + z^2 + 4xy + 6yz + 2zx$ to canonical form. (5)

15. (*a*) Discuss the convergence of the series $1 + 2x + 3x^2 + 4x^3 + \ldots\ldots\ldots$ (5)

(*b*) Test the convergence of the series $\frac{1}{3} + \left(\frac{2}{5}\right)^2 + \left(\frac{3}{7}\right)^3 + \ldots\ldots\ldots$ (5)

16. (*a*) State and prove Cauchy's mean value theorem. (5)

(*b*) Find the value of n such that $r^n \bar{r}$ is solenoidal. (5)

17. (*a*) Find the rank of the matrix $A = \begin{pmatrix} 1 & -1 & 2 & 3 \\ 0 & 2 & 4 & 3 \\ -2 & 1 & 5 & 2 \end{pmatrix}$. (5)

(*b*) Test whether the series $-1 + \frac{1}{2^2} - \frac{1}{3^2} + \frac{1}{4^2}$ .... is absolutely convergent or conditionally convergent. (5)

FACULTY OF ENGINEERING AND INFORMATICS

B.E. I-Year (Main) Examination, May/June 2015

Subject: Mathematics–I

(Code No. 9002/M)

Time: 3 hours *Max. Marks: 75*

Note: Answer All questions from Part–A. Answer any Five questions from Part–B.

PART-A (25 Marks)

1. Let $f'(x) = \dfrac{1}{3 - x^2}$ and $f(0) = 1$. Find an interval in which $f(1)$ lies. (2)

2. Find the equation of the envelope of the family of straight lines $y = cx + c^2$, where c is a parameter. (3)

3. Prove that $f(x, y) = \begin{cases} \dfrac{x^2 + xy + x + y}{x + y}; & (x, y) \neq (2, 2) \\ 4 \quad ; & (x, y) = (2, 2) \end{cases}$ (2)

is discontinuous at the point (2, 2).

4. If $f(x, y) = \tan^{-1}(x, y)$, find an approximate value of $f(1.1, 0.8)$ using the Taylor series linear approximation. (3)

5. Evaluate the double integral $\iint_R xy\, dx\, dy$, where 'R' is the region bounded by the x-axis, the line $y = 2x$ and the parabola $x^2 = 4ay$. (2)

6. If $\bar{a}$ is a constant vector and $\bar{r} = xi + yj + zk$, then prove that $\nabla X(a x \bar{r}) = 2\bar{a}$. (3)

7. Test whether the vectors (1, 0, 0), (0, 2, 0), (0, 0, 3) are linearly independent or not. (2)

8. Find all values of λ for which rank of the matrix. (3)

$$A = \begin{bmatrix} \lambda & -1 & 0 & 0 \\ 0 & \lambda & -1 & 0 \\ 0 & 0 & \lambda & -1 \\ -6 & 11 & -6 & 1 \end{bmatrix}$$

is equal to 3.

9. Test the convergence of the series $\sum \left[\dfrac{(1 + nx)^n}{n^n}\right]$. (2)

10. Show by an example that every convergent series need not be absolute convergent. (3)

PART–B (50 Marks)

11. (*a*) State and prove Lagrange's mean value theorem. (5)

(*b*) Find the evolute of $x^2 = 4ay$. (5)

12. (*a*) Find the shortest distance between the line $y = 10 - 2x$ and the ellipse $\frac{x^2}{4} + \frac{y^2}{9} = 1$. (5)

(*b*) Show that $\lim\limits_{(x,y)\to(0,0)} \frac{xy}{x^2 + y^2}$ does not exist. (5)

13. (*a*) Show that the vector field defined by the vector function $\bar{V} = xyz\,(yzi + xzj + xyk)$ is conservative. (5)

(*b*) Show that $\int_C (yz - 1)\,dx + (z + xz + z^2)\,dy + (y + xy + 2yz)\,dz$ is independent of the path of integration from (1, 2, 2) to (2, 3, 4). Evaluate the integral. (5)

14. (*a*) Prove that eigen values of (*i*) an Hermitian matrix are real (*ii*) a Skew-Hermitian matrix are zero or purely imaginary. (5)

(*b*) Examine $A = \begin{bmatrix} 1 & 0 & i \\ 0 & 1 & 0 \\ -i & 0 & 3 \end{bmatrix}$ is positive definite. (5)

15. (*a*) Discuss the convergence of the series $\sum \left[\frac{1.35...(2n-1)}{2.46...(2n)} \frac{x^{2n}}{2n}\right]$. (5)

(*b*) Test the convergence of the series $1 + 3x + 5x^2 + 7x^3 + ...$ (5)

16. (*a*) State and prove Cayley Hamilton theorem. (5)

(*b*) Find the eigen values and the corresponding eigen vectors. (5)

$$A = \begin{bmatrix} 1 & 2 & 2 \\ 0 & 2 & 1 \\ -1 & 2 & 2 \end{bmatrix}$$

17. (*a*) If $x = r\cos\theta$, $y = r\sin\theta$, then find $\left(\frac{\partial r}{\partial x}\right)^2 + \left(\frac{\partial r}{\partial y}\right)^2$. (5)

(*b*) If $u = \log\,[x^2 + xy + y^2]$, then find $x\frac{\partial u}{\partial x} + y\frac{\partial u}{\partial y}$. (5)

Index